W9-AOZ-842

OXFORD MATHEMATICAL MONOGRAPHS

Editors

TO THE GEOMETERS

W. L. Edge
T. G. Room
B. Segre

PROJECTIVE GEOMETRIES OVER FINITE FIELDS

BY

J. W. P. HIRSCHFELD

CLARENDON PRESS · OXFORD
1979

Oxford University Press, Walton Street, Oxford OX2 6DP
OXFORD LONDON GLASGOW
NEW YORK TORONTO MELBOURNE WELLINGTON
IBADAN NAIROBI DAR ES SALAAM LUSAKA CAPE TOWN
KUALA LUMPUR SINGAPORE JAKARTA HONG KONG TOKYO
DELHI BOMBAY CALCUTTA MADRAS KARACHI

ISBN 0 19 853526 0

Published in the United States by
Oxford University Press, New York

British Library Cataloguing in Publication Data

Hirschfeld, J W P
Projective geometries over finite fields. -
(Oxford mathematical monographs).
1. Geometry, Projective 2. Modular fields
I. Title II. Series
516'.5 QA471 78-40487

ISBN 0-19 853526-0

Printed in Great Britain
by Thomson Litho Ltd
East Kilbride

PREFACE

Purpose

The study of finite projective spaces was at one time no more than an adjunct to algebraic geometry over the real and complex numbers. But, more recently, finite spaces have been studied both for their application to practical topics such as coding theory and design of experiments, and for their illumination of more abstract mathematical topics such as finite group theory and graph theory.

There have been many papers but few books. This work is an attempt to collect the results of these papers, and to present a self-contained and comprehensive account of the subject, assuming no more knowledge than the group theory and linear algebra taught in a first degree course, as well as a little projective geometry, and a very little algebraic geometry.

Historical remarks

The first reference I can find on finite geometry in the literature is von Staudt [1856], pp. 86-91. He already has the idea of 'real' and 'complex' points of a finite space as defined in §2.6, but he restricts his attention to dimensions two and three. Then Fano [1892] defined $PG(n,p)$ synthetically for an arbitrary prime p, while Hessenberg [1902/3] did it analytically. The first systematic account of $PG(n,q)$ for arbitrary n and an arbitrary prime power q is due to Veblen and Bussey [1906], although the group of projectivities $PGL(n+1,q)$ goes back to Jordan [1870].

The growth of abstract group theory diverted attention from the geometries on which the linear groups act. Research continued on modular invariants as well as on plane algebraic geometry strictly analogous to classical geometry. This was followed by much deeper work in number theory by Hasse [1934], and by Weil [1948a] on estimates for the number of points on an algebraic curve.

Another stimulus was the realization that finite spaces

had applications in statistics. The central problem is the determination of the numbers $m(l;\ r,\ s;\ n,\ q)$ defined in §3.3. Also, several Finnish astronomers tried to use finite spaces as an approximation to real Euclidean spaces.

Several of these strands were brought together by the beautiful result of Segre [1955a], theorem 8.2.4 in the text, that in $PG(2,q)$ with q odd a set of $q+1$ points, no three of which are collinear, is a conic. This stimulated a considerable amount of research, which still leaves many easily stated, but apparently difficult, problems unsolved.

However, finite geometry is still an immature subject. There is a wide gap in knowledge between the many specialized results proved in the text and the very deep results of the Weil [1949] conjectures proved recently by Deligne [1974]. Finite geometry is still a rich field for applicable research.

Outline

The two chapters of Part I consist of more or less familiar results on finite fields and on projective spaces, of which very few are proved. Although §2.1 should be read before the rest of the book, the other sections of Part I are there for later reference.

Part II deals with elementary properties of $PG(n,q)$ for arbitrary n. Part III is a detailed exposition of the line and the plane.

In a further volume, Part IV will deal in great detail with $PG(3,q)$ and Part V with further properties of $PG(n,q)$ for $n \geq 4$.

Even though some topics could be handled more economically by considering the n-dimensional case before the particular properties of low dimensions, the policy has been the reverse: to deal thoroughly with low dimensions in order to ease the understanding of the general case.

The three themes of this work are

(i) properties of algebraic varieties,

(ii) combinatorial properties of subsets of points,

(iii) group theoretical properties of configurations.

The first two intertwine in theorems on the characterization of varieties such as Segre's theorem above. The third theme, although less emphasized, appears in proving isomorphisms between groups by constructing a suitable mapping between two configurations which the respective groups fix.

An elementary course could begin with §2.1 and §3.1, and then continue from Chapter 6.

References and numeration

No references to the Bibliography are given in the body of the chapters: they are all gathered in the last section of each chapter.

In places, such as the first two chapters, where results are not proved, the results are numbered by small Roman or Arabic numerals. In the rest of the book, the theorems and lemmas are together numbered consecutively throughout a section: thus theorem 4.3.5 is the fifth theorem or lemma in §4.3, which is the third section of Chapter 4. Statements are numbered, when necessary, consecutively throughout a chapter: thus (8.6) is the sixth labelled statement in Chapter 8. The end of the proof of a theorem or lemma is indicated by the symbol □.

Acknowledgements

This project was begun on a visit to Italy under the European Programme sponsored by the Royal Society and the Accademia Nazionale dei Lincei. I am most grateful to them and to the University of Perugia and of Rome for their hospitality, as well as to the University of Sussex for its continued support. The original typescript was mainly done by Mrs. Jill Foster, whose skill made it a pleasure for me to read.

Beniamino Segre

As this volume was being completed, the mathematician with the greatest influence on it died. As the Bibliography reveals, he published the most as well as the deepest papers in the subject. His enormous knowledge of classical algebraic geometry enabled him to identify those results which could be applied in finite spaces. His theorem on the characterization of conics, mentioned above, not only stimulated a great deal of research but also made many mathematicians realize that finite spaces were worth studying. His long paper, Segre

[1959a], gave a comprehensive list of results and methods, and is to my mind the seminal paper in the subject. This whole work was conceived while I was editing for Professor Segre the monograph, Segre [1967], which contains applications of the Hasse-Weil theorem to arcs and caps and which shows his ingenuity and originality. His most recent paper, Segre and Korchmáros [1977], on which he delivered a lecture at the British Combinatorial Colloquium in July, 1977, was a final gem.

University of Sussex J.W.P.H.
October, 1977

CONTENTS

PART I. INTRODUCTION

PART II. ELEMENTARY PROPERTIES OF $PG(n,q)$

PART I
INTRODUCTION

1
FINITE FIELDS

1.1. Definitions and existence

(i) A *field* is a set K closed under two operations +, × such that

(a) $(K, +)$ is an abelian group with identity 0;

(b) $(K_0, \times)$ is an abelian group with identity 1, where $K_0 = K \setminus \{0\}$;

(c) $x(y + z) = xy + xz$, $(x + y)z = xz + yz$ for all x, y, z in K.

(ii) A *finite field* is a field with only a finite number of elements.

(iii) The *characteristic* of a finite field K is the smallest positive integer (and hence a prime) p such that $px = 0$ for all x in K.

(iv) $GF(p) = \mathbf{Z}/p\mathbf{Z}$, p prime, is a finite field of p elements and consists of the residue classes of the integers modulo p.

(v) A finite field K of characteristic p has a subfield isomorphic to $GF(p)$ and has p^h elements for some h in $\mathbf{N}$.

(vi) If $F(x)$ is an irreducible polynomial of degree h over $GF(p)$, then

$$GF(p^h) = GF(p)[x]/(F(x))$$

$$= \{a_0 + a_1 t + \ldots + a_{h-1} t^{h-1} \mid a_i \in GF(p),\ F(t) = 0\}.$$

(vii) The elements of $GF(q)$, $q = p^h$, satisfy

$$x^q - x = 0,$$

and there exists s in $GF(q)$ such that

$$GF(q) = \{0, 1, s, \ldots, s^{q-2} \mid s^{q-1} = 1\}:$$

s is called a *primitive element* or *primitive root* of $GF(q)$.

(viii) Any field of q elements is isomorphic to $GF(q)$, which will also be denoted by γ.

1.2. Automorphisms

(i) The group of automorphisms of $\gamma = GF(q)$, $q = p^h$, is cyclic of order h. It is generated by the automorphism ϕ, where

$t\phi = t^p$: so $t\phi^i = t^{p^i}$.

Equivalently, if σ is any automorphism of γ, then $\sigma = \phi^i$ for some i. It is often convenient to write t^σ instead of $t\sigma$.

(ii) Since ϕ is an automorphism

$(s + t)^p = s^p + t^p$ for all s, t in γ.

(iii) If $F(x)$ is an irreducible polynomial of degree n over γ, and θ is a root of $F(x)$ in $GF(q^n)$, then the set of roots of $F(x)$ in $GF(q^n)$ is $\{\theta, \theta^q, \ldots, \theta^{q^{n-1}}\}$.

1.3. Functions

(i) With $\gamma[x]$ the ring of polynomials in x over $\gamma = GF(q)$, let $\Gamma[x] = \gamma[x]/(x^q - x)$ and $\mathcal{G}[x] = \{F \in \gamma[x] \mid \deg F < q\}$.

Then any two polynomials in $\gamma[x]$ with the same image in $\Gamma[x]$ take the same value for each element of γ. Thus there is a bijection $\alpha : \mathcal{G}[x] \to \Gamma[x]$ given by

$F\alpha = F(x) + (x^q - x)\gamma[x]$.

The inverse map $\beta : \Gamma[x] \to \ [x]$ is given by selecting from any class of $\Gamma[x]$ the unique polynomial of degree less than q.

(ii) Any function $f\colon \gamma \to \gamma$ is defined by a polynomial $F \in \mathcal{G}[x]$. In fact, $tf = F(t) = -\sum_{s\in\gamma} sf \cdot (t^q - t)/(t - s)$. This is Lagrange's classical interpolation formula.

(iii) If f in (ii) is a bijection, then $\deg F \leq q - 2$ when $q \geq 2$; for, the coefficient of t^{q-1} is $\sum_{s\in\gamma} sf = \sum_{s\in\gamma} s = 0$. When $q = 2$, $F(t) = t + 0f$.

(iv) For $q > 2$, suppose the function f is given by a polynomial F of degree at most $q - 2$ and $f|\gamma_0$ is a bijection; then $0f = F(0) = 0$ and f is a bijection. For,

$$tf = F(t) = 0f.(1 - t^{q-1}) - \sum_{s\in\gamma_0} sf.\ (t^q - t)/(t - s).$$

Since $f|\gamma_0$ is a bijection, $\sum_{s\in\gamma_0} sf = \sum_{s\in\gamma_0} s = 0$. So the co-

efficient of t^{q-1} is $-0f$. Since $\deg F \leq q - 2$, $0f = 0$ and f is a bijection.

(v) Let us call an element F of $\mathcal{G}[x]$, whose corresponding function f on γ is a bijection, a *permutation polynomial* and write $\mathscr{P}(q;x)$ for the set of permutation polynomials in the indeterminate x.

(vi) There is a useful criterion for a polynomial to be a permutation polynomial.

DICKSON'S CRITERION. *If* $F \in \mathscr{G}[x]$, *then* $F \in \mathscr{P}(q;x)$ *if and only if*

(a) *for* $r \not\equiv 0 \pmod p$ *and* $r \le q-2$, *the degree of* $F(x)^r$ *reduced modulo* $x^q - x$ *is at most* $q-2$;

(b) $F(t) = 0$ *has exactly one solution in* γ.

(vii) In the particular case that $p = 2$ and $F(0) = 0$, this becomes the following. *If* $F \in \mathscr{G}[x]$, *then* $F \in \mathscr{P}(q;x)$ *if and only if*

(a) *for* r *odd and* $r \le q-2$, *the degree of* $F(x)^r$ *reduced modulo* $x^q - x$ *is at most* $q-2$;

(b) $F(t) = 0 \Rightarrow t = 0$.

1.4. Quadratic equations

To solve, over $\gamma = GF(q)$ with $q = p^h$, the equation

$$ax^2 + bx + c = 0, \qquad a \ne 0, \tag{1.1}$$

the cases of odd and even characteristic are considered separately.

(i) Let $p \ne 2$; the *discriminant* $\Delta = b^2 - 4ac$. If $\Delta = 0$, then eqn (1.1) has one solution $x = -b/(2a)$. If Δ is a non-square, then (1.1) has no solutions.

(ii) Let $p = 2$. If $b = 0$, then eqn (1.1) has one solution $x = \surd(c/a)$. If $b \ne 0$, put $y = ax/b$ and $\partial = ac/b^2$. Then eqn (1.1) becomes

$$y^2 + y + \partial = 0 ; \tag{1.2}$$

$\partial = ac/b^2$ is the *S-invariant* of the equation.

Let $D(t) = D_2(t) = t + t^2 + t^4 + \dots + t^{2^{h-1}}$.

Then $D(t)^2 + D(t) = 0$, all t. So $D(\partial) = 0$ or 1.

If $D(\partial) = 0$, then (1.2) and so (1.1) has two solutions.

If $D(\partial) = 1$, then (1.2) and so (1.1) has no solutions.

If $y = s$ is one solution of (1.2), then $y = s + 1$ is the other.

If $D(\partial) = 0$ and $D(k) = 1$, then a solution of (1.2) is

$$y = k\,\partial^2 + (k + k^2)\partial^4 + \ldots + (k + k^2 + \ldots + k^{2^{h-2}})\partial^{2^{h-1}}. \tag{1.3}$$

If $q = 2^{2m+1}$ and $D(\partial) = 0$, then a solution of (1.2) is

$$y = \partial + \partial^{2^2} + \partial^{2^4} + \ldots + \partial^{2^{2m}}.$$

(iii) From (ii), for q even,

$$\gamma = \mathscr{C}_0 \cup \mathscr{C}_1,$$

where $\mathscr{C}_0 = \{t \in \gamma \mid D(t) = 0\}$ and $\mathscr{C}_1 = \{t \in \gamma \mid D(t) = 1\}$. $\mathscr{C}_0$ is the set of elements of *category zero* and $\mathscr{C}_1$ the set of elements of *category one*.

(a) $0 \in \mathscr{C}_0$;

(b) $q = 2^{2m} \Rightarrow 1 \in \mathscr{C}_0$;

(c) $q = 2^{2m+1} \Rightarrow 1 \in \mathscr{C}_1$;

(d) $t \in \mathscr{C}_i \Rightarrow t^\sigma \in \mathscr{C}_i$ for any automorphism σ of γ ;

(e) $s \in \mathscr{C}_i,\ t \in \mathscr{C}_j \Rightarrow \begin{cases} s + t \in \mathscr{C}_0 & \text{if } i = j \\ s + t \in \mathscr{C}_1 & \text{if } i \neq j. \end{cases}$

(f) $|\mathscr{C}_0| = |\mathscr{C}_1| = q/2$.

1.5. *n*-th roots

To solve the equation $x^n = c$ in $\gamma = GF(q)$, let $d = (n, q-1)$, $e = (q-1)/d$ and s be a primitive element of γ.

(i) $x^n = 1$ has d solutions in γ, namely $x = 1, s^e, s^{2e}, \ldots, s^{(d-1)e}$.

(ii) $x^n = 1$ has the unique solution $x = 1$ if $d = 1$.

(iii) $x^n = 1$ has n solutions if $n|(q-1)$, namely $x = 1, s^{(q-1)/n}, \ldots, s^{(n-1)(q-1)/n}$.

(iv) $x^n = c$ has a unique solution if $d = 1$, namely $x = c^r$ where $r, r' \in \mathbf{Z}$ and $rn + r'(q-1) = 1$.

(v) $x^n = c$ has n solutions if $n|(q-1)$ and $c^{(q-1)/n} = 1$.

(vi) $x^2 = c$ has a unique solution in $GF(2^h)$, namely

$x = c^{2^{h-1}}$.

(vii) $x^2 = c$ has two solutions for exactly half the non-zero values of c and no solutions for the other half.

(viii) For $\gamma = GF(q)$, the following are equivalent:

(a) q is even;

(b) $(q-1,2) = 1$;

(c) $(q-1,4) = 1$;

(d) $x^2 = 1$ has exactly one solution in γ;

(e) $x^4 = 1$ has exactly one solution in γ.

(ix) For $\gamma = GF(q)$, the following are equivalent:

(a) $(q-1,4) = 4$;

(b) $q \equiv 1 \pmod 4$;

(c) -1 is a square in γ;

(d) $x^4 = 1$ has four solutions in γ.

(x) For $\gamma = GF(q)$, the following are equivalent:

(a) $(q-1, 4) = 2$;

(b) $q \equiv -1 \pmod 4$;

(c) -1 is a non-square in γ;

(d) $x^4 = 1$ has exactly two solutions in γ;

(e) The squares in γ are the fourth powers in γ.

(xi) For $\gamma = GF(q)$, the following are equivalent:

	q odd	*q even*
(a)	$(q-1, 3) = 3$	$(q-1, 3) = 3$
(b)	$q \equiv 1 \pmod 3$	$q \equiv 1 \pmod 3$
(c)	$x^2 + x + 1$ has two distinct roots in γ	$x^2 + x + 1$ has two distinct roots in γ
(d)	-3 is a non-zero square in γ	$1 \in \mathscr{C}_0$
(e)	$x^3 = 1$ has three solutions in γ	$x^3 = 1$ has three solutions in γ
(f)		$q = 2^{2m}$

(xii) For $\gamma = GF(q)$, the following are equivalent:

	q odd	*q even*
(a)	$q \equiv -1 \pmod 3$	$q \equiv -1 \pmod 3$
(b)	$x^2 + x + 1$ has no root in γ	$x^2 + x + 1$ has no root in γ

(c) -3 is a non-square in γ	$1 \in \mathscr{C}_1$
(d) $x^3 = 1$ has one solution in γ and three in $\gamma' = \mathrm{GF}(q^2)$	$x^3 = 1$ has one solution in γ and three in γ'
(e)	$q = 2^{2m+1}$

(xiii) For $\gamma = GF(q)$, the following are equivalent:

(a) $q \equiv 0 \pmod 3$;

(b) $x^2 + x + 1$ has one root in γ;

(c) $x^3 = 1$ has a triple solution in γ;

(d) $q = 3^h$.

1.6. Primitive and subprimitive roots of polynomials

Let $\mathscr{N}(m,q)$ be the *set of monic irreducible polynomials over* $GF(q)$ *of degree* m. Write $N(m,q) = \mathscr{N}|(m,q)|$.

(i) If $F \in \mathscr{N}(m,q)$, then F has *exponent* e if e is the smallest positive integer such that $F(x)$ divides $x^e - 1$. The exponent e always divides $q^m - 1$. If $e = q^m - 1$, then F is *primitive* and has a primitive root in $GF(q^m)$. So if α is a root in $GF(q^m)$ of a primitive F, then α has order $q^m - 1$.

(ii) If α is a primitive root of $\mathrm{GF}(q^m)$, then $\beta = \alpha^{\theta(m-1)}$ is a primitive root of the subfield $GF(q)$, where $\theta(n) = (q^{n+1} - 1)/(q - 1)$. More generally, if $r|m$, then $\kappa = \alpha^{\theta(m-1)/\theta(r-1)}$ is a primitive root of the subfield $GF(q^r)$.

(iii) In particular, if α is a primitive root of $GF(q^2)$, then $\beta = \alpha^{q+1}$ is a primitive root of the subfield $GF(q)$. So $GF(q) = \{0,1,\beta \ldots,\beta^{q-2}\}$. Further, if $t \in GF(q^2)$, then $t^{q+1} \in GF(q)$. Thus the equation $x^{q+1} = c$ has $q + 1$ solutions or none according as c is or is not in $GF(q)$.

(iv) Let $\mathscr{N}^e(m,q) = \{F \in \mathscr{N}(m,q) \mid \text{exponent of } F \text{ is } e\}$. Then $\mathscr{N}(m,q) = \bigcup_e \mathscr{N}^e(m,q)$, where e runs over all divisors of $q^m - 1$ which are not divisors of $q^r - 1$ for $r < m$. Then $|\mathscr{N}^e(m,q)| = \phi(e)/m$, where ϕ is the Euler totient function. For $e = q^m - 1$, write $P(m,q) = |\mathscr{N}^e(m,q)|$.

(v) For projective spaces, there is a larger subset of $\mathscr{N}(m,q)$ than the primitive polynomials which is pertinent. If $F \in \mathscr{N}(m,q)$ then F has *subexponent* e if e is the smallest positive integer such that $F(x)$ divides $x^e - c$ for some c in $GF(q)$. The subexponent e always divides $\theta(m - 1) = (q^m - 1)/(q - 1)$. If $e = \theta(m - 1)$, then F is *subprimitive* and has an *m-subprimitive* (or, usually, *subprimitive*) root in $GF(q^m)$.

(vi) If α is a root of a subprimitive F, then the smallest positive integer n such that α^n is in the subfield $GF(q)$ of $GF(q^m)$ is $\theta(m - 1)$. So, a subprimitive polynomial for which α is a primitive root of $GF(q)$ is primitive.

(vii) Let $R(m,q)$ be the number of subprimitive polynomials in $\mathscr{N}(m,q)$. Then

$$P(m,q) \leq R(m,q) \leq N(m,q).$$

If the prime decomposition of m is $m = p_1^{k_1} \ldots p_s^{k_s}$, then

$$mN(m,q) = q^m - \Sigma'' q^{m/p_i} + \Sigma'' q^{m/p_i p_j} \ldots + (-1)^s q^{m/p_1 \ldots p_s}.$$

In particular, if $m = p^k$, then

$$N(m,q) = (q^m - q^{m/p})/m$$

and, if m is a prime,

$$N(m,q) = (q^m - q)/m.$$

Thus $N(2,q) = (q^2 - q)/2$, $N(3,q) = (q^3 - q)/3$, $N(4,q) = (q^4 - q^2)/4$.

(viii) $P(m,q) = \phi(q^m - 1)/m$.

Thus $P(2,q) = \phi(q^2 - 1)/2$, $P(3,q) = \phi(q^3 - 1)/3$, $P(4,q) = \phi(q^4 - 1)/4$.

(ix) If α is a primitive root of $GF(q^m)$, then α^n is primitive if and only if $(n, q^m - 1) = 1$. However, α^n is m-subprimitive if and only if $(n, \theta(m - 1)) = 1$. Hence $mR(m,q)$ is the number of integers less than $q^m - 1$ and prime to $\theta(m)$. The number of integers less than and prime to $\theta(m - 1)$ is $\phi(\theta(m - 1))$: if $0 < x < \theta(m - 1)$, then $x + k\theta(m - 1)$ is or is not prime to $\theta(m - 1)$, according as x is or is not prime to $\theta(m - 1)$. Thus

$$R(m,q) = (q-1)\,\phi(\theta(m-1))/m.$$

So $R(2,q) = (q-1)\phi(q+1)/2$, $R(3,q) = (q-1)\phi(q^2+q+1)/3$, $R(4,q) = (q-1)\phi((q+1)(q^2+1))/4$.

(x) For small values of m and q, $N(m,q)$, $R(m,q)$, and $P(m,q)$ are compared in Table 1.1.

TABLE 1.1

q	2	3	4	5	7	8	9
$N(2,q)$	1	2	6	10	21	28	36
$R(2,q)$	1	2	6	4	12	21	16
$P(2,q)$	1	2	4	4	8	18	16
$N(3,q)$	2	8	20	40	112	168	240
$R(3,q)$	2	8	12	40	72	168	192
$P(3,q)$	2	4	12	20	36	144	96
$N(4,q)$	3	18	60	150	588	1008	1620
$R(4,q)$	2	8	48	48	240	504	640
$P(4,q)$	2	8	32	48	160	432	640

1.7. Notation for small finite fields

$GF(2) = \{0,1 \mid 2 = 0\}$

$GF(3) = \{0,1,-1 \mid 3 = 0\}$

$GF(4) = \{0,1,\omega,\omega^2 \mid 2 = \omega^2 + \omega + 1 = \omega^3 + 1 = 0\}$

$GF(5) = \{0,1,-1,2,-2 \mid 5 = 0\}$

$GF(7) = \{0,1,-1,2,-2,3,-3 \mid 7 = 0\}$

$GF(8) = \{0,1,\varepsilon,\varepsilon^2,\varepsilon^3,\varepsilon^4,\varepsilon^5,\varepsilon^6 \mid 2 = \varepsilon^3 + \varepsilon^2 + 1 = \varepsilon^5 + \varepsilon + 1 = \varepsilon^6 + \varepsilon^4 + 1 = \varepsilon^7 + 1 = 0\}$

$GF(9) = \{0,1,-1,\sigma,-\sigma,\sigma^2,-\sigma^2,\sigma^3,-\sigma^3 \mid 3 = \sigma^2 - \sigma - 1 = \sigma^3 + \sigma - 1 = \sigma^3 + \sigma^2 + 1 = \sigma^4 + 1 = 0\}$

$GF(16) = \{0,1,\eta^i \mid i \in \mathbf{N}_{14}, 2 = \eta^4 + \eta + 1 = \eta^{15} + 1 = 0\}$

$GF(32) = \{0,1,\chi^i \mid i \in \mathbf{N}_{30}, 2 = \chi^5 + \chi^2 + 1 = \chi^{31} + 1 = 0\}$

$GF(64) = \{0,1,\phi^i | i \in \mathbf{N}_{62}, 2 = \phi^6 + \phi + 1 = \phi^{63} + 1 = 0\}$

$GF(128) = \{0,1,\beta^i | i \in \mathbf{N}_{126}, 2 = \beta^7 + \beta + 1 = \beta^{127} + 1 = 0\}$

$GF(256) = \{0,1,\rho^i | i \in \mathbf{N}_{254}, 2 = \rho^8 + \rho^4 + \rho^3 + \rho^2 + 1 = \rho^{255} + 1 = 0\}$

Complete addition tables for $GF(16)$, $GF(32)$, $GF(64)$, $GF(128)$, and $GF(256)$ are given in Appendix II.

For an example of subprimitivity, let us consider $GF(16)$ which can be regarded both as a quadratic extension of $GF(4)$ and as a quartic extension of $GF(2)$.

(i) $\mathcal{N}(2,4)$: $N(2,4) = R(2,4) = 6$, $P(2,4) = 4$

Elements of $\mathcal{N}(2,4)$	Roots in $GF(16)$
$t^2 + t + \omega$	η, η^4
$t^2 + \omega t + 1$	η^6, η^9
$t^2 + \omega^2 t + \omega^2$	η^{11}, η^{14}
$t^2 + t + \omega^2$	η^2, η^8
$t^2 + \omega^2 t + 1$	η^3, η^{12}
$t^2 + \omega t + \omega$	η^7, η^{13}

where ω and η are as above, and $\omega = \eta^5$.

The 12 elements $\eta, \eta^2, \eta^3, \eta^4, \eta^6, \eta^7, \eta^8, \eta^9, \eta^{11}, \eta^{12}, \eta^{13}, \eta^{14}$ are the 2-subprimitive roots of $GF(16)$.

(ii) $\mathcal{N}(4,2)$: $N(4,2) = 3$, $R(4,2) = P(4,2) = 2$

Elements of $\mathcal{N}(4,2)$	Roots in $GF(16)$
$t^4 + t + 1$	$\eta, \eta^2, \eta^4, \eta^8$
$t^4 + t^3 + 1$	$\eta^7, \eta^{11}, \eta^{13}, \eta^{14}$
$t^4 + t^3 + t^2 + t + 1$	$\eta^3, \eta^6, \eta^9, \eta^{12}$
$= (t^2 + \omega t + 1)(t^2 + \omega^2 t + 1)$	

again with ω, η as above and $\omega = \eta^5$. The eight elements $\eta, \eta^2, \eta^4, \eta^7, \eta^8, \eta^{11}, \eta^{13}, \eta^{14}$ are the 4-subprimitive roots of $GF(16)$.

1.8. Invariants of polynomials

Let F be in $\gamma[x]$ and of degree n, where $\gamma = GF(q)$ and $q = p^h$. So

$$F(x) = a_0 x^n + a_1 x^{n-1} + \ldots + a_n, \quad a_0 \neq 0.$$

Let α_i, $i \in \mathbf{N}_n$, be the roots of F in some extension of γ. The *discriminant* $\Delta = \Delta(F) = d^2$, where

$$d = a_0^{n-1} \prod_{i<j} (\alpha_i - \alpha_j).$$

When $p = 2$, d is a symmetric function of the roots and so is a form of degree $2n - 2$ in the coefficients of F.

1.8.1. LEMMA: *If q is odd and F is irreducible over γ of degree $n>1$, then the following are equivalent:*

(i) $d^q = d$;

(ii) *n is odd*;

(iii) *Δ is a square in γ*.

Proof. If $\alpha_1, \alpha_2, \ldots, \alpha_n$ are the roots of F, then $\gamma' = \gamma(\alpha_n)$ is a simple extension of γ, in which the other roots of F are conjugates of α_n. So we may put $\alpha_i = \alpha_n{}^{q^i}$, $i \in \mathbf{N}_{n-1}$. Hence $\alpha_i{}^q = \alpha_{i+1}$, all i, where $\alpha_{n+1} = \alpha_1$. So $d \in \gamma'$ and $d^q = a_0{}^{n-1} \prod_{i<j} (\alpha_{i+1} - \alpha_{j+1}) = (-1)^{n-1} d$.

Therefore (i) and (ii) are equivalent.

Now, $d^q = d$ if and only if $d \in \gamma$. Since $\Delta = d^2$, (i) and (iii) are equivalent.□

If F and G are in $\gamma[x]$ and of degree n and m respectively, then $F = \Sigma a_i x^{n-i}$ and $G = \Sigma b_i x^{m-i}$ with $a_0 b_0 \neq 0$. The *resultant* $R = R(F, G)$ of F and G is

$$R = a_0{}^m b_0{}^n \prod (\alpha_i - \beta_j),$$

where $\alpha_1, \ldots, \alpha_n$ are the roots of F and $\beta_1, \ldots, \beta_m$ are the roots of G.

Also

$$R = a_0^{\ m} \ \Pi G(\alpha_i) = (-1)^{mn} \ b_0^{\ n} \ \Pi F(\beta_j)$$

and R is a form of degree $m + n$ in the coefficients of F and G.

1.8.2. LEMMA: *If* $F = F_1 F_2 \ldots F_k$ *and* $\Delta(F) \neq 0$, *where each* F_i *is in* $\gamma[x]$, *then*

$$\Delta[F] = \Delta_1 \Delta_2 \ldots \Delta_k \left(\prod_{i<j} R_{ij} \right)^2,$$

where $\Delta_i = \Delta(F_i)$ and $R_{ij} = R(F_i, F_j)$.□

1.8.3. LEMMA: *If* q *is odd and* $F = F_1 F_2 \ldots F_k$ *with* $\Delta(F) \neq 0$ *and each* F_i *in* $\gamma[x] \backslash \gamma$ *and irreducible, then* Δ *is a square or a non-square according as the number of irreducible factors* F_i *of even order is even or odd.*

Proof. This follows directly from lemmas 1.8.1 and 1.8.1 and 1.8.2.□

A comparable result is required for q even. As above, let F in $\gamma[x]$ have roots $\alpha_1, \alpha_2, \ldots, \alpha_n$ with $\alpha_i \neq \alpha_j$ for $i \neq j$, so that $\Delta(F) \neq 0$. The *S-invariant* $\partial = \partial(F)$ is

$$\partial = \Sigma'' \ \alpha_i \ \alpha_j \ / \ (\alpha_i + \alpha_j)^2;$$

also write

$$\delta = \Sigma'' \ \alpha_i \ / \ (\alpha_i + \alpha_j).$$

For completeness, define $\partial = 0$ when $\deg F = 1$.

1.8.4. LEMMA: *If* q *is even and* F *is irreducible over* γ *of degree* $n > 1$, *then the following are equivalent:*

(i) $\delta^q = \delta$;

(ii) n *is odd;*

(iii) $\partial \in \mathscr{C}_0$.

Proof. As in lemma 1.8.1, let $\alpha_1, \alpha_2, \ldots, \alpha_n$ be the roots of F, which may be chosen so that $\alpha_i = \alpha_n^{q^i}$, $i \in \mathbf{N}_{n-1}$; hence $\alpha_i^q = \alpha_{i+1} = \alpha_1$. So

$$\delta^q = \Sigma'' \alpha_{i+1}/(\alpha_{i+1} + \alpha_{j+1}) = (n - 1) + \delta.$$

Hence (i) and (ii) are equivalent.

Also $\delta^2 + \delta = \partial$. So $\partial \in \mathscr{C}_0$ or $\mathscr{C}_1$ according as δ is or is not in γ; that is, (i) and (iii) are equivalent.□

1.8.5. LEMMA: *If q is even and $F = F_1F_2\ldots F_k$, where $\Delta(F) \neq 0$ and* each F_i is in $\gamma[x]$, *then*

$$\partial(F) = \partial_1 + \partial_2 + \ldots + \partial_k + R,$$

where $\partial_i = \partial(F_i)$ and $R \in \mathscr{C}_0$.

Proof. It suffices to prove the result when $F = F_1F_2$. Suppose $\alpha_1, \alpha_2, \ldots, \alpha_n$ are the roots of F_1 and $\beta_1, \ldots, \beta_m$ the roots of F_2. Then

$$\partial(F) = \partial_1 + \partial_2 + R,$$

where $R = \Sigma\alpha_i\beta_j/(\alpha_i^2 + \beta_j^2)$.

Put $r = \Sigma\alpha_i/(\alpha_i + \beta_j)$. Then $R = r^2 + r$. However, r is a symmetric function both of the roots of F_1 and of the roots of F_2. So $r \in \gamma$ and hence $R \in \mathscr{C}_0$, §1.4.□

1.8.6. LEMMA: *If q is even and $F = F_1F_2\ldots F$ with $\Delta(F) \neq 0$ and each F_i in $\gamma[x]\backslash\gamma$ and irreducible, then $\partial(F)$ is in $\mathscr{C}_0$ or $\mathscr{C}_1$ according as the number of irreducible factors of even order is even or odd.*

Proof. This is a consequence of lemmas 1.8.4 and 1.8.5.□

To unite even and odd characteristic, we extend the notion

of category zero and one to γ with q odd. If $t \in \gamma$ with q odd, then t is of *category zero or one* according as t is a square or a non-square. So we have a function $C : \gamma \to \{0,1\}$, where $C(t) = 0$ for t of category zero and $C(t) = 1$ for t of category one. In fact,

$$\text{for } q = 2^h,\ C(t) = D_2(t) = t + t^2 + \ldots + t^{2^{h-1}};$$
$$\text{for } q = p^h,\ p > 2,\ C(t) = (1 - t^{(q-1)/2})/2.$$

1.8.7. LEMMA.

(i) For q even, $C(s + t) \equiv C(s) + C(t) \bmod 2$;
(ii) for q odd, $C(st) \equiv C(s) + C(t) \bmod 2$.□

A polynomial F in $\gamma[x]$ of discriminant $\Delta \neq 0$ and S-invariant ∂ has *category* $C(F)$ given by

$$C(F) = C(\Delta) \text{ for } q \text{ odd},$$
$$C(F) = C(\partial) \text{ for } q \text{ even}.$$

Lemmas 1.8.3 and 1.8.6 can now be stated together.

1.8.8. THEOREM. *If $F = F_1F_2\ldots F_k$ with $\Delta(F) \neq 0$ and each F_i in $\gamma[x]\backslash\gamma$ and irreducible, then $C(F)$ is 0 or 1 according as the number of irreducible factors F_i of even order is even or odd.*□

COROLLARY 1.

(i) deg $F = 2$: *if $\Delta \neq 0$, then F has roots in γ if and only if $C(F) = 0$.*
(ii) deg $F = 3$: *if $\Delta \neq 0$, then F has exactly one root in γ if and only if $C(F) = 1$.*
(iii) deg $F = 4$: *if $\Delta \neq 0$, then F has one root or four roots in γ or is the product of two irreducible quadratics if and only if $C(F) = 0$.*□

COROLLARY 2. *If F in $\gamma[x]$ is irreducible, then $C(F)$ is zero or one according as deg F is even or odd.*□

COROLLARY 3. *If $F_1, F_2 \in \gamma[x]$*, then $C(F_1F_2) \equiv C(F_1) + C(F_2) \bmod 2$.□

We record Δ, d, and ∂ for deg F = 2, 3 and 4. In fact, for q even, $\partial = \tilde{\omega}/\Delta = \tilde{\omega}/d^2$, where $\tilde{\omega}$ is a symmetric polynomial in the roots of F. So $\tilde{\omega}$ rather than ∂ is given.

(i) deg $F=2$: $\Delta=a_1^2-4a_0a_2$; for $p=2$, $d=a_1$, $\tilde{\omega}=a_0a_2$.

(ii) deg $F=3$: $\Delta=a_1^2a_2^2-4a_0a_2^3-4a_1^3a_3-27a_0^2a_3^2+18a_0a_1a_2a_3$;

for $p=2$, $d=a_1a_2+a_0a_3$, $\tilde{\omega}=a_0a_2^3+a_1^3a_3+a_0^2a_3^2+a_0a_1a_2a_3$.

(iii) deg $F=4$:

(a) $p=3$, $\Delta=(a_0a_4-a_1a_3)^3 + (a_0a_4-a_1a_3)^2a_2^2$
$-(a_0a_3^2+a_1^2a_4-a_0a_2a_4)a_2^3$;

(b) $p\neq 3$, $\Delta=(4I^3-J^2)/27$,

$I=a_2^2-3a_1a_3+12a_0a_4$,

$J=2a_2^3+27a_1^2a_4+27a_0a_3^2-72a_0a_2a_4-9a_1a_2a_3$;

for $p=2$, $d=J=a_1^2a_4+a_0a_3^2+a_1a_2a_3$,

$\tilde{\omega} = a_1^4a_4^2+a_1^3a_3^3+a_1^3a_2a_3a_4+a_1^2a_2^3a_4$
$+a_0a_1a_2a_3^3+a_0a_2^3a_3^2+a_0^2a_3^4$.

1.9. Some equations with few terms

We wish to find roots of polynomials of the form

$$F(x) = x^{q'} - bx - c\ , \tag{1.5}$$

where $q' = p^r$ with $h = mr$ and, as usual, $F \in \gamma[x]$ with $\gamma = GF(q)$, $q = p^h$. Here we take $\gamma' = GF(q')$ as a subfield of γ.

Firstly, we consider the case with $b = 1$:

$$G(x) = x^{q'} - x - d\ . \tag{1.6}$$

Define the element $D_{q'}$ of $\gamma[x]$ as follows:

$$D_{q'}(x) = x + x^{q'} + x^{q'^2} + \ldots + x^{q'^{m-1}}\ . \tag{1.7}$$

In particular,

$$D_p(x) = x + x^p + \ldots + x^{p^{h-1}}.$$

$D_{q'}$ is just a generalization of D_2 from §1.4.

1.9.1. THEOREM. *G has q' or no roots in γ according as $D_{q'}(d)$ is zero or not.*

Proof. $D_{q'}(t)^{q'} = D_{q'}(t)$ for any t in γ. So $D_{q'}$ only takes values in γ'. Therefore $D_{q'} - e$ has exactly $q'^{m-1} = q/q'$ roots in γ for any e in γ'.

If α and β are two roots of G, then $(\alpha - \beta)^{q'} = \alpha^{q'} - \beta^{q'} = (\alpha + d) - (\beta + d) = \alpha - \beta$. So $\alpha - \beta \in \gamma'$ and, if G has one root α in γ, it has q' roots in γ obtained by adding the elements of γ' to α. So G has q' roots in γ for q/q' values of d and no roots in γ for the others.

Finally, if α is a root of G, then

$$0 = \sum_{i=0}^{m-1} G(\alpha)^{q'^i} = \alpha^q - \alpha - D_{q'}(\alpha).$$

So $D_{q'}(\alpha) = 0$ if and only if $\alpha \in \gamma$.□

COROLLARY. *If* $D_{q'}(d) = 0$ *and* k *is an element of* γ *with* $D_{q'}(k) = -1$, *then the roots of* G *are*

$$\alpha = \alpha' - kd^{q'} - (k + k^{q'})d^{q'^2} \dots - (k + k^{q'} + \dots + k^{q'^{m-2}})d^{q'^{m-1}},$$

where α' *varies in* γ'.□

Now we return to the original eqn (1.5).

1.9.2. LEMMA. *If* α *is a root of* F *in* γ, *then*

(i) $\alpha^{q'^i} = b_i\alpha + c_i$, $i \in \mathbf{N}_m$,

where $b_1 = b$, $c_1 = c$, *and, in general,*

$$b_i = b^{1+q'+q'^2+\dots+q'^{i-1}},$$

$$c_i = cb^{q'+q'^2+\dots+q'^{i-1}} + c^{q'}b^{q'^2+\dots q'^{i-1}} + \dots + c^{q'^{i-1}};$$

(ii) $(b_m - 1)\alpha + c_m = 0$.□

1.9.3. THEOREM. *If* $\gamma = GF(q)$ *with* $q = q'^m$, *then* $F(x) = x^{q'} - bx - c \in \gamma[x]$ *has zero, one, or* q' *distinct roots in* γ:

(i) *when* $b_m \neq 1$, F *has the one root* $\alpha = c_m/(1 - b_m)$;

(ii) *when* $b_m = 1$ *and* $c_m \neq 0$, F *has no roots;*

(iii) *when* $b_m = 1$ *and* $c_m = 0$, F *has* q' *distinct roots.*

Proof. Part (i) follows from part (ii) of the previous lemma. If $b_m = 1$, then $b^{(q+1)/(q'-1)} = 1$. So, by §1.5(iii), there exists s in γ such that $b = s^{q'-1}$. Hence putting

$x = sy$ and $c = ds^{q'}$, we obtain

$$F(x) = F(sy) = s^{q'}y^{q'} - bsy - ds^{q'} = s^{q'}G(y)$$

and

$$c_m = sD_{q'}(d).$$

So (ii) and (iii) follow from theorem 1.9.1.□

COROLLARY 1. *Over* $\gamma = GF(3^h)$, $F(x) = x^3 - bx - c$ *has zero, one, or three distinct roots.*

(i) *If* $b = s^2$ *with* s *in* γ_0, *then*

(a) *if* $D_3(c/s^3) = 0$, F *has three roots in* γ;

(b) *if* $D_3(c/s^3) \neq 0$, F *has no roots in* γ.

(ii) *If* b *is zero or a non-square,* F *has one root* α in γ:

(a) *if* $b = 0$, *then* $\alpha = c^{3^{h-1}}$;

(b) *if* $b \neq 0$, *then*

$$\alpha = -cb^{(3^h-3)/2} - c^3b^{(3^h-3^2)/2} - \ldots - c^{3^{h-1}}.$$

Proof. In the theorem, $q' = 3$ and $m = h$. So $b_h = b^{(3^h-1)/2}$. Hence $b_h = 0$, 1, or -1 as b is zero, a non-zero square, or a non-square.□

COROLLARY 2. *Over* $\gamma = GF(q)$, q *square,* $F(x) = \bar{x} - bx - c$ *has zero, one or* $\sqrt{q}$ *distinct roots in* γ, *where* $\bar{x} = x^{\sqrt{q}}$.

(i) *If* $b\bar{b} = 1$, *then*

(a) *if* $\bar{c} = -c\bar{b}$, F *has* $\sqrt{q}$ *distinct roots in* γ;

(b) *if* $\bar{c} \neq -c\bar{b}$, F *has no roots in* γ.

(ii) *If* $b\bar{b} \neq 1$, *then* F *has the unique root* α *in* γ *where* $\alpha = (c\bar{b} + \bar{c})/(1 - b\bar{b})$.

(iii) *When* $b = -1$, F *has zero or* $\sqrt{q}$ *roots in* γ *as* c *is in* $GF(\sqrt{q})$ *or not.*□

1.9.4. THEOREM. *Over* $\gamma = GF(2^h)$, $G(x) = x^4 + x + c$ *has zero, two, or four distinct roots.*

(i) *When* h *is even,* G *has four or zero roots in* γ *as* $D_4(c)$ *is zero or not; if* α *is a root of* G *in* γ, *the others are* $\alpha + 1$, $\alpha + \omega$, $\alpha + \omega^2$, *where* $\omega^2 + \omega + 1 = 0$.

(ii) *When* h *is odd,* G *has two or zero roots in* γ *as* $D_2(c)$ *is zero or not; if* α *is a root of* G *in* γ, *the other is* $\alpha + 1$.

Proof. Part (i) is simply a corollary to theorem 1.9.1 with $q' = 4$ and $q = 4^m = 2^h$.

For part (ii), if α and β are roots of G in γ, then $(\alpha + \beta)^4 = \alpha + \beta$, whence $\alpha = \beta$ or $\alpha = \beta + 1$; since h is odd, $x^2 + x + 1$ is irreducible over γ and there are no other alternatives. So G has two roots or none in γ. Then, if α is a root of G in ,

$$0 = \sum_{i=0}^{h-1} G(\alpha)^{2^i} = D_2(c). \square$$

COROLLARY 1. *The roots* α *of* G *are as follows:*

(i) $h = 2m$: *let* k *in* γ *satisfy* $D_4(k) = 1$. *Then*

$$\alpha = \lambda + kc^4 + (k + k^4)\, c^{4^2} + \ldots + (k + k^4 + \ldots + k^{4^{m-2}})\, c^{4^{m-1}},$$

where $\lambda = 0, 1, \omega, \omega^2$.

(ii) $h = 2m + 1$: *since* $D_2(c) = 0$, *there exists* s *in* γ *with* $s^2 + s = c$. *Then*

$$\alpha = \lambda + s^2 + s^{2^3} + \ldots + s^{2^{h-2}},$$

where $\lambda = 0, 1$.

More explicitly, there is the following:

(a) $m = 2n$ *and* $h = 4n + 1$:

$$\alpha = \lambda + (c^2 + c^{2^2}) + (c^{2^5} + c^{2^6}) + \ldots + (c^{2^{h-4}} + c^{2^{h-3}});$$

(b) $m = 2n + 1$ *and* $h = 4n + 3$:

$$\alpha = \lambda + (c + c^2) + (c^{2^4} + c^{2^5}) + \ldots + (c^{2^{h-3}} + c^{2^{h-2}}).$$

In both (a) *and* (b), $\lambda = 0, 1$.

Proof. All these solutions can be verified by substitution although (a) and (b) are deducible from the first case in (ii) by use of the formula (1.4) from §1.4 for s; namely, $s = c + c^{2^2} + c^{2^4} + \ldots + c^{2^{h+1}}$. $\square$

COROLLARY 2. *Let* $\gamma = GF(q)$ *with* $q = 4^m$ *and let* $c \in \gamma$. *Then three or one of* $x^2 + x + c$, $x^2 + x + \omega c$, $x^2 + x + \omega^2 c$ *is reducible or irreducible over* γ.

Proof. If $\lambda^3 = 1$, then

$$D_2(\lambda c) = \sum_{i=0}^{2m-1} (\lambda c)^{2^i} = \sum_{i=0}^{m-1}\left[(\lambda c)^{2^{2i}} + (\lambda c)^{2^{2i+1}}\right]$$

$$= D_4(\lambda c) + D_4(\lambda c)^2 = \lambda D_4(c) + \lambda^2 D_4(c)^2 .$$

If $x^4 + x + c$ is reducible, then by the theorem $D_4(c) = 0$; so $D_2(\lambda c) = 0$ and all three quadratics are reducible. If $D_4(c) = \mu$ with $\mu^3 = 1$, then $D_2(\lambda c) = \lambda\mu + \lambda^2\mu^2$. So $D_2(\lambda c) = 0$ for exactly one non-zero value of λ, namely $\lambda = 1/\mu$.

1.10. Cubic Equations

In this section we describe how to solve cubic equations over $\gamma = GF(q)$, $q = p^h$, and begin with those for which $p \neq 3$.

1.10.1. THEOREM. *Let* $F(x) = a_0x^3 + a_1x^2 + a_2x + a_3 \in \gamma[x]$, $p \neq 3$, *and let* $\Delta(F) \neq 0$. *Then*

$$F(x) = b_1(x - \beta_1)^3 + b_2(x - \beta_2)^3 ,$$

where $b_1 = F(\beta_2)/(\beta_2 - \beta_1)^3$, $b_2 = F(\beta_1)/(\beta_1 - \beta_2)^3$ *and* β_1 β_2 *are the roots of the* Hessian H *of* F, *where*

$$H(x) = (3a_0a_2 - a_1^2)\, x^2 + (9a_0a_3 - a_1a_2)\, x + (3a_1a_3 - a_2^2)$$

$$= A_0x^2 + A_1x + A_2 .$$

Proof. By equating the two forms for F, we obtain

$$a_0:a_1:a_2:a_3 = (b_1 + b_2) : -3(b_1\beta_1 + b_2\beta_2) : 3(b_2\beta_1^2 + b_2\beta_2^2) : -(b_1\beta_1^3 + b_2\beta_2^3).$$

Hence
$$-\frac{b_2}{b_1} = \frac{a_1 + 3a_0\beta_1}{a_1 + 3a_0\beta_2} = \frac{a_2 - 3a_0\beta_1}{a_1 - 3a_0\beta_2^2} = \frac{a_3 + a_0\beta_1^3}{a_3 + a_0\beta_2^3} . \tag{1.8}$$

So
$$-\frac{3}{a_1} = \frac{3(\beta_1 + \beta_2)}{a_2 + 3a_0\beta_1\beta_2} = \frac{\beta_1^2 + \beta_1\beta_2 + \beta_2^2}{-a_3 + a_0\beta_1\beta_2(\beta_1 + \beta_2)} ,$$

which equations give the required values for $\beta_1 + \beta_2$ and $\beta_1\beta_2$.

It is assumed in the proof that $\beta_1 \neq \beta_2$. But, as is explained in theorem 1.10.4, this is precisely the condition $\Delta(F) \neq 0$.□

COROLLARY 1. *If* $\Delta(F) \neq 0$, *and* $\alpha_1', \alpha_2', \alpha_3'$, *are the roots of* $y^3 - F(\beta_1)/F(\beta_2)$, *then the roots* $\alpha_1, \alpha_2, \alpha_3$ *of* F *are*

$$\alpha_i = (\beta_2\,\alpha_i' - \beta_1)/(\alpha_i' - 1), \qquad i = 1, 2, 3.$$

Proof. By the theorem, if $F(x) = 0$, the substitution $y = (x - \beta_1)/(x - \beta_2)$ gives $y^3 = e$, where $e = -b_2/b_1 = F(\beta_1)/F(\beta_2)$. Hence the inverse substitution gives the result.□

COROLLARY 2. *Let* $\Delta(F) = 0$. *Then*

(i) *if* F *has roots* $\alpha_1, \alpha_1, \alpha_2$ $(\alpha_2 \neq \alpha_1)$, *they are both in* γ *and* α_1 *is the common root of* $a_1x^2 + 2a_2x + 3a_3$ *and* $3a_0\,x^2 + 2a_1\,x + a_2$, *while* $\alpha_2 = -a_1/a_0 - 2\alpha_1$;

(ii) *if* F *has roots* $\alpha_1, \alpha_1, \alpha_1$, *we have* $\alpha_1 = -a_1/3a_0$, *which occurs when* $A_0 = A_1 = A_2 = 0$.□

The roots of a cubic must now be described qualitatively. We continue with the case $p \neq 3$. The field γ has *character* χ, where

$$\chi = \chi(\gamma) = 0 \quad \text{if } q \equiv 1 \pmod 3,$$

$$\chi = \chi(\gamma) = 1 \quad \text{if } q \equiv -1 \pmod 3.$$

1.10.2. LEMMA. $\chi = C(-3)$ *if* $p > 3$; $\chi = C(1)$ *if* $p = 2$.

Proof. $q \equiv 1$ or $-1 \pmod 3$ as $G(x) = x^2 + x + 1$ has two roots or none in γ. For $p \neq 2$, $\Delta(G) = -3$; for $p = 2$, $\partial(G) = 1$. Then, from §1.8, the definition of $C(G)$ gives the result.□

1.10.3. LEMMA. *If* $F \in \gamma[x]$, $p \neq 3$, *and* deg $F = 3$, *with* H *the Hessian of* F, *then*

$$C(H) \equiv C(F) + \chi \pmod 2.$$

Proof. If $p \neq 2$, then from theorem 1.10.1, $\Delta(H) = A_1^2 - 4A_0A_2$. By evaluating this and comparing the formula for $\Delta(F)$ in §1.8, we obtain that $\Delta(H) = -3\Delta(F)$. By the previous lemma, $\chi = C(-3)$. So

$$\begin{aligned} C(H) = C(\Delta(H)) &= C(-3\Delta(F)) \\ &\equiv C(-3) + C(\Delta(F)) \pmod 2 \\ &\equiv \chi + C(F) \pmod 2. \end{aligned}$$

If $p = 2$, then a similar evaluation gives $\partial(H) = \partial(F) + 1$. As $\chi = C(1)$, so

$$\begin{aligned} C(H) = C(\partial(H)) &= C(\Delta(F) + 1) \\ &\equiv C(\partial(F)) + 1 \pmod 2 \\ &\equiv C(F) + \chi \pmod 2. \square \end{aligned}$$

COROLLARY. *The roots* β_1, β_2 *of* H *are in* γ *or not as* $C(F) + \chi \equiv 0$ *or* 1 (*mod* 2).□

1.10.4. **THEOREM**. *Let* F *be a cubic polynomial over* $\gamma = GF(q)$, $q = p^h$ *and* $p \neq 3$, *such that* $\Delta(F) \neq 0$. *Let* α_1, α_2, α_3 *be the roots of* F *and* β_1, β_2 *the roots of its Hessian* H; *also* $e = F(\beta_1)/F(\beta_2)$. *Then these five roots are distinct and related by Table 1.2, where* γ', γ'', *and* γ''' *are respectively quadratic, cubic, and sextic extensions of* γ, *and where* K, K', K_1, K_2, K_3 *are subsets of fields.*

TABLE 1.2

$q \equiv c \pmod 3$	$\beta_1, \beta_2 \in K$	$e = s^3,\ s \in K'$	$\alpha_i \in K_i$		
c	K	K'	K_1	K_2	K_3
1	γ	γ	γ	γ	γ
1	γ	$\gamma''\backslash\gamma$	$\gamma''\backslash\gamma$	$\gamma''\backslash\gamma$	$\gamma''\backslash\gamma$
1	$\gamma'\backslash\gamma$	γ'	γ	$\gamma'\backslash\gamma$	$\gamma'\backslash\gamma$
-1	γ	γ	γ	$\gamma'\backslash\gamma$	$\gamma'\backslash\gamma$
-1	$\gamma'\backslash\gamma$	γ'	γ	γ	γ
-1	$\gamma'\backslash\gamma$	$\gamma'''\backslash\gamma'$	$\gamma''\backslash\gamma$	$\gamma''\backslash\gamma$	$\gamma''\backslash\gamma$

Proof. One can ensure, by a linear substitution for x in $F(x)$ if necessary, that $A_0 = 3a_0a_2 - a_1^2 \neq 0$. So neither $\beta_1 = \infty$ nor $\beta_2 = \infty$.

From §1.8, $R(F,H) = A_0{}^3 F(\beta_1)\ F(\beta_2) = -A_0{}^3 b_1 b_1\ (\beta_1 - \beta_2)^6$. When b_1 and b_2 are calculated from eqn (1.6) in the proof of theorem 1.10.1, we obtain that $9b_1 b_2\ (\beta_1 - \beta_2) = A_0$. Hence

$$R(F,H) = -A_0{}^4\ (\beta_1 - \beta_2)^4/9 = -[\Delta(H)/3]^2 = -\Delta(F)^2.$$

As $\Delta(F) \neq 0$, all five of α_1, α_2, α_3, β_1, β_2 are distinct.

We now apply lemma 1.10.3. When $c = 1$, $\chi = 0$ and so $C(H) = C(F)$. When $\beta_1, \beta_2 \notin \gamma$, then $C(H) = C(F) = 1$; by theorem 1.8.8, corollary 1 (ii), F has then exactly one root in γ. Similarly, when $\beta_1, \beta_2 \in \gamma$, $C[H] = C[F] = 0$ and F has zero or three roots in γ: three roots if e is a cube in γ and no roots if e is a non-cube in γ. This establishes the first three rows of the table.

When $c = -1$, $\chi = 1$ and so $C(F)$ is 0 or 1 as $C(H)$ is 1 or 0. When $\beta_1, \beta_2 \in \gamma$, then $C(H) = 0$ and $C(F) = 1$. So F has just one root in γ. When $\beta_1, \beta_2 \notin \gamma$, then $C(H) = 1$ and $C(F) = 0$. So F has zero or three roots in γ. This time, $e \in \gamma'$.

As γ' has q^2 elements, and $q^2 \equiv 1 \pmod 3$, all or none of the cube roots α_1', α_2', α_3' of e lie in γ'. If all lie in γ', then α_2, α_2, and α_3 are in γ' and so in γ: if none lie in γ', they lie in γ''' whence α_1, α_2, and α_3 lie in γ''.□

Now we turn to the case $p = 3$. So, for the remainder of this section, $\gamma = GF(q)$ with $q = 3^h$. The essential result is theorem 1.9.3, corollary 1; the other cases are either trivial or reduce to this one.

Let $F(x) = a_0 x^3 + a_1 x^2 + a_2 x + a_3$, $a_0 \neq 0$.

(i) If $a_3 = 0$, $F(x) = x(a_0 x^2 + a_1 x + a_2)$ and quadratics were dealt with in §1.4.

(ii) If $a_1 = 0$, $F(x) = a_0(x^3 - bx - c)$ with $a_2 = -a_0 b$ and $a_3 = -a_0 c$. This is case (v).

(iii) If $a_3 \neq 0$ and $a_2 = 0$, then F has no zero roots and $F(x^{-1}) = a_3 x^{-3}(x^3 - bx - c)$ with $a_1 = -a_3 b$ and $a_0 = a_3 c$. This is also case (v).

(iv) If $a_1\ a_2\ a_3 \neq 0$, then, with $\alpha = a_2/a_1$, $F(x + \alpha) = a_0 x^3 + a_1 x^2 + F(\alpha)$. So, either α is a double root of F or we are reduced to case (iii).

(v) $F(x) = x^3 - bx - c$. Then, by theorem 1.9.3, corollary 1, F has zero, one, or three roots in γ. If b is a non-zero square in γ, namely if $b = s^2$, then F has three roots or none in γ as $D_3(c/s^3)$ is zero or not. If b is zero, F has three identical roots in γ. If b is a non-square, F has just one root in γ. If F has just one root α, then

$$\alpha = -cb^{(3^h-3)/2} - c^3 b^{(3^h-3^2)/2} \dots - c^{3^{h-1}}.$$

If F has three roots in γ, then with $b = s^2$, $c = ds^3$, and $x = sy$, $G(y) = y^3 - y - d$ has three roots β in γ, where, for a fixed k such that $D_3(k) = -1$,

$$\beta = \beta' - kd^3 - (k + k^3)\, d^{3^2} \dots -(k + k^3 \dots + k^{3^{h-2}})d^{3^{h-1}}$$

with $\beta' = 0, 1, -1$. So F has three roots $s\beta$ in γ.

1.11. Quartic equations

It is convenient to treat quartic equations at this stage, although several geometrical notions of Chapters 6 and 7 are used: this section might appear more properly after §7.2.

Let $F(x) = a_0x^4 + a_1x^3 + a_2x^2 + a_3x + a_4$.
The invariants I and J of F from §1.8 are

$$I = a_2^2 - 3a_1a_3 + 12a_0a_4,$$
$$J = 2a_2^3 + 27a_1^2a_4 + 27a_0a_3^2 - 72a_0a_2a_4 - 9a_1a_2a_3. \qquad (1.9)$$
$$\Delta = (4I^3 - J^2)/27\,, \qquad p \neq 3\,,$$
$$\Delta = (a_0a_4 - a_1a_3)^3 + (a_0a_4 - a_1a_3)^2\, a_2^2 - (a_0a_3^2 + a_1^2a_4 - a_0a_2a_4)a_2^3, \quad p = 3\,.$$

When $\Delta \neq 0$ and $I = 0$, F is *equianharmonic*. When $\Delta \neq 0$ and $J = 0$, F is *harmonic*. These terms apply when $p = 2$ or 3, although the geometrical meaning of §6.1 no longer applies for $p = 2$; for $p = 3$, they are equivalent.

$$p = 2 : I = a_2^2 + a_1a_3, \quad J = a_1^2a_4 + a_0a_3^2 + a_1a_2a_3\ ;$$
$$p = 3 : I = a_2^2\,, \qquad J = -a_2^3.$$

The *cross-ratio* $\{\alpha_1, \alpha_2; \alpha_3, \alpha_4\}$ of the four roots

$\alpha_1, \alpha_2, \alpha_3, \alpha_4$ is

$$\frac{(\alpha_1 - \alpha_3)(\alpha_2 - \alpha_4)}{(\alpha_1 - \alpha_4)(\alpha_2 - \alpha_3)} .$$

If λ is this value, then permutations of the α_i give just the six values

$$\lambda,\ 1 - \lambda,\ 1/\lambda,\ 1/(1-\lambda),\ \lambda/(\lambda-1),\ (\lambda-1)/\lambda .$$

The geometrical significance of the cross-ratio is discussed in §6.1.

1.11.1. THEOREM. *Let $F \in \gamma[x]$, $\gamma = GF(q)$ and $q = p^h$, with $\deg F = 4$. The sextic equation satisfied by the six cross-ratios of the roots of F is*

$$\frac{(x^2 - x - 1)^3}{x^2(x - 1)^2} = \frac{I^3}{\Delta} .$$

If $p \neq 3$, an equivalent equation is

$$\frac{(x^2 - x + 1)^3}{[(x + 1)(x - 2)(2x - 1)]^2} = \frac{I^3}{J^2} ,$$

where I and J are given in (1.9) *and* $\Delta = \Delta(F)$.

Proof. The roots of the required sextic polynomial G are (1.10). So, pairing the cross-ratios with sum 1, we obtain

$$G = \left[x^2 - x + \lambda(1 - \lambda)\right]\left[x^2 - x + (\lambda - 1)/\lambda^2\right]\left[x^2 - x - \lambda/(1 - \lambda)^2\right]$$

$$= (x^2 - x)^3 + (x^2 - x)^2 \left[\frac{-\lambda^3(\lambda - 1)^3 + (\lambda - 1)^3 - \lambda^3}{\lambda^2(\lambda - 1)^2}\right] + 3(x^2 - x) + 1$$

$$= (x^2 - x + 1)^3 - (x^2 - x)^2 \left[(\lambda^2 - \lambda + 1)^3 / \lambda^2(\lambda - 1)^2\right].$$

Therefore the equation is

$$\frac{(x^2 - x + 1)^3}{(x^2 - x)^2} = \frac{(\lambda^2 - \lambda + 1)^3}{\lambda^2(\lambda - 1)^2} .$$

If, for the next few lines, we write $\alpha_i - \alpha_j = (ij)$,
then $\lambda = (13)(24)/[(14)(23)]$, $\lambda - 1 = (12)(34)/[(14)(23)]$.

So $\lambda^2 - \lambda + 1 = [(13)^2(24)^2 - (12)(34)(14)(23)]/[(14)(23)]^2$

$= T/[(14)(23)]^2.$

Then
$$\frac{(\lambda^2 - \lambda + 1)^3}{\lambda^2(\lambda - 1)^2} = \frac{T^3}{[\Delta/a_0{}^6]} .$$

Finally,
$$\begin{aligned} T &= \Sigma'' \alpha_i{}^2 \alpha_j{}^2 - \Sigma'' \alpha_i{}^2 \alpha_j \alpha_k + \Pi\alpha_i \\ &= \left(\Sigma'' \alpha_i \alpha_j\right)^2 - 3\Sigma'' \alpha_i{}^2 \alpha_j \alpha_k \\ &= \left(\Sigma'' \alpha_i \alpha_j\right)^2 - 3\Sigma'' \alpha_i \alpha_j \alpha_k \Sigma\alpha_\ell + 12\,\Pi\,\alpha_i \\ &= I/a_0{}^2, \end{aligned}$$

which gives the required equation.

When $p \neq 3$, $\Delta = (4I^3 - J^2)/27$.
So $J^2 = 4I^3 - 27\Delta$. Since
$$4(x^2 - x + 1)^3 - 27x^2(x - 1)^2 = [(x + 1)(x - 2)(2x - 1)]^2,$$
the second form of the equation appears.□

For the remainder of this section, we rely heavily on §7.2 on conics. The intention is to find the roots of F.

1.11.2. LEMMA. *The roots of F correspond to the base points of the pencil of quadrics $\mathscr{G} + \lambda\mathscr{H}$ in $PG(2,q)$, where $\mathscr{G}$ and $\mathscr{H}$ are given by*

$$G = a_0x_1{}^2 + a_1x_0x_1 + a_2x_1x_2 + a_3x_0x_2 + a_4x_2{}^2,$$

$$H = x_1x_2 - x_0{}^2 .$$

Proof. The points of $\mathscr{H}$ are $\mathbf{P}(t, t^2, 1)$, $t \in \gamma^+$.
Such a point lies on $\mathscr{G}$ exactly when $F(t) = 0$.□

1.11.3. LEMMA. *The quadric $\mathscr{G}_\lambda = \mathbf{V}(G + \lambda H)$ is singular if the resolvent cubic*

$$S(\lambda) = \lambda^3 + 2a_2\lambda^2 + (a_1a_3 + a_2{}^2 - 4a_0a_4)\lambda + a_1a_2a_3 - a_0a_3{}^2 - a_1{}^2a_4 = 0.$$

If $\Delta(F) \neq 0$, the singular quadric $\mathscr{G}_\lambda$ is not a repeated line.□

1.11.4. LEMMA. *The canonical forms for the resolvent cubic are as follows:*

(i) *if* $p \neq 3$, $R(x) = x^3 - 3Ix + J$;

(ii) *if* $p = 3$ *and* $a_2 \neq 0$, $R(x) = -\Delta x^3 + Ix + 1$;

(iii) *if* $p = 3$ *and* $a_2 = 0$, $R(x) = x^3 + (a_1a_3 - a_0a_4)x - (a_0a_3^2 + a_1^2a_4)$.

Proof.

(i) Substitute $\lambda = -x/3 - 2a_2/3$ in $S(\lambda)$.

(ii) Substitute $\lambda = [(a_1a_3 + a_2^2 - a_0a_4)x + 1]/(2a_2x)$.

(iii) Substitute $\lambda = x$ and put $a_2 = 0$ in $S(\lambda)$.□

COROLLARY 1. *If* S *is the resolvent cubic of the quartic* F, *then* $\Delta(S) = \Delta(F)$.

Proof. For $p \neq 3$, $\Delta(R) = 27(4I^3 - J^2)$, §1.8. So $\Delta(S) = \Delta(R)/3^6 = (4I^3 - J^2)/27 = \Delta(F)$.

For $p = 3$, if $S = x^3 + b_1x^2 + b_2x + b_3$, then $\Delta(S) = b_1^2b_2^2 - b_2^3 - b_1^3b_3$. Hence

$$\begin{aligned}\Delta(S) &= (a_0a_4 - a_1a_3)^3 + (a_0a_4 - a_1a_3)^2a_2^2 \\ &\quad - (a_0a_3^2 + a_1^2a_4 - a_0a_2a_4)a_2^3 \\ &= \Delta(F),\end{aligned}$$

as in §1.8.□

COROLLARY 2. *If* S *is irreducible, then* $F = F_1F_3$ *or* $F = F_2F_2'$, *where* deg F_i = deg F_i' = i.

Proof. This follows from corollary 1 and theorem 1.8.8, corollary 1, (ii) and (iii).□

COROLLARY 3. *When* F *is equianharmonic,*

(i) $p \neq 3$, $R(x) = x^3 + J$;

(ii) $p = 3$, $R(x) = x^3 + (a_1a_3 - a_0a_4)x - (a_0a_3^2 + a_1^2a_4)$.□

1.11.5. THEOREM. *The roots of the quartic* F *are the values of* t *for which* $\mathbf{P}(t, t^2, 1)$ *lies on the degenerate quadrics* $\mathscr{G}_\lambda$.□

1.11.6. THEOREM. *When* $\Delta(F) \neq 0$, *the number of roots of* F *in* γ

is one more than the number of roots of R in γ, except when F has no roots in γ.

	Roots in γ			
F	0	1	2	4
R	0,1,3	0	1	3

Equivalently,

R	0	1	3
F	0,1	0,2	0,4

Proof. Since $\Delta(F) \neq 0$, $\mathscr{G}$ and $\mathscr{H}$ meet in four distinct points lying in γ or a quadratic extension γ' or a quartic extension γ''. The three degenerate quadrics $\mathscr{G}_\lambda$ are therefore the three pairs of opposite sides of the tetrastigm $\mathscr{Q}$ with the four points as vertices. If for such a $\mathscr{G}_\lambda$, λ is in γ, then $\mathscr{G}_\lambda$ is either a pair of lines over γ or a conjugate pair over γ' with point of intersection over γ. In the classification of theorem 7.2.1, $\mathscr{G}_\lambda$ is respectively of type (i)(*b*) or (*c*).

If R has three roots in γ, choose the diagonal points of the quadrangle as $\mathbf{U}_0$, $\mathbf{U}_1$, $\mathbf{U}_2$. then the three $\mathscr{G}_\lambda$ are $\mathbf{V}(x_1^2 - e_0x_2^2)$, $\mathbf{V}(x_2^2 - e_1x_0^2)$, $\mathbf{V}(x_0^2 - e_2x_1^2)$ with $e_0e_1e_2 = 1$. So either one or three of the e_i are square. In the former case, one $\mathscr{G}_\lambda$ is of type (*b*) and two of type (*c*); so no vertex of $\mathscr{Q}$ lies over γ and F has no roots in γ.

If R has one root in γ, the corresponding $\mathscr{G}_\lambda$ is pair of lines over γ. The vertices of $\mathscr{Q}$ either consist of a pair of conjugate points on each line of $\mathscr{G}_\lambda$ in which case F is the product of two irreducible quadratics or consist of two points over γ on one line of $\mathscr{G}_\lambda$ and two conjugate points on the other line. F has correspondingly zero or two roots in γ. Finally, when R has no roots in γ, F has zero or one root.□

1.11.7. THEOREM. *For an equianharmonic quartic F, the possible number of roots of F is given in Table 1.3, where $q \equiv m$ (mod 3) and $\Delta(F) = s^6$. Equivalently, listing the number of roots of F against the number r of roots in γ of the resolvent R gives Table 1.4.*

TABLE 1.3

m	1	-1	0
$s \in \gamma$	0, 4	0, 2	0, 1, 4
$s \in \gamma$	0, 1	0, 2	0, 2

TABLE 1.4

r	$m=1$	$m=-1$	$m=0$
0	0, 1		0, 1
1		0, 2	0, 2
3	0, 4		0, 4

Proof. When $p \neq 3$, $R(x) = x^3 + J$, lemma 1.11.4, corollary 3. So when $q \equiv 1 \pmod 3$, R has three roots or none in γ as J is a cube or not. Since $q \equiv 1 \pmod 3$, $x^2 + x + 1$ is reducible whence -3 is a square, and -27 a sixth power. As $\Delta = \Delta(F) = -J^2/27$, so Δ is a sixth power if and only if J is a cube.

When $q \equiv -1 \pmod 3$, R has exactly one root in γ.

When $q \equiv 0 \pmod 3$, then

$$R(x) = x^3 + (a_1 a_3 - a_0 a_4)\, x - (a_0 a_3{}^2 + a_1{}^2 a_4).$$

So, from theorem 1.9.3, corollary 1, R has one root in γ when $b = a_0 a_4 - a_1 a_3$ is zero or a non-square, and zero or three roots in γ when b is a non-zero square. But, from §1.8, since $I = a_2{}^2 = 0$, $\Delta = \Delta(F) = b^3$. So Δ is a non-zero sixth power or not as b is a non-zero square or not. Now, the previous theorem allows the tables to be filled in.□

1.12. Notes and references

§§1.1, 1.2, 1.5. For general properties of finite fields, see Albert [1956], Dickson [1901]. See also the assorted works on coding theory in the Bibliography.

§1.3. There are many unsolved problems on permutation polynomials, and the Bibliography contains many papers on the subject. See particularly Dickson [1897], [1901] Chapter 5, Hayes [1967], Redéi [1973], and the papers of Carlitz.

An important application of permutation polynomials is the still unsolved problem of the classification of ovals for q even : see §§8.4, 8.8.

§1.4. This is based on Segre [1960a], pp. 103-108 .

§1.6. The numbers of monic irreducible and primitive polynomials are taken from Dickson [1901], pp. 18, 20. For subprimitivity, see Hirschfeld [1976].

§§1.8-1.11. These are based on Segre [1964a].

See the Bibliography for assorted other references on properties of polynomials, invariants, solution of equations and miscellaneous properties of finite fields.

2
PROJECTIVE SPACES

2.1. $PG(n,K)$

(i) Let $V = V(n+1,K)$ be an $(n+1)$-dimensional vector space over the field K with origin 0. Then consider the equivalence relation on the points of $V\backslash\{0\}$ whose equivalence classes are the one-dimensional subspaces of V with the origin deleted; that is, if $X,Y \in V\backslash\{0\}$ and for some basis $X = (x_0,\ldots,x_n)$, $Y = (y_0,\ldots, y_n)$, X is equivalent to Y if, for some t in K_0, $y_i = tx_i$ for all i. Then the set of equivalence classes is the *n-dimensional projective space over K* and is denoted by $PG(n,K)$ or, if $K = GF(q)$, by $PG(n,q)$. The elements of $PG(n,K)$ are called *points*. If the point $\mathbf{P}(X)$ is the equivalence class of the vector X, then we will say that X is a vector *representing* $\mathbf{P}(X)$: in this case, tX for t in K_0 also represents $\mathbf{P}(X)$.

The points $\mathbf{P}(X_1),\ldots,\mathbf{P}(X_r)$ are *linearly independent* if a set of vectors $X_1,\ldots, X_r$ representing them are linearly independent.

A *subspace of dimension m, or m-space,* of $PG(n,K)$ is a set of points all of whose representing vectors form (together with the origin) a subspace of dimension $m+1$ of $V(n+1,K)$. A subspace of dimension zero has already been called a point. Subspaces of dimensions one, two, and three are respectively called a *line*, a *plane*, and a *solid*. Subspaces of dimension $n-1$ and $n-2$ are called a *prime* and a *secundum*. The term *hyperplane* is commonly used elsewhere instead of *prime*. So a prime is the set of points $\mathbf{P}(X)$ whose vectors $X = (x_0,\ldots, x_n)$ satisfy an equation $a_0x_0 + a_1x_1 +\ldots+ a_nx_n = 0$. Thus an m-space Π_m is the set of points represented by the vectors $t_0X_0 + t_1X_1 +\ldots+ t_mX_m$, where $X_0,\ldots, X_m$ are $m+1$ linearly independent vectors and $(t_0,t_1,\ldots, t_m) \in K^{m+1}\backslash\{(0,\ldots,0)\}$; or, equivalently, Π_m is the set of points whose representing vectors $X = (x_0,\ldots, x_n)$ satisfy the equations $XA = 0$, where A is an $(n+1) \times (n-m)$ matrix of rank $n-m$ with coefficients in K. If a point P lies in a prime Π, then P is *incident* with Π or, equally well, Π is *incident* with P.

(a) If Π_r and Π_s are subspaces of $PG(n,K)$, then the *meet* or *intersection* of Π_r and Π_s written $\Pi_r \cap \Pi_s$, is the set of points common to Π_r and Π_s and is also a subspace.

(b) The *join* of Π_r and Π_s, written $\Pi_r \Pi_s$, is the smallest subspace containing Π_r and Π_s.

(c) If $\Pi_r \cap \Pi_s = \Pi_t$ and $\Pi_r \Pi_s = \Pi_m$, then $r + s = m + t$.

(d) If Π_r and Π_r' are both r-spaces in $PG(n,K)$ and $\Pi_r' \subset \Pi_r$, then $\Pi_r' = \Pi_r$.

(e) A subspace Π_m is the join of $m+1$ linearly independent points or the intersection of $n-m$ linearly independent primes.

If S and S' are two spaces $PG(n,K)$, then a *collineation* $\mathfrak{T} : S \to S'$ is a bijection which preserves incidence; that is, if $\Pi_r \subset \Pi_s$, then $\Pi_r \mathfrak{T} \subset \Pi_s \mathfrak{T}$. It is sufficient that $\mathfrak{T}$ is a bijection such that if $\Pi_0 \subset \Pi_1$ then $\Pi_0 \mathfrak{T} \subset \Pi_1 \mathfrak{T}$.

A *projectivity* $\mathfrak{T} : S \to S'$ is a bijection given by a matrix T: if $\mathbf{P}(X') = \mathbf{P}(X)\mathfrak{T}$, then $tX' = XT$ where X' and X are coordinate vectors for $\mathbf{P}(X')$ and $\mathbf{P}(X)$, and $t \in K_0$. Write $\mathfrak{T} = \mathbf{M}(T)$: then $\mathfrak{T} = \mathbf{M}(\lambda T)$ for any λ in K_0. The matrix T is non-singular.

A projectivity is a collineation. We shall mostly be concerned with the case that $S = S'$.

With respect to a fixed basis of $V(n+1, K)$, an automorphism σ of K can be extended to an *automorphism* σ of $S = PG(n,K)$: this is a collineation given by $\mathbf{P}(X)\sigma = \mathbf{P}(X\sigma)$, where $X\sigma = (x_0\sigma, x_1\sigma, \ldots, x_n\sigma)$. Sometimes it will be more convenient to write X^σ and x^σ for $X\sigma$ and $x\sigma$.

(ii) *Fundamental theorem of projective geometry*

(a) If $\mathfrak{T}' : S \to S$ is a collineation, then $\mathfrak{T}' = \sigma\mathfrak{T}$ where σ is an automorphism and $\mathfrak{T}$ is a projectivity. In particular if $K = GF(p^h)$ and $\mathbf{P}(X') = \mathbf{P}(X)\mathfrak{T}'$, then there exists m in N_h, t_{ij} in K for (i,j) in $\bar{N}_n^2$, and t in K_0 such that

$$tX' = X^{p^m} T \text{ where } X^{p^m} = (x_0^{p^m}, \ldots, x_n^{p^m})$$

and
$$T = (t_{ij}),\ i,j \in \bar{N}_n;$$

that is,
$$tx_i' = x_0^{p^m} t_{0i} + \ldots + x_n^{p^m} t_{ni} .$$

(b) If $\{P_1, \ldots, P_{n+2}\}$, $\{P'_1, \ldots, P'_{n+2}\}$ are sets of $n+2$ points of $PG(n,K)$ such that no $n+1$ points chosen from the same set lie in a prime (or, in the language of Chapter 8, the two sets form $(n+2)$-arcs), then there exists a unique projectivity $\mathfrak{T}$ such that $P'_i = P_i\mathfrak{T}$, all i in $\mathbf{N}_{n+2}$.

(c) For $n=1$, (b) simplfies: there is a unique projectivity transforming any three distinct points on a line to any other three.

(d) When $K = GF(2)$, it suffices in (b) to give the images of $P_1, \ldots, P_{n+1}$ to determine $\mathfrak{T}$. Similarly, in (c), the images of two points determine the projectivity.

(iii) *Principle of duality*

To any $S = PG(n,K)$, there is a *dual* space S^*, whose *points* and *primes* are respectively the primes and points of S. For any theorem true in S, there is an equivalent theorem true in S^*. In particular, if T is a theorem in S stated in terms of points, primes, and incidence, the same theorem is true in S^* and gives a dual theorem T^* in S by interchanging 'point' and 'prime' whenever they occur. Thus 'join' and 'meet' are dual. Hence the dual of an r-space in $PG(n,K)$ is an $(n-r-1)$-space.

In particular, in $PG(2,K)$, point and line are dual; in $PG(3,K)$ point and plane are dual, whereas the dual of a line is a line.

(iv) *Coordinate Systems*

Part (b) of the fundamental theorem of projective geometry emphasizes a basic difference between $V(n+1,K)$ and $PG(n,K)$. In the former, linear transformations are determined by the images of $n+1$ points: in the latter, projectivities are determined by the images of $n+2$ points.

Let $\{P_0, \ldots, P_{n+1}\}$ be any set of $n+2$ points in $PG(n,q)$, no $n+1$ in a prime. If P is any other point of the space, then a coordinate vector for P is determined in the following manner. P_i is represented by the vector t_iX_i for some vector X_i and any t_i in K. Since X_{n+1} is linearly dependent on $X_0, \ldots, X_n$, for any given t in K there exist a_i in K for all i in $\bar{\mathrm{N}}_n$ such that

$$tX_{n+1} = a_0X_0 + \ldots + a_nX_n .$$

So, for variable t, the ratios a_i/a_j remain fixed. Thus, if P is any point with $P = \mathbf{P}(X)$, then

$$X = t_0a_0X_0 + \ldots + t_na_nX_n .$$

So, with respect to $\{P_0, \ldots, P_{n+1}\}$, P is given by $(t_0, \ldots, t_n)$ where the t_i are determined up to a common factor. $\{P_0, \ldots, P_n\}$ is the *simplex of reference* and P_{n+1} the *unit point*. Together the $n+2$ points form a *basis for the coordinate system*. In particular, let $E_i = (0, \ldots, 0, 1, 0, \ldots, 0)$ be the vector with one in the $(i+1)$-th place and zeros elsewhere, and let $E = (1, \ldots, 1)$. Write

$$\mathbf{U}_i = \mathbf{P}(E_i) \quad , \quad \mathbf{U} = \mathbf{P}(E) .$$

Then $\{\mathbf{U}_0, \ldots, \mathbf{U}_n\}$ is the simplex of reference and $\mathbf{U}$ the unit point in a coordinate system with basis $\{\mathbf{U}_0, \ldots, \mathbf{U}_n, \mathbf{U}\}$.

Thus, in $V(n+1,K)$, a basis is a set of $n+1$ linearly independent points, and, in $PG(n,K)$, a basis for a coordinate system is a set of $n+2$ points, no $n+1$ in a prime (that is, every set of $n+1$ points is linearly independent). Dually, a coordinate system is determined by $n+2$ primes, no $n+1$ of which have a point in common. The faces of the simplex of reference will be written $\mathbf{u}_0, \ldots, \mathbf{u}_n$ and the unit prime $\mathbf{u}$. So $\mathbf{u}_i$ is given by $x_i = 0$ and $\mathbf{u}$ by $\Sigma x_i = 0$.

Again from the fundamental theorem, if two coordinate systems are given by the vectors $X = (x_0, \ldots, x_n)$ and $Y = (y_0, \ldots, y_n)$, we can change from one system to the other by $Y = XA$, where A is non-singular. If a projectivity $\mathfrak{T}$ in the one system is given by $X' = XT$, then, since $Y' = X'A$, in the other system it is given by $Y' = X'A = XTA = YA^{-1}TA$.

(v) *Polarities*

A prime is the set of points $\mathbf{P}(X)$ with $X = (x_0, \ldots, x_n)$ satisfying a relation $u_0x_0 + u_1x_1 + \ldots + u_nx_n = 0$. The vector $U = (u_0, \ldots, u_n)$ is the coordinate vector of the prime, which will be denoted $\pi(U)$: the u_i are called *prime* or *tangential* coordinates.

Let S be a space $PG(n,K)$ and S' its dual space superimposed on S: that is, the points of S' are the primes of S and the primes of S' are the points of S. Consider a function $\mathfrak{T} : S \to S'$. If $\mathfrak{T}$ is a collineation, it is called a *reciprocity* of S and induces a collineation, also called $\mathfrak{T}$, of S' to S; that is, if the points of S are transformed to primes, then, since $\mathfrak{T}$ preserves incidence, primes are transformed to points. If $\mathfrak{T}$ is a projectivity, then it is called a *correlation* of S. In either case, if $\mathfrak{T}$ is involutory, that is, $\mathfrak{T}^2 = \mathfrak{J}$, where $\mathfrak{J}$ is the identity, then it is called a *polarity* of S. Thus, if P and P' are points and π a prime such that $P\mathfrak{T} = \pi$ and $\pi\mathfrak{T} = P'$, then, in a polarity, $P = P'$.

Suppose $\mathfrak{T}$ is a reciprocity: then it is the product of an automorphism σ and a projectivity of S to S' given by the matrix T. Let us write $x\sigma = \tilde{x}$ for any x in K. If $X = (x_0, \ldots, x_n)$, write $\tilde{X} = (\tilde{x}_0, \ldots, \tilde{x}_n)$, and if $T = (t_{ij})$ write $\tilde{T} = (\tilde{t}_{ij})$. Also T^* will denote the transpose of T, that is, $T^* = (t_{ji})$.

Suppose $\mathbf{P}(X)\mathfrak{T} = \pi(V)$ and $\pi(U)\mathfrak{T} = \mathbf{P}(Y)$. Then the first relation is given by $tV = \tilde{X}T$. Now, $\mathbf{P}(X)$ lies in $\pi(U)$ if and only if $\pi(V)$ contains $\mathbf{P}(Y)$: therefore $UX^* = 0 \Leftrightarrow VY^* = 0$. Hence $\tilde{U}\tilde{X}^* = 0 \Leftrightarrow UX^* = 0 \Leftrightarrow VY^* = 0 \Leftrightarrow \tilde{X}TY^* = 0 \Leftrightarrow YT^*\tilde{X}^* = 0$. So $\tilde{U} = sYT^*$ for some s in K and $sY = \tilde{U}T^{*-1}$.

If $\mathfrak{T}$ is a correlation, these equations become $tV = XT$, $sY = UT^{*-1}$. Suppose the reciprocity $\mathfrak{T}$ is a polarity, then $\mathbf{P}(X) \to \pi(\tilde{X}T) \to \mathbf{P}(\tilde{\tilde{X}}\tilde{T}T^{*-1}) = \mathbf{P}(X)$. So $tX = \tilde{\tilde{X}}\tilde{T}T^{*-1}$. Hence $\tilde{\tilde{X}} = X$ and $\tilde{T}T^{*-1} = tI$. There are two possibilities: either σ is the identity and $x\sigma = x$ for all x in K or σ has period two and we will write $x\sigma = x$ and similarly $\bar{X}$ for $\tilde{X}$ and $\bar{T}$ for $\tilde{T}$.

In a polarity $\mathfrak{T}$, if $\mathbf{P}(X)\mathfrak{T} = \pi(V)$ and $\pi(U)\mathfrak{T} = \mathbf{P}(Y)$, then $\pi(V)$ is the *polar* (prime) of $\mathbf{P}(X)$ and $\mathbf{P}(Y)$ is the *pole* of $\pi(U)$. Since $\mathfrak{T}^2 = \mathfrak{J}$, the converse is also true. If $\mathbf{P}(Y)$ lies in $\pi(V)$, the polar of $\mathbf{P}(X)$, then $\mathbf{P}(X)$ lies in $\pi(U)$, the polar of $\mathbf{P}(Y)$. In this case, $\mathbf{P}(X)$ and $\mathbf{P}(Y)$ are *conjugate* points and $\pi(U)$ and $\pi(V)$ are *conjugate* primes. $\mathbf{P}(X)$ is *self-conjugate* if it lies in its own polar prime: $\pi(U)$ is *self-conjugate* if it contains its own pole.

Thus $\mathbf{P}(X)$ and $\mathbf{P}(Y)$ are conjugate if $VY^* = 0$; that is, $\tilde{X}TY^* = 0$. The self-conjugate points $\mathrm{P}(X)$ are given by $\tilde{X}TX^* = 0$.

Let us now examine the different types of polarity.

I. *σ is the identity*

Then $TT^{*-1} = tI$, $T = tT^*$, $T^* = tT$, $T = t^2T$. Since T is non-singular, $t^2 = 1$. $\mathbf{P}(X)$ is conjugate to $\mathbf{P}(Y)$ when $XTY^* = 0$.

(a) K has characteristic $p \neq 2$.

(1) $t = 1$, $T = T^*$.

$$XTY^* = t_{00}x_0y_0 + t_{01}(x_0y_1 + x_1y_0) + \dots \quad .$$

$$XTX^* = t_{00}x_0^2 + 2t_{01}x_0x_1 + \dots \quad .$$

$\mathfrak{T}$ is called an *ordinary polarity* or a *polarity with respect to a quadric.*

(2) $t = -1$, $T = \ T^*$: so $t_{ii} = 0$ and $t_{ij} = -t_{ji}$.

$$XTY^* = \mathbf{\Sigma}'' t_{ij}(x_iy_j - x_jy_i)$$

$$= t_{01}(x_0y_1 - x_1y_0) + \dots \quad .$$

$$XTX^* = 0 \ .$$

Hence every point $\mathbf{P}(X)$ is self-conjugate. $\mathfrak{T}$ is called a *null polarity* or *symplectic polarity* or a *polarity with respect to a linear complex*. Since T is non-singular and skew-symmetric, n must be odd.

(b) K has characteristic $p = 2$.

Now $t = 1$, $T = T^*$, $t_{ij} = t_{ji}$.

(1) $t_{ii} = 0$, all i. This is similar to the case (a)(2).

$$XTY^* = \mathbf{\Sigma}'' t_{ij}(x_iy_j + x_jy_i)$$

$$= t_{01}(x_0y_1 + x_1y_0) + \dots \quad .$$

$$XTX^* = 0 \ .$$

Every point $\mathbf{P}(X)$ is self-conjugate. $\mathfrak{T}$ is called a *null polarity* or a *symplectic polarity* or a *polarity with respect to a linear complex*. It can only occur for n odd.

(2) $t_{ii} \neq 0$, some i.

$$XTY^* = t_{00}x_0y_0 + t_{01}(x_0y_1 + x_1y_0) + \dots .$$

$$XTX^* = \Sigma t_{ii}x_i^2$$

$$= (\Sigma\sqrt{t_{ii}}x_i)^2$$

The self-conjugate points of $\mathfrak{T}$ therefore comprise a prime. $\mathfrak{T}$ is called a *pseudo polarity*.

II. *σ has period two*

$\overline{T}T^{*-1} = tI$, $\overline{T} = tT^*$, $T = t\overline{T}^*$, $T = t^2T$. Since T is non-singular, $t^2 = 1$. So, either $T^* = \overline{T}$ or $T^* = -\overline{T}$. If $T^* = -\overline{T}$ and so K has characteristic $p \neq 2$ to be distinct from the previous case, then there exists λ in K with $\overline{\lambda} = -\lambda$ as follows. Since σ has period two, the set $K' = \{t \in K \mid \overline{t} = t\}$ is a subfield of K, and K is an extension of degree two over K'. So, there exists c in K' such that c is not a square in K'; for, if every element in K' were a square, every polynomial $x^2 + ax + b$ would be reducible. So there exists λ in $K\backslash K'$ such that $\lambda^2 = c$. Both $\overline{\lambda}$ and $-\lambda$ are also solutions in K of $x^2 = c$ and since neither can be equal to λ, $\overline{\lambda} = -\lambda$. So $-1 = \overline{\lambda}/\lambda$. Hence $T^* = (\overline{\lambda}/\lambda)\overline{T}$, $\lambda T^* = \overline{\lambda T}$, $(\lambda T)^* = \overline{\lambda T}$. As λT gives the same polarity as T, we need only consider the case $T^* = \overline{T}$: $t_{ij} = \overline{t}_{ji}$. The points $\mathbf{P}(X)$ and $\mathbf{P}(Y)$ are conjugate when $\overline{X}TY^* = 0$.

$$\overline{X}TY^* = \Sigma t_{ij}\,\overline{x}_iy_j$$

$$= t_{00}\,\overline{x}_0y_0 + (t_{01}\overline{x}_0y_1 + \overline{t}_{01}\overline{x}_1y_0) + \dots .$$

$$\overline{X}TX^* = t_{00}\overline{x}_0x_0 + (t_{01}\overline{x}_0x_1 + \overline{t}_{01}\overline{x}_1x_0) + \dots .$$

$\mathfrak{T}$ is called a *Hermitian polarity* or a *unitary polarity*.

TABLE 2.1

Number	Name	Characteristic p of K	Equations	Conditions on T	Locus of self-conjugate points	Occurrence in $PG(n,K)$
Ia(1)	*Ordinary*	$p \neq 2$	$tV=XT, sY=UT^{*-1}$	$t_{ij} = t_{ji}$	$\mathbf{V}(XTX^*)$	*all* n
Ia(2)	*Null*	$p \neq 2$	"	$t_{ij} = -t_{ij}$ $t_{ii} = 0$	$PG(n,K)$	*odd* n
Ib(1)	*Null*	$p = 2$	"	$t_{ij} = t_{ji}$ $t_{ii} = 0$	$PG(n,K)$	*odd* n
Ib(2)	*Pseudo*	$p = 2$	"	$t_{ij}=t_{ji}$, $t_{kk}\neq 0$ *some* k	$\mathbf{V}(\Sigma\sqrt{t_{ii}}x_i)$	*all* n
II	*Hermitian*	p *arbitrary* K *quadratic extension of* K'	$tV=\bar{X}T$, $sY=\bar{U}T^{*-1}$	$t_{ij} = -\bar{t}_{ji}$	$\mathbf{V}(\bar{X}TX^*)$	*all* n

To summarize, a polarity of $PG(n,K)$ is a function $\mathfrak{T}$ which transforms points to primes, primes to points, preserves incidence and is involutory. The five types are given in Table 2.1. See §2.6 for the notation $\mathbf{V}(F)$.

2.2. Incidence structures

If Π_{n-1} is any prime in $PG(n,K)$, then $AG(n,K) = PG(n,K)\backslash\Pi_{n-1}$ is an *n-dimensional affine space over* K. When $K = GF(q)$, write $AG(n,K) = AG(n,q)$. The *subspaces* of $AG(n,K)$ are the subspaces of $PG(n,K)$ with the points of Π_{n-1} deleted.

An *incidence structure* $\mathcal{S}$ is a triple $(\mathcal{P}, \mathcal{B}, I)$, where $\mathcal{P}$ is a set whose elements are called *points*, $\mathcal{B}$ is a set whose elements are called *blocks* (or *lines* in several specific cases), and $I \subset \mathcal{P} \times \mathcal{B}$. If $(P,L) \in I$, then we say that P *is incident with* L, or L *is incident with* P, or P *lies on* L, or L *contains* P. Frequently, the blocks are sets of points, and we write '$P \in L$' instead of '$(P,L) \in I$'. Let $|P| = v$, $|B| = b$.

An incidence structure $\mathcal{S}$ is a *tactical configuration* if

(i) every block is incident with k points;

(ii) every point is incident with r blocks.

Then

$$b\,k = v\,r\,. \tag{2.1}$$

So $\mathcal{S}$ is occasionally called a (v_r, b_k) configuration.

A tactical configuration $\mathcal{S}$ is a t-(v,k,λ) *design* or, briefly, a *t-design* if

(iii) any set of t points is incident with exactly λ blocks, $\lambda > 0$.

Then, if λ_i is the number of blocks containing a given set of i points with $0 \leqslant i \leqslant t$,

$$\lambda_i\, \mathbf{c}(k-i,\ t-i) = \lambda \mathbf{c}(v-i,\ t-i). \tag{2.2}$$

So $\mathcal{S}$ is an i-design for $0 \leqslant i \leqslant t$. Also $\lambda_0 = b$, $\lambda_1 = r$. A tactical configuration is just a 1-design.

A tactical configuration is a *generalized quadrangle* if

(iv) $k \geqslant 2$, $r \geqslant 2$ and any two points are contained in at most one block;

(v) for a point P and a block L which are not incident,

there exists a unique point P' and a unique block L' such that L contains P' and L' contains P and P'.

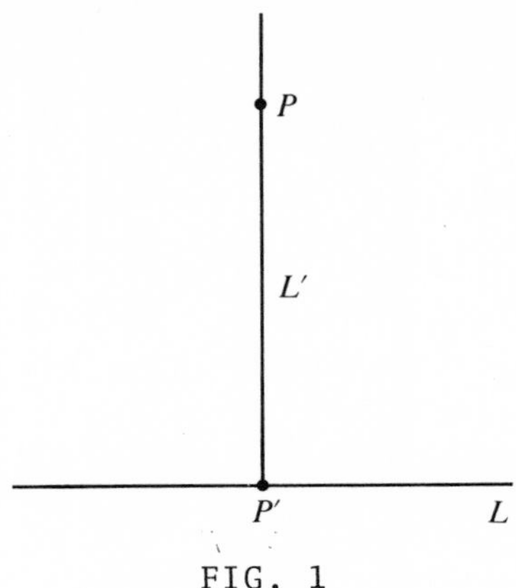

FIG. 1

Now, we wish to give sets of axioms which characterize $PG(n,q)$ and $AG(n,q)$, both for $n \geq 2$. Firstly, we consider the case $n = 2$. In these incidence structures, call the blocks *lines*, and regard them as sets of points.

Let $\mathcal{S}$ be an incidence structure such that

(i) any two points are incident with exactly one line;

(ii) any two lines are incident with exactly one point;

(iii) there exist four points no three collinear;

(iv) some line comprises $q+1$ points;

(v) the theorem of Desargues (theorem 7.5.5) holds,

or (v') the theorem of Pappus (theorem 7.5.4) holds.

Then $\mathcal{S} = PG(2,q)$.

Let $\mathcal{S}$ be an incidence structure. Then two lines l_1, l_2 are called *parallel* if $l_1 = l_2$ or $l_1 \cap l_2 = \emptyset$;

(i) any two points are incident with exactly one line;

(ii) for every point P and line l, there is a unique line l' parallel to l and containing P;

(iii) there exist three non-collinear points;

(iv) some line contains q points.

Parallelism can now be shown to be an equivalence relation. Let the equivalence classes be called *parallel classes* or *ideal points*, and call the set of ideal points the *ideal line*. The incidence structure $\mathcal{S}'$ obtained by including the ideal points and the ideal line satisfies axioms (i) - (iv) for

$PG(2,q)$. The final axiom for $\mathcal{S}$ is

(vi) the theorem of Desargues holds in $\mathcal{S}'$,

or (vi') the theorem of Pappus holds in $\mathcal{S}'$.

Then $\mathcal{S} = AG(2,q)$.

Next we give a set of axioms for $PG(n,q)$, $n > 2$, which correspond to the properties (a) - (e) of §2.1(i).

Let $\mathcal{S}$ be a finite set of *points* with assigned subsets, called *subspaces*. To each subspace is associated an integer d, called the *dimension* satisfying $-1 \leqslant d \leqslant n$, where n is a fixed integer greater than two. A subspace of dimension d is written Π_d.

(i) For each $d = -1, 0, 1, \ldots, n$ there exists a subspace Π_d and

(a) there is a unique Π_{-1}, namely the empty set;

(b) the subspaces Π_0 are the points of $\mathcal{S}$;

(c) there is a unique Π_n, namely $\mathcal{S}$.

(ii) If $\Pi_r \subset \Pi_s$, then $r \leqslant s$, with $r = s$ if and only if $\Pi_r = \Pi_s$.

(iii) $\Pi_r \cap \Pi_s = \Pi_t$.

(iv) If Π_m is the intersection of all subspaces containing Π_r and Π_s, where $\Pi_r \cap \Pi_s = \Pi_t$, then

$$m + t = r + s .$$

(v) Some line contains $q + 1 \geqslant 3$ points.

Then $\mathcal{S} = PG(n,q)$, $n \geqslant 3$.

There are many variations on this set of axioms for $PG(n,q)$. Also there are several ways of axiomatizing $AG(n,q)$ by relating it to $PG(n,q)$. Now we give a set of axioms for $AG(n,q)$, $n > 2$, in which n is not specified but q is.

Let $\mathcal{S}$ be an incidence structure with an equivalence relation *parallelism* on its lines (blocks).

(i) Any two points P_1, P_2 are incident with exactly one line P_1P_2.

(ii) For every point P and line l, there is a unique line l' parallel to l and containing P.

(iii) If P_1P_2 and P_3P_4 are parallel lines and P is a point on P_1P_3 distinct from P_1 and P_3, then there is a point P' on

PP_2 and P_3P_4 (see Fig. 2).

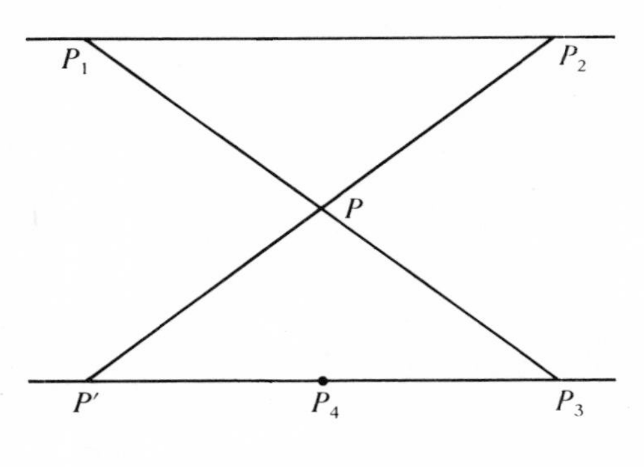

FIG. 2

(iv) If no line contains more than two points and P_1, P_2, P_3 are distinct points, then the line l_3 through P_3 parallel to P_1P_2 and the line l_2 through P_2 parallel to P_1P_3 have a point P in common (see Fig. 3).

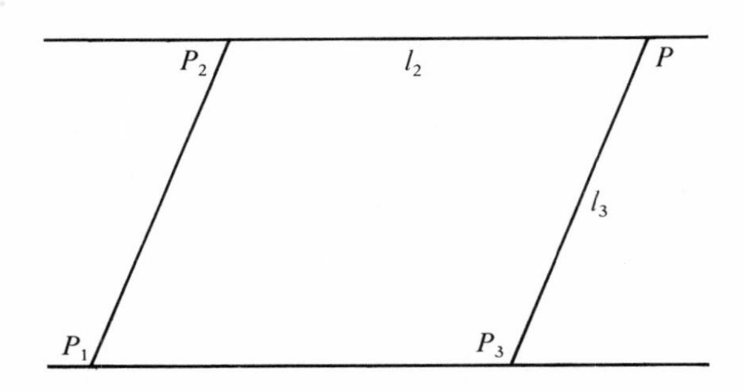

FIG. 3

(v) Some line contains exactly $q \geqslant 2$ points.

(vi) There exist two lines neither parallel nor with a common point.

Then $\mathscr{S} = AG(n,q)$ for some $n \geqslant 3$.

2.3. Canonical forms for projectivities in $PG(n-1,q)$

The *projective group* $PGL(n,q)$ is the group of projectivities of $PG(n-1,q)$. The *collineation* group $P\Gamma L(n,q)$ is the group of collineations of $PG(n-1,q)$.

The *general linear group* $GL(n,q)$ is the group of all bijective linear transformations of $V(n,q)$. The group $\Gamma L(n,q)$ is the group of all bijective semi-linear transformations of $V(n,q)$. The *special linear group* $SL(n,q)$ is the subgroup of $GL(n,q)$ comprising those linear transformations of determinant one.

Write $Z(G)$ for the centre of the group G. Let $Z = Z(GL(n,q)) = Z(\Gamma L(n,q))$. Then $Z = \{tI_n \mid t \in \gamma_0, I_n$ the identity of $GL(n,q)\}$. Write $G \cong H$ when the two groups G and H are isomorphic. Then $PGL(n,q) \cong GL(n,q)/Z$. The *projective special linear* group $PSL(n,q) \cong SL(n,q)/\{Z \cap SL(n,q)\}$. Then $PGL(n,q) \cong PSL(n,q)$ if and only if $(n,q-1) = 1$. Let $p(n,q) = |PGL(n,q)|$.

Let
$$[r,s]_- = \prod_{i=r}^{i=s} (q^i - 1) \quad \text{for } s \geqslant r$$
$$= 1 \quad \text{for } s < r.$$

2.3.1. THEOREM.

(i) $p(n,q) = |PGL(n,q)| = q^{n(n-1)/2}\,[2,\ n]_-$;

(ii) $|P\Gamma L(n,q)| = hp(n,q)$;

(iii) $|GL(n,q)| = q^{n(n-1)/2}\,[1,\ n]_-$;

(iv) $|\Gamma L(n,q)| = hq^{n(n-1)/2}\,[1,\ n]_-$;

(v) $|SL(n,q)| = p(n,q)$;

(vi) $|PSL(n,q)| = p(n,q)/(n,q-1)$.

Proof. Each projectivity $\mathfrak{T}$ of $PG(n-1,\ q)$ is given by $q-1$ matrices λT, $\lambda \in \gamma_0$. So $p(n,q) = |GL(n,q)|/(q-1)$. As $GL(n,q)$ can be considered as the set of $n \times n$ non-singular matrices over γ,

$$|GL(n,q)| = (q^n - 1)(q^n - q)\ \ldots\ (q^n - q^{n-1})$$
$$= q^{n(n-1)/2}\ [1,\ n]_-\ .$$

Hence (i) and (iii) follow. As the automorphism group of γ has order h, (ii) and (iv) follow by the fundamental theorem, §2.1(ii).

$SL(n,q)$ is the kernel of the determinant map from $GL(n,q)$ to γ_0. So $|SL(n,q)| = |GL(n,q)|/|\gamma_0| = p(n,q)$. From §1.5(i), $t^n = 1$ has $(n, q-1)$ solutions in γ.
So $|Z \cap SL(n,q)| = |\{tI_n \mid t^n = 1\}| = (n, q-1)$. This gives (vi).□

For a direct geometrical proof of the value of $p(n,q)$, see §4.3.

By a change of coordinate system in $PG(n-1,q)$ as in §2.1, two projectivities $\mathfrak{A}$ and $\mathfrak{B}$ in the same conjugacy class of $PGL(n,q)$ act on $PG(n-1, q)$ in the same manner. If $\mathfrak{A}$ and $\mathfrak{B}$ are given by the matrices A and B, there exists a non-singular matrix T and some t in γ_0 such that $tB = T^{-1}AT$. If $\mathfrak{A}$ and $\mathfrak{B}$ give conjugate elements of $GL(n,q)$, then there exists a non-singular matrix T such that $T^{-1}AT = B$.

Let $E(n,q)$ and $e(n,q)$ be the respective numbers of conjugacy classes in $GL(n,q)$ and $PGL(n,q)$. The generating functions are as follows:

$$E(x) = 1 + \sum_{n=1}^{\infty} E(n,q)\, x^n = \prod_{n=1}^{\infty} \frac{1 - x^n}{1 - qx^n}; \tag{2.3}$$

$$e(x) = 1 + \sum_{n=1}^{\infty} e(n,q)\, x^n = (q-1)^{-1} \sum_{d|(q-1)} \phi(d)\, E(x^d). \tag{2.4}$$

In particular,

$$E(2,q) = q^2 - 1,\quad e(2,q) = q + (q-1,\ 2), \tag{2.5}$$
$$E(3,q) = q^3 - q,\quad e(3,q) = q^2 + q - 1 + (q-1,\ 3), \tag{2.6}$$
$$E(4,q) = q^4 - q,\quad e(4,q) = q^3 + q^2 + (q-1,\ 2)q - 1 + (q-1,\ 4); \tag{2.7}$$

see theorems 6.3.1 and 7.4.1.

Once canonical forms are found for the equivalence classes of $GL(n,q)$, the canonical forms for the equivalence classes of $PGL(n,q)$ can be found by selecting, in each case, one from a set of at most $q-1$ matrices, which represent non-conjugate linear transformations, but conjugate projectivities. That is, if a canonical form is given by the matrix A, there are $q-1$

distinct matrices tA as t varies in γ_0: all $q-1$ matrices tA represent the same projectivity of $PG(n-1,q)$, but up to $q-1$ non-conjugate linear transformations of $V(n,q)$.

For any commutative ring R, let $\mathcal{M}(n,R)$ be the ring of $n \times n$ matrices with coefficients in R under the usual addition and multiplication of matrices.

If A, $B \in \mathcal{M}(n, \gamma)$, then A is *similar* to B, written $A \approx B$, if there exists a non-singular T in $\mathcal{M}(n, \gamma)$ such that $T^{-1}AT = B$.

An *elementary operation* on an element $A(x)$ of $\mathcal{M}(n, \gamma[x])$ is a procedure of one of the following types:

(I) It interchanges any two rows (or columns) of the matrix;

(II) It multiplies the matrix by an element of γ_0;

(III) It adds to any row (resp. column) a multiple by any element of $\gamma[x]$ of any other row (resp. column).

If $A(x)$, $B(x) \in \mathcal{M}(n, \gamma[x])$, then $A(x)$ is *equivalent* to $B(x)$, written $A(x) \sim B(x)$, if there is a finite sequence of elementary operations which reduces $A(x)$ to $B(x)$.

The following are then classical results, and, for the purposes of this book, it will suffice to take A and B non-singular. I is the identity of $\mathcal{M}(n, \gamma)$.

2.3.2. THEOREM. (i) $A \approx B \Leftrightarrow xI-A \sim xI-B$

(ii) $xI - A \sim \text{diag}\,(F_1(x), \ldots, F_n(x))$ *where*

(a) *if* $G_i(x)$ *is the greatest common divisor of all the* $i \times i$ *minors of* $xI - A$, *then*

$$F_i(x) = G_i(x)/G_{i-1}(x), i = 2, \ldots, n, \quad F_1(x) = G_1(x);$$

(b) $F_i | F_{i+1}$;

(c) F_n *is the minimum polynomial of* A;

(d) $C(x) = F_1(x) \ldots F_n(x)$ *is the characteristic polynomial of* A;

(e) $F_1, \ldots, F_n$ *are the invariant factors of* $xI-A$;

(f) *if* $F_i = E_1^{n_{i1}} \ldots E_m^{n_{im}}$ *where each* E_j *is irreducible in* $\gamma[x]$, *then the polynomials* $E_i^{n_{ij}}$ *are the elementary divisors of* $xI-A$.

(iii) $A \approx B$ *if and only if* $xI-A$ *and* $xI-B$ *have the same set of invariant factors or, equivalently, the same set of invariant factors or, equivalently, the same set of elementary divisors.*

(iv) *If* $F(x) = x^n - a_{n-1}x^{n-1} \ldots - a_0$ *is any monic polynomial, then its companion matrix,* $C(F)$, *is given by the* $n \times n$ *matrix*

$$C(F) = \begin{bmatrix} 0 & 1 & 0 & \ldots & 0 \\ 0 & 0 & 1 & 0 \ldots & 0 \\ \vdots & & & & \\ 0 & 0 \ldots & & 0 & 1 \\ a_0 & a_1 & \ldots & & a_{n-1} \end{bmatrix}.$$

(v) *With* F *as in* (iv) *and irreducible, the hyper-companion matrix,* $H(F^m)$, *of the monic polynomial* $G = F^m$ *is the* $mn \times mn$ *matrix.*

$$H(G) = H(F^m) = \begin{bmatrix} C(F) & L & 0 & 0 & \ldots & 0 & 0 & 0 \\ 0 & C(F) & L & 0 & \ldots & 0 & 0 & 0 \\ \vdots & & & & & & & \\ 0 & 0 & \ldots & \ldots & \ldots & 0 & C(F) & L \\ 0 & 0 & \ldots & \ldots & \ldots & 0 & 0 & C(F) \end{bmatrix}$$

where $L = (l_{ij})$ *is the* $n \times n$ *matrix with one in its lower left-hand corner and all other elements zero; that is,* $l_{n1} = 1$, $l_{ij} = 0$ *otherwise. When* $m = 1$, $H(F) = C(F)$.

(vi) *If* $xI - A$ *has invariant factors* $F_1, \ldots, F_n$, *then* $A \approx \operatorname{diag}(C(F_1), \ldots, C(F_n))$.

(vii) *If* $xI - A$ *has elementary divisors* $E_1, \ldots, E_k$, *then* $A \approx \operatorname{diag}(H(E_1), \ldots, H(E_k))$. *These are the canonical forms that will be used here.*

(viii) *If* $xI - A$ *has invariant factors* $F_1(x), \ldots, F_n(x)$ *and elementary divisors* $E_1(x), \ldots, E_k(x)$ *then* $xI - tA$ *has invariant factors* $F_1(x/t), \ldots, F_n(x/t)$ *and elementary divisors* $E_1(x/t), \ldots, E_k(x/t)$.

(ix) *If* $E(x)$ *is an elementary divisor of* $xI - A$ *and* $E(x) = (x-t)^m$, *write* $C_m(t) = H(E)$. *Thus*

$$C_m(t) = \begin{bmatrix} t & 1 & 0 & 0 & \cdots & 0 & 0 & 0 \\ 0 & t & 1 & 0 & \cdots & 0 & 0 & 0 \\ \vdots & & & & & & & \\ 0 & 0 & \cdots & & & 0 & t & 1 \\ 0 & 0 & \cdots & & & 0 & 0 & t \end{bmatrix}$$

If $E(x)$ is irreducible of degree n and α is a root of $E(x)$ in $GF(q^n)$, write $D_n(\alpha) = \mathrm{diag}\,(\alpha, \alpha^q, \ldots, \alpha^{q^{n-1}})$. So, over $GF(q^n)$, $H(E) \sim D_n(\alpha)$. Further, let

$$D_n^m(\alpha) = \begin{bmatrix} D_n(\alpha) & L & 0 & 0 \cdots 0 & 0 & 0 \\ 0 & D_n(\alpha) & L & 0 \cdots 0 & 0 & 0 \\ \vdots & & & & & \\ 0 & 0 & \cdots & \cdots 0 & D_n(\alpha) & L \\ 0 & 0 & \cdots & \cdots 0 & 0 & D_n(\alpha) \end{bmatrix}$$

with L as in (v). Then, over $GF(q^n)$, $H(E^n) \sim D_n^m(\alpha)$. □

Although the list of elementary divisors identifies a projectivity, a briefer notation is required. Both the *symbol* and the *extended symbol* will be defined.

Let $\mathfrak{T} = \mathbf{M}(T)$ be a projectivity of $PG(n-1,q)$. Each elementary divisor of T has a *symbol*

$$(r)m,$$

where m is the number of conjugates over γ of the corresponding eigenvalue λ and mr is the degree of the elementary divisor. To each eigenvalue λ of T, we associate the symbol

$$(r_1, r_2, \ldots, r_k)m,$$

where $r_1 m, \ldots, r_k m$ are the degrees of the elementary divisors with λ as eigenvalue. Finally, if T has eigenvalues $\lambda_1, \ldots, \lambda_s$, it has the *symbol*

$$\left[(r_1^{(1)}, \ldots, r_{k_1}^{(1)})m_1, \ldots, (r_1^{(s)}, \ldots, r_{k_s}^{(s)})m_s \right].$$

Sometimes it is desirable to include the eigenvalues; then T has the *extended symbol*

$$\left[\ \lambda_1(r_1^{(1)},\ldots,\ r_{k_1}^{(1)})m_1,\ldots,\ \lambda_s(r_1^{(s)},\ldots,r_{k_s}^{(s)})m_s\ \right].$$

Note that

$$\Sigma_{i,j}\ r_i^{(j)}m_j\ =\ n.$$

The *symbol* for $\mathfrak{T}$ is the same as the symbol for T. The *extended symbol* for $\mathfrak{T}$ is the extended symbol of tT, any t in γ_0. If $m_j = 1$, then we may take $\lambda_j = 1$, and we usually select that λ_j for which $\Sigma_i\ r_i^{(j)}$ is the greatest.

Examples:

Elementary divisors of T	Symbol for $\mathfrak{T}$	Extended symbol for $\mathfrak{T}$
$(x-\lambda)^r$	$[(r)1]$	$[\lambda(r)1]$
$(x-\lambda)^{r_1},\ldots,(x-\lambda)^{r_k}$	$[(r_1,\ldots,r_k)1]$	$[\lambda(r_1,\ldots,r_k)1]$
$(x-\lambda_0)^r,(x-\lambda_1)^s,\ \lambda_0\neq\lambda_1$	$[(r)1,\ (s)1]$	$[\lambda_0(r)1,\ \lambda_1(s)1]$
$F(x) = x^m - a_1x^{m-1} - \ldots - a_0$, $F(\alpha) = 0$ in $\bar{\gamma}$	$[(1)m]$	$[\alpha(1)m]$
$F(x)^r$	$[(r)m]$	$[\alpha(r)m]$

2.4. Projectivities with subprimitive characteristic polynomial

From theorem 2.3.2(viii), if $xI-A$ has invariant factors $F_1(x),\ldots,\ F_n(x)$, then $xI-tA$ has invariant factors $F_1(x/t),\ldots,$ $F_n(x/t)$. In particular, if $F(x) = x^n - a_{n-1}x^{n-1}\ \ldots\ -a_0$ is irreducible, then $G(x) = t^nF(x/t) = x^n - a_{n-1}tx^{n-1}\ \ldots\ -a_0t^n$. Thus $H(F)$ and $H(G)$ are matrices of conjugate projectivities. So, if $a_{n-1}\neq 0$, the projectivities given by $H(G)$ as t varies in γ_0 are all in the same conjugacy class, which may be represented by $H(G)$ with $G = x^n - x^{n-1} - b_{n-2}x^{n-2}\ldots -b_0$ where $a_{n-1}t = 1$ and $b_i = a_it^{n-i}$ for i in $\bar{\mathrm{N}}_{n-2}$; if $(q-1,n) = 1$, the form $G =$ $x^n - b_{n-1}x^{n-1}\ \ldots\ -b_1x-1$ could be used instead. It should be noted that $t^nF(x/t)$ does not always give q-1 distinct polynomials as t varies in γ_0.

2.4.1. LEMMA. *If F is a subprimitive polynomial of degree n, $t^nF(x/t)$ gives $q-1$ distinct polynomials as t varies in λ_0.*

Proof. Suppose the roots in $GF(q^n)$ of F are $\beta, \beta^q, \ldots, \beta^{q^{n-1}}$ where β is a subprimitive root of $GF(q^n)$. If there exists t in γ_{01} such that $t^nF(x/t) = F(x)$, then $\{\beta, \beta^q, \ldots, \beta^{q^{n-1}}\} = \{t\beta, t\beta^q, \ldots, t\beta^{q^{n-1}}\}$. So $t\beta = \beta^{q^i}$ and therefore $t = \beta^{q^i-1}$ for some i with $0 < i \leqslant n-1$. Since β is subprimitive, the smallest positive power of it in $GF(q)$ is $\theta(n-1)$ by §1.6(vi). However, for $i \leqslant n-1$, $q^i-1 < q^{n-1} + q^{n-2} + \ldots + 1 = \theta(n-1)$, giving a contradiction.□

COROLLARY. *If F is subprimitive, $H(F)$ is not similar to $tH(F)$ for any t in γ_{01}.*□

2.4.2. LEMMA. *If F is irreducible of degree n, then the number of elements of $GL(n,q)$ which commute with $H(F)$ is q^n-1.*

Proof. Consider $H(F)$ as $D_n(\beta)$. Then, if $TD_n(\beta) = D_n(\beta)T$, T must take the form diag $(\alpha^r, \alpha^{rq}, \ldots, \alpha^{rq^{n-1}})$ where α is a primitive root of $GF(q^n)$. Hence, there are q^n-1 possibilities for T as r varies in $\mathbf{N}_{q^n-2}$.□

COROLLARY. *If F is a subprimitive polynomial of degree n and $\mathfrak{H}$ is the element of $PGL(n,q)$ given by the matrix $H(F)$, then $\mathfrak{H}$ commutes with $\theta(n-1)$ elements of $PGL(n,q)$.*

Proof. By the corollary to lemma 2.4.1, there is no T in $GL(n,q)$ such that $TH(F) = tH(F)T$ for some t in γ_{01}. So, for $\mathfrak{T}\mathfrak{H} = \mathfrak{H}\mathfrak{T}$, it is only necessary to consider the corresponding matrix equation $TH = tHT$ for $t = 1$. Hence, from the lemma, $\mathfrak{H}$ commutes with $(q^n-1)/(q-1) = \theta(n-1)$ elements of $PGL(n,q)$.□

2.4.3. LEMMA. *If $\mathfrak{H}$ is a projectivity of $PG(n,q)$ whose characteristic polynomial is subprimitive, then the conjugacy class of $\mathfrak{H}$ in $PGL(n+1,q)$ has $p(n+1,q)/\theta(n) = q^{n(n+1)/2}\,[1,n]_-$ elements.*□

2.4.4. LEMMA. *The number of conjugacy classes of $PGL(n+1,q)$ whose elements have subprimitive characteristic polynomial is $\phi(\theta(n))/(n+1)$.*

Proof. If F is subprimitive, then $G = t^{n+1}F(x/t)$ is also subprimitive and distinct from F for t in γ_{01}. Since $H(F)$ and $H(G)$ define conjugate projectivities, and, from §1.6, the number of subprimitive polynomials of degree $n+1$ is $R(n+1,q) = (q-1)\phi(\theta(n))/(n+1)$, the required answer is $R(n+1,q)/(q-1) = \phi(\theta(n))/(n+1)$, where ϕ is the Euler function.□

2.4.5. THEOREM. *The number, $\sigma(n,q)$, of projectivities of $PG(n,q)$ with subprimitive polynomial is*

$$\sigma(n,q) = q^{n(n+1)/2}\ [1,\ n]_-\ \phi(\theta(n))/(n+1).\square$$

2.5. Orders of projectivities

We will calculate the order of the projectivities given by the matrix $H(F)$ where $F = E^m$ and E is irreducible. There are three separate cases:

(i) $E = x-t$: $H(F) = C_m(t)$;

(ii) $m = 1$, E has degree n and a root α in $GF(q^n)$: $H(F) \sim D_n(\alpha)$ over $GF(q^n)$;

(iii) $m \geqslant 1$, E has degree n and a root α in $GF(q^n)$: $H(F) \sim D_n^m(\alpha)$ over $GF(q^n)$.

Let $\mathfrak{T} = \mathrm{M}(T)$; then the *order of T*, $o(T)$, is the least positive integer k such that $T^k = I$; the *order of* $\mathfrak{T}$, $o(\mathfrak{T})$ is the least positive integer k such that $T^k = cI$ for some c in γ_0.

(i) If $T = C_m(t)$, then $o(\mathfrak{T}) = o(T) = p^r$ where $p^{r-1} < m \leqslant p^r$ and $q = p^h$. In this case $c = t^{p^r}$.

(ii) If $T = D_n(\alpha)$ where α is a root in $GF(q^n)$ of $E(x) = x^n - a_{n-1}x^{n-1} \ldots - a_1x - a_0$, then

(a) $o(\mathfrak{T})$ is the smallest positive integer k such that $\alpha^k \in GF(q)$;

(b) by §1.6 (v) and (vi), $o(\mathfrak{T}) \mid \theta(n-1)$;

(c) if α is an n-subprimitive root of $GF(q^n)$, that is, E is subprimitive, $o(\mathfrak{T}) = \theta(n-1)$.

(d) $c = \alpha^{o(\mathfrak{T})}$, $\alpha^{\theta(n-1)} = (-1)^{n-1}a_0 = c^{\theta(n-1)/o(\mathfrak{T})}$.

(iii) If $T = D_n^m(\alpha)$, then $o(\mathfrak{T}) = p^r k_0$, where

(a) $p^{r-1} < m \leqslant p^r$;

(b) k_0 is the smallest integer k such that $\alpha^k \in GF(q)$;

(c) $k_0 \mid \theta(n-1)$;

(d) if α is an n-subprimitive root of $GF(q^n)$, $k_0 = \theta(n-1)$;

(e) $c = \alpha^{o(\mathfrak{T})}$ and $\alpha^{p^r \theta(n-1)} = (-1)^{n-1}a_0^{p^r} = c^{\theta(n-1)/k_0}$.

2.6. Varieties

If F in $K[x_0,\ldots, x_n]$ is homogeneous, it is a *form*. As in §2.1, let $X = (x_0,\ldots, x_n)$. Also, write $K[X] = K[x_0,\ldots, x_n]$; if $F \in K[X]$, then $F(X) = F(x_0,\ldots, x_n)$. There is a danger of confusion as the x_i are used both as indeterminates and as elements of K; however, it should be clear from the context which is meant.

(i) A subset $\mathscr{F}$ of $PG(n,K)$ is a *variety* (*over* K) if there exist forms $F_1, F_2,\ldots, F_r$ in $K[X]$ such that

$$\begin{aligned}\mathscr{F} &= \{\mathbf{P}(A) \in PG(n,K) \mid F_1(A) = F_2(A) = \ldots = F_r(A) = 0\} \\ &= \mathbf{V}(F_1,\ldots, F_r).\end{aligned}$$

The points $\mathbf{P}(A)$ are *points of* $\mathscr{F}$. Occasionally, we write $\mathscr{F} = \mathbf{V}_n(F_1,\ldots, F_r)$ or $\mathscr{F} = \mathbf{V}_{n,K}(F_1,\ldots, F_r)$ if n or K is doubtful. When $K = GF(q)$, the notation $\mathbf{V}_{n,q}(F_1,\ldots, F_r)$ is used.

$\mathbf{V}(\lambda_1 F_1,\ldots, \lambda_r F_r) = \mathbf{V}(F_1,\ldots, F_r)$ for all $(\lambda_1,\ldots, \lambda_r)$ in K_0^r. Also

$$\mathbf{V}(F_1,\ldots, F_r) = \mathbf{V}(F_1) \cap \mathbf{V}(F_2) \cap \ldots \cap \mathbf{V}(F_r). \tag{2.8}$$

A prime $\pi(A)$ can also be written $\mathbf{V}(AX^*)$.

(ii) A variety $\mathbf{V}(F)$ is a *primal*; elsewhere, the term *hypersurface* is common. A primal in $PG(2,K)$ is a (*plane*) (*algebraic*) *curve*; a primal in $PG(3,K)$ is a *surface*.

(iii) The *order* or *degree* of a primal $\mathbf{V}(F)$ is the degree m of F. A projectivity of the space does not change the order of $\mathbf{V}(F)$.

(iv) $\mathbf{P}(A)$ is a *complex point* of $\mathscr{F} = \mathbf{V}_{n,K}(F_1, \ldots, F_r)$ if $F_1(A) = \ldots = F_r(A) = 0$ and $A \in \overline{K}^{n+1}$, where $\overline{K}$ is the algebraic closure of K. An *imaginary point* $\mathbf{P}(A)$ of $\mathscr{F}$ is a complex point of $\mathscr{F}$ which is not a point of $\mathscr{F}$; that is, $A \in \overline{K}^{n+1}\backslash K^{n+1}$ and there is no λ in $\overline{K}_0$ such that $\lambda A \in K^{n+1}$. More precisely, $\mathbf{P}(A)$ is a *k-complex point of* $\mathscr{F}$ if K' is the extension of K of least degree k such that $\mathbf{P}(A)$ is a point of $\mathbf{V}_{n,K'}(F_1, \ldots, F_r)$. Sometimes it is convenient to describe the points of $\mathscr{F}$ as *real points*.

(v) If the primals $\mathbf{V}(F_1)$ and $\mathbf{V}(F_2)$ in $PG(n,K)$ have the same set of complex points, then F_1 and F_2 are not necessarily proportional. However, over $\overline{K}$, they will have the same set of irreducible factors with perhaps different multiplicities. When it is necessary to take account of these multiplicities, we write $\mathbf{V}^*(F)$ for $\mathbf{V}(F)$. Thus $\mathbf{V}^*(F)$ and $\mathbf{V}(F)$ have the same set of complex points but $\mathbf{V}^*(F_1) = \mathbf{V}^*(F_2)$ if and only if $F_1 = \lambda F_2$ for some λ in K_0.

(vi) The meet $\mathscr{F} \cap \Pi_r$ of a primal $\mathscr{F}$ of order m and an r-space Π_r not lying on $\mathscr{F}$ is a primal of Π_r of order m.

If $\mathscr{F}$ is a primal order m in $PG(1,K)$, then $\mathscr{F}$ has a set G_m of exactly m complex points, some of which may be repeated. Similarly, if $\mathscr{F}$ is a primal of order m in $PG(n,K)$ and Π_1 is a line not lying on $\mathscr{F}$, then $\mathscr{F} \cap \Pi_1$ has a set of m complex points.

(vii) A primal $\mathscr{F} = \mathbf{V}(F)$ in $PG(n,K)$ is *irreducible* if F is irreducible over K; $\mathscr{F}$ is *absolutely irreducible* if F is irreducible over $\overline{K}$. If, over $\overline{K}$, $F = F_1 F_2 \ldots F_s$, where each F_i has degree at least one and is irreducible over $\overline{K}$, then

$$\mathbf{V}_{n,\overline{K}}(F) = \mathbf{V}_{n,\overline{K}}(F_1) \cup \ldots \cup \mathbf{V}_{n,\overline{K}}(F_s). \qquad (2.9)$$

The sets $\mathscr{G}_i = \mathbf{V}_{n,\overline{K}}(F_i) \cap PG(n,K)$ are the *components* of $\mathscr{F}$. We call $\mathscr{G}_i$ a *regular* component of $\mathscr{F}$ if $\mathscr{G}_i = \mathbf{V}(F_i)$; that is, $\lambda F_i \in K[X]$ for some λ in $\overline{K}_0$. See §10.1.

(viii) If $\mathscr{F}$ is an irreducible primal of order m in $PG(1,K)$, then the complex points of $\mathscr{F}$ are *m-complex conjugate points* of $\mathscr{F}$. This definition can also be applied to $\mathscr{F} = l \cap F'$, where $\mathscr{F}'$ is a variety and l a line of $PG(n,K)$. See Chapter 21 in Volume II on twisted cubics.

(ix) Let $\mathscr{F} = \mathbf{V}_{n,K}(F_1, \ldots, F_r)$ and $\bar{\mathscr{F}} = \mathbf{V}_{n,\bar{K}}(F_1, \ldots, F_r)$. Then $\bar{\mathscr{F}}$ is *reducible* if there exist varieties $\mathscr{G}_1$, $\mathscr{G}_2$ over $\bar{K}$ with $\mathscr{G}_1$, $\mathscr{G}_2$ proper subsets of $\bar{\mathscr{F}}$. Otherwise $\bar{\mathscr{F}}$ is *irreducible*. Then $\bar{\mathscr{F}}$ has a unique decomposition into irreducible components

$$\bar{\mathscr{F}} = \mathscr{G}_1 \cup \mathscr{G}_2 \cup \ldots \cup \mathscr{G}_s \tag{2.10}$$

such that $\mathscr{G}_i \not\subset \mathscr{G}_j$ for $i \neq j$.

If $s = 1$ in (2.10), then $\mathscr{F}$ is *absolutely irreducible*. If $s > 1$, suppose that $\mathscr{G}_i = \bar{\mathscr{H}}_i$ for $i \leqslant s_0$, where $\mathscr{H}_i$ is a variety over K, and that there exists no such $\mathscr{H}_i$ for $i > s_0$. Then, still with $s > 1$, we define $\mathscr{F}$ to be *irreducible* if $s_0 = 0$.

The *components* of $\mathscr{F}$ are $\mathscr{G}_i \cap PG(n,K)$, $i \in \mathbf{N}_s$. So, if $s_0 = s$, then the components of $\mathscr{F}$ are just $\mathscr{G}_1, \mathscr{G}_2, \ldots, \mathscr{G}_s$.

(x) The *dimension* of an irreducible variety $\mathscr{F}$ in $PG(n,K)$ is $n - k - 1$ where k is the maximum of the dimensions of the subspaces containing no complex points of $\mathscr{F}$. Thus the dimension of a primal is $n - 1$; in particular, a curve in $PG(2,K)$ and a surface in $PG(3,K)$ have dimension one and two respectively. More generally, varieties of dimension one and two are called *curves* and *surfaces* respectively. Naturally, as a variety, Π_k has dimension k.

(xi) The order of a primal was defined in (iii). If an irreducible variety $\mathscr{F}$ has dimension e, then the *order* or *degree* of $\mathscr{F}$ is the maximum number of complex points in which the subspaces of dimension $n-e$ not lying on $\mathscr{F}$ meet $\mathscr{F}$. Thus the order of a primal $\mathscr{F}$ is the maximum number of complex points in which a line not on $\mathscr{F}$ can meet $\mathscr{F}$.

(xii) If $\mathscr{F}$ is a variety contained in a subspace Π_r of $PG(n,K)$ and Π_s is a subspace of $PG(n,K)$ skew to Π_r, then the *cone* $\Pi_s\mathscr{F}$ consists of the points on the lines PQ with P in Π_s and Q in $\mathscr{F}$: it is a variety.

(xiii) Let $P = \mathbf{P}(A)$ be a point of the irreducible primal $\mathscr{F} = \mathbf{V}(F)$ of order d. Let $l = \mathbf{P}(A)\mathbf{P}(B)$. Then

$$f(t) = F(A + tB) = F^{(0)} + F^{(1)}t + \ldots + F^{(d)}t^d. \tag{2.11}$$

Since $\mathbf{P}(A) \in \mathscr{F}$, so $F^{(0)} = F(A) = 0$; also $F^{(d)} = F(B)$.

Suppose l is not on $\mathscr{F}$; that is, not all the $F^{(i)}$ are zero. The *intersection multiplicity of l and $\mathscr{F}$ at* $\mathbf{P}(A)$, denoted $m_P(l,\mathscr{F})$, is the multiplicity of the root $t=0$ of f.

When $\mathscr{F} = \mathbf{V}(F_1, F_2,\ldots, F_r)$, then define $m_P(l,\mathscr{F}) = (m_1, m_2,\ldots, m_r)$, where $m_i = m_P(l,\mathscr{F}_i)$, $\mathscr{F}_i = \mathbf{V}(F_i)$ and $(m_1, m_2,\ldots, m_r)$ is the greatest common divisor of the m_i.

If l lies on $\mathscr{F}$, write $m_P(l,\mathscr{F}) = \infty$.

The *multiplicity of P on $\mathscr{F}$*, denoted $m_P(\mathscr{F})$, is the minimum of $m_P(l,\mathscr{F})$ for all lines l through P. Then P is a *singular* point of $\mathscr{F}$ if $m_P(\mathscr{F}) > 1$ and a *simple* point of $\mathscr{F}$ if $m_P(\mathscr{F}) = 1$.

A line l is a *tangent line* to $\mathscr{F}$ at P or *touches $\mathscr{F}$ at P* if $m_P(l,\mathscr{F}) > m_P(\mathscr{F})$.

See §10.1 for a different but equivalent definition of singular points for plane curves.

(xiv) As in (xiii), let $P = \mathbf{P}(A)$ be a point of the irreducible primal $\mathscr{F}$. The *tangent space* of $\mathscr{F}$ at P, denoted $T_P(\mathscr{F})$, is the set of points on all the tangent lines. It is formed in the following way. Consider

$$f(t) = F(A + tX) = F^{(0)}(X) + F^{(1)}(X)t + \ldots + F^{(d)}(X)t^d. \tag{2.12}$$

Over any field,

$$F^{(1)}(X) = \frac{\partial F}{\partial a_0}x_0 + \frac{\partial F}{\partial a_1}x_1 + \ldots + \frac{\partial F}{\partial a_n}x_n, \tag{2.13}$$

where $\dfrac{\partial F}{\partial a_i} = \dfrac{\partial F}{\partial x_i}(A)$ and $A = (a_0, a_1,\ldots, a_n)$. So P is singular if $\dfrac{\partial F}{\partial a_0} = \dfrac{\partial F}{\partial a_1} = \ldots = \dfrac{\partial F}{\partial a_n} = 0$.

If P is simple, then $T_P(\mathscr{F}) = \mathbf{V}(F^{(1)})$. If $m_P(\mathscr{F}) = m$, then $T_P(\mathscr{F}) = \mathbf{V}(F^{(m)})$.

If $\mathscr{F} = \mathbf{V}(F_1,\ldots, F_r)$ is irreducible, then $T_P(\mathscr{F}) = T_P(\mathscr{F}_1) \cap\ldots\cap T_P(\mathscr{F}_r)$, where $\mathscr{F}_i = \mathbf{V}(F_i)$. Hence P is a simple point of $\mathscr{F}$ if and only if $T_P(\mathscr{F})$ is a prime of $PG(n,K)$.

2.6.3. LEMMA. *If Π_r is a subspace of $PG(n,K)$ lying on the irreducible variety $\mathscr{F}$ and containing the point P of $\mathscr{F}$, then $\Pi_r \subset T_P(\mathscr{F})$.*

Proof. It suffices to prove this for $r = 1$ and $\mathscr{F}$ a primal, since if the result is true for all lines in Π_r through P and $\mathscr{F}$ a primal, it is true for Π_r; if the result is true for a primal, it is true for an intersection of primals.

For $\mathscr{F} = \mathbf{V}(F)$, we have $T_P(\mathscr{F}) = \mathbf{V}(F^{(m)})$ as in (xiv). If l lies on $\mathscr{F}$, then $F^{(i)}(B) = 0$ for all i and any point $\mathbf{P}(B)$ of l. So, in particular, $F^{(m)}(B) = 0$, whence $\mathbf{P}(B) \in T_P(\mathscr{F})$ and so l is in $T_P(\mathscr{F})$.

2.7. Linear systems of primals

(i) The set of primals of order m of $PG(n,K)$ form a projective space $PG(N,K)$ where $N = \mathbf{h}(n+1, m) - 1$.

(ii) The set of primals forming a d-space Π_d of $PG(N,K)$ is also called a *linear system* $\mathscr{L}_d$ *of dimension* d. Thus if a point of Π_d represents a primal $\mathbf{V}(F)$, then

$$F = \lambda_0 F_0 + \ldots + \lambda_d F_d,$$

where $F_1, \ldots, F_d$ are linearly independent forms such that $\mathbf{V}(F_i) \in \mathscr{L}_d$ and $\lambda_i \in K$, all i. To avoid confusion we speak of "primals of the system $\mathscr{L}_d$" rather than 'points of the subspace Π_d'. Similarly, instead of 'subspaces of Π_d', we refer to 'linear subsystems of $\mathscr{L}_d$'. A linear system of dimension one is called a *pencil* and of dimension two is called a *net*.

(iii) A point common to all the primals of a linear system $\mathscr{L}_d$ is a *base point*. All the base points of $\mathscr{L}_d$ are obtained as the intersection of any $d + 1$ linearly independent primals of $\mathscr{L}_d$.

(iv) All the primals of $\mathscr{L}_d$ which pass through a fixed point P_0, which is not a base point of $\mathscr{L}_d$, form a linear system $\mathscr{L}_{d-1}$. More generally, the primals of $\mathscr{L}_d$ through $P_1, \ldots, P_k$, none of which are base points, form a system $\mathscr{L}_e$ where $d - k < e < d$. Also, $e = d - k$ if $P_1, \ldots, P_k$ impose independent conditions.

(v) The intersection of a linear system $\mathscr{L}_d$ of primals of order m of $PG(n,q)$ with a subspace Π_r is a linear system $\mathscr{L}_e$ of primals of order m of Π_r, where $e = d - s$ is the number of linearly independent primals of $\mathscr{L}_d$ containing Π_r.

(vi) A linear system $\mathscr{L}_r$ of primals of order m of $PG(1,K)$ is a *linear series of order m and dimension r* and is denoted by g_r^m. Then we have that $r \leqslant m$. In fact $r = m$ if and only if g_r^m is the *complete series* of all sets G_m consisting of m complex points.

If $\{\mathbf{U}_0, \mathbf{U}_1, \mathbf{U}\}$ is the basis for a coordinate system on $PG(1,K)$, then the (*non-homogeneous*) *coordinate* of a complex point $\mathbf{P}(t_0, t_1)$ is t_0/t_1. The product of the coordinates of the elements of a G_m not containing $\mathbf{U}_0$ or $\mathbf{U}_1$, where $G_m = \mathbf{V}_{1,\bar{K}}(c_0x_0^m + \ldots + c_1x_1^m)$, is $(-1)^m c_1/c_0$. This product lies in K (since c_0 and c_1 do), even if some of the points of G_m are imaginary.

2.7.1. THEOREM. *If $G_{m+1} = \{P_0, \ldots, P_m\}$ is a set of m+1 distinct points of $PG(1,K)$, the m+1 sets $G_m^i = G_{m+1}\backslash\{P_i\}$ for i in $\bar{\mathbf{N}}_m$ are linearly independent. Thus the g_r^m containing all m+1 sets G_m^i is the complete series g_m^m of $PG(1,K)$.*

Proof. Let t_i be the coordinate of P_i, $i \in \bar{\mathbf{N}}_m$. Let $f(x) = \prod_{i=0}^{m}(x - t_i)$ and let $f_i(x) = f(x)/(x - t_i)$. If the G_m^i are dependent, there exist a_i in K not all zero such that $\sum_{i=0}^{m} a_i f_i(x) = 0$.

Suppose $a_0 \neq 0$. Put $x = t_0$. Then $\prod_{i=1}^{m}(t_0 - t_i) = 0$. So $t_0 = t_j$ for some j, which is impossible since all points P_i are distinct.□

2.8. A generalization of the theorem of Menelaus

An *n-gram* is a set of n lines in $PG(2,K)$, no three of which are concurrent, together with the set of $n(n-1)/2$ points of intersection of pairs of the lines. The points and lines are called *vertices* and *sides* of the n-gram. The n sides of the n-gram form an *n-side*.

Dually, an *n-stigm* is a set of points in $PG(2,K)$, no three of which are collinear, together with the set of $n(n-1)/2$ joins of pairs of the points. The points and lines are again called *vertices* and *sides* of the n-stigm. The n vertices form an *n-point* or an *n-arc*.

An n-gon is a cycle $(P_1P_2 \ldots P_n)$ of n points no three of which are collinear: the P_i are called the *vertices* and the n lines $P_1P_2, P_2P_3, \ldots, P_{n-1}P_n, P_nP_1$ the *sides* of the n-gon.

For n-grams, n-stigms and n-gons with arbitrary n, the respective terms are *polygram*, *polystigm*, and *polygon*. For small n, the following terms, mostly classical, are used:

	n-gram	n-stigm	n-gon
$n = 3$	triangle	triangle	triangle
$n = 4$	tetragram	tetrastigm	quadrangle, quadrilateral
$n = 5$	pentagram	pentastigm	pentagon
$n = 6$	hexagram	hexastigm	hexagon

For *tetragram* and *tetrastigm*, it is common to see *complete quadrilateral* and *complete quadrangle* respectively. A triangle with vertices P_0, P_1, P_2 and sides l_0, l_1, l_2 is denoted $P_0P_1P_2$ or $l_0l_1l_2$, when there is no risk of confusing it with the entire plane.

Let $\mathscr{C}_n$ be an algebraic curve of order n in $PG(2,K)$.

Certain problems on the intersection of two curves $\mathscr{C}_n$ and $\mathscr{C}_k$ will now be considered, particularly when $\mathscr{C}_k$ is a k-side $\mathscr{K}$, the set of sides of a k-gram $\mathscr{K}'$.

Let $\mathscr{K} = \{\alpha_0, \ldots, \alpha_{k-1}\}$ with $\alpha_i = \mathbf{V}(L_i)$, let $A_{ij} = \alpha_i \cap \alpha_j$, let $\mathscr{A} = \{A_{ij}\}$ and let $\mathscr{A}_i = \{A_{ij} \mid j \neq i\}$ be the set of vertices of $\mathscr{K}'$ on α_i. Consider on each α_i a set $G_n^{\,i}$ of n points such that $\mathscr{A}_i \cap G_n^{\,i} = \emptyset$. Let $G = \cup G_n^{\,i}$. For any $\mathscr{K}$ such that $\mathscr{C}_n \cap \mathscr{A} = \emptyset$, $\mathscr{C}_n \cap \mathscr{K}$ is such a set G. Conditions are required under which a given G is the complete intersection of $\mathscr{K}$ with some $\mathscr{C}_n$. In fact, given a G on $\mathscr{K}$, necessary and sufficient conditions will be found for the set $\mathscr{M}_G = \{\mathscr{C}_n \mid \mathscr{C}_n \cap \mathscr{K} = G\}$ to be non-empty.

Let $\mathscr{L}$ be the linear system of all $\mathscr{C}_n$ containing the points G; then $\mathscr{L} \supset \mathscr{M}_G$. The dimension of $\mathscr{L}$ will be found, as well as the number of conditions for a $\mathscr{C}_n$ in the plane to belong to $\mathscr{L}$.

2.8.1. THEOREM.

(i) *$\mathscr{C}_n$ meets $\mathscr{K}$ in G and some further point P if and only if all the lines α_i of $\mathscr{K}$ are part of $\mathscr{C}_n$.*

(ii) *The set* $\mathcal{L}\backslash\mathcal{M}_G$ *of curves* $\mathcal{C}_n$ *containing* $\mathcal{K}$ *forms a linear subsystem* $\mathcal{L}_K$ *of* $\mathcal{L}$ *and has dimension* $\mathbf{h}(3, n-k) - 1$.

Proof. If $P \in \alpha_0$, then $\mathcal{C}_n \cap \alpha_0$ consists of $n+1$ points $\{P\} \cup G_n^{\ 0}$. So $\mathcal{C}_n$ contains α_0. However, any other line α_i meets α_0 in A_{0i}. So each α_i meets $\mathcal{C}_n$ in at least $n+1$ points. Hence $\mathcal{C}_n$ contains all α_i. Thus any curve of $\mathcal{L}$ not in $\mathcal{M}_G$ contains $\mathcal{K}$, and $\mathcal{L}_{\mathcal{K}}$ consists of all curves $\mathbf{V}(L_0 L_1 \ldots L_{k-1} F_{n-k})$, where F_{n-k} is a form of order $n-k$; such a curve comprises the k lines α_i and $\mathcal{C}_{n-k} = \mathbf{V}(F_{n-k})$. Thus $\mathcal{L}_{\mathcal{K}}$ is a linear subsystem of dimension $\mathbf{h}(3, n-k) - 1$.□

2.8.2. THEOREM. *Suppose* G *lies on* $\mathcal{C}_n{}^* = \mathbf{V}(F^*)$.

(i) *If* $k > n$, $\mathcal{L}_{\mathcal{K}} = \emptyset$ *and* $\mathcal{L} = \mathcal{M}_G = \{\mathcal{C}_n{}^*\}$.

(ii) *If* $\mathcal{L}_{\mathcal{K}} \neq \emptyset$, $\mathcal{M}$ *is not a linear subsystem of* $\mathcal{L}$. *However, if* $\mathcal{C}_n \in \mathcal{L}$ *and* $\mathcal{C}_n{}^* \in \mathcal{M}_G$, *there is a* $\mathcal{C}_{n-k} = \mathbf{V}(F_{n-k})$ *such that* $\mathcal{C}_n = \mathbf{V}(F^* + L_0 L_1 \ldots L_{k-1} F_{n-k})$.

(iii) *If* d *is the dimension of* $\mathcal{L}$, *then* $d = \mathbf{h}(3, n-k)$.

Proof. (i) If $k > n$, no $\mathcal{C}_n$ can have $\mathcal{K}$ as a component. So $\mathcal{L}_{\mathcal{K}} = \emptyset$ and $\mathcal{L} = \mathcal{M}_G$ and is the single curve $\mathcal{C}_n{}^*$.

(ii) If $\mathcal{L}_{\mathcal{K}} \neq \emptyset$, then $\mathcal{L}_{\mathcal{K}}$ and $\mathcal{M}_G$ form a partition of $\mathcal{L}$. So $\mathcal{M}_G$ cannot be a linear subsystem. If $\mathcal{C}_n = \mathbf{V}(F) \in \mathcal{L}$ and $\mathcal{C}_n{}^* \in \mathcal{M}_G$, then some member of the pencil $\mathbf{V}(sF + tF^*)$ contains the point P in $\mathcal{K}\backslash G$ and so is in $\mathcal{L}_{\mathcal{K}}$. Hence $\lambda F = F^* + L_0 L_1 \ldots L_{k-1} F_{n-k}$ for some λ and some F_{n-k}.

(iii) From (ii), $d = \dim \mathcal{L}_{\mathcal{K}} + 1 = \mathbf{h}(3, n-k)$.□

COROLLARY. (i) *If there is a non-empty family* $\mathcal{M}_G$ *of curves* $\mathcal{C}_n$ *such that* $\mathcal{C}_n \cap \mathcal{K} = G$, *then* G *imposes* $\mathbf{h}(3, n) - 1 - \mathbf{h}(3, n-k) = k(2n-k+3)/2 - 1$ *conditions on a* $\mathcal{C}_n$ *to contain it.*

(ii) *If* $n \geqslant k$, *then* $\mathcal{K}$ *imposes* $\mathbf{h}(3, n) - \mathbf{h}(3, n-k) = k(2n-k+3)/2$ *conditions on a* $\mathcal{C}_n$ *to contain it.*□

2.8.3$_1$. **THEOREM.** *If* $k = 1$, $\mathcal{M}_G$ *is never empty and the* n *conditions imposed by* G *are independent.*

Proof. If $\mathcal{K} = \{\alpha_0\}$, $G = \{P_1, \ldots, P_n\}$, $P \notin \alpha_0$ and

$PP_i = \mathbf{V}(L_i')$, then $\mathbf{V}(L_1'L_2'\dots L_n')$ is in $\mathcal{M}_G$.□

2.8.3$_2$. THEOREM. *If $k = 2$, $\mathcal{M}_G$ is never empty and the $2n$ conditions imposed by G are independent.*

Proof. If $\mathcal{K} = \{\alpha_0, \alpha_1\}$, $G_n^i = \{P_1^i, \dots, P_n^i\}$ and $P_j^0 P_j^1 = \mathbf{V}(L_j')$ then $\mathbf{V}(L_1'L_2'\dots L_n')$ is in $\mathcal{M}_G$.□

Let the coordinate system in $PG(2,K)$ be fixed. If the point P lies on a side of $\mathbf{U}_0\mathbf{U}_1\mathbf{U}_2$ and is not a vertex, it is the meet of that side and one of $\mathbf{V}(x_1 - cx_2)$, $\mathbf{V}(x_2 - cx_0)$, $\mathbf{V}(x_0 - cx_1)$: then c is called the *coordinate* of P.

2.8.3$_3$. THEOREM. *Let $\mathcal{K}$ be the triangle $\mathbf{U}_0\mathbf{U}_1\mathbf{U}_2$. Then $\mathcal{M}_G \neq \emptyset$ if and only if $\Pi c = (-1)^n$ where the product ranges over all $3n$ points P of G and c is the coordinate of P. If $\mathcal{M}_G \neq \emptyset$, then G imposes $3n-1$ conditions on the curves of order n to contain it; if $n = 1$ or 2, $\mathcal{M}_G$ consists of a single $\mathcal{C}_n$.*

Proof. If $\mathcal{C}_n \in \mathcal{M}_G$ and $\mathcal{C}_n = \mathbf{V}(c_0x_0^n + c_1x_1^n + c_2x_2^n + \dots)$ then $c_0c_1c_2 \neq 0$ since $\mathcal{C}_n$ contains no vertex of $\mathbf{U}_0\mathbf{U}_1\mathbf{U}_2$. So

$$G_n^0 = \mathbf{V}(x_0,\ c_1x_1^n + \dots + c_2x_2^n),$$
$$G_n^1 = \mathbf{V}(x_1,\ c_0x_0^n + \dots + c_2x_2^n),$$
$$G_n^2 = \mathbf{V}(x_2,\ c_0x_0^n + \dots + c_1x_1^n).$$

Hence $\Pi c = (-1)^n c_2/c_1$, $(-1)^n c_0/c_2$ or $(-1)^n c_1/c_0$ according as the product is taken respectively over G_n^0, G_n^1, or G_n^2. So $\Pi c = (-1)^n$ where the product is over all points of G.

Now suppose that the condition is satisfied by the points of G. Then $3n-1$ points of G determine the remaining one P, say. Let $\mathcal{L}_P$ be the linear system of $\mathcal{C}_n$ through the points $G\backslash\{P\}$. Then $\mathcal{L} \subset \mathcal{L}_P$ and, if d is the dimension of $\mathcal{L}_P$, $0 \leqslant \mathbf{h}(3,n)-1-(3n-1) = \mathbf{h}(3,n-3) \leqslant d \leqslant \mathbf{h}(3,n)-1$. Since the dimension of $\mathcal{L}_{\mathcal{K}}$ is $\mathbf{h}(3, n-3)-1 < d$, then $\mathcal{L}_P \setminus \mathcal{L}_{\mathcal{K}} \neq \emptyset$. Now, $\mathcal{M}_G = \mathcal{L} \setminus \mathcal{L}_{\mathcal{K}} \subset \mathcal{L}_P \setminus \mathcal{L}_{\mathcal{K}}$. However, if $\mathcal{C}_n^* \in \mathcal{L}_P \setminus \mathcal{L}_{\mathcal{K}}$, it meets $\mathcal{K}$ in G' where $G' = (G\backslash\{P\}) \cup \{Q\}$ and $Q \in \mathcal{K}\backslash\mathcal{A}$. But G' also satisfies the given condition. So $Q = P$ and $G' = G$. So

$\mathscr{C}_n^* \in \mathscr{M}_G$ and $\mathscr{M}_G = \mathscr{L}_P \setminus \mathscr{L}_{\mathscr{K}} \neq \emptyset$.□

For $n = 1$, this is the theorem of Menelaus and, for $n = 2$, the theorem of Carnot.

We now go on to the case that $k \geqslant 4$. The symbols $\mathscr{L}$, $\mathscr{L}_{\mathscr{K}}$, $\mathscr{M}_G$ are as above. Also, let $\mathscr{K}_i = \mathscr{K}\setminus\{\alpha_i\}$, $\mathscr{K}_{ij} = \mathscr{K}\setminus\{\alpha_i,\alpha_j\}$, $G_i = G \setminus G_n{}^i$, $G_{ij} = G \setminus (G_n{}^i \cup G_n{}^j)$. Then $\mathscr{L}^i$, $\mathscr{L}_{\mathscr{K}}{}^i$, $\mathscr{M}_G{}^i$ have the same significance for $\mathscr{K}_i$ as $\mathscr{L}$, $\mathscr{L}_{\mathscr{K}}$, $\mathscr{M}_G$ have for $\mathscr{K}$; similarly for $\mathscr{L}^{ij}$, $\mathscr{L}_{\mathscr{K}}{}^{ij}$, $\mathscr{M}_G{}^{ij}$ with regard to $\mathscr{K}_{ij}$.

2.8.3$_k$. THEOREM. (i) *For $k \geqslant 4$, $\mathscr{M}_G$ is not empty if and only if the condition $\Pi c = (-1)^n$ is satisfied for each of the* $\mathbf{c}(k,3)$ *triangles of $\mathscr{K}$.*

(ii) *If $\mathscr{M}_G$ is not empty, then G imposes* $\mathbf{h}(3,n)-1-\mathbf{h}(3,n-k) = k(2n-k+3)/2-1$ *conditions on the curves of order n to contain it.*

(iii) *If $\mathscr{M}_G$ is not empty and $k > n$, then $\mathscr{M}_G$ consists of a unique curve $\mathscr{C}_n$.*

Proof. The necessity of (i) follows from the previous theorem, where $k = 3$. Part (iii) is theorem 2.8.2(i) and part (ii) is corollary (i) to the same theorem.

The sufficiency of (i) will be proved by induction. Assume the theorem true for the values $k-1$ and $k-2$. So $\mathscr{L}^0$, the set of all $\mathscr{C}_n$ containing G_0, contains $\mathscr{M}_G{}^0$, the set of all $\mathscr{C}_n$ of $\mathscr{L}^0$ meeting $\mathscr{K}_0$ in exactly G_0, and is not empty. Therefore, $\mathscr{L}^0$ cuts out on α_0 a series $g_d{}^n$. It suffices to show that $g_d{}^n$ contains $G_n{}^0$. For, if there is a curve $\mathscr{C}_n$ in $\mathscr{L}^0$ that meets α_0 in $G_n{}^0$, then such a $\mathscr{C}_n$ cannot belong to $\mathscr{L}_{\mathscr{K}}{}^0$, so belongs to $\mathscr{M}_G{}^0$ and so to $\mathscr{M}_G$.

Now, by the theorem for the value $k-1$, $\mathscr{M}_G{}^j$ is not empty and every $\mathscr{C}_n$ in $\mathscr{M}_G{}^j$ passes through $G_n{}^0$ on α_0, for j in $\mathbf{N}_{k-1}$. Since $\mathscr{M}_G{}^j \subset \mathscr{L}^j \subset \mathscr{L}^{0j}$, it follows that $\mathscr{L}^{0j}$ contains curves meeting α_0 in $G_n{}^0$. If the series cut out on α_0 by $\mathscr{L}^{0i}$ is $g_{d_j}{}^n$, then $g_{d_j}{}^n$ contains $G_n{}^0$ for j in $\mathbf{N}_{k-1}$. Since $\mathscr{L}^0 \subset \mathscr{L}^{0j}$, so $g_d{}^n \subset g_{d_j}{}^n$ and $g_d{}^n \subset \bigcap_{j=1}^{k-1} g_{d_j}{}^n$. If the last inequality can be shown to be an equality, then $g_d{}^n$ will contain $G_n{}^0$ and the theorem will be proved.

In fact, the dimension d is required. The curves of $\mathscr{L}^0$

which contain α_0 necessarily contain all of $\mathscr{K}$ and so form the linear system $\mathscr{L}_{\mathscr{K}}$ of dimension $\mathbf{h}(3, n-k)-1$. So there exist $\mathbf{h}(3,n-k)$ but no more linearly independent curves of $\mathscr{L}^0$ through α_0. Since the dimension of $\mathscr{L}^0$ is $\mathbf{h}(3,n-k+1)$, $d =$

$$\mathbf{h}(3,n-k+1) - \mathbf{h}(3,n-k) = \begin{cases} n-k+2, & n \geq k-1 \\ 0, & n < k-1. \end{cases}$$

Similarly, for each j in $\mathbf{N}_{k-1}$, $g_{d_j}^{\;n}$ has dimension $d_j =$

$$\mathbf{h}(3,n-k+2) - \mathbf{h}(3,n-k+1) = \begin{cases} n-k+3, & n \geq k-2 \\ 0, & n < k-2. \end{cases}$$

Therefore, if $n < k-2$, then $d = d_1 = \ldots = d_{k-1} = 0$. So $g_{d_j}^{\;n} = g_d^{\;n}$ and $g_d^{\;n} = \bigcap g_{d_j}^{\;n}$.

Now suppose that $n \geq k-2$. Since $k \geq 4$, we have that $n > n-k+3 = d_j = d+1$. Since $g_d^{\;n} \subset \bigcap g_{d_j}^{\;n}$ and $d = d_j - 1$, so $g_d^{\;n} = \bigcap g_{d_j}^{\;n}$ unless $g_{d_1}^{\;n} = \ldots = g_{d_{k-1}}^{\;n}$.

It will be shown that, if $g_{d_j}^{\;n} = g$ for j in $\mathbf{N}_{k-1}$, then g is the complete series $g_n^{\;n}$ on α_0, whence $d_j = n$ for all j, contradicting the above statement that $n > d_j$. The proof will then be complete.

Choose on α_0 a set $E = \{P_1, \ldots, P_{n-k+2}\}$ of distinct points such that $E \cap \mathscr{A}_0 = \emptyset$. Let $E' = E \cup \mathscr{A}_0 = \{Q_1, \ldots, Q_{n+1}\}$ and write $E_i = E' \backslash \{Q_i\}$ for i in $\mathbf{N}_{n+1}$. If E_i is in g for all i in $\mathbf{N}_{n+1}$, then by theorem 2.7.1, the dimension of g is n.

If Q_i is a point A_{0j} of $\mathscr{A}_0$, then $E_i = E' \backslash \{A_{0j}\} = (\mathscr{A}_0 \backslash \{A_{0j}\}) \cup E$ can be obtained as the intersection of α_0 with a $\mathscr{C}_n$ of $\mathscr{L}^{0j}$ consisting of $\mathscr{K}_{0j}$ and a suitable $\mathscr{C}_{n-k+2}$ such as $\mathbf{V}(L_1 L_2 \ldots L_{n-k+2})$, where $\mathbf{V}(L_i) = PP_i$ and P is any point not on α_0. So E_i belongs to $g_{d_j}^{\;n} = g$.

If Q_i is a point P_j of E, the set $E_i = E' \backslash \{P_j\}$ can be obtained as the intersection of α_0 with a $\mathscr{C}_n$ of $\mathscr{L}_0$ consisting of $\mathscr{K}_0$ and a suitable $\mathscr{C}_{n-k+1}$ such as $\mathbf{V}(L_1 \ldots L_{j-1} L_{j+1} \ldots L_{n-k+2})$. So E_i belongs to $g_d^{\;n} \subset g$.

Thus, in all cases E_i belongs to g, which therefore has dimension n.□

The theorems $2.8.3_k$ can be dualized and it is in this form that they will in fact be utilized. A line in the plane through a vertex of $\mathbf{U}_0\mathbf{U}_1\mathbf{U}_2$ other than a side is one of $\mathbf{V}(x_1 - dx_2)$, $\mathbf{V}(x_2 - dx_0)$, $\mathbf{V}(x_0 - dx_1)$, with $d \neq 0$: d is the *coordinate* of such a line.

In the correlation given by $\mathbf{U}_0 \to \mathbf{u}_0$, $\mathbf{U}_1 \to \mathbf{u}_1$, $\mathbf{U}_2 \to \mathbf{u}_2$, and $\mathbf{U} \to \mathbf{u}$, the point $\mathbf{P}(0, c, 1)$ becomes the line $\pi(0, c, 1) = \mathbf{V}(cx_1 + x_2)$. Thus the dual of the coordinate c of a point on $\mathbf{u}_i$ is the coordinate $-1/c$ of a line through $\mathbf{U}_i$.

The dual of a plane algebraic curve $\mathscr{C}_n$ is a plane algebraic envelope Γ_n of class n. This means that any point P, such that the pencil of lines through it does not belong to Γ_n, lies on at most n lines of Γ_n or exactly n if K is algebraically closed and the lines properly counted.

Now consider a k-stigm $\mathscr{K}'$ whose vertices $B_0, \ldots, B_{k-1}$ form the k-arc $\mathscr{K}$ and whose sides $b_{ij} = B_iB_j$ form the set $\mathscr{B}$. Through each B_i there is a set h_n^i of n distinct lines none of which is a side of $\mathscr{K}'$. Write $H = \bigcup_{i=0}^{k-1} h_n^i$. Let $\mathscr{N}_H$ be the totality of algebraic envelopes of class n containing H but no line of $\mathscr{B}$.

2.8.4$_3$. THEOREM. *Let $\mathscr{K}$ be the triangle $\mathrm{u}_0\mathrm{u}_1\mathrm{u}_2$. Then $\mathscr{N}_H \neq \emptyset$ if and only if $\Pi d = 1$ where the product ranges over all $3n$ lines l of H, and d is the coordinate of l. If $\mathscr{N}_H \neq \emptyset$, then H imposes $3n-1$ conditions on the envelopes of class n to contain it; if $n = 1$ or 2, $\mathscr{N}_H$ consists of a single Γ_n.* □

For $n = 1$, this is Ceva's theorem.

2.8.4$_k$. THEOREM. (i) *For $k \geq 4$, $\mathscr{N}_H \neq \emptyset$ if and only if the condition $\Pi d = 1$ is satisfied for each of the $\mathrm{c}(k, 3)$ triangles of $\mathscr{K}$.*

(ii) *If $\mathscr{N}_H \neq \emptyset$, then H imposes $k(2n-k+3)/2-1$ conditions on the envelopes of class n to contain it.*

(iii) *If $\mathscr{N}_H \neq \emptyset$ and $k > n$, then $\mathscr{N}_H$ consists of a single envelope Γ_n.* □

2.9. Notes and references

§2.1. For books on finite projective spaces, see particularly Artin [1957], Baer [1952], Barlotti [1965], Biggs [1971], Bumcrot [1969], Dembowski [1968], Kaplansky [1974], Kárteszi [1976], Levi [1942], Segre [1948], [1960a], as well as the papers Segre [1959a], [1967].

For background on projective geometry, see also Baker

[1921], Dorwart [1966], Gruenberg and Weir [1967], Hartshorne [1967], Horadam [1970], Pedoe [1963], Rosenbaum [1963], Semple and Kneebone [1952].

§2.2. For general results on incidence structures, see particularly Dembowski [1968], as well as Biggs [1971], Cameron [1976], Cameron and van Lint [1975], Hall [1967], Street and Wallis [1977].

For non-Desarguesian projective plans, see Albert and Sandler [1968], Hughes and Piper [1973], Pickert [1955]. For generalized quadrangles, see Buekenhout and Lefèvre [1974], and the papers of Payne and of Thas. The axioms for $AG(n,q)$ are taken from Lenz [1954].

§2.3 For books on finite linear groups, see Artin [1957], Biggs [1971], Blichfeldt [1917], Burnside [1911], Carmichael [1937], Davis [1969], Dembowski [1968], Dickson [1901], Dieudonné [1971], Huppert [1967], van der Waerden [1935].

The generating function $E(x)$ for the number of conjugacy classes in $GL(n,q)$ is given by Feit and Fine [1960]; another form is given by Green [1955]. The formula for $e(x)$ was communicated separately by G.E. Wall and I.G. Macdonald. Wall and Macdonald have also independently worked out that if $e^*(n,q)$ is the number of conjugacy classes in $SL(n,q)$, then the generating function is

$$e^*(x) = 1 + \sum_{n=1}^{\infty} e^*(n,q)x^n = (q-1)^{-1} \sum_{d|(q-1)} \phi_2(d)E(x^d) - q + 2,$$

where

$$\phi_2(d) = d^2 \prod_{\substack{p \text{ prime} \\ p|d}} (1 - p^{-2}).$$

Macdonald has also shown that if $e^\dagger(n,q)$ is the number of conjugacy classes in $PSL(n,q)$, then

$$e^\dagger(n,q) = (n,q-1)^{-1} \sum_{\substack{d_1|(q-1) \\ d_2|(q-1) \\ (d_1d_2)|n}} \phi(d_1)\phi_2(d_2)E(n/(d_1d_2),q).$$

Also,

$$e(n,q) = (q-1)^{-1} \sum_{d|(n,q-1)} \phi(d)E(n/d,q),$$

$$e^*(n,q) = (q-1)^{-1} \sum_{d|(n,q-1)} \phi_2(d)E(n/d,q).$$

For the linear algebra involved in theorem 2.3.2, see, for example, Cullen [1972].

§2.4. This is taken from Hirschfeld [1976].

§2.6. For background on algebraic geometry, see particularly W. Fulton [1969], Shafarevich [1974], as well as Baker [1921], Hodge and Pedoe [1947], Mumford [1976], Segre [1972a], Semple and Roth [1949].

What is here called a variety is now more usually called a projective algebraic set. In fact, the definition of variety given is not terribly satisfactory for several reasons. Firstly, in a finite space, *every* subset is a variety. Secondly, two different ideals in $K[x_0,\ldots, x_n]$ may define the same subset of $PG(n,K)$, especially for K finite. Thirdly, the definitions of singular point, tangent space, etc. depend entirely on the forms defining the variety. So we really require a variety to be a pair $(\mathscr{P}, \mathscr{I})$ where $\mathscr{P}$ is a subset of $PG(n,K)$ and $\mathscr{I}$ is a homogeneous ideal of $K[x_0,\ldots, x_n]$. In other words, the theory of schemes is needed: see, for example, Shafarevich [1974]. However, that is well beyond the scope of this book and not really essential here, as only very simple types of varieties are considered in detail. See also §10.5.

§§2.7, 2.8. See Segre [1959a].

PART II
ELEMENTARY PROPERTIES OF $PG(n, q)$

3
SUBSPACES OF *PG (n, q)*

3.1. Numbers

Let $PG^{(r)}(n,q)$ be the set of r-spaces of $PG(n,q)$ and let $\phi(r;\ n,q) = |PG^{(r)}(n,q)|$. Thus $PG^{(0)}(n,q) = PG(n,q)$. Let $\chi(s,r;\ n,q)$ be the number of r-spaces through an s-space in $PG(n,q)$.

In §1.6, the notation $\theta(n) = (q^{n+1} - 1)/(q - 1)$ was introduced. From §2.3, $[r,\ s]_- = \prod_{i=r}^{i=s} (q^i - 1)$, for $s \geqslant r$, and equals one otherwise.

3.1.1. THEOREM.

$$\text{(i)}\quad \phi(0;\ n,q) = \theta(n) = (q^{n+1} - 1)/(q - 1) = q^n + q^{n-1} + \ldots + q + 1.$$

$$\text{(ii)}\quad \phi(r;\ n,q) = [n - r + 1,\ n + 1]_-/[1,\ r + 1]_-.$$

$$\text{(iii)}\quad \chi(s,r;\ n,q) = [r - s + 1,\ n - s]_-/[1,\ n - r]_-.$$

Proof. (i) From §2.1, the points of $PG(n,q)$ are $\mathbf{P}(X)$, where $X = (x_0, x_1, \ldots, x_n)$ with the x_i in $\gamma = GF(q)$ and some x_i in γ_0. Also $\mathbf{P}(X) = \mathbf{P}(Y)$ if and only if $X = tY$ for some t in γ_0. So

$$\phi(0;\ n,q) = (q^{n+1} - 1)/(q - 1) = \theta(n).$$

(ii) $\phi(r;\ n,q)$ is the number of ways of choosing r+1 linearly independent points in $PG(n,q)$ divided by the number of ways of choosing such a set of points in a particular r-space. Hence

$$\phi(r;\ n,q) = \theta(n)\{\theta(n)-\theta(0)\}\{\theta(n)-\theta(1)\}\ldots\{\theta(n)-\theta(r-1)\}/[\theta(r)\{\theta(r)-\theta(0)\}\ldots\{\theta(r)-\theta(r-1)\}]$$

$$= (q^{n+1}-1)(q^{n+1}-q)\ldots(q^{n+1}-q^r)/[(q^{r+1}-1)(q^{r+1}-q)\ldots(q^{r+1}-q^r)]$$

$$= [n-r+1,\ n+1]_-/[1,\ r+1]_-.$$

(iii) By the Principle of Duality, §2.1(iii),

$$\chi(s,r;\ n,q) = \phi(n - r - 1;\ n - s - 1,\ q)$$
$$= [r - s + 1,\ n - s]_-/[1,\ n - r]_-.\square$$

COROLLARY. *In particular,*

(i) $\phi(0;\ 1,q) = q + 1;$

(ii) $\phi(0;\ 2,q) = \phi(1;\ 2,q) = q^2 + q + 1;$

(iii) $\phi(0;\ 3,q) = \phi(2;\ 3,q) = (q + 1)(q^2 + 1);$

(iv) $\phi(1;\ 3,q) = (q^2 + 1)(q^2 + q + 1);$

(v) $\chi(0,1;\ 3,q) = q^2 + q + 1;$

(vi) $\chi(1,2;\ 3,q) = \chi(0,1;\ 2,q) = q + 1;$

(vii) $\chi(s,s+1;\ n,q) = \theta(n-s-1).\square$

Some further numbers of subspaces are of interest. In $PG(n,q)$, let $\psi_{12}(t,s,r;\ n,q)$ be the number of spaces Π_r meeting a fixed Π_s in a fixed Π_t; let $\psi_2(t,s,r;\ n,q)$ be the number of spaces Π_r meeting a fixed Π_s in some Π_t; let $\psi_1(t,s,r;\ n,q)$ be the number of ordered pairs $(\Pi_s,\ \Pi_r)$ meeting in a fixed Π_t; let $\psi(t,s,r;\ n,q)$ be the number of ordered pairs $(\Pi_s,\ \Pi_r)$ meeting in some Π_t. Thus the suffix indicates which of the following integers t and s refer to a fixed subspace of that dimension.

3.1.2. THEOREM.

(i) $\psi_{12}(t,s,r;\ n,q) = q^{(r-t)(s-t)}[n-r-s+t+1,\ n-s]_-/[1,\ r-t]_-;$

(ii) $\psi_2(t,s,r;\ n,q) = q^{(r-t)(s-t)}[n-r-s+t+1,\ n-s]_-[s-t+1,\ s+1]_- \div ([1,\ r-t]_-[1,\ t+1]_-);$

(iii) $\psi_1(t,s,r;\ n,q) = q^{(r-t)(s-t)}[n-r-s+t+1,\ n-s]_-[s-t+1,\ n-t]_- \div ([1,\ r-t]_-[1,\ n-s]_-);$

(iv) $\psi(t,s,r;\ n,q) = q^{(r-t)(s-t)}[n-r-s+t+1,\ n+1]_- \div ([1,\ r-t]_-[1,\ s-t]_-[1,\ t+1]_-).$

Proof. Firstly, we determine ψ_{12}. Given Π_t and Π_s, the number of ways of choosing points $P_1,\ldots,\ P_{r-t}$ outside Π_s so that $\Pi_t P_1 P_2 \ldots P_{r-t}$ is an r-space Π_r such that $\Pi_r \cap \Pi_s = \Pi_t$ is

$$N_1 = \{\theta(n) - \theta(s)\}\{\theta(n) - \theta(s+1)\}\ldots\{\theta(n) - \theta(s+r-t-1)\}.$$

Given one of the Π_r, the number of ways of choosing $P_1,\ldots,P_{r-t}$ in $\Pi_r\backslash\Pi_t$ is

$$N_2 = \{\theta(r) - \theta(t)\}\{\theta(r) - \theta(t+1)\}\ldots\{\theta(r) - \theta(r-1)\}.$$

Then $\psi_{12}(t,s,r;\ n,q) = N_1/N_2$, which immediately gives (i).
Then,

$$\begin{aligned}
\psi_2(t,s,r;\ n,q) &= \phi(t;\ s,q)\ \psi_{12}(t,s,r;\ n,q);\\
\psi_1(t,s,r;\ n,q) &= \chi(t,s;\ n,q)\ \psi_{12}(t,s,r;\ n,q);\\
\psi(t,s,r;\ n,q) &= \phi(t;\ n,q)\ \psi_1(t,s,r;\ n,q)\\
&= \phi(s;\ n,q)\ \psi_2(t,s,r;\ n,q).
\end{aligned}$$

So ψ_2 and ψ_1 are calculated from ψ_{12} and the results of theorem 3.1.1; finally, ψ is calculated from ψ_1 or ψ_2.□

3.2. Characterization of subspaces

3.2.1. THEOREM. *If $\mathcal{L}$ is a set of points in $PG(n,q)$ which has non-empty intersection with every line of $PG(n,q)$, then* $|\mathcal{L}| \geq \theta(n-1)$. *Also,* $|\mathcal{L}| = \theta(n-1)$ *if and only if $\mathcal{L}$ is a prime.*

Proof. If $\mathcal{L}$ is a prime, then it meets every line of the space and $|\mathcal{L}| = \theta(n-1)$.

Suppose now that $\mathcal{L}$ meets every line. Let P be any point not in $\mathcal{L}$. There are $\chi(0,\ 1;\ n,q) = \theta(n-1)$ lines m_i through P, each of which meets $\mathcal{L}$ in at least one point Q_i. If $Q_i = Q_j$, then $m_i = PQ_i = PQ_j = m_j$. So $|\mathcal{L}| \geq \theta(n-1)$.

If $\mathcal{L}$ meets every line of the space and $|\mathcal{L}| = \theta(n-1)$, then each line through a point P not in $\mathcal{L}$ contains exactly one point of $\mathcal{L}$. Thus, if a line contains two points of $\mathcal{L}$, it lies entirely in $\mathcal{L}$. So $\mathcal{L}$ is a subspace. As $|\mathcal{L}| = \theta(n-1)$, $\mathcal{L}$ is a prime.□

3.2.2. THEOREM. *If $\mathcal{L}$ is a set of points of $PG(n,q)$ which has non-empty intersection with every r-space, then* $|\mathcal{L}| \geq \theta(n-r)$. *Also,* $|\mathcal{L}| = \theta(n-r)$ *if and only if $\mathcal{L}$ is an $(n-r)$-space.*

Proof. If $\mathcal{L}$ is an $(n-r)$-space, then $\mathcal{L}$ meets every r-space of $PG(n,q)$ and $|\mathcal{L}| = \theta(n-r)$.

The theorem is true by definition for $r=0$ and by the

previous theorem for $r=1$. So we prove the theorem by induction on r and assume it true for all values less than r.

So, suppose that $\mathscr{L}$ meets every r-space and that $|\mathscr{L}| \leqslant \theta(n-r)$. Since $|\mathscr{L}| < \theta(n-r+1)$, the inductive hypothesis says that there exists an $(r-1)$-space Π_{r-1} such that $\mathscr{L} \cap \Pi_{r-1} = \emptyset$. There are $\theta(n-r)$ distinct r-spaces Π_r^i, $i \in \mathbf{N}_{\theta(n-r)}$, each meeting $\mathscr{L}$ in at least a point P_i. If $P_i = P_j$, then $\Pi_r^i = P_i\Pi_{r-1} = P_j\Pi_{r-1} = \Pi_r^j$. So $|\mathscr{L}| \geqslant \theta(n-r)$.

Now, suppose that $\mathscr{L}$ meets every r-space and that $|\mathscr{L}| = \theta(n-r)$. Then, from above $\mathscr{L} = \{P_i \mid i \in \mathbf{N}_{\theta(n-r)}\}$. Let Q be any point of Π_{r-1} and consider the lines QP_i. Since, for $j \neq i$, P_j is not in Π_r, all the lines QP_i are distinct. Let $\mathscr{M}$ be the set of points on all the lines QP_i; then $|\mathscr{M}| = q\,\theta(n-r) + 1 = \theta(n-r+1)$.

Suppose that $\mathscr{M}$ is not an $(n-r+1)$-space. Then, by the inductive hypothesis, there is an $(r-1)$-space Π'_{r-1} such that $\mathscr{M} \cap \Pi'_{r-1} = \emptyset$. So Q is not in Π'_{r-1}. Therefore, let Π'_r be the r-space $Q\Pi'_{r-1}$.

Since $\mathscr{L}$ meets every r-space, $\mathscr{L}$ meets Π'_r in some point P_k. So $\mathscr{M}$ and Π'_r have this point P_k in common. But QP_k and Π'_{r-1} are both subspaces of Π'_r and so have a common point P. As QP_k is contained in $\mathscr{M}$, so P lies in $\mathscr{M} \cap \Pi'_{r-1}$, contradicting that $\mathscr{M} \cap \Pi'_{r-1} = \emptyset$. Therefore $\mathscr{M}$ is an $(n-r+1)$-space.

Let Q' be a point of Π_{r-1} distinct from Q and let $\mathscr{M}'$ be the set of points on all the lines $Q'P_i$. Then, as above, $\mathscr{M}'$ is also an $(n-r+1)$-space. Further, $\mathscr{M}$ and $\mathscr{M}'$ are distinct, since if Q were in $\mathscr{M}'$, it would lie on a line $Q'P_i = QQ'$ and Q would be in Π_{r-1}. Now, $\mathscr{M} \cap \mathscr{M}'$ is a subspace of dimension at most $n-r$. As $\mathscr{M} \cap \mathscr{M}'$ contains all $\theta(n-r)$ points of $\mathscr{L}$, so $\mathscr{M} \cap \mathscr{M}'$ is an $(n-r)$-space and equals $\mathscr{L}$.□

COROLLARY. *If $\mathscr{L}$ is a set of points in $PG(n,q)$ such that, for any r-space Π_r, there is an s-space in $\mathscr{L} \cap \Pi_r$, then $|\mathscr{L}| \geqslant \theta(n-r+s)$. Also $|\mathscr{L}| = \theta(n-r+s)$ if and only if $\mathscr{L}$ is an $(n-r+s)$-space.*

Proof. Let Π_r be an r-space and Π_s an s-space contained in $\mathscr{L} \cap \Pi_r$. Since every $(r-s)$-space in Π_r meets Π_s in at least a point, $\mathscr{L}$ meets every $(r-s)$-space in Π_r. Since every $(r-s)$-

space is contained in some r-space, $\mathscr{L}$ meets every $(r-s)$-space. So the corollary follows by replacing r by $r-s$ in the theorem.□

3.2.3. THEOREM. *If $\mathscr{L}$ is a set of points in $PG(n,q)$ such that, for any r-space Π_r, $|\mathscr{L}\cap\Pi_r| \geqslant \theta(s)$, then $|\mathscr{L}| \geqslant \theta(n-r+s)$. Also $|\mathscr{L}| = \theta(n-r+s)$ if and only if $\mathscr{L}$ is an $(n-r+s)$-space.*

Proof. If $\mathscr{L}$ is an $(n-r+s)$-space, then $|\mathscr{L}| = \theta(n-r+s)$ and, if Π_r is any r-space, $\mathscr{L} \cap \Pi_r$ is a subspace of dimension at least s so that $|\mathscr{L}\cap\Pi_r| \geqslant \theta(s)$.

For $s=0$, the theorem becomes the previous theorem. So we prove the theorem by induction on s and assume it true for all values less than s.

Assume the result false so that one of the following holds:

(i) $|\mathscr{L}| < \theta(n-r+s)$;

(ii) $|\mathscr{L}| = \theta(n-r+s)$ and $\mathscr{L}$ is not an $(n-r+s)$-space.

Then, by the inductive hypothesis for $s-1$, there exists an $(r-1)$-space Π_{r-1} such that $N = |\Pi_{r-1} \cap \mathscr{L}| < \theta(s-1)$. There are $\chi(r-1, r; n,q) = \theta(n-r)$ distinct r-spaces through Π_{r-1}, the intersection of any two of which is exactly Π_{r-1}. Since every r-space has at least $\theta(s)$ points in common with $\mathscr{L}$,

$$\begin{aligned}|\mathscr{L}| &= \theta(n-r)\ [\theta(s) - N] + N \\ &> \theta(n-r)\ [\theta(s) - \theta(s-1)] + \theta(s-1) \\ &= \theta(n-r+s),\end{aligned}$$

contradicting (i) and (ii).□

3.3. Sets of subspaces

A recurring theme of the book is the characterization of algebraic varieties in $PG(n,q)$ as finite sets of points with certain combinatorial properties.

A *(k,l)-set* in $PG(n,q)$ is a set of k spaces Π_l. A *k-set* is a $(k,0)$-set, that is a set of k points. The most general type of (k,l)-set that will be considered is a *$(k,l;\ r,s;\ n,q)$-set*, that is, a (k,l)-set in $PG(n,q)$ at at most r spaces Π_l of which lie in any Π_s.

It is of great interest for applications to find maximal and particularly maximum such sets as k varies but the other parameters remain fixed. Thus a *complete* $(k,l;\ r,s;\ n,q)$-set is one not contained in any $(k+1,l;\ r,s;\ n,q)$-set. Let $m(l;\ r,s;\ n,q)$ be the maximum value of k that such a set can have. This number is known for only a few cases, although some bounds are always available.

The definition of $(k,l;\ r,s;\ n,q)$-set is specialized as follows:

a $(k;\ r,s;\ n,q)$-*set* is a $(k,0;\ r,s;\ n,q)$-set;

a $(k,r;\ n,q)$-*set* is a $(k;\ r,r-1;\ n,q)$-set;

a $(k;\ r)$-*cap* is a $(k;\ r,1;\ n,q)$-set with $n \geqslant 3$;

a k-*cap* is a $(k,2)$-cap;

a k-*arc* is a $(k;\ n,n-1;\ n,q)$-set;

a *plane* k-*arc* is a $(k;\ 2,1;\ 2,q)$-set;

a *plane* $(k;\ r)$-*arc* is a $(k;\ r,1;\ 2,q)$-set;

a (k,l)-*span* is a $(k,l;\ 1,2l;\ n,q)$-set with $l \geqslant 1$.

Thus a plane k-arc (or just k-*arc* if there is no ambiguity) is a set of k points in $PG(2,q)$, no three of which are collinear: a maximum plane k-arc is an *oval*. A k-cap is a set of k points in $PG(n,q)$ with $n \geqslant 3$ such that no three are collinear. A maximum k-cap in $PG(3,q)$ is an *ovaloid*. We denote $m(0;\ 2,1;\ n,q)$ by $m_2(n,q)$ and $m(0;\ n,n-1;\ n,q)$ by $m(n,q)$.

A (k,l)-span is a set of k spaces Π_l no two of which intersect: in particular cases, §4.1, a maximum (k,l)-span is a spread.

If $\mathscr{K}$ is any k-set, then a *d-secant* of $\mathscr{K}$ is a line l such that $|l \cap \mathscr{K}| = d$. In particular, the following terminology is also used:

1-secant	*unisecant*
2-secant	*bisecant*
3-secant	*trisecant*.

A *tangent* to a k-set $\mathscr{K}$ is a line which is a tangent when $\mathscr{K}$ is considered as an algebraic variety, as in §2.6(xiii). Also when $\mathscr{K}$ is a variety, a *chord* is a line meeting $\mathscr{K}$ in two complex points, possible coincident; that is, a line meeting $\mathscr{K}$ in a pair of real points, or a line meeting $\mathscr{K}$ in a pair of 2-complex conjugate points, or a unisecant tangent line. See, for example, the chapter on twisted cubics in Volume II. It follows that any line not lying on a non-singular quadric $\mathscr{Q}_n$ in $PG(n,q)$ is a chord of $\mathscr{Q}_n$.

3.4. Notes and references

§3.2. This is taken from Bose and Burton [1966]. For a related characterization of subspaces, see Rothschild and van Lint [1974], as well as MacWilliams [1961].

4
PARTITIONS OF *PG* (n, q)

4.1. Partitions of $PG(n,q)$ by r-spaces

A *spread* $\mathscr{F}$ of r-spaces of $PG(n,q)$ is a set of r-spaces which partitions $PG(n,q)$; that is, every point of $PG(n,q)$ lies in some r-space of $\mathscr{F}$ and every two r-spaces of $\mathscr{F}$ are disjoint. Here it is convenient to write $\theta(n,q)$ for $\phi(0;\ n,q)$.

4.1.1. THEOREM. *The following are equivalent:*

(i) *there exists a spread* $\mathscr{F}$ *of* r*-spaces of* $PG(n,q)$;

(ii) $\theta(r,q) \mid \theta(n,q)$;

(iii) $(r + 1) \mid (n + 1)$.

Proof. If $\mathscr{F}$ exists, then the number of points in $PG(r,q)$ divides the number of points in $PG(n,q)$; that is, (i) implies (ii).

$$\theta(n,q)/\theta(r,q) = (q^{n+1} - 1)/(q^{r+1} - 1),$$

which is an integer if and only if $r + 1$ divides $n + 1$. So (ii) is equivalent to (iii).

It remains to show that (iii) implies (i). Let s be given by

$$n + 1 = (r + 1)(s + 1), \tag{4.1}$$

Take an algebra of dimension $r + 1$ over $GF(q)$. For simplicity, let us use $GF(q^{r+1})$. If F is an irreducible polynomial of degree $r + 1$ over $GF(q)$ and α is a root of F in $GF(q^{r+1})$, then every element χ of $GF(q^{r+1})$ can be written

$$\chi = x_0 + x_1\alpha + \ldots + x_r\alpha^r,$$

where $x_i \in GF(q)$, for all i in $\mathbf{\bar{N}}_r$. If we take the $s + 1$ elements $\chi_0,\ldots,\chi_s$ of $GF(q^{r+1})$, they can be written

$$\chi_i = x_{i0} + x_{i1}\alpha + \ldots + x_{ir}\alpha^r,$$

where $x_{ij} \in GF(q)$, $i \in \mathbf{\bar{N}}_s$. The $n + 1$ elements x_{ij} in, say, lexicographical order, will be interpreted as coordinates of a point in $PG(n,q)$. Thus each point of $PG(n,q)$ is given by an $(s+1)$-ple $(\chi_0,\ldots,\chi_s)$ of elements of $GF(q^{r+1})$.

Let $\tau_0,\ldots,\tau_s$ be any elements, not all zero, of $GF(q^{r+1})$. Then the equations

$$\chi_0/\tau_0 = \chi_1/\tau_1 = \ldots = \chi_s/\tau_s, \tag{4.2}$$

which could more properly be written

$$\tau_i\chi_j = \tau_j\chi_i \quad \text{for } i,\ j \text{ in } \overline{\mathbf{N}}_s,$$

define an r-space Π_r in $PG(n,q)$. For, the equations (4.2) give $s(r+1)$ linearly independent equations in the x_{ij}, and so define a subspace of dimension $n - s(r+1) = r$.

Each $(s+1)$-ple $\tau = (\tau_0,\ldots,\tau_s)$ corresponds to a point $P(\tau)$ in $PG(s,\ q^{r+1})$. As $P(\tau)$ varies in $PG(s,\ q^{r+1})$, so Π_r varies through a partition of $PG(n,q)$.

This may be seen in two ways. Every point of $PG(n,q)$ lies in one of the Π_r thus defined. If Π_r' is given by

$$\chi_0/\tau_0' = \chi_1/\tau_1' = \ldots = \chi_s/\tau_s', \tag{4.3}$$

then $\Pi_r \cap \Pi_r'$ is given by both (4.2) and (4.3). Since $\Pi_r \neq \Pi_r'$, there exist j, k in $\overline{\mathbf{N}}_s$ such that $\tau_j/\tau_j' \neq \tau_k/\tau_k'$. Hence $\chi_j = \chi_k = 0$. Hence $\chi_i = 0$ for i in $\overline{\mathbf{N}}_s$. Since $\chi_i = 0$, so $x_{ij} = 0$ all j in $\overline{\mathbf{N}}_r$. So $\Pi_r \cap \Pi_r' = \Pi_{-1}$. Thus the spaces Π_r form a partition of $PG(n,q)$.

Alternatively, since $PG(s,\ q^{r+1})$ contains $\theta(s,\ q^{r+1}) = (q^{n+1}-1)/(q^{r+1}-1)$ points by (4.1), this number of spaces Π_r is obtained. Thus the number of points in all these spaces Π_r is $\theta(s,\ q^{r+1})\ \theta(r,\ q) = (q^{n+1}-1)/(q-1) = \theta(n,q)$. Since $PG(n,q)$ contains this number of points and every point is in some Π_r, no two spaces Π_r can intersect. So, again the spaces Π_r form a partition of $PG(n,q)$. □

The simplest non-trivial example of a spread occurs for $r = s = 1$ and $n = 3$. These spreads have been much studied lately, particularly for their application to non-Desarguesian planes. They will be considered again in Volume II. For another proof of the theorem, see theorem 4.2.7 and its corollary.

Now we show how to use a spread to reconstruct a plane. In (4.1), let

$$n = 2k - 1, \quad r = k - 1, \quad s = 1.$$

Then, as in the proof of theorem 4.1.1, a spread $\mathscr{F}$ of spaces Π_{k-1} of $PG(2k - 1,q)$ corresponds to a line $PG(1,q^k)$. A plane $PG(2,q^k)$ with this $PG(1,q^k)$ as one of its lines can now be constructed.

Let us write $\Pi = PG(2k - 1,q)$ and embed Π in $PG(2k, q)$. An incidence structure $\mathscr{S}$ is defined as follows.

(i) (a) The *proper* points of $\mathscr{S}$ are the points of $PG(2k,q)\backslash\Pi$. Thus $\mathscr{S}$ has $\theta(2k, q) - \theta(2k - 1, q) = q^{2k}$ proper points.

(b) The *ideal* points of $\mathscr{S}$ are the Π_{k-1} of $\mathscr{F}$. Thus $\mathscr{S}$ has $q^k + 1$ ideal points.

(ii) (a) The *proper* lines of $\mathscr{S}$ are the Π_k of $PG(2k, q)$ which meet Π in a Π_{k-1} of $\mathscr{F}$. No Π_k can contain two Π_{k-1} of $\mathscr{F}$. So the number of proper lines of $\mathscr{S}$

$$\begin{aligned} &= |\mathscr{F}|\ [\theta(2k, q) - \theta(2k - 1, q)]/[\theta(k, q) - \theta(k - 1, q)] \\ &= (q^k + 1)q^{2k}/q^k \\ &= q^k(q^k + 1). \end{aligned}$$

(b) The *ideal* line of $\mathscr{S}$ is $\mathscr{F}$.

Then it may be shown that $\mathscr{S}$ satisfies all the axioms of §2.2 for $PG(2,q^k)$. So $\mathscr{S} = PG(2,q^k)$.

4.2. Cyclic projectivities

A projectivity $\mathfrak{T}$ which permutes the $\theta(n)$ points of $PG(n,q)$ in a single cycle is called a *cyclic projectivity*.

4.2.1. THEOREM. *A projectivity $\mathfrak{T}$ of $PG(n,q)$ is cyclic if and only if the characteristic polynomial of an associated matrix is subprimitive.*

Proof. Suppose $\mathfrak{T} = M(T)$, where T has subprimitive characteristic polynomial F, and suppose F has a root α in $GF(q^{n+1})$. Then, by §1.6(vi), the smallest power of α which lies in the subfield $GF(q)$ of $GF(q^{n+1})$ is $\theta(n)$. Thus T^m has no eigenvalues in $GF(q)$ for $1 \leqslant m < \theta(m)$. Hence $\mathfrak{T}^m$ has no fixed points for $1 \leqslant m < \theta(n)$. Also $o(\mathfrak{T}) = \theta(n)$. So $\mathfrak{T}$ acts as a single cycle on the $\theta(n)$ points of $PG(n,q)$.

Conversely, if $\mathfrak{T}$ is cyclic, then no power of $\mathfrak{T}$ less than its order $\theta(n)$, can have fixed points. If the characteristic polynomial F of an associated matrix T were reducible, then $xI - T$ would have an elementary divisor E, either irreducible of degree less than n+1 or reducible of degree n+1 (and so a power of some other polynomial). In either case, $T \approx \text{diag}\ (H(E), \ldots)$ where $H(E)$ is the hypercompanion matrix as in theorem 2.3.2(v). Then $\mathfrak{H}$, the projectivity associated with $H(E)$ has order $o(\mathfrak{H})$, which is less than $\theta(n)$ by §2.5. So $\mathfrak{T}^{o(\mathfrak{H})}$ has a fixed point. Thus F is irreducible. Since $o(\mathfrak{T}) = \theta(n)$, so F is subprimitive.□

COROLLARY 1. *By duality, a cyclic projectivity permutes the primes of $PG(n,q)$ in a single cycle.*□

COROLLARY 2. *The number of cyclic projectivities in $PG(n,q)$ is given by*

$$\sigma(n,q) = q^{n(n+1)/2}\ [1,n]_-\ \phi(\theta(n))/(n+1),$$

where ϕ is the Euler function.

Proof. The number was determined in theorem 2.4.5.□

COROLLARY 3. *In particular,*

$\sigma(1,q) = q(q-1)\ \phi(q+1)/2$;

$\sigma(2,q) = q^3(q+1)(q-1)^2\ \phi(q^2+q+1)/3$;

$\sigma(3,q) = q^6(q^2+q+1)\ (q+1)\ (q-1)^3\phi((q+1)(q^2+1))/4$.□

COROLLARY 4. *The number of conjugacy classes of*

PGL$(n+1, q)$ *consisting of cyclic projectivities is* $\phi(\theta(n))/(n+1)$.□

An example of a cyclic projectivity $\mathfrak{T}$ occurs when the associated matrix $T = H(F)$, where $F(x) = x^{n+1} - a_n x^n \ldots - a_0$ and F is subprimitive.

The existence of cyclic projectivities enables us to set up a correspondence between the points of $PG(n,q)$ and the elements of $GF(q^{n+1})$. Let $\mathfrak{T}$ be the cyclic projectivity just given and let F have the root α in $GF(q^{n+1})$. So

$$\alpha^{n+1} = a_n\alpha^n + \ldots\ldots + a_1\alpha + a_0 \quad .$$

Let $P(0)$ be the point $\mathbf{U}_0$ and let $P(i) = P(0)\mathfrak{T}^i$. Then for i in $\overline{\mathbf{N}}_{\theta(n)-1}$, there exist $y_0^{(i)}, y_1^{(i)}, \ldots, y_n^{(i)}$ in $GF(q)$ such that

$$\alpha^i = y_0^{(i)} + y_1^{(i)}\alpha + \ldots\ldots + y_n^{(i)}\alpha^n.$$

Then

$$\alpha^{i+1} = \alpha \cdot \alpha^i = y_0^{(i)}\alpha + y_1^{(i)}\alpha^2 + \ldots\ldots + y_n^{(i)}\alpha^{n+1}$$

$$= y_n^{(i)}a_0 + (y_n^{(i)}a_1 + y_0^{(i)})\alpha + \ldots\ldots +$$

$$+ (y_n^{(i)}a_n + y_{n-1}^{(i)})\alpha^n.$$

But also

$$\alpha^{i+1} = y_0^{(i+1)} + y_1^{(i+1)}\alpha + \ldots\ldots\ldots + y_n^{(i+1)}\alpha^n.$$

Thus $(y_0^{(i+1)}, \ldots, y_n^{(i+1)}) = (y_0^{(i)}, \ldots, y_n^{(i)})\,T$,
since $T = \begin{bmatrix} 0 & 1 & 0 & \ldots & 0 \\ 0 & 0 & 1 & 0\ldots & 0 \\ \cdot & & & & \\ \cdot & & & & \\ 0 & \ldots & \ldots & 0 & 1 \\ a_0 & \ldots & \ldots & a_{n-1} & a_n \end{bmatrix}$.

So $(y_0^{(i)}, \ldots\ldots, y_n^{(i)})$ is the vector of $P(i)$, and $\alpha^i \leftrightarrow P(i)$ gives a bijection between the sets $\{\alpha^i \mid i \in \overline{\mathbf{N}}_{\theta(n)-1}\}$ and $PG(n,q)$

If α is primitive, this bijection can be extended to a correspondence between $GF(q^{n+1})\backslash\{0\}$ and $PG(n,q)$. For, then $GF(q)\backslash\{0\} = \{\alpha^{i\theta(n)} | i \in \overline{\mathbf{N}}_{q-2}\}$ and $\alpha^i = t\,\alpha^j$ with t in $GF(q)$ if and only if $i \equiv j \pmod{\theta(n)}$. In this case

$$(y_0^{(i)}, \ldots\ldots, y_n^{(i)}) = (ty_0^{(j)}, \ldots\ldots, ty_n^{(j)}).$$

So, in the correspondence

$$\{\alpha^i, \alpha^{i+\theta(n)}, \ldots\ldots, \alpha^{i+(q-2)\theta(n)}\} \leftrightarrow P(i),$$

α^i and α^j correspond to the same point $P(i)$ and we have a $(1, q-1)$ correspondence between $GF(q^{n+1})\backslash\{0\}$ and $PG(n,q)$.

In any case, the points $P(i)$, $P(j)$, $P(k)$ are collinear if and only if there exist a_i, a_j, a_k in $GF(q)\backslash\{0\}$ such that

$$a_i\,\alpha^i + a_j\,\alpha^j + a_k\,\alpha^k = 0,$$

and, more generally, $P(i_0), \ldots, P(i_m)$ are linearly dependent if and only if there exist $a_{i_0}, \ldots, a_{i_m}$ in γ such that $a_{i_0}\,\alpha^{i_0} + \ldots\ldots + a_{i_m}\,\alpha^{i_m} = 0$.

The existence of a cyclic projectivity affords an attractive representation of the points and lines of $PG(2,q)$. If $\mathfrak{T}$ acts cyclically on the points of $PG(2,q)$ then dually it acts cyclically on the lines of $PG(2,q)$. The plane can always be represented by an array of $q+1$ rows and q^2+q+1 columns, where each element of the array is a point, and each column consists of the points on a line. We now show that it can be represented by a regular array; that is, each row is a cyclic permutation of the first.

Suppose the points collinear with $P(0)$ and $P(1)$ are those $P(i)$ with indices $i=d_2, d_3, \ldots, d_q$. Let us write d_0 for 0 and d_1 for 1. Then consider the array

$$\begin{array}{lllll} d_0 & d_0+1 & d_0+2 & \ldots\ldots & d_0+q^2+q \\ d_1 & d_1+1 & d_1+2 & \ldots\ldots & d_1+q^2+q \end{array}$$

$$d_q \quad d_q+1 \quad d_q + 2 \; \ldots\ldots \; d_q + q^2 + q$$

Let $\mathscr{M}'$ be the above array with each integer reduced if necessary modulo $q^2 + q + 1$ so that each entry x satisfies $0 \leqslant x \leqslant q^2 + q$.

4.2.2. THEOREM. *$\mathscr{M}'$ is a regular array and represents the points and lines of $PG(2,q)$.*

Proof. The first row of $\mathscr{M}'$ represents all the points $P(0), P(1), \ldots P(q^2 + q)$ of $PG(2,q)$. Since $\mathfrak{T}$ acts cyclically on the lines of the plane and the first column represents the points of a line, each successive column is the transformation by $\mathfrak{T}$ of the previous one, and so each column represents the points of a line. By construction, each row is a cyclic permutation of the first, and so $\mathscr{M}'$ is regular.□

COROLLARY. *The integers $d_0, d_1, \ldots\ldots, d_q$ form a perfect difference set; that is, the $q^2 + q$ integers $d_i - d_j$ with $i \neq j$ are all distinct.*

Proof. Consider the array $\mathscr{N}$ formed by the columns of $\mathscr{M}'$ which contain $d_0 = 0$. They may be written, modulo $q^2 + q + 1$, as

$$\begin{array}{cccc} d_0 - d_0 & d_0 - d_1 & \ldots\ldots & d_0 - d_q \\ d_1 - d_0 & d_1 - d_1 & \ldots\ldots & d_1 - d_q \\ d_2 - d_0 & d_2 - d_1 & \ldots\ldots & d_2 - d_q \\ \vdots & & & \\ d_q - d_0 & d_q - d_1 & \ldots\ldots & d_q - d_q \end{array}$$

The columns of $\mathscr{N}$ represent the lines of $PG(2,q)$ through $P(0)$. Apart from $P(0)$, each point of the plane lines on just one of these lines. Hence the $q^2 + q$ integers in the array $\mathscr{N}$ apart from those on the main diagonal are distinct and consist of all the differences $d_i - d_j$. So any column of $\mathscr{M}'$ forms a perfect difference set.□

Conversely, it may be shown that given a perfect difference set, a projective plane may be formed. It is conjectured that all such planes are Desarguesian.

Perfect difference sets for the initial values of q are given in the Table 4.1.

TABLE 4.1

q	Perfect different set									
2	0	1	3							
3	0	1	3	9						
4	0	1	4	14	16					
5	0	1	3	8	12	18				
7	0	1	3	13	32	36	43	52		
8	0	1	3	7	15	31	36	54	63	
9	0	1	3	9	27	49	56	61	77	81

$PG(2,2)$						
0	1	2	3	4	5	6
1	2	3	4	5	6	0
3	4	5	6	0	1	2

$PG(2,3)$												
0	1	2	3	4	5	6	7	8	9	10	11	12
1	2	3	4	5	6	7	8	9	10	11	12	0
3	4	5	6	7	8	9	10	11	12	0	1	2
9	10	11	12	0	1	2	3	4	5	6	7	8

This type of representation of $PG(2,q)$ can be extended to higher dimensions. As before, we have a cyclic projectivity $\mathfrak{T}$ and $P(i) = P(0)\mathfrak{T}^i$ for $i \in \overline{\mathbf{N}}_{\theta(n)-1}$. Then, with the correspondence $P(i) \leftrightarrow i$, the points are denoted by integers modulo $\theta(n)$.

4.2.3. LEMMA. *If the points of an m-space in $PG(n,q)$ are represented by the integers $c_0, c_1, \ldots, c_{\theta(m)-1}$, then the integers $c_0 + k, c_1+k, \ldots, c_{\theta(m)-1} + k$, modulo $\theta(n)$,* also *represents the points of an m-space.*

Proof. The point $c_i + k$ is the point c_i transformed by the projectivity $\mathfrak{T}^k$. So, as the set $\{c_i\}$ represents the points of an m-space, so does the set $\{c_i + k\}$ for any k.□

COROLLARY. *Each m-space of PG(n,q) occurs as one of a cycle of order N, where N divides* $\theta(n)$.□

4.2.4. LEMMA. *If the points of an m-space in* $PG(n,q)$, $q = p^h$, *are represented by the integers* $c_0, c_1, \ldots, c_{\theta(m)-1}$, *then the integers* $q^r c_0, q^r c_1, \ldots, q^r c_{\theta(m)-1}$ *modulo* $\theta(n)$ *also represent the points of an m-space for any r in* $\overline{\mathbf{N}}_n$.

Proof. From above, $P(i)$ is also represented by α^i, where α is a primitive root of $GF(q^{n+1})$; also $P(i)$, $P(j)$, $P(k)$ are collinear if and only if there exist a_i, a_j, a_k in γ_0 such that $a_i\alpha^i + a_j\alpha^j + a_k\alpha^k = 0$. Under an automorphism $t \to tq^r$ of $GF(q^{n+1})$, which leaves every element of $GF(q)$ fixed, this relation becomes

$$a_i\ \alpha^{iq^r} + a_j\ \alpha^{jq^r} + a_k\ \alpha^{kq^r} = 0\ :$$

so any relation of linear dependency among the points given by $c_0, c_1, \ldots, c_{\theta(m)-1}$ is satisfied for the corresponding points among $q^r c_0, q^r c_1, \ldots, q^r c_{\theta(m)-1}$.□

4.2.5. LEMMA. *If an m-space* Π_m *is one of a cycle of order N, its points can be written in the form, modulo* $\theta(n)$,

$$\begin{array}{llll} c_0, & c_0 + N, & \ldots, & c_0 + (r-1)N \\ c_1, & c_1 + N, & \ldots, & c_1 + (r-1)N \\ \cdot & & & \\ \cdot & & & \\ c_s, & c_s + N, & \ldots, & c_s + (r-1)N \end{array}$$

where $c_i \not\equiv c_j \bmod N$, $\theta(n) = rN$ *and* $r = \theta(l)$ *for some integer l. Also, a necessary condition that* $N < \theta(n)$ *is that* $(\theta(m), \theta(n)) > 1$.

Proof. Since Π_m has cycle N, so $N \mid \theta(n)$. Let $\theta(n) = Nr$. Also, if c_0 represents a point on Π_m, so does $c_0 + kN$. Let

$$\mathscr{C}_0 = \{c_0 + kN \mid k \in \overline{\mathbf{N}}_{r-1}\}$$

and suppose that

$$\mathscr{C}'_0 = \{c_0 + kN \mid k \in \overline{\mathbf{N}}_l\}$$

is the largest subset of $\mathscr{C}_0$ representing independent points of $\mathscr{C}_0$. Suppose c represents a point of Π_m linearly dependent on those represented by $\mathscr{C}'_0$ and b a point of Π_m linearly independent of those represented by $\mathscr{C}'_0$. Then c is dependent on

$$c_0,\ c_0 + N,\ \ldots, c_0 + lN.$$

So, applying $\mathfrak{T}^{b-c_0}$, we have that $b + c - c_0$ is dependent on $b,\ b + N,\ \ldots,\ b + lN$. Since all $b + kN$ represent points of Π_m, so does $b + c - c_0$. Suppose Π_m is generated by the points given by $c_0, \ldots, c_0 + lN, b_1, b_2, \ldots, b_{m-l}$; then the m-space generated by points given by $c, \ldots, c + lN$, $b_1 + c - c_0, \ldots, b_{m-l} + c - c_0$ is also Π_m, since all these points lie in Π_m and are independent. Hence Π_m has cycle $c - c_0$. Thus $c - c_0 = iN$. Therefore all points on the l-space Π_l defined by $\mathscr{C}'_0$ have the representation $c_0 + jN$. But r points have such a representation. Hence $r = \theta(l)$, $\theta(m) =$ rs and $\theta(n) = rN$. Hence $r \mid (\theta(m), \theta(n))$ and so $(l+1) \mid (m+1, n+1)$.□

COROLLARY. *Every prime in* $PG(n,q)$ *has cycle* $\theta(n)$ *exactly*.

Proof. This was also corollary 1 to theorem 4.2.1, but it may also be shown as follows. For a prime to have cycle N, the lemma implies that $(l+1) \mid (n, n+1)$. Hence $l = 0$ and $r = \theta(l) = 1$. So $N = \theta(n) / r = \theta(n)$.□

4.2.6. **LEMMA**. *A necessary and sufficient condition that there exists an* m*-space* Π_m *in* $PG(n,q)$ *of cycle* N *less than* $\theta(n)$ *is that* $(m+1, n+1) > 1$.

Proof. Lemma 4.2.5 proves the necessity. Suppose $l+1 = (m+1, n+1)$ and $l > 0$. Let $r = \theta(l)$ and $N = \theta(n)/r$. Then, by lemma 4.2.5, the points represented by $0, N, \ldots, (r-1)N$ lie in an l-space Π_l. Suppose a generating set for Π_l is given by the set

$$\mathcal{L}_1 = \{iN \mid i \in \mathrm{N}_{l-1}\} .$$

Let b_1 represent a point independent of those given by $\mathcal{L}_1$. Let

$$\mathcal{L}_2 = \{b_1 + iN \mid i \in \mathrm{N}_{l-1}\} .$$

Suppose the points represented by $\mathcal{L}_1 \cup \mathcal{L}_2$ are dependent. Then there exist $A_0, \ldots, A_l, B_0, \ldots, B_l$ in γ such that

$$A_0 + A_1\alpha^N + \ldots + A_l\alpha^{lN} + B_0\alpha^{b_1} + B_1\alpha^{b_1+N} + \ldots + B_l\alpha^{b_1+lN} = 0.$$

Hence $\quad A + B\alpha^{b_1} = 0,$

where $\quad A = A_0 + A_1\alpha^N + \ldots + A_l\alpha^{lN}$

and $\quad B = B_0 + B_1\alpha^N + \ldots + B_l\alpha^{lN}.$

But B^{-1} can be written as a polynomial in α^N and hence as a polynomial in α^N of degree at most l. So there exist $C_0, \ldots, C_l$ in γ such that

$$\alpha^{b_1} = -AB^{-1} = C_0 + C_1\alpha^N + \ldots + C_l\alpha^{lN},$$

contradicting the choice of b_1.

Now take b_2 representing a point independent of those given by $\mathcal{L}_1 \cup \mathcal{L}_2$ and let

$$\mathcal{L}_3 = \{b_2 + iN \mid i \in \overline{\mathrm{N}}_{l-1}\} .$$

Then, exactly as above, $\mathcal{L}_1 \cup \mathcal{L}_2 \cup \mathcal{L}_3$ represent independent points. Continuing this process, we obtain the set $\mathcal{L}_1 \cup \mathcal{L}_2 \cup \ldots \cup \mathcal{L}_{s+1}$, where $s+1 = (m+1)/(l+1)$; this set generates Π_m. By construction, Π_m has cycle N and $b_1, \ldots, b_s$ may be chosen so that the cycle is not less than N; for example, $b_i = i$ for i in $\mathbf{N}_s$.□

In $PG(n,q)$, a set $\mathscr{S}$ of m-spaces is a *k-fold spread* if every point of $PG(n,q)$ lies in k of the m-spaces of $\mathscr{S}$. In the terminology of §4.1, a 1-fold spread is a spread. So we can now give a generalization of theorem 4.1.1.

4.2.7. THEOREM. *If* $(l+1)\,|\,(m+1,\ n+1)$, *then there exists a k-fold spread of m-spaces in* $PG(n,q)$, *where* $k = \theta(m)/\theta(l)$.

Proof. By lemma 4.2.5, there exists an m-space Π_m of cycle $N = \theta(n)/\theta(l)$, whose points are given by

$$\begin{array}{l} c_0,\ c_0 + N,\ \ldots\ ,\ c_0 + (r-1)N, \\ \vdots \\ c_{k-1},\ c_{k-1} + N, \ldots,\ c_{k-1} + (r-1)N\ , \end{array}$$

where $r = \theta(l)$. From Π_m, we can choose N-1 other m-spaces by successively adding 1,2,..., N-1 and reducing modulo $\theta(n)$. These N spaces then constitute a k-fold spread of m-spaces.□

COROLLARY. *A spread of m-spaces exists in* $PG(n,q)$ *if and only if* $(m+1)\ |\ (n+1)$.

Proof. If $(m+1)\,|\,(n+1)$, then in the theorem, take $l = m$. Then $k = 1$ and $N = \theta(n)/\theta(m)$.□

The corollary is precisely theorem 4.1.1.

Example. $x^4 + x + 1$ is primitive over $GF(2)$. So the projectivity $\mathfrak{T}$ given by

$$T = H(x^4 + x + 1) = \begin{bmatrix} 0 & 1 & 0 & 0 \\ 0 & 0 & 1 & 0 \\ 0 & 0 & 0 & 1 \\ 1 & 1 & 0 & 0 \end{bmatrix}$$

is cyclic on $PG(3,2)$. $P(i) = P(0)\mathfrak{T}^i$:

$P(0) = \mathbf{P}(1,0,0,0)$, $P(1) = \mathbf{P}(0,1,0,0)$, $P(2) = \mathbf{P}(0,0,1,0)$, $P(3) = \mathbf{P}(0,0,0,1)$, $P(4) = \mathbf{P}(1,1,0,0)$, $P(5) = \mathbf{P}(0,1,1,0)$, $P(6) = \mathbf{P}(0,0,1,1)$, $P(7) = \mathbf{P}(1,1,0,1)$, $P(8) = \mathbf{P}(1,0,1,0)$, $P(9) = \mathbf{P}(0,1,0,1)$, $P(10) = \mathbf{P}(1,1,1,0)$, $P(11) = \mathbf{P}(0,1,1,1)$, $P(12) = \mathbf{P}(1,1,1,1)$, $P(13) = \mathbf{P}(1,0,1,1)$, $P(14) = \mathbf{P}(1,0,0,1)$.

The 35 lines are given by two cycles of fifteen and one of five (Table 4.2).

TABLE 4.2

0	1	2	3	4	5	6	7	8	9	10	11	12	13	14
1	2	3	4	5	6	7	8	9	10	11	12	13	14	0
4	5	6	7	8	9	10	11	12	13	14	0	1	2	3
0	1	2	3	4	5	6	7	8	9	10	11	12	13	14
2	3	4	5	6	7	8	9	10	11	12	13	14	0	1
8	9	10	11	12	13	14	0	1	2	3	4	5	6	7
0	1	2	3	4										
5	6	7	8	9										
10	11	12	13	14										

The existence of these lines follows from the fact that $P(0)$, $P(1)$, $P(4)$ are collinear. These themselves give the first cycle of 15. Then, by lemma 4.2.4, $P(0)$, $P(2)$, $P(8)$ are collinear giving the second cycle of 15. From these two cycles, there are six lines through $P(0)$. Hence the seventh must be $P(5)P(10)$, and thus we have the cycle of five. Alternatively, since $m=1$ and $n=3$, lemma 4.2.6 gives the existence of a cycle of order N less than $\theta(3) = 15$. In fact, since $l+1 = (m+1, n+1)$, so $l = 1$, $r = \theta(l) = 3$ and $N = \theta(n)/r = 5$. So, again $P(0)$, $P(5)$, $P(10)$ are collinear and the line is part of a cycle of five. The cycle of five lines is a spread of $PG(3,2)$. The 15 planes are given in Table 4.3 by a cycle of 15, where the first member represents $\mathbf{u}_3$.

TABLE 4.3

0	1	2	3	4	5	6	7	8	9	10	11	12	13	14
1	2	3	4	5	6	7	8	9	10	11	12	13	14	0
2	3	4	5	6	7	8	9	10	11	12	13	14	0	1
4	5	6	7	8	9	10	11	12	13	14	0	1	2	3
5	6	7	8	9	10	11	12	13	14	0	1	2	3	4
8	9	10	11	12	13	14	0	1	2	3	4	5	6	7
10	11	12	13	14	0	1	2	3	4	5	6	7	8	9

Since the primes of $PG(n,q)$ have cycle $\theta(n)$, we can generalize the plane array and give an n-dimensional array $\mathcal{M}$ representing the subspaces of $PG(n,q)$. The numbers $0, 1, \ldots, \theta(n)-1$ represent the points of the space. The integers in any $(n-1)$-dimensional layer of $\mathcal{M}$ represent the points of an $(n-1)$-space and are obtained as follows. Take the integers in the base layer as representing the points of an $(n-1)$-space; to obtain the next layer add one to each integer in the base and reduce if necessary modulo $\theta(n)$, then this process is repeated for the next layer, and so on. The integers in an $(n-2)$-dimensional layer of an $(n-1)$-dimensional layer represent the points of an $(n-2)$-space, and so on. The array $\mathcal{M}$ has $\prod_{i=1}^{n} \theta(i)$ entries. The elements in a layer of $\mathcal{M}$ representing an $(n-1)$-space provide a difference set of $\theta(n-1)$ integers with the property that their differences are congruent modulo $\theta(n)$ to the elements of $N_{\theta(n)}$, each integer of which is congruent to $\theta(n-2)$ of the differences.

Example. $PG(3,2)$ is represented by a three-dimensional array $\mathcal{M}$ of $3\cdot 7\cdot 15$ entries. The 15 faces are given in Table 4.4; the seven columns of each face represent the lines of the corresponding plane.

4.3. Partitions of $PG(n,q^k)$ by subgeometries $PG(n,q)$

Cyclic projectivities have a further application in elucidating the structure of projective spaces over finite fields when

TABLE 4.4

0	1	2	4	5	8	10
1	2	4	5	10	0	8
4	5	10	8	0	2	1
1	2	3	5	6	9	11
2	3	5	6	11	1	9
5	6	11	9	1	3	2
2	3	4	6	7	10	12
3	4	6	7	12	2	10
6	7	12	10	2	4	3
3	4	5	7	8	11	13
4	5	7	8	13	3	11
7	8	13	11	3	5	4
4	5	6	8	9	12	14
5	6	8	9	14	4	12
8	9	14	12	4	6	5
5	6	7	9	10	13	0
6	7	9	10	0	5	13
9	10	0	13	5	7	6
6	7	8	10	11	14	1
7	8	10	11	1	6	14
10	11	1	14	6	8	7
7	8	9	11	12	0	2
8	9	11	12	2	7	0
11	12	2	0	7	9	8
8	9	10	12	13	1	3
9	10	12	13	3	8	1
12	13	3	1	8	10	9

TABLE 4.4 (cont.)

9	10	11	13	14	2	4
10	11	13	14	4	9	2
13	14	4	2	9	11	10
10	11	12	14	0	3	5
11	12	14	0	5	10	3
14	0	5	3	10	12	11
11	12	13	0	1	4	6
12	13	0	1	6	11	4
0	1	6	4	11	13	12
12	13	14	1	2	5	7
13	14	1	2	7	12	5
1	2	7	5	12	14	13
13	14	0	2	3	6	8
14	0	2	3	8	13	6
2	3	8	6	13	0	14
14	0	1	3	4	7	9
0	1	3	4	9	14	7
3	4	9	7	14	1	0

the order of the field is not prime.

Firstly, since $GF(q)$ is a subfield of $GF(q^k)$ for k in N, $PG(n,q)$ is naturally embedded in $PG(n,q^k)$ once the coordinate system is fixed. Any $PG(n,q)$ embedded in $PG(n,q^k)$ is a *subgeometry* of $PG(n,q^k)$. Let $\mathscr{S}$ be the set of such subgeometries $PG(n,q)$ and let $s(n,q,q^k) = |\mathscr{S}|$.

4.3.1. LEMMA. $s(n, q, q^k) = p(n+1,q^k)/p(n+1,q)$

$$= q^{n(n+1)(k-1)/2} \prod_{i=2}^{n+1} [(q^{ki}-1)/(q^i-1)].$$

Proof. Any $n+2$ points, no $n+1$ of which lie in a prime, determine a $PG(n,q)$. Hence $PGL(n+1,q^k)$ acts transitively on the set $\mathcal{S}$ of $PG(n,q)$, each of which has a stabilizer isomorphic to $PGL(n+1,q)$. Therefore, $s(n, q, q^k) = p(n+1,q^k)/p(n+1,q)$.

Alternatively, let $M(n,q)$ be the number of sets of $n+2$ points in $PG(n,q)$, no $n+1$ of which lie in a prime. Then

$$(n+2)!\, M(n,q) = \theta(n)\, [\theta(n) - \theta(0)][\theta(n) - \theta(1)] \ldots [\theta(n) - \theta(n-1)](q-1)^n$$
$$= q^{n(n+1)/2}[2, n+1]_- = p(n+1,q).$$

In $PG(n,q^k)$, since any such set determines a $PG(n,q)$ so $s(n, q, q^k) = M(n,q^k)/M(n,q) = p(n+1,q^k)/p(n+1,q)$.□

COROLLARY 1. *$s(n, q, q^k)$ is multiplicative in the sense that $s(n, q, q^k)\, s(n, q^k, q^{kl}) = s(n, q, q^{kl})$*.□

COROLLARY 2.

(i) $s(1, q, q^k) = q^{k-1}\, (q^{2k} - 1)/(q^2 - 1)$;

(ii) $s(1, q, q^2) = q(q^2 + 1)$.□

COROLLARY 3. $s(2, q, q^2) = q^3(q^2 + 1)(q^3 + 1)$.□

In particular, $s(2, 2, 4) = 2^3 \cdot 3^2 \cdot 5 = 360$;
$s(2, 3, 9) = 2^3 \cdot 3^3 \cdot 5 \cdot 7 = 7{,}560$;
$s(2, 4, 16) = 2^6 \cdot 5 \cdot 13 \cdot 17 = 70{,}720$;
$s(2, 2, 16) = 360 \cdot 70{,}720 = 25{,}459{,}200$.

It is now natural to ask if there exists a partition of $PG(n,q^k)$ into disjoint subgeometries $PG(n,q)$. Firstly, we require some lemmas.

4.3.2. LEMMA. *If r, s and x are positive integers with $x > 1$, then*

$$N = \frac{(x^{rs} - 1)(x - 1)}{(x^r - 1)(x^s - 1)}$$

is an integer if and only if $(r,s) = 1$.

Proof.

$$N = \frac{x^{s(r-1)} + x^{s(r-2)} + \ldots + x^{s} + 1}{x^{r-1} + x^{r-2} + \ldots + x + 1}$$

Let $kr \leqslant m$. Then,

$$x^{m} - x^{m-kr} = x^{m-kr}(x^{kr} - 1),$$

which is divisible by $x^{r} - 1$ and so by $x^{r-1} + x^{r-2} + \ldots + x + 1$. Hence

$$\frac{x^{m}}{x^{r-1} + \ldots + x + 1} - \frac{x^{m-kr}}{x^{r-1} + \ldots + x + 1}$$

is an integer.

Let $(r,s) = d$; then the set of residues of $s(r-1)$, $s(r-2), \ldots, s$, 0 modulo r is, in some order, $r-d$, $r-2d, \ldots, d$, 0, each of which occurs d times. Hence, in the expression for N, if x^{si} is replaced by x^{j} where $si \equiv j \pmod{r}$ and $0 \leqslant j < r$, then N is an integer if and only if M is, where

$$M = \frac{d(x^{r-d} + \ldots + x^{d} + 1)}{x^{r-1} + \ldots + x + 1}$$

If $d = 1$, $M = 1$ and so N is an integer. If $d > 1$, $M < 1$ and so N is not an integer.□

COROLLARY. *$\theta(n,q)$ divides $\theta(n,q^{k})$ if and only if $(k, n+1) = 1$.*□

Proof.

$$\theta(n,q^{k}) \;/\; \theta(n,q) = (q^{k(n+1)} - 1)(q - 1)/[(q^{k} - 1)(q^{n+1} - 1)].$$

So $x = q$, $r = k$, $s = n+1$ gives the result.□

4.3.3. LEMMA. *If F is an irreducible polynomial of degree r over $GF(q)$ and $(k,r) = 1$, then F is irreducible over $GF(q^k)$.*

Proof. Suppose that α is a root of F in $GF(q^r)$ and that F is reducible over $GF(q^k)$. Then F has a factor G over $GF(q^k)$ such that G is irreducible over $GF(q^k)$, α is a root of G, G has degree s, and the coefficients of G lie in a subfield $GF(q^m)$ of $GF(q^k)$ with $m > 1$; so $m \mid k$. The roots of G lie in $GF(q^{sm})$ which must be a subfield of $GF(q^r)$; so $m \mid r$. Hence $m \mid (k,r)$, giving a contradiction.□

4.3.4. LEMMA. *Let A, B, C be three positive integers such that $(A, B) = 1$. Then there exists an integer m with $0 \leqslant m \leqslant C - 1$ such that $(A + mB, C) = 1$.*

Proof. If $C = 1$, let $m = 0$. Suppose the primes dividing C are $P_1, \ldots, P_a, Q_1, \ldots, Q_b, R_1, \ldots, R_c$, where $P_1, \ldots, P_a$ divide A, where $Q_1, \ldots, Q_b$ divide B and where $R_1, \ldots, R_c$ divide neither A nor B. Let $m = R_1R_2 \ldots R_c$ if such primes R_i exist, and let $m = 1$ otherwise. Then, if $P_i \mid (A + mB, C)$, P_i divides A, C, and so mB; if $Q_i \mid (A + mB, C)$, Q_i divides B, C, and so A; if $R_i \mid (A + mB, C)$, R_i divides m, C, and so A. In each case there is a contradiction. So $(A + mB, C) = 1$.□

For the remainder of this section, write

$$\begin{aligned}
w &= \theta(n,q^k) = (q^{k(n+1)} - 1)/(q^k - 1),\\
v &= \theta(n,q) = (q^{n+1} - 1)/(q - 1),\\
z &= (q^{k(n+1)} - 1)/(q^{n+1} - 1),\\
y &= (q^k - 1)/(q - 1),\\
u &= w/v = z/y = (q^{k(n+1)} - 1)(q - 1)/[(q^k - 1)(q^{n+1} - 1)].
\end{aligned}$$

4.3.5. THEOREM. *If $(k, n+1) = 1$ and $\mathfrak{S}$ is a projectivity of $PG(n,q^k)$ which acts as a cyclic projectivity on some $PG(n,q)$ embedded in $PG(n,q^k)$, then*

(i) *there exists a cyclic projectivity $\mathfrak{T}$ of $PG(n,q^k)$ such that $\mathfrak{T}^u = \mathfrak{S}$;*

(ii) *every orbit of $\mathfrak{S}$ is a space $PG(n,q)$.*

Proof. Select the simplex of reference and the unit point in the given $PG(n,q)$. Then $\mathfrak{S}$ is given by a matrix S whose characteristic polynomial F has coefficients in $GF(q)$ and is subprimitive. By lemma 4.3.3, F is irreducible over $GF(q^k)$. So all orbits of $\mathfrak{S}$ have length v, the order of $\mathfrak{S}$, and $\mathfrak{S}$ has $w/v = u$ orbits.

Suppose (i) is true and there exists a cyclic projectivity $\mathfrak{S}$ of $PG(n,q)$ with $\mathfrak{T}^u = \mathfrak{S}$. Now, $\mathfrak{T}$ has order w and the points of $PG(n,q^k)$ are $P(0)$, $P(1),\ldots, P(w-1)$, where $P(0)$ is any point and $P(i) = P(0)\mathfrak{T}^i$. Then $\mathfrak{T}^u$ has u orbits $\mathcal{O}(i)$, i in $\overline{\mathbf{N}}_{u-1}$, where $\mathcal{O}(i) = \{P(i + ju) \mid j \in \overline{\mathbf{N}}_{v-1}\}$. All the orbits are projectively equivalent, since $\mathcal{O}(i) = \mathcal{O}(0)\mathfrak{T}^i$: so every orbit is a subgeometry $PG(n,q)$.

It remains to find $\mathfrak{T}$. In fact, we will find conjugates $\mathfrak{S}'$ of $\mathfrak{S}$ and $\mathfrak{T}'$ of $\mathfrak{T}$ such that $\mathfrak{T}'^u = \mathfrak{S}'$.

Suppose now that the roots of F in $GF(q^{n+1})$ are β, β^q, ..., β^{q^n}. Then, since F is also irreducible over $GF(q^k)$, the roots can also be written (in a different order) β, $\beta^{q^k},\ldots, \beta^{q^{nk}}$. So there exist projectivities $\mathfrak{S}'$ and $\mathfrak{A}$, where $\mathfrak{S}'$ has matrix $S' = \text{diag}\ (\beta, \beta^{q^k}, \ldots, \beta^{q^{nk}})$ and $\mathfrak{S} = \mathfrak{A}\ \mathfrak{S}'\mathfrak{A}^{-1}$.

Let κ be a primitive root of $GF(q^{k(n+1)})$. Then $\kappa^i \in GF(q^k)$ if and only if $i = fw$ and $\kappa^i \in GF(q^{n+1})$ if and only if $i = gz$ for some integers f and g. Further, κ^w is a primitive root of $GF(q^k)$ and κ^z is a primitive root of $GF(q^{n+1})$. Since β is a subprimitive root of $GF(q^{n+1})$, $\beta = \kappa^{rz}$ where $(r, v) = 1$, §1.6(v). So

$$S' = \text{diag}\ (\kappa^{rz}, \kappa^{rzq^k}, \ldots., \kappa^{rzq^{nk}}).$$

Since $(k, n+1) = 1$, so also $(y, v) = 1$. Hence, since $(r, v) = 1$, we have $(ry, v) = 1$. So, by lemma 4.3.4, there is an integer d such that $(ry + dv, w) = 1$. Now $\lambda = \kappa^{duv} = \kappa^{dw}$ is in $GF(q)$. Since $(ry + dv, w) = 1$, $\alpha = \kappa^{ry+dv}$ is an $(n+1)$-subprimitive root of $GF(q^{k(n+1)})$. So the projectivity $\mathfrak{T}'$ with matrix

$$T' = \text{diag}\ (\alpha, \alpha^{q^k}, \ldots, \alpha^{q^{nk}})$$

is a cyclic projectivity of $PG(n,q^k)$. However,

$$\alpha^{uq^{sk}} = \kappa^{(ry + dv)uq^{sk}} \cdot \kappa^{dvuq^{sk}} = \kappa^{rzq^{sk}} \cdot \lambda^{q^{sk}} = \kappa^{rzq^{sk}}\lambda .$$

So
$$T'^u = \text{diag}\,(\alpha^u, \alpha^{uq^k}, \ldots, \alpha^{uq^{nk}})$$
$$= \lambda\,\text{diag}\,(\kappa^{rz}, \kappa^{rzq^k}, \ldots, \kappa^{rzq^{nk}})$$
$$= \lambda S'.$$

Thus $\mathfrak{T}'^u = \mathfrak{S}'$. Hence, let $\mathfrak{T} = \mathfrak{A}\mathfrak{T}'\mathfrak{A}^{-1}$. Then $\mathfrak{T}^u = \mathfrak{S}$.□

4.3.6. THEOREM. *The following are equivalent:*

(i) *there exists a partition of $PG(n,q^k)$ into subgeometries $PG(n,q)$;*

(ii) $\theta(n,q) \mid \theta(n,q^k)$;

(iii) $(k, n+1) = 1$.

Proof. The theorem states that a partition of $PG(n,q^k)$ into $PG(n,q)$ exists if and only if the number of points in $PG(n,q)$ divides the number of points in $PG(n,q^k)$. Thus (i) implies (ii). By the corollary to lemma 4.3.2, (ii) and (iii) are equivalent. By the previous theorem, (iii) implies (i).□

This theorem should be compared with Theorem 4.1.1.

COROLLARY 1. *$PG(1,q^k)$ can be partitioned into subgeometries $PG(1,q)$ if and only if $(k, 2) = 1$.*□

COROLLARY 2. *$PG(2,q^k)$ can be partitioned into subgeometries $PG(2,q)$ if and only if $(k, 3) = 1$.*□

COROLLARY 3. *$PG(2,q^2)$ can be partitioned into $q^2 - q + 1$ subgeometries $PG(2,q)$.*□

Let $\theta(n, q, q^k) = \theta(n,q^k)/\theta(n,q)$. Then, if $(k, n+1) = 1$, it has been shown that $PG(n,q^k)$ can be partitioned into $\theta(n, q, q^k)$ spaces $PG(n,q)$, where

$$\theta(n, q, q^k) = (q^{k(n+1)} - 1)(q - 1)/[(q^k - 1)(q^{n+1} - 1)].$$

4.3.7. LEMMA. *$\theta(n, q, q^k)$ is multiplicative in the sense that* $\theta(n, q, q^k) \cdot \theta(n, q^k, q^{kl}) = \theta(n, q, q^{kl})$.□

In particular, from corollary 3 to theorem 4.3.6, $\theta(2, q, q^2) = q^2 - q + 1$ and so

$$\theta(2, 2, 4) = 3,$$
$$\theta(2, 3, 9) = 7,$$
$$\theta(2, 4, 16) = 13,$$
$$\theta(2, 2, 16) = 39.$$

If $(k, n+1) = 1$, the number $\sigma(n, q, q^k)$ of projectivities $\mathfrak{S}$ which act cyclically on a $PG(n,q)$ of $PG(n,q^k)$ can be determined.

4.3.8. LEMMA. If $(k, n+1) = 1$,

$$\sigma(n, q, q^k) = q^{kn(n+1)/2} \prod_{i=1}^{n} (q^{ki} - 1)\, \phi(\theta(n,q))/(n+1).$$

Proof. $\sigma(n, q, q^k) = \sigma(n, q)\, s(n, q, q^k)/\theta(n, q, q^k)$. But $\sigma(n,q)$ was given in theorem 4.2.1, corollary 2, and $s(n, q, q^k)$ in lemma 4.3.1.

Alternatively, the number of elements of $PGL(n+1,q^k)$ in the conjugacy class of $\mathfrak{S}$ is $p(n+1,q^k)/\theta(n,q^k)$ by lemma 2.4.3. The number of such conjugacy classes is $R(n+1,q)/(q-1) = \phi(\theta(n,q))/(n+1)$. Hence

$$\begin{aligned}\sigma(n, q, q^k) &= p(n+1,q^k)\, \phi(\theta(n,q))/[(n+1)\theta(n,q^k)]\\ &= q^{kn(n+1)/2} \prod_{i=1}^{n} (q^{ki} - 1)\, \phi(\theta(n,q))/(n+1).\end{aligned}$$ □

Now let $p(n, q, q^k)$ be the number of partitions of $PG(n,q^k)$ into spaces $PG(n,q)$ when $(k, n+1) = 1$ and let $\sigma(n, q, q^k)$ be the number of partitions in which a given $PG(n,q)$ lies. Then

$$p(n, q, q^k)\, \theta(n, q, q^k) = \sigma(n, q, q^k)\, s(n, q, q^k).$$

4.3.9. THEOREM.

(i) $p(n, q, q^k) \geqslant q^{kn(n+1)/2} \prod_{i=1}^{n} (q^{ki} - 1)/(n+1)$

$$= p_0(n, q, q^k);$$

(ii) $\rho(n, q, q^k) \geq q^{n(n+1)/2} \prod_{i=1}^{n} (q^i - 1)/(n+1) =$

$= \rho_0(n, q).$

Proof.

(i) Each cyclic projectivity $\mathfrak{T}$ of $PG(n, q^k)$ determines a partition. However, if $\mathfrak{T}^i$ is cyclic, it determines the same partition. Hence

$$p(n, q, q^k) \geq \sigma(n,q^k) / \phi(\theta(n,q^k)) = p_0(n, q, q^k).$$

(ii) Similarly, if $\mathfrak{S}$ is a projectivity of $PG(n,q^k)$ which acts cyclically on a fixed $PG(n,q)$, the orbits of $\mathfrak{S}$ form a partition. If $\mathfrak{S}^i$ is also cyclic on the fixed $PG(n,q)$, it gives the same partition. Hence

$$\rho(n, q, q^k) \geq \sigma(n,q) / \phi(\theta(n,q)) = \rho_0(n,q).$$

This argument also shows that

$$p(n, q, q^k) \geq \theta(n, q, q^k) / \phi(\theta(n,q)) = p_0(n, q, q^k).\square$$

COROLLARY.
$p_0(n, q, q^k)\ \theta(n, q, q^k) = \rho_0(n, q)\ s(n, q, q^k).\square$

In general, the inequality signs of the theorem will not be equalities; for

$$\begin{aligned} \rho(n, q, q^{kl}) &\geq \rho(n, q, q^k)\ \rho(n, q^k, q^{kl}) \\ &\geq \rho_0(n,q)\ \rho_0(n,q^k) \\ &> \rho_0(n,q). \end{aligned}$$

So $p(n, q, q^{kl}) > p_0(n, q, q^{kl})$.

Examples.

(i) The plane of order four consists of three disjoint planes of order two:

$k = q = n = 2,\ u = y = 3,\ v = 7,\ z = 9,\ w = 21.$

TABLE 4.5

PG(2,4)																				
0	1	2	3	4	5	6	7	8	9	10	11	12	13	14	15	16	17	18	19	20
1	2	3	4	5	6	7	8	9	10	11	12	13	14	15	16	17	18	19	20	0
4	5	6	7	8	9	10	11	12	13	14	15	16	17	18	19	20	0	1	2	3
14	15	16	17	18	19	20	0	1	2	3	4	5	6	7	8	9	10	11	12	13
16	17	18	19	20	0	1	2	3	4	5	6	7	8	9	10	11	12	13	14	15

Since $u = 3$, we can immediately pick out a $PG(2,2)$ by taking just the multiples of three. Hence, from the second, third, and fifth rows of Table 4.5, we have, beginning with the last column

(A)	0	3	6	9	12	15	18
	3	6	9	12	15	18	0
	15	18	0	3	6	9	12;

adding one and two to each entry, we have

(B_1)	1	4	7	10	13	16	19	(C_1)	2	5	8	11	14	17	20
	4	7	10	13	16	19	1		5	8	11	14	17	20	2
	16	19	1	4	7	10	13		17	20	2	5	8	11	14

If (B_2), (C_2) is another pair of planes $PG(2,2)$ into which the remaining 14 points can be divided, (B_2) cannot have a quadrangle in common with (B_1) or (C_1). However, any five points of $PG(2,2)$ contain a quadrangle. So (B_2) must have four points in common with one of the planes, (B_1) say, three of which are collinear and three non-collinear points in common with (C_1). In fact, if (B_2) has three given collinear points in common with (B_1), the fourth common point is determined. Thus there are seven other pairs (B_i), (C_i) of planes apart from (B_1), (C_1) into which the 14 points are partitioned. The relation between the eight pairs (B_i), (C_i) is symmetrical.
So $\rho(2, 2, 4) = 8 = \rho_0(2, 2)$. But $\theta(2, 2, 4) = 3$ and $s(2, 2, 4) = 360$. Thus $p(2, 2, 4) = p_0(2, 2, 4) = 960$.

(ii) $PG(1,8)$ consists of three disjoint $PG(1,2)$:
$n = 1$, $q = 2$, $k = v = u = 3$, $y = 7$, $w = 9$, $z = 21$.
$\theta(1, 2, 8) = 3$, $s(1, 2, 8) = 84$, $\rho_0(1, 2) = 1$ and

$p_0(1, 2, 8) = 28$.

In fact, $p(1, 2, 8)$ is the number of ways nine points can be partitioned into three sets of three. So

$$p(1, 2, 8) = \mathbf{c}(9,3)\mathbf{c}(6,3)/(3!) = 280,$$

and

$$\rho(1, 2, 8) = \mathbf{c}(6,3)/(3!) = 10.$$

A similar argument shows that

$$p(1, 2, 2^k) = (2^k + 1)! \; / \; \{6^{(2^k+1)/3} \, [(2^k+1)/3]\,!\},$$

for k odd.

4.4. Notes and references.

§4.1. For q a prime, theorem 4.1.1 was given by Burnside [1911], p. 459, as a theorem on abelian groups. It was rediscovered by André [1954], Segre [1964b], Bruck and Bose [1964]. See Carmichael [1937] for the connection between abelian groups and finite geometries.

§4.2. The idea of the correspondence between $GF(q^{n+1})\backslash\{0\}$ and $PG(n,q)$ is due to Singer [1938], as is theorem 4.2.2. See also Snapper [1950]. The other theorems of §4.2 are taken from Rao [1969]. Many authors have taken up the themes of cyclic projectivities and difference sets. See, for example, Baumert [1971], Hirschfeld [1976], Lunelli, Lunelli and Sce [1959], Mazumdar [1967], Raktoe [1967], Rao [1945], [1946].

§4.3. Theorem 4.3.6 was first proved by Yang [1949a], although his proof of lemma 4.3.2 is incorrect. The proof of the lemma is due to J.A. Tyrrell. The whole section follows Hirschfeld [1976]. See also Yang [1949b], Yff [1977].

5
CANONICAL FORMS FOR VARIETIES AND POLARITIES

5.1. Quadric and Hermitian varieties

A *quadric (variety)* $\mathcal{Q}$ in $PG(n,q)$ is a primal of order two. So $\mathcal{Q} = \mathbf{V}(Q)$, where Q is a quadratic form; that is,

$$Q = \sum_{i,j=0}^{n}{}' a_{ij}x_ix_j$$

$$= a_{00}x_0^2 + a_{01}x_0x_1 + \ldots .$$

A *Hermitian variety* $\mathcal{U}$ in $PG(n,q)$ is a variety $\mathbf{V}(H)$, where H is a Hermitian form; that is,

$$H = \sum_{i,j=0}^{n} t_{ij}\bar{x}_ix_j , \qquad \text{where} \quad t_{ji} = \bar{t}_{ij},$$

$$= t_{00}\bar{x}_0x_0 + t_{01}\bar{x}_0x_1 + \bar{t}_{01}x_0\bar{x}_1 + \ldots .$$

Here $x \to \bar{x}$ is an involutory automorphism of $\gamma = GF(q)$. So q is a square and $\bar{x} = x^{\sqrt{q}}$.

In each case, the form and the variety are *degenerate* if there is a change of coordinate system which reduces the form to one in fewer variables: otherwise, the form and the variety are *non-degenerate*.

5.1.1. LEMMA. *A quadric or a Hermitian variety is degenerate if and only if it is singular.*

Proof. In either case, if x_0, say, is eliminated by a change of coordinate system, then in the new system $\mathbf{U}_0$ is a singular point of the variety. Conversely, if $\mathbf{U}_0$ is a singular point, then consider $\mathcal{Q}$ first. Since $\mathbf{U}_0 \in \mathcal{Q}$, so $a_{00} = 0$. Since $\mathbf{U}_0$ is singular, $a_{01} = a_{02} = \ldots = a_{0n} = 0$, § 2.6. Therefore $\mathcal{Q}$ is degenerate. For $\mathcal{U}$, since $\mathbf{U}_0$ lies on it, $t_{00} = 0$. Since $\mathbf{U}_0$ is singular, $t_{01} = t_{02} = \ldots = t_{0n} = 0$. Since H is Hermitian, $t_{10} = t_{20} = \ldots = t_{n0} = 0$. So $\mathcal{U}$ is degenerate.□

We wish to find canonical forms for $\mathcal{Q}$ and $\mathcal{U}$. When $p \neq 2$, we may write $t_{ij} = t_{ji} = (a_{ij} + a_{ji})/2$, so that $Q = \sum_{i,j=0}^{n} t_{ij}x_ix_j = XTX^*$ where $T = (t_{ij})$. For any characteristic, $H = \bar{X}TX^*$. So we begin by excluding quadrics when $p = 2$ and thus deal with the cases that $\mathcal{Q}$ and $\mathcal{U}$ are the sets of self-conjugate points of a polarity as in §2.1(v).

Let $F = \tilde{X}TX^*$, where $T = (t_{ij})$, $\tilde{t}_{ji} = t_{ij}$, $\tilde{X} = (\tilde{x}_0, \ldots, \tilde{x}_n)$, and $x \to \tilde{x}$ is an automorphism σ of γ such that $\tilde{\tilde{x}} = x$; thus σ has order one or two. When σ has order one and so $\tilde{x} = x$ all x, then $F = XTX^* = Q$. When σ has order two and so $\tilde{x} = \bar{x}$ all x, then $F = \bar{X}TX^* = H$.

If two varieties V_1, V_2 are projectively equivalent, write $V_1 \sim V_2$.

5.1.2. LEMMA. $\mathbf{V}(F) \sim \mathbf{V}(a_0\tilde{x}_0x_0 + a_1\tilde{x}_1x_1 + \ldots + a_r\tilde{x}_rx_r)$, where $\tilde{a}_i = a_i$.

Proof. If some $t_{kk} \neq 0$, F can be transformed to a form with $t_{00} \neq 0$. If every $t_{kk} = 0$ and, say, $t_{01} \neq 0$, the transformation $x_0 = y_0$, $x_1 = b\, y_0 + y_1$, $x_i = y_i$ for $i > 1$ gives a form $\Sigma\, s_{ij}\, \tilde{y}_i\, y_j$ with $s_{00} = t_{01}\, b + t_{10}\, \tilde{b}$. So, if $p \neq 2$ and we put $b = t_{01}{}^{-1}$, then $s_{00} = 2$. If $p = 2$, choose b such that $t_{01}\, b \notin GF(\sqrt{q})$; then, also, $s_{00} \neq 0$.

So consider $F = \Sigma\, t_{ij}\, \tilde{x}_ix_j$ with $t_{00} \neq 0$. Then

$$F = t_{00}\, \tilde{x}_0\, x_0 + \tilde{x}_0\, (t_{01}\, x_1 + \ldots + t_{0n}\, x_n)$$

$$+ x_0(t_{10}\, \tilde{x}_1 + \ldots + t_{1n}\, \tilde{x}_n) + G(x_1, \ldots, x_n)$$

$$= t_{00}{}^{-1}\, (t_{00}\, \tilde{x}_0 + t_{10}\, \tilde{x}_1 + \ldots + t_{n0}\, \tilde{x}_n)(t_{00}\, x_0 + t_{01}\, x_1 +$$

$$+ \ldots + t_{0n}\, x_n) + G'(x_1, \ldots, x_n)\,.$$

Put $x_0 = y_0 - t_{00}{}^{-1}\, (t_{01}\, y_1 + \ldots + t_{0n}\, y_n)$, $x_i = y_i$ for $i > 1$. Then $y_0 = t_{00}{}^{-1}\, (t_{00}\, x_0 + t_{01}\, x_1 + \ldots + t_{0n}\, x_n)$, $y_i = x_i$ for $i > 1$. Thus $F = t_{00}\, \tilde{y}_0\, y_0 + G'(y_1, \ldots, y_n)$.

Therefore, by continuing this process, we obtain the canonical form

$$F = a_0 \tilde{x}_0 x_0 + \ldots + a_r \tilde{x}_r x_r \quad \text{with } \tilde{a}_i = a_i \text{ and } r \leq n. \square$$

COROLLARY 1.

(i) *For* $p \neq 2$ *and* $\tilde{x} = x$, $\mathbf{V}(Q) \sim \mathbf{V}(a_0 x_0^2 + \ldots + a_r x_r^2)$.

(ii) *For* p *arbitrary and* $\tilde{x} = \bar{x} = x^{\sqrt{q}}$,

$$\mathbf{V}(H) \sim V(a_0 \bar{x}_0 x_0 + \ldots + a_r \bar{x}_r x_r), \qquad \textit{where } \bar{a}_i = a_i. \square$$

COROLLARY 2. *A quadric* $\mathbf{V}(XTX^*)$ *when* q *is odd or a Hermitian variety* $\mathbf{V}(\bar{X}TX^*)$ *is singular if and only if the corresponding matrix* T *is singular.* □

5.1.3. LEMMA. *If* q *is a square, then every element of the subfield* $GF(\sqrt{q})$ *of* $GF(q)$ *can be written in the form* $c^{\sqrt{q}+1} = c\, c^{\sqrt{q}} = c\,\bar{c}$.

Proof. If α is a primitive element of $GF(q)$, then $\alpha^{\sqrt{q}+1}$ is a primitive element of $GF(\sqrt{q})$, §1.6(iii). So if $t \in GF(\sqrt{q})$, $t = 0$ or $t = \alpha^{m(\sqrt{q}+1)} = \alpha^m \cdot \alpha^{m\sqrt{q}}$. So, if $c = \alpha^m$, then $t = c\, c^{\sqrt{q}}$. □

5.1.4. LEMMA. *Given any non-square* ν *in* $\gamma = GF(q)$ *with* q *odd, there exist* c, d *in* γ *such that* $c^2 + d^2 = \nu$.

Proof. If $c^2 + d^2 = \nu$, and μ is any other non-square, there exists b in γ such that $\mu = \nu b^2$; then $(bc)^2 + (bd)^2 = \mu$. So, if the lemma is true for one non-square, it is true for all. Suppose the lemma is false; then $1 + x^2$ is a square for each x in γ. Let N be the number of solutions of $1 + x^2 = y^2$. Then, when -1 is a non-square, $1 + x^2$ is never zero and $N = 2q$: when -1 is a square, $N = 2(q-2) + 2 = 2q - 2$. However, if $1 + x^2 = y^2$, then $1 = y^2 - x^2 = (y - x)(y + x)$. So $y - x = e$, $y + x = 1/e$. For each non-zero e, there is a unique solution. Hence $N = q - 1$. Thus we have a contradiction. □

5.1.5. THEOREM. *A non-singular Hermitian variety in $PG(n,q)$ with q square has the canonical form*

$$\mathcal{U}_n = V(\bar{x}_0 x_0 + \ldots + \bar{x}_n x_n)$$

$$= V(x_0^{\sqrt{q}+1} + \ldots + x_n^{\sqrt{q}+1}).$$

Proof. From lemma 5.1.2, corollary 1, a non-singular Hermitian form F can be reduced to $F = a_0 \bar{x}_0 x_0 + \ldots + a_n \bar{x}_n x_n$, where $a_i = \bar{a}_i = a_i^{\sqrt{q}}$. So each a_i is in the subfield $GF(\sqrt{q})$. Therefore, for each a_i, there exists b_i with $b_i \bar{b}_i = a_i$. If we now put $y_i = b_i x_i$, then

$$F = \bar{y}_0 y_0 + \ldots\ldots + \bar{y}_n y_n = y_0^{\sqrt{q}+1} + \ldots\ldots + y_n^{\sqrt{q}+1}. \square$$

5.1.6. THEOREM. *A non-singular quadric in $PG(n,q)$ with q odd has one of the following canonical forms:*

(i) *n even,* $\mathcal{Q}_n = V(x_0^2 + \ldots + x_n^2)$;

(ii) *n odd,*

(a) $\mathcal{Q}_n = V(x_0^2 + \ldots + x_n^2)$

or (b) $\mathcal{Q}'_n = V(x_0^2 + \ldots + x_{n-1}^2 + \nu x_n^2)$,

where ν is any non-square.

Proof. From lemma 5.1.2, corollary 1, a non-singular quadratic form F can be reduced to $F = a_0 x_0^2 + \ldots + a_n x_n^2$. Suppose $a_0, \ldots, a_s$ are squares $b_0^2, \ldots, b_s^2$ and $a_{s+1}, \ldots, a_n$ are non-squares $\nu b_{s+1}^2, \ldots, \nu b_n^2$, where ν is any particular non-square. Then, putting $y_i = b_i x_i$ for each i, we have $F = F_s$ where

$$F_s = y_0^2 + \ldots + y_s^2 + \nu\,(y_{s+1}^2 + \ldots + y_n^2).$$

However, by lemma 5.1.4, there exist c, d in γ such that $c^2 + d^2 = \nu$. Then, if z_i, z_j are given by

$$z_i = c\, y_i - d\, y_j$$

and

$$z_j = d\, y_i + c\, y_j,$$

$$z_i^2 + z_j^2 = (c^2 + d^2)(y_i^2 + y_j^2) = \nu(y_i^2 + y_j^2).$$

So F_s can be transformed into F_{s+2}. Hence there are two canonical forms F_{n-1} and F_n. But for n even, $F_{n-1} = y_0^2 + \ldots + y_{n-1}^2 + \nu y_n^2$ and can be transformed into $\nu(y_0^2 + \ldots + y_n^2)$. So, for n even, we have the single canonical form $\mathcal{Q}_n = \mathbf{V}(F_n)$ and, for n odd, the two forms $\mathcal{Q}_n = \mathbf{V}(F_n)$ and $\mathcal{Q}'_n = \mathbf{V}(F_{n-1})$. □

There do exist other useful canonical forms for quadrics when the characteristic is odd. But firstly, we will find canonical forms when the characteristic is two.

5.1.7. THEOREM. *A non-singular quadric $\mathcal{Q}_n$ in $PG(n,q)$ with q even has one of the following forms:*

(i) *n even,* $\mathcal{Q}_n = \mathbf{V}(x_0{}^2 + x_1x_2 + x_3x_4 + \ldots + x_{n-1}x_n)$;

(ii) *n odd,* (a) $\mathcal{Q}_n = \mathbf{V}(x_0x_1 + x_2x_3 + \ldots + x_{n-1}x_n)$

or (b) $\mathcal{Q}_n = \mathbf{V}(dx_0{}^2 + x_1{}^2 + x_0x_1 + x_2x_3 + \ldots + x_{n-1}x_n)$,

where d is any particular element in $\mathscr{C}_1$.

Proof. Let Q be the non-degenerate quadratic form over $GF(2^h)$ given by

$$Q = \sum_{i,j=0}^{n}{}' a_{ij}\, x_i\, x_j \,.$$

Suppose $n \geq 2$. Firstly, not all a_{ij} with $i \neq j$ are zero, since otherwise

$$Q = (\surd a_{00}\, x_0 + \ldots + \surd a_{nn}\, x_n)^2.$$

Let us now transform Q so that $a_{00} = 0$. Suppose $a_{00} \neq 0$. Let $a_{12} \neq 0$ and further suppose $a_{11} \neq 0$; for, if $a_{11} = 0$, the transformation

$$y_0 = x_1,\ y_1 = x_0,\ y_i = x_i \qquad \text{for} \quad i > 1$$

will produce a form with no term in y_0^2. So

$$Q = a_{11}\, x_1^2 + x_1(a_{01}x_0 + a_{12}x_2 + \ldots + a_{1n}\, x_n) + G(x_0, x_2, \ldots, x_n).$$

Now, put $y_2 = a_{01}\, x_0 + a_{12}\, x_2 + \ldots + a_{1n}\, x_n$, $y_i = x_i$ for $i \neq 2$.

Then

$$Q = a_{11}\, y_1^2 + y_1 y_2 + G(y_0, y_2, \ldots, y_n), \text{ where } G(y_0, 0, \ldots 0) = b.$$

Put $z_1 = y_1 + ty_0$, $z_i = y_i$ for $i \neq 1$. Then $Q = cz_0^2 + \ldots$, where $c = a_{11}t^2 + b$: so t may be chosen so that $c = 0$. Therefore, in the original form of Q, suppose that $a_{00} = 0$. Since not all a_{0j} are zero, suppose $a_{01} \neq 0$. Put $y_1 = a_{01}x_1 + a_{02}x_2 + \ldots + a_{0n}x_n$, $y_i = x_i$ for $i \neq 1$. Then

$$Q = y_0 y_1 + \sum_{i=1}^{n}{}' \, b_{ij} y_i y_j .$$

Now put

$$x_0 = y_0 + b_{11}y_1 + b_{12}y_2 + \ldots + b_{1n}y_n, \quad x_i = y_i \text{ for } i \neq 0.$$

Then

$$Q = x_0 x_1 + \sum_{i=1}^{n}{}' \, b_{ij} x_i x_j .$$

This process can now be continued to obtain

$$Q = x_0 x_1 + x_2 x_3 + \ldots + x_{n-2}x_{n-1} + x_n^2 \text{ for } n \text{ even}$$

and

$$Q = x_0 x_1 + x_2 x_3 + \ldots + x_{n-3}x_{n-2} + f(x_{n-1}, x_n) \text{ for } n \text{ odd},$$

where f is a quadratic form. Either f is the product of two non-proportional linear factors or is irreducible. In the former case, it may be transformed to $x_{n-1}x_n$: in the latter case, to $x_{n-1}x_n + x_{n-1}^2 + dx_n^2$, where $d \in \mathscr{C}_1$, §1.4. Now, put $y_i = x_{n-i}$, all i.□

The canonical form for $\mathscr{U}_n$ enables us to calculate the number of points $\mu(n,q)$ on it.

5.1.8. THEOREM. $\mu(n,q) = [q^{(n+1)/2} + (-1)^n][q^{n/2} - (-1)^n]/(q-1)$.

Proof. In $GF(q)$ with q square, the equation $x^{\sqrt{q}+1} = c$

with c in $GF(\sqrt{q})$ has $\sqrt{q}+1$ solutions, §1.5(v). Therefore since ${}_n\mathcal{U}_1 = \mathbf{V}(x_0^{\sqrt{q}+1} + x_1^{\sqrt{q}+1})$, so $\mu(1,q) = \sqrt{q}+1$. Let $H_n = \sum_0^n \bar{x}_i x_i$: then $H_n = H_{n-1} + \bar{x}_n x_n$. So $|\mathcal{U}_n \cap \mathbf{u}_n| = \mu(n-1, q)$ and $|\mathcal{U}_n \backslash \mathbf{u}_n| = [\theta(n-1, q) - \mu(n-1, q)](\sqrt{q}+1)$. Hence

$$\mu(n,q) = \mu(n-1,q) + [\theta(n-1,q) - \mu(n-1,q)](\sqrt{q}+1)$$

$$= (\sqrt{q}+1)\theta(n-1,q) - \sqrt{q}\,\mu(n-1,q). \tag{5.1}$$

Applying (5.1) for $n = n, n-1, \ldots, n-r+1$ gives

$$\mu(n,q) = (\sqrt{q}+1)[\theta(n-1,q) - \sqrt{q}\theta(n-2,q) \ldots + $$
$$+ (-1)^{r-1}q^{(r-1)/2}\theta(n-r,q)] + (-1)^r q^{r/2}\mu(n-r,q).$$

So, for $r = n-1$,

$$\mu(n,q) = (\sqrt{q}+1)[\theta(n-1,q) - \sqrt{q}\theta(n-2,q) \ldots +$$
$$+ (-1)^n q^{(n-2)/2}\theta(1,q)] + (-1)^{n-1}q^{n/2}(\sqrt{q}+1)$$
$$= \frac{\sqrt{q}+1}{q-1}\begin{bmatrix} q^n - q^{(2n-1)/2} + \ldots + (-1)^{n-1}q^{(n+2)/2} \\ -1 + q^{1/2} + \ldots + (-1)^{n-1}q^{(n-2)/2} \end{bmatrix}$$
$$+ (-1)^{n-1}q^{n/2}(\sqrt{q}+1)$$
$$= (\sqrt{q}-1)^{-1}[q^n - q^{(2n-1)/2} + \ldots + (-1)^{n-1}q^{(n+1)/2}$$
$$-1 + q^{1/2} - \ldots\ldots\ (-1)^n q^{(n-1)/2}]$$
$$= (\sqrt{q}-1)^{-1}[(-1)^{n-1}q^{(n+1)/2} - 1][(-\sqrt{q})^n - 1]/$$
$$(-\sqrt{q}-1)$$
$$= [q^{(n+1)/2} + (-1)^n][q^{n/2} - (-1)^n]/(q-1). \square$$

COROLLARY. $\mu(2,q) = q\sqrt{q}+1$, $\mu(3,q) = (q+1)(q\sqrt{q}+1)$. □

5.2. Projective classification of quadrics

We require canonical forms for the quadrics which are similar whether or not the characteristic of the field is two.

5.2.1. LEMMA. *Let* $F = x_0^2 + x_1^2 + \ldots + x_{2s+1}^2$ *be a quadratic form over* $GF(q)$ *with* q *odd.*

(i) *If* $q \equiv 1 \pmod 4$, *F can be transformed to*

$$F = x_0x_1 + x_2x_3 + \ldots + x_{2s}x_{2s+1}.$$

(ii) *If* $q \equiv -1 \pmod 4$, *F can be transformed to*

(a) *if s is odd,*

$$F = x_0x_1 + x_2x_3 + \ldots + x_{2s}x_{2s+1};$$

(b) *if s is even,*

$$F = x_0^2 + x_1^2 + x_2x_3 + \ldots + x_{2s}x_{2s+1}.$$

Proof. (i) By §1.5(ix), $q \equiv 1 \pmod 4$ means that there exists t in γ with $t^2 = -1$. Then put $y_{2i} = x_{2i} + tx_{2i+1}$, $y_{2i+1} = x_{2i} - tx_{2i+1}$ for i in $\overline{\mathbf{N}}_s$. Hence $F = y_0y_1 + y_2y_3 + \ldots + y_{2s}y_{2s+1}$.

(ii) Since $q \equiv -1 \pmod 4$, -1 is a non-square in γ, by §1.5(x). Hence by lemma 5.1.4, there exist c, d in γ such that $c^2 + d^2 = -1$. Then, if $x_{2i-1} = cy_{2i-1} + dy_{2i+1}$, $x_{2i+1} = dy_{2i-1} - cy_{2i+1}$, we have $x_{2i-1}^2 + x_{2i+1}^2 = (c^2+d^2)(y_{2i-1}^2+y_{2i+1}^2) = -(y_{2i-1}^2+y_{2i+1}^2)$. Hence, if s is odd and so the number of terms in F is a multiple of four, F becomes

$$F = y_0^2 - y_1^2 + y_2^2 - y_3^2 + \ldots + y_{2s}^2 - y_{2s+1}^2.$$

Then, putting $z_{2i} = y_{2i} + y_{2i+1}$, $z_{2i+1} = y_{2i} - y_{2i+1}$, we have

$$F = z_0z_1 + z_2z_3 + \ldots + z_{2s}z_{2s+1}.$$

In exactly the same way, if s is even and so the number of terms in F is two more than a multiple of four, F may be transformed to

$$F = z_0^2 + z_1^2 + z_2z_3 + z_4z_5 + \ldots + z_{2s}z_{2s+1}. \square$$

5.2.2. LEMMA. *Let* $F = x_0^2 + x_1^2 + \ldots + x_{2s}^2 + \nu x_{2s+1}^2$ *be a quadratic form over* $GF(q)$ *with* q *odd and* ν *a non-square.*

(i) *If* $q \equiv 1 \pmod 4$, F *can be transformed to*

$$F = \nu x_0^2 + x_1^2 + x_2x_3 + \ldots + x_{2s}x_{2s+1}.$$

(ii) *If* $q \equiv -1 \pmod 4$, F *can be transformed to*

(a) *if* s *is odd,*

$$F = x_0^2 + x_1^2 + x_2x_3 + \ldots + x_{2s}x_{2s+1};$$

(b) *if* s *is even,*

$$F = x_0x_1 + x_2x_3 + \ldots + x_{2s}x_{2s+1}.$$

Proof. (i) By suitable transposition of variables,

$$F = \nu x_0^2 + x_1^2 + \ldots + x_{2s+1}^2.$$

s in lemma 5.2.1 (i), $x_2^2 + \ldots + x_{2s+1}^2$ can be transformed to $_2x_3 + x_4x_5 + \ldots + x_{2s}x_{2s+1}$. So F can be transformed to

$$F = x_0^2 + x_1^2 + x_2x_3 + x_4x_5 + \ldots + x_{2s}x_{2s+1}.$$

(ii) If m is a multiple of four, then by emma 5.2.1(ii)(a), $\sum_1^m y_i^2$ can be transformed to $y_1y_2 + y_3y_4 + \ldots + y_{m-1}y_m$. So, for (a), we must examine $\nu x_0^2 + x_1^2 + x_2^2 + x_3^2$ nd, for (b), $\nu x_0^2 + x_1^2$. Firstly, since $q \equiv -1 \pmod 4$, -1 is non-square. So there exists b in γ with $\nu = -b^2$. Hence $x_0^2 + x_1^2 = -b^2x_0^2 + x_1^2 = (x_1 + bx_0)(x_1 + bx_0)$, which can be ransformed to x_0x_1. This completes (b). Similarly, $x_0^2 + x_1^2 + x_2^2 + x_3^2$ can be transformed to $x_0x_1 + x_2^2 + x_3^2$, which an in turn be transformed to $x_0^2 + x_1^2 + x_2x_3$. This completes a). $\square$

5.2.3. LEMMA. *Let* $F = x_0^2 + \ldots + x_{2s}^2$ *be a quadratic form over* $GF(q)$ *with* q *odd. Then* $\mathbf{V}(F)$ *is projectively equivalent to*

$$\mathbf{V}(x_0^2 + x_1x_2 + \ldots + x_{2s-1}x_{2s}).$$

Proof. If s is even, then $2s$ is a multiple of four and so, by lemma 5.2.1 parts (i) and (ii) (a), $x_1^2 + \ldots + x_{2s}^2$ can be transformed to $x_1x_2 + \ldots + x_{2s-1}x_{2s}$, giving the required result.

If s is odd, the number of terms in F is three more than a multiple of four. So $\mathbf{V}(F)$

$$\sim \mathbf{V}(x_0^2 + x_1^2 + x_2^2 + x_3x_4 + \ldots + x_{2s-1}x_{2s})$$

$$\sim \mathbf{V}(x_0^2 - x_1^2 - x_2^2 + x_3x_4 + \ldots + x_{2s-1}x_{2s})$$

$$\sim \mathbf{V}(x_0x_1 - x_2^2 + x_3x_4 + \ldots + x_{2s-1}x_{2s})$$

$$\sim \mathbf{V}(-x_0^2 + x_1x_2 + x_3x_4 + \ldots + x_{2s-1}x_{2s})$$

$$\sim \mathbf{V}(x_0^2 + x_1x_2 + x_3x_4 + \ldots + x_{2s-1}x_{2s}).\square$$

From the three lemmas of the section and theorem 5.1.7, the projective classification of non-singular quadrics is given by the following.

5.2.4. THEOREM. *In* $PG(n,q)$, *the number of projectively distinct non-singular quadrics is one or two as* n *is even or odd. They have the following canonical forms.*

(i) $n = 2s$, $s \geqslant 0$:

$$\mathscr{P}_{2s} = \mathbf{V}(x_0^2 + x_1x_2 + x_3x_4 + \ldots + x_{2s-1}x_{2s});$$

(ii) $n = 2s - 1$, $s \geqslant 1$:

$$\mathscr{H}_{2s-1} = \mathbf{V}(x_0x_1 + x_2x_3 + \ldots + x_{2s-2}x_{2s-1}),$$

$$\mathscr{E}_{2s-1} = \mathbf{V}(f(x_0,x_1) + x_2x_3 + x_4x_5 + \ldots + x_{2s-2}x_{2s-1}),$$

where f is an irreducible binary quadratic form. □

$\mathscr{P}_{2s}$, $\mathscr{H}_{2s-1}$, $\mathscr{E}_{2s-1}$ are respectively called *parabolic, hyperbolic, elliptic.* The form f can be chosen as $f(x_0,x_1) = dx_0^2 + x_0x_1 + x_1^2$, where d is category one for q even and $1-4d$ is a non-square for q odd. In particular, $\mathscr{P}_2$ is a *conic.*

The maximum dimension $g = g(\mathscr{Q})$ of subspaces lying on a quadric $\mathscr{Q}$ is called the *projective index* of $\mathscr{Q}$.

COROLLARY 1. *The projective index has the following values*

$\mathscr{Q}$	$\mathscr{P}_{2s}$	$\mathscr{H}_{2s-1}$	$\mathscr{E}_{2s-1}$
g	$s-1$	$s-2$	$s-2$

Proof. A suitable subspace in each case is the following.

$$\mathscr{P}_{2s} : \mathbf{V}(x_0, x_2, x_4, \ldots, x_{2s});$$

$$\mathscr{H}_{2s-1} : \mathbf{V}(x_0, x_2, x_4, \ldots, x_{2s-2});$$

$$\mathscr{E}_{2s-1} : \mathbf{V}(x_0, x_1, x_3, \ldots, x_{2s-1}). \square$$

The *character* $w = w(\mathscr{Q})$ of a non-singular quadric in $PG(n,q)$ is

$$w = 2g - n + 3.$$

COROLLARY 2.

$\mathscr{Q}$	$\mathscr{P}_{2s}$	$\mathscr{H}_{2s-1}$	$\mathscr{E}_{2s-1}$
w	1	2	0

□

These values of w are the motivation for the names of the quadrics. In Volume II, the character of singular quadrics is also defined.

COROLLARY 3. *If* $\mathscr{Q}=\mathbf{V}(a_0x_0^2 + a_1x_1^2 + \ldots + a_{2s-1}x_{2s-1}^2)$ *is a* *non-singular quadric in* $PG(2s-1,q)$, *with* q *odd, then its type is given by Table* 5.1, *where* m *is the number of squares (or, equally well, non-squares) among the* a_i, $q \equiv e \pmod 4$, *and* E *or* H *indicates elliptic or hyperbolic respectively.*

TABLE 5.1

s	odd	odd	odd	odd	even	even	even	even
m	even	even	odd	odd	even	even	odd	odd
e	1	-1	1	-1	1	-1	1	-1
$\mathscr{Q}$	H	E	E	H	H	H	E	E

Proof. This follows from lemmas 5.2.1 and 5.2.2.□

Note. Let $f = (e+1)/2$ so that $f = 1$ when $e = 1$, and $f = 0$ when $e = -1$. Then, by a little Boolean algebra, we obtain the function $B = s(f+1) + m + 1$. Then, we have the result:

B	odd	even
$\mathscr{Q}$	H	E

In fact, if s is even, then $B \equiv m + 1 \pmod 2$. So, if s is even, $\mathscr{Q}$ is hyperbolic or elliptic as m is even or odd.

The canonical forms for quadrics enable us to readily determine the number of points on a non-singular quadric in $PG(n,q)$. Let this number be $\psi(n) = \psi(n,q)$ and, in particular write $\psi(2s,q)$, $\psi_+(2s-1,q)$, $\psi_-(2s-1,q)$ for the respective numbers of points on $\mathscr{P}_{2s}$, $\mathscr{H}_{2s-1}$, $\mathscr{E}_{2s-1}$.

5.2.5. LEMMA. (i) $\psi(0,q) = 0$; (ii) $\psi_+(1,q) = 2$; (iii) $\psi_-(1,q) = 0$; (iv) $\psi(2,q) = q + 1$.

Proof. (i) $\mathscr{P}_0 = \mathbf{V}(x_0{}^2)$. But $PG(0,q) = \{\mathbf{P}(1)\}$. So $\psi(0,q) = 0$.

(ii) $\mathscr{H}_1 = \mathbf{V}(x_0x_1) = \{\mathbf{U}_0, \mathbf{U}_1\}$. So $\psi_+(1,q) = 2$.

(iii) $\mathscr{E}_1 = \mathbf{V}(f(x_0,x_1))$, where f is irreducible. So $\psi_-(1,q) = 0$.

(iv) $\mathscr{P}_2 = \mathbf{V}(x_0{}^2 + x_1x_2) = \{\mathbf{P}(t, -t^2, 1) \mid t \in \gamma\} \cup \{\mathbf{U}_1\}$. So $\psi(2,q) = q + 1$.□

5.2.6. THEOREM. (i) $\psi(2s,q) = (q^{2s} - 1)/(q - 1)$;

(ii) $\psi_+(2s-1,q) = (q^{s-1} + 1)(q^s - 1)/(q - 1)$;

(iii) $\psi_-(2s-1,q) = (q^s + 1)(q^{s-1} - 1)/(q - 1)$.

Proof. A non-singular quadric $\mathscr{Q}_n$ in $PG(n,q)$ can be written $\mathscr{Q}_n = \mathbf{V}(F_n)$, where, for $n \geqslant 2$,

$$F_n = F_{n-2} + x_{n-1}x_n ,$$

theorem 5.2.4. So, a point of $\mathscr{Q}_n$ lies in the intersection of $\mathscr{Q}_{n-2}$ with one of the primes $\mathbf{u}_{n-1}$ or $\mathbf{u}_n$, or it does not. There are

$$2\{q\psi(n-2) + 1\} - \psi(n-2) = (2q - 1)\psi(n-2) + 2$$

of the former type, and

$$(q-1)\ \{\theta(n-2) - \psi(n-2)\} = q^{n-1} - 1 - (q-1)\psi(n-2)$$

of the latter. Hence

$$\psi(n) = q^{n-1} + 1 + q\,\psi\,(n-2). \tag{5.2}$$

Applying (5.2) for $n = n, n-2, \ldots, n-2r+2$ gives

$$\psi(n) = (q^{n-1} + q^{n-2} + \ldots + q^{n-r}) + (q^{r-1} + q^{r-2} + \ldots + 1) + $$
$$+ q^r\psi(n-2r). \tag{5.3}$$

Now we use (5.3) in the three cases.

(i) $n = 2s$, $r = s$, $\psi(0) = 0$:

$$\psi(2s) = (q^{2s-1} + \ldots + q^s) + (q^{s-1} + \ldots + 1) + 0$$
$$= (q^{2s} - 1)/(q - 1).$$

(ii) $n = 2s - 1$, $r = s - 1$, $\psi_+(1) = 2$;

$$\psi_+(2s - 1) = (q^{2s-2} + \ldots + q^s) + (q^{s-2} + \ldots + 1) + 2q^{s-1}$$
$$= (q^{2s-2} + \ldots + q^{s-1}) + (q^{s-1} + \ldots + 1)$$
$$= (q^{s-1} + 1)(q^s - 1)/(q - 1).$$

(iii) $n = 2s - 1$, $r = s - 1$, $\psi_-(1) = 0$:

$$\psi_-(2s - 1) = (q^{2s-2} + \ldots + q^s) + (q^{s-2} + \ldots + 1)$$
$$= (q^s + 1)(q^{s-1} - 1)/(q - 1).\square$$

COROLLARY. *Let the non-singular quadric* $\mathcal{Q}_n$ *in* $PG(n,q)$ *have character* w. *Then*

$$\psi(n,q) = (q^n - 1)/(q - 1) + (w - 1)\, q^{(n-1)/2}.\square$$

Further properties of quadrics and Hermitian varieties will be investigated in later chapters and in Volume II. For now, we name the groups of projectivities leaving the varieties fixed.

$$\mathcal{U}_n \quad : \quad PGU(n+1,\, q)$$
$$\mathcal{P}_{2s} \quad : \quad PGO(2s+1, q)$$
$$\mathcal{H}_{2s-1} \quad : \quad PGO_+(2s,\, q)$$
$$\mathcal{E}_{2s-1} \quad : \quad PGO_-(2s,\, q).$$

For the orders of these groups, see Appendix I.

5.3. Polarities

In the cases that the self-conjugate points of a polarity $\mathfrak{T}$ form a quadric or Hermitian variety, the canonical form for $\mathfrak{T}$ may be taken from the canonical form for the variety, since if the variety is $V(\tilde{X}TX^*)$, then $\mathbf{P}(X)$ and $\mathbf{P}(Y)$ are conjugate in the corresponding polarity if $\tilde{X}TY^* = 0$, §2.1(v). The polar prime $\mathbf{V}(\tilde{X}TY^*)$ of a point $\mathbf{P}(Y)$ on the variety is the tangent prime at $\mathbf{P}(Y)$.

5.3.1. THEOREM. *The canonical forms for ordinary polarities in* $PG(n,q)$ *with* q *odd and Hermitian polarities with* q *square are determined by the variety* $\mathcal{V}$ *of self-conjugate points.*

If $\mathbf{P}(X)$ *and* $\mathbf{P}(Y)$ *are conjugate, they are connected by the following equations:*

(i) $\mathscr{V} = \mathscr{U}_n$: $x_0{}^q y_0 + \ldots + x_n{}^q y_n = 0$;

(ii) $\mathscr{V} = \mathscr{P}_{2s}$: $2x_0y_0 + (x_1y_2 + x_2y_1) + (x_3y_4 + x_4y_3) + \ldots +$
$+ (x_{2s-1}y_{2s} + x_{2s}y_{2s-1}) = 0$;

(iii) $\mathscr{V} = \mathscr{H}_{2s-1}$: $(x_0y_1 + x_1y_0) + (x_2y_3 + x_3y_2) + \ldots +$
$+ (x_{2s-2}y_{2s-1} + x_{2s-1}y_{2s-2}) = 0$;

(iv) $\mathscr{V} = \mathscr{E}_{2s-1}$: $2tx_0y_0 + 2x_1y_1 + (x_2y_3 + x_3y_2) + \ldots +$
$+ (x_{2s-2}y_{2s-1} + x_{2s-1}y_{2s-2}) = 0$,

where $f(x_0, x_1) = tx_0^2 + x_1^2$ is irreducible.□

COROLLARY 1. *If* $\mathscr{V} = \mathbf{V}(a_0x_0{}^2 + \ldots + a_nx_n{}^2)$ *in* $PG(n,q)$ *with* q *odd, and* $\Pi a_i \neq 0$*, then* $\mathbf{P}(X)$ *and* $\mathbf{P}(Y)$ *are conjugate if*

$$a_0x_0y_0 + \ldots + a_nx_ny_n = 0.\square$$

5.3.2. THEOREM. *A null polarity in* $PG(n,q)$ *with* n *odd has the canonical bilinear form* F*, where*

$$F = (x_0y_1 - x_1y_0) + (x_2y_3 - x_3y_2) + \ldots + (x_{n-1}y_n - x_ny_{n-1}).$$

Proof. From §2.1(v), $F = \Sigma_0^n{}'' t_{ij}(x_iy_j - x_jy_i) = x_0(t_{01}y_1 + \ldots + t_{0n}y_n) - y_0(t_{01}x_1 + \ldots + t_{0n}x_n) + \Sigma_1^n{}'' t_{ij}(x_iy_j - x_jy_i)$.
Suppose $t_{01} \neq 0$. Then put $x_1' = t_{01}x_1 + \ldots + t_{0n}x_n$, $y_1' = t_{01}y_1 + \ldots + t_{0n}y_n$ and, for $i \neq 1$, $x_i' = x_i$, $y_i' = y_i$. Then

$$\begin{aligned} F &= x_0'y_1' - x_1'y_0' + \Sigma_1^n{}'' b_{ij}(x_i'y_j' - x_j'y_i') \\ &= (x_0' - b_{12}x_2' \ldots - b_{1n}x_n')y_1' - x_1'(y_0' - b_{12}y_2' \ldots - b_{1n}y_n') \\ &\quad + \Sigma_2^n{}'' b_{ij}(x_i'y_j' - x_j'y_i'). \end{aligned}$$

Now, put $x_0 = x_0' - b_{12}x_2' \ldots - b_{1n}x_n'$, $y_0 = y_0' - b_{12}y_2' \ldots - b_{1n}y_n'$ and, for $i > 0$, $x_i = x_i'$, $y_i = y_i'$. Then

$$F = x_0y_1 - x_1y_0 + \Sigma_2^n{}'' b_{ij}(x_iy_j - x_jy_i).$$

Continuing this process, we finally have

$$F = (x_0y_1 - x_1y_0) + (x_2y_3 - x_3y_2) + \ldots + (x_{n-1}y_n - x_ny_{n-1}).\square$$

5.3.3. THEOREM. *A pseudo polarity in $PG(n,q)$ with q even has the canonical bilinear forms F as follows.*

(i) $n = 2s : F = x_0y_0 + (x_1y_2 + x_2y_1) + (x_3y_4 + x_4y_3) + \ldots + (x_{2s-1}y_{2s} + x_{2s}y_{2s-1})$;

(ii) $n = 2s-1 : F = x_0y_0 + (x_0y_1 + x_1y_0) + (x_2y_3 + x_3y_2) + \ldots + (x_{2s-2}y_{2s-1} + x_{2s-1}y_{2s-2})$.

Proof. A pseudo polarity in $PG(n,q)$ with q even has bilinear form.

$$F = \sum_0^n t_{ii}x_iy_i + \sum_0^n{}'' t_{ij}(x_iy_j + x_jy_i).$$

Suppose $t_{00} \neq 0$. Then, put $x_0' = \sqrt{t_{00}}x_0 + \ldots + \sqrt{t_{nn}}x_n$, $y_0' = \sqrt{t_{00}}y_0 + \ldots + \sqrt{t_{nn}}y_n$ and, for $i > 0$, $x_i' = x_i$, $y_i' = y_i$. Then,

$$F = x_0'y_0' + \sum_0^n{}'' a_{ij}(x_i'y_j' + x_j'y_i').$$

Let $G = \sum_1^n{}'' a_{ij}(x_iy_j + x_jy_i)$. Then, if n is even, G determines a null polarity in $PG(n-1,q)$ and can be transformed as in theorem 5.3.2 to

$$G = (x_1y_2 + x_2y_1) + \ldots + (x_{n-1}y_n + x_ny_{n-1}).$$

So
$$\begin{aligned} F &= x_0y_0 + a_{01}(x_0y_1 + x_1y_0) + \ldots + a_{0n}(x_0y_n + x_ny_0) + G \\ &= x_0y_0 + a_{01}(x_0y_1 + x_1y_0) + a_{02}(x_0y_2 + x_2y_0) + (x_1y_2 + x_2y_1) + \ldots \\ &= x_0y_0 + (x_1 + a_{02}x_0)(y_2 + a_{01}y_0) + (x_2 + a_{01}x_0)(y_1 + a_{02}y_0) + \ldots . \end{aligned}$$

Put $x_0' = x_0$, $y_0' = y_0$, $x_{2i-1}' = x_{2i-1} + a_{0,2i}x_0$, $x_{2i}' = x_{2i} + a_{0,2i-1}x_0$ and similarly for y_{2i-1}', y_{2i}'. Then, omitting the dashes, for $n = 2s$, we have

$$F = x_0y_0 + (x_1y_2 + x_2y_1) + (x_3y_4 + x_4y_3) + \ldots + (x_{2s-1}y_{2s} + x_{2s}y_{2s-1}).$$

For $n = 2s - 1$, we obtain similarly

$$F = x_0y_0 + (x_2y_3 + x_3y_2) + (x_4y_5 + x_5y_4) + \ldots + (x_{2s-2}y_{2s-1} + x_{2s-1}y_{2s-2}) + \sum''_{j \neq 1} a_{1j}(x_1y_j + x_jy_1).$$

We note that $a_{10} \neq 0$, for otherwise the polarity would have a singular matrix.

$$\begin{aligned} F &= x_0y_0 + a_{10}(x_0y_1 + x_1y_0) + (x_2y_3 + x_3y_2) + a_{12}(x_1y_2 + x_2y_1) + \\ &\quad + a_{13}(x_1y_3 + x_3y_1) + \ldots \\ &= x_0y_0 + a_{10}(x_0y_1 + x_1y_0) + (x_2 + a_{13}x_1)(y_3 + a_{12}y_1) + \\ &\quad + (x_3 + a_{12}x_1)(y_2 + a_{13}y_1) + \ldots \quad . \end{aligned}$$

So put $x_1' = a_{10}x_1$, $x_{2i}' = x_{2i} + a_{1,2i+1}x_1$, $x_{2i+1}' = x_{2i+1} + a_{1,2i}x_1$ and similarly for y_1', y_{2i}', y_{2i+1}'. Then, omitting the dashes, for $n = 2s - 1$ we have

$$F = x_0y_0 + (x_0y_1 + x_1y_0) + (x_2y_3 + x_3y_2) + \ldots + (x_{2s-2}y_{2s-1} + x_{2s-1}y_{2s-2}). \square$$

For q odd, the groups of projectivities fixing (that is, commuting with) the various polarities other than the null polarity are just the groups of the corresponding varieties of self-conjugate points. This also applies for q even and a Hermitian polarity.

For any q, the group of projectivities in $PG(n,q)$ fixing a null polarity is called the *projective symplectic group* and denoted $PGSp(n+1, q)$.

For q even, the group of projectivities fixing a pseudo polarity in $PG(n,q)$ is called the *projective pseudo symplectic group* and is denoted $PGPs(2s+1, q)$ for $n = 2s$ and $PGPs^*(2s, q)$ for $n = 2s - 1$. See Appendix I for the orders of these groups.

It remains to comment on the polarities derived from non-singular quadrics when q is even.

5.3.4. THEOREM. *In $PG(n,q)$ with q even, a quadric $\mathscr{Q}$ defines a null polarity if n is odd but no polarity if n is even.*

Proof. Suppose $\mathscr{Q} = \mathbf{V}(F)$, where F is a quadratic form. Then, as in (2.11) and (2.12),

$$F(X + tY) = F(X) + tG(X,Y) + t^2F(Y). \tag{5.4}$$

If $F(X) = \Sigma a_{ii}x_i + \Sigma'' a_{ij}x_ix_j$, then G is a bilinear form:

$$\begin{aligned} G(X,Y) &= 2\Sigma\, a_{ii}x_iy_i + \Sigma''\, a_{ij}(x_iy_j + x_jy_i) \\ &= \Sigma''\, a_{ij}(x_iy_j + x_jy_i), \end{aligned} \tag{5.5}$$

since q is even.

When $\mathscr{Q} = \mathscr{P}_{2s}$,

$$G(X,Y) = (x_1y_2 + x_2y_1) + (x_3y_4 + x_4y_3) + \ldots + (x_{2s-1}y_{2s} + y_{2s}y_{2s-1}). \tag{5.6}$$

As there are no terms in x_0 or y_0, the matrix of the form is singular and G does not define a polarity.

When $\mathscr{Q} = \mathscr{H}_{2s-1}$ or $\mathscr{E}_{2s-1}$, then

$$G(X,Y) = (x_0y_1 + x_1y_0) + (x_2y_3 + x_3y_2) + \ldots + (x_{2s-2}y_{2s-1} + x_{2s-1}y_{2s-2}), \tag{5.7}$$

where, for $\mathscr{E}_{2s-1}$ in theorem 5.2.4, $f(x_0, x_1) = dx_0^{\,2} + x_0x_1 + x_1^{\,2}$. So, from theorem 5.3.2, $\mathscr{Q}$ defines a null polarity.□

COROLLARY. *The tangent primes to $\mathscr{P}_{2s}$ all have the point $\mathbf{U}_0$ in common.*

Proof. Since there is no term in x_0 or y_0 in $G(X,Y)$ as in (5.6), $\mathbf{U}_0$ lies in each tangent prime. The tangent prime at $\mathbf{P}(Y)$ is $\mathbf{V}(G(X,Y))$.□

The meet of the tangent primes of a parabolic quadric $\mathscr{Q}$ is called the *nucleus* and denoted $N(\mathscr{Q})$.

The polarity defined by a variety will be denoted by the corresponding German letter:

$$\begin{array}{ll} \mathcal{U}_n & \mathfrak{U}_n \\ \mathcal{P}_{2s} & \mathfrak{P}_{2s} \\ \mathcal{H}_{2s-1} & \mathfrak{H}_{2s-1} \\ \mathcal{E}_{2s-1} & \mathfrak{E}_{2s-1}. \end{array}$$

5.4. Notes and references

§5.1. In the reduction to canonical form of quadratic and Hermitian forms, we follow Dickson [1901], [1926].

§5.2. For quadrics and Hermitian varieties, see Arf [1940], Bose and Chakravarti [1966], Châtelet [1948], Primrose [1951], Ramanujacharyulu [1965], Ray-Chaudhuri [1962a], [1962b], Segre [1959a], [1965a], Witt [1937]. Further references will be given in later chapters in Volume II dealing in more detail with their properties.

§5.2. For pseudo polarities or, equivalently, non-alternating, symmetric, bilinear forms, see Segre [1960a], pp.242-245, Pless [1964], [1965a], [1965b]. In discussions of quadratic and bilinear forms this is a topic often neglected.

PART III

$PG(1, q)$ and $PG(2, q)$

6

$PG(1,q)$

6.1. Harmonic tetrads

The $q+1$ points of $PG(1,q)$ are $\mathbf{P}(x_0,x_1)$, $x_i \in \gamma$. So $PG(1,q) = \{\mathbf{U}_0 = \mathbf{P}(1,0)\} \cup \{\mathbf{P}(x_0,1) \mid x_0 \in \gamma\}$. Each point $\mathbf{P}(x_0,x_1)$ with $x_1 \neq 0$ is determined by the non-homogeneous coordinate x_0/x_1. The coordinate for $\mathbf{U}_0$ is ∞. Then, with $\gamma^+ = \gamma \cup \{\infty\}$, each point of $PG(1,q)$ is represented by a single element in γ^+. Let $\mathbf{P}(\infty,1) = \mathbf{P}(1,0)$. Then $PG(1,q) = \{\mathbf{P}(t,1) \mid t \in \gamma^+\}$.

A projectivity $\mathfrak{T} = \mathbf{M}(T)$ of $PG(1,q)$ is given by $Y = XT$ where $Y = (y_0,y_1)$, $X = (x_0,x_1)$ and $T = \begin{bmatrix} a & b \\ c & d \end{bmatrix}$. Let $s = y_0/y_1$ and $t = x_0/x_1$; then $s = (at + c)/(bt + d)$. (This is a slight disadvantage to writing mappings on the right: if the projectivity was given by $Y^* = TX^*$, then $s = (at + b)/(ct + d)$.) If $Q_i = P_i\mathfrak{T}$ for $i = 2,3,4$, and P_i and Q_i have the respective coordinates t_i and s_i, then $\mathfrak{T}$ is given by

$$\frac{(s - s_3)(s_2 - s_4)}{(s - s_4)(s_2 - s_3)} = \frac{(t - t_3)(t_2 - t_4)}{(t - t_4)(t_2 - t_3)}.$$

So, the *cross-ratio* of four points P_1,P_2,P_3,P_4 with coordinates t_1,t_2,t_3,t_4 is

$$\{P_1,P_2;\ P_3,P_4\} = \{t_1,t_2;\ t_3,t_4\} = \frac{(t_1 - t_3)(t_2 - t_4)}{(t_1 - t_4)(t_2 - t_3)}\ ;\ \text{see §1.11.}$$

A projectivity is determined by the images of three points. Therefore there exists a projectivity $\mathfrak{T}$ such that $Q_i = P_i\mathfrak{T}$ for i in $\mathbf{N}_4$ if and only if the cross-ratios of the two sets of four points in the corresponding order are equal.

We may note that

$$\lambda = \{t_1,t_2;\ t_3,t_4\} = \{t_2,t_1;\ t_4,t_3\} = \{t_3,t_4;\ t_1,t_2\} = \{t_4,t_3;\ t_2,t_1\}. \qquad (6.1)$$

So $\{P_1,P_2;P_3,P_4\}$ is invariant under a projective group of order four isomorphic to $\mathbf{Z}_2 \times \mathbf{Z}_2 = \{1, (P_1P_2)(P_3P_4), (P_1P_3)(P_2P_4), (P_1P_4)(P_2P_3)\}$. Under all permutations of $\{P_1, P_2, P_3, P_4\}$, the cross-ratio takes just the six values

$$\lambda,\ 1/\lambda,\ 1-\lambda,\ 1/(1-\lambda),\ \lambda/(\lambda-1),\ (\lambda-1)/\lambda,$$

as follows:

$$\begin{aligned}
1/\lambda &= \{t_1,t_2;\ t_4,t_3\} = \{t_2,t_1;\ t_3,t_4\} = \{t_4,t_3;\ t_1,t_2\} = \\ &= \{t_3,t_4;\ t_2,t_1\},\\
1-\lambda &= \{t_1,t_3;\ t_2,t_4\} = \{t_3,t_1;\ t_4,t_2\} = \{t_2,t_4;\ t_1,t_3\} = \\ &= \{t_4,t_2;\ t_3,t_1\},\\
1/(1-\lambda) &= \{t_1,t_3;\ t_4,t_2\} = \{t_3,t_1;\ t_2,t_4\} = \{t_4,t_2;\ t_1,t_3\} = \\ &\quad \{t_2,t_4;\ t_3,t_1\},\\
(\lambda-1)/\lambda &= \{t_1,t_4;\ t_2,t_3\} = \{t_4,t_1;\ t_3,t_2\} = \{t_2,t_3;\ t_1,t_4\} = \\ &\quad \{t_3,t_2;\ t_4,t_1\}\\
\lambda/(\lambda-1) &= \{t_1,t_4;\ t_3,t_2\} = \{t_4,t_1;\ t_2,t_3\} = \{t_3,t_2;\ t_1,t_4\} = \\ &\quad \{t_2,t_3;\ t_4,t_1\}.
\end{aligned}$$

Also, $\{t_1,t_2;\ t_3,t_4\}$ takes the values ∞, 0, or 1 if and only if two of the t_i are equal.

If P_1, P_2, P_3, P_4 are distinct points, then P_1 and P_2 *separate* P_3 and P_4 *harmonically*, written $H(P_1,P_2;\ P_3,P_4)$, if

$$\{t_1,t_2;\ t_3,t_4\} = -1. \tag{6.2}$$

So,

$$H(P_1,P_2;\ P_3,P_4) \Leftrightarrow H(P_2,P_1;\ P_3,P_4).$$

In this case, the permutations of the points only give two other values of the cross-ratio, namely 2 and 1/2. Equation (6.2) is equivalent to

$$(t_1 + t_2)(t_3 + t_4) = 2(t_1t_2 + t_3t_4). \tag{6.3}$$

If $t_1 = \infty$, then $2t_2 = t_3 + t_4$. If $t_1 = \infty$ and $t_2 = 0$, then

$t_3 + t_4 = 0$. This means that $\mathbf{P}(X)$ and $\mathbf{P}(Y)$ harmonically separate $\mathbf{P}(tX + sY)$ and $\mathbf{P}(tX - sY)$.

Four points form a *harmonic tetrad* if two of them harmonically separate the other two. A tetrad is harmonic if $\lambda = 1/\lambda$ or $\lambda = \lambda/(\lambda - 1)$ or $\lambda = 1 - \lambda$.

When $p = 2$, there are no harmonic tetrads. When $p = 3$, then $2 = 1/2 = -1$. So every pair of points in a harmonic tetrad harmonically separates the other pair: (6.3) becomes

$$\Sigma'' \, t_i t_j = 0. \tag{6.4}$$

A tetrad is *equianharmonic* if $\lambda = 1/(1 - \lambda)$ or, equivalently, $\lambda = (\lambda - 1)/\lambda$; that is,

$$\lambda^2 - \lambda + 1 = 0. \tag{6.5}$$

So, equianharmonic tetrads exist if and only if $\lambda^3 + 1 = 0$ has three solutions in γ, or $\lambda = -1$ is a solution of (6.5); that is, when $q \equiv 1 \pmod 3$ or $q \equiv 0 \pmod 3$, §1.5(xi), (xiii).

A tetrad is *superharmonic* if it is both harmonic and equianharmonic. Thus superharmonic tetrads exist if and only if $p = 3$.

6.1.1. LEMMA. *On* $PG(1,q)$, $q = p^h$,

(i) *the number of harmonic tetrads and the projective group of each one is as follows:*

$p = 3$	$q(q^2 - 1)/24$	$\mathbf{S}_4$
$p > 3$	$q(q^2 - 1)/8$	$\mathbf{D}_4$;

(ii) *the number of equianharmonic tetrads and the projective group of each one is as follows:*

$p = 3$	$q(q^2 - 1)/24$	$\mathbf{S}_4$
$q \equiv 1 \pmod 3$	$q(q^2 - 1)/12$	$\mathbf{A}_4$.

Proof. If $H(P_1, P_2; P_3, P_4)$ and $\lambda = \{t_1, t_2; t_3, t_4\}$, then $\lambda = 1/\lambda$, whence for $p > 3$ there are eight permutations of $\{P_1, P_2, P_3, P_4\}$ which are projectively equivalent; for $p = 3$, all 24 permutations are equivalent. In the equianharmonic case, $\lambda = 1/(1-\lambda) = (\lambda-1)/\lambda$, whence there are 12 projectively equivalent permutations for $p \neq 3$.□

6.2. Quadric and Hermitian varieties

Let $Q(1,q)$ be the set of quadrics in $PG(1,q)$. Then $Q(1,q) = \mathbf{V}\{a_0x_0^2 + a_{01}x_0x_1 + a_1x_1^2 \mid a_0, a_{01}, a_1 \in \gamma\}$. So $|Q(1,q)| = q^2 + q + 1$. Then, if $PGL(2,q)$ is considered as acting on $Q(1,q)$, there are three orbits:

(i) $q+1$ repeated points $\mathbf{V}((x_0 + tx_1)^2)$, $t \in \gamma^+$;

(ii) $q(q+1)/2$ point pairs $\mathbf{V}((x_0 + sx_1)(x_0 + tx_1))$, $s, t \in \gamma^+$, $s \neq t$; that is hyperbolic quadrics $\mathscr{H}_1$;

(iii) $q(q-1)/2$ elliptic quadrics $\mathscr{E}_1$ containing no points and given by $\mathbf{V}(f(x_0, x_1))$, where f is irreducible.

Hence $|PGO_+(2,q)| = p(2,q)/[q(q+1)/2] = 2(q-1)$; $|PGO_-(2,q)| = p(2,q)/[q(q-1)/2] = 2(q+1)$.

An elliptic quadric may be considered as comprising a pair of 2-complex conjugate points; that is, a pair of points of $PG(1, q^2)$ conjugate over $GF(q)$. So, if $\mathscr{E}_1 = \mathbf{V}(f)$ with $f = a_0x_0^2 + a_{01}x_0x_1 + a_1x_1^2$, then $f = a_0(x_0 + \alpha x_1)(x_0 + \alpha^q x_1)$ where $\alpha \in GF(q^2)\backslash GF(q)$. The q^2+1 points of $PG(1,q^2)$ fall into two sets:

(i) $q+1$ in a subgeometry $PG(1,q)$;

(ii) $q^2 - q$ in $PG(1,q^2)\backslash PG(1,q)$ comprising $q(q-1)/2$ conjugate pairs.

When q is square let $U(1,q)$ be the set of Hermitian varieties in $PG(1,q)$. Then $U(1,q) = \{\mathbf{V}(a_0\bar{x}_0x_0 + b\bar{x}_0x_1 + \bar{b}x_0\bar{x}_1 + a_1\bar{x}_1x_1 \mid a_0, a_1 \in GF(\sqrt{q}),\ b \in GF(q)\}$, where $\bar{x} = x^{\sqrt{q}}$. So $|U(1,q)| = (q+1)(\sqrt{q}+1)$. By lemma 5.1.2, if $\mathscr{F} \in U(1,q)$, then $\mathscr{F} \sim \mathbf{V}(\bar{x}_0x_0 + e\bar{x}_1x_1)$, where $e = 0$ or 1. When $e = 0$, then F comprises the single point $\mathbf{U}_1$. When $e = 1$, $\mathscr{F} = \mathscr{U}_1 = V(\bar{x}_0x_0 + \bar{x}_1x_1)$. By §1.5(v), $x^{\sqrt{q}+1} = -1$ has $\sqrt{q}+1$ solutions in $GF(q)$. So $\mathscr{U}_1$ comprises $\sqrt{q}+1$ points. Thus, under the action of $PGL(2,q)$, the set $U(1,q)$ has two orbits:

(i) $q+1$ points, each counted $\sqrt{q}+1$ times;

(ii) $\sqrt{q}(q+1)$ varieties $\mathscr{U}_1$, each comprising $\sqrt{q}+1$ points.

Hence, $|PGU(2,q)| = p(2,q)/[\sqrt{q}(q+1)] = \sqrt{q}(q-1)$.

It should be noted that in both $Q(1,q)$ and $U(1,q)$, the variety $\mathbf{V}(0)$, that is the whole line $PG(1,q)$, has not been included.

6.2.1. LEMMA. *Each non-singular Hermitian variety $\mathscr{U}_{1,q}$ is a subline $PG(1,\sqrt{q})$ and conversely.*

Proof. By §1.5(v), there exists α in $GF(q)$, q square, such that $\alpha^{\sqrt{q}-1} = -1$, that is $\bar{\alpha} = -\alpha$. By theorem 5.1.5 and lemma 5.1.2, corollary 2,

$$
\begin{aligned}
\mathscr{U}_1 &= V(\bar{x}_0 x_0 + \bar{x}_1 x_1) \\
&\sim V(\alpha\bar{x}_0 x_1 + \bar{\alpha} x_0 \bar{x}_1) \\
&\doteq V(x_0 x_1 (x_0^{\sqrt{q}-1} - x_1^{\sqrt{q}-1})) \\
&= \{U_0, U_1\} \cup \{P(t, 1) \mid t \in GF(\sqrt{q})_0\} \\
&= PG(1,\sqrt{q}). \square
\end{aligned}
$$

COROLLARY. *For q square,*
(i) $PGU(2,q) \cong PGL(2,\sqrt{q})$;
(ii) $PSU(2,q) \cong PSL(2,\sqrt{q})$. □

Corresponding to $\mathscr{U}_1$, for q square, and to $\mathscr{H}_1$ and $\mathscr{E}_1$, for q odd, are the respective polarities. It remains to consider a null polarity for any q and a pseudo polarity for q even.

A null polarity has bilinear form $x_0 y_1 - x_1 y_0$, §2.1(v), and is therefore the identity mapping. The canonical bilinear form for a pseudo polarity is $x_0 y_0 + x_0 y_1 + x_1 y_0$, theorem 5.3.3. It has the one self-polar point U_1.

6.2.2. LEMMA. (i) $PGSp(2,q) = PGL(2,q)$;
(ii) $PGPs^*(2,q) \cong Z_q$.

(i) Since the only null polarity is the identity, every projectivity fixes it.

(ii) The projectivities commuting with the pseudo polarity given are those with matrix $\begin{bmatrix} 1 & b \\ 0 & 1 \end{bmatrix}$. Hence they form a group isomorphic to the additive group of γ.□

COROLLARY. *There are $q^2 - 1$ pseudo polarities on $PG(1,q)$.*□

6.3. *PGL*(2,*q*)

From §6.1, $PGL(2,q)$ acts sharply triply transitively on $PG(1,q)$. The order $p(2,q) = q(q^2 - 1)$. We would like to calculate $e(2,q)$, the number of conjugacy classes of $PGL(2,q)$. This number is the number of distinct projectivities of $PG(1,q)$ under any change of coordinate system.

Firstly, as a basis for comparison, we calculate $E(2,q)$, the number of conjugacy classes of $GL(2,q)$. In Table 6.1, each canonical matrix T with a distinct symbol, §2.3, we list the elementary divisors of $xI - T$ (each of which is the power of an irreducible polynomial), any special conditions on the divisors, and the number of distinct forms with this symbol.

TABLE 6.1

Symbol	Elementary divisors	Conditions	Number
[(1,1)1]	$x - t,\ x - t$	$t \in \gamma_0$	$q - 1$
[(1)1, (1)1]	$x - t_0,\ x - t_1$	$t_i \in \gamma_0,\ t_0 \neq t_1$	$\mathbf{c}(q - 1, 2)$
[(2)1]	$(x - t)^2$	$t \in \gamma_0$	$q - 1$
[(1)2]	$x^2 - a_1 x - a_0$		$N(2,q)$

So
$$\begin{aligned} E(2,q) &= 2(q - 1) + \mathbf{c}(q - 1,2) + N(2,q) \\ &= 2(q - 1) + (q - 1)(q - 2)/2 + q(q - 1)/2, \ \S1.6(\text{vii}), \\ &= q^2 - 1; \end{aligned}$$
see (2.5) in §2.3.

The elementary divisors for projectivities can then be listed in Table 6.2.

To count the number of distinct types, it must be remembered that the set of elementary divisors $F_1(x), \ldots, F_r(x)$ give the same projectivity as $F_1(x/t), \ldots, F_r(x/t)$ for t in γ_0. The cases of q even (Table 6.3) and odd (Table 6.4) need to be distinguished. Here $o(t)$ is the order of the element t. So, for q even,
$$\begin{aligned} e(2,q) &= 1 + (q - 2)/2 + 1 + N(2,q)/(q - 1) \\ &= 1 + (q - 2)/2 + 1 + q/2 \\ &= q + 1. \end{aligned}$$
In this case, $e(2,q) = E(2,q)/(q - 1)$.

TABLE 6.2

Symbol	Elementary divisors	Conditions
[(1,1)1]	$x - 1,\ x - 1$	
[(1)1, (1)1]	$x - 1,\ x - t$	$t \in \gamma_{01}$
[(2)1]	$(x - 1)^2$	
[(1)2]	$x \quad - a_1x - a_0$	$a_1 = 0$ *or* $a_0 = 1$

TABLE 6.3

q even

Symbol	Elementary divisors	Conditions	Order of projectivity	Number of non-conjugate projectivities
[(1,1)1]	$x + 1,\ x + 1$		1	1
[(1)1, (1)1]	$x + 1,\ x + t$	$t \in \gamma_{01}$	$o(t)$	$(q - 2)/2$
[(2)1]	$(x + 1)^2$		2	1
[(1)2]	$x^2 + a_1x + 1$	*root* α *in* $GF(q^2)$ *subexponent* m	m	$N(2,q)/(q - 1)$

TABLE 6.4

q odd

Symbol	Elementary divisors	Conditions	Order of projectivity	Number of non-conjugate projectivites
[(1,1)1]	$x - 1,\ x - 1$		1	1
[(1)1, (1)1]	$x - 1,\ x + 1$		2	1
[(1)1, (1)1]	$x - 1,\ x - t$	$t^3 \neq t$	$o(t)$	$(q - 3)/2$
[(2)1]	$(x - 1)^2$		p	1
[(1)2]	$x^2 - \nu$	ν *non-square*	2	1
[(1)2]	$x^2 - x - a_0$	*root* α *in* $GF(q^2)$ *subexponent* $m>2$	m	$[N(2,q) - -(q-1)/2]/(q-1)$

For q odd, $e(2,q) = 1 + 1 + (q-3)/2 + 1 + 1 + (q-1)/2 = q + 2$.

6.3.1. THEOREM. (i) $E(2,q) = q^2 - 1$;
(ii) $e(2,q) = q + (q-1, 2)$.□

Let $\mathfrak{T} = \mathrm{M}(T)$ be a projectivity given by $Y = XT$ with $T = \begin{bmatrix} a & b \\ c & d \end{bmatrix}$. In non-homogeneous coordinates $s = y_0/y_1$ and $t = x_0/x_1$, this becomes

$$bst + ds - at - c = 0 \tag{6.6}$$

with $ad - bc \neq 0$. When $a + d = 0$ so that (6.6) becomes

$$bst + d(s + t) - c = 0, \tag{6.7}$$

$\mathfrak{T}$ is called an *involution*. So an involution has order one or two, and is accordingly called *improper* or *proper*.

A proper involution has two, one, or no fixed points as

$$bt^2 + 2dt - c = 0 \tag{6.8}$$

has two, one, or no solutions; it is respectively called *hyperbolic*, *parabolic*, or *elliptic*. Since, on a line, a prime is the same thing as a point, each proper involution must be a polarity and, conversely, each polarity must be an involution, not necessarily proper. From (6.7), a proper involution is determined by two pairs of corresponding points.

For q odd, there are two types of proper involution. From Table 6.4, they have extended symbols $[1(1)1, -1(1)1]$ and $[\nu(1)2]$; the corresponding matrices are

$$T = \begin{bmatrix} 1 & 0 \\ 0 & -1 \end{bmatrix}, \begin{bmatrix} 0 & 1 \\ \nu & 0 \end{bmatrix}.$$

In non-homogeneous coordinates, the equations are

$$s = -t, \qquad s = \nu/t.$$

The involution $s = -t$ has fixed points $t = 0$ and $t = \infty$, and is therefore hyperbolic. Since, for any pair of corresponding points, the coordinates s and t satisfy $s + t = 0$, the pair is harmonically separated by the fixed points, §6.1. The involution $s = \nu/t$ has no fixed points and is therefore elliptic. These hyperbolic and elliptic involutions are nothing else than the polarities determined by the respective hyperbolic and elliptic quadrics; namely $\mathscr{H}_1 = V(x_0x_1)$ and $\mathscr{E}_1 = V(x_0^2 - \nu x_1^2)$.

For q even, there is just one type of proper involution. It has the symbol [(2)1] and matrix

$$T = \begin{bmatrix} 1 & 1 \\ 0 & 1 \end{bmatrix}.$$

In non-homogeneous coordinates, its equation is

$$s = t + 1 .$$

There is just the one fixed point $t = \infty$, and the involution is therefore parabolic. It is in fact a pseudo polarity with bilinear form $x_1y_1 + x_0y_1 + x_1y_0$.

6.3.2. LEMMA. *On* $PG(1,q)$ *the number of involutions is* q^2:

(i) *when* q *is even,* $q^2 - 1$ *are parabolic and one is the identity, the only improper involution;*

(ii) *when* q *is odd,* $q(q + 1)/2$ *are hyperbolic and* $q(q - 1)/2$ *are elliptic.*

Proof. The possible types for q odd and even follow from the above analysis of canonical forms, or equally well, from (6.8). For q even, all points are fixed when $b = c = 0$, and so by (6.7) the involution is the identity: otherwise there is just the one fixed point $t = \surd(c/b)$. For q odd, the involution is hyperbolic or elliptic as $\Delta = d^2 + bc$ is a square or not: the discriminant Δ cannot be zero as otherwise $\mathfrak{T}$ would not be bijective.

The number of involutions is $|\{\mathbf{P}(b,d,c) \mid d^2 + bc \neq 0\}| = |PG(2,q)\backslash\mathscr{P}_2|$, the number of points in the plane off a conic, which is $(q^2 + q + 1) - (q + 1) = q^2$, §7.2.

For q even, the only improper involution is the identity.

The other $q^2 - 1$ involutions are parabolic; alternatively, each is a pseudo polarity as in §6.2.

For q odd, there are as many hyperbolic and elliptic involutions as corresponding quadrics, §6.2.□

6.3.3. LEMMA. *In* $PG(2,q)$, *q even and $q \geqslant 4$, each of the three proper involutions determined by a tetrad* $\{P_1,P_2,P_3,P_4\}$ *of points has the same fixed point* P.

Proof. If $P_1\mathfrak{I} = P_2$ and $P_3\mathfrak{I} = P_4$ in an involution $\mathfrak{I}$ given by (6.7), then if P_i has coordinate t_i,

$$b\ t_1t_2 + d(t_1 + t_2) + c = 0,$$

$$b\ t_3t_4 + d(t_3 + t_4) + c = 0,$$

whence

$$\lambda b = t_1 + t_2 + t_3 + t_4\ ,$$

$$\lambda d = t_1t_2 + t_3t_4,$$

$$\lambda c = t_1t_2t_3 + t_1t_2t_4 + t_1t_3t_4 + t_2t_3t_4.$$

So the fixed point P has parameter $t = \sqrt{(c/b)}$. As this is invariant under any permutation of the t_i, the lemma is proved.□

In this lemma, P is called the *associated* point of the tetrad $\{P_1,P_2,P_3,P_4\}$. If the points of the tetrad have coordinates ∞, 0, 1, r, then P has coordinate $\sqrt{r}$.

6.3.4. LEMMA. *If* $H(P_1,P_2;\ P_3,P_4)$, *then the involution* $\mathfrak{I}$ *in which* $\{P_1,P_2\}$ *and* $\{P_3,P_4\}$ *are pairs is hyperbolic or elliptic as* $q \equiv 1$ *or* $q \equiv -1 \pmod 4$.

Proof. If P_1, P_2, P_3, P_4 have respective coordinates ∞, 0, 1, -1, then $\mathfrak{I}$ is given by $st + 1 = 0$. So the fixed points are given by $t^2 + 1 = 0$. Hence $\mathfrak{I}$ is hyperbolic or elliptic as -1 is a square or not, whence the result, §1.5.□

6.4. *PG*(1,*q*) for $q = 2,3,4,5,7,8,9$

(i) *PG*(1,2). The line consists of three points, any permutation of which can be performed by a projectivity. So $PGL(2,2) \cong S_3$ and $p(2,2) = 6$. Thus $PGL(2,2)$ acts sharply triply transitively on $PG(1,2)$.

(ii) *PG*(1,3). If t_1, t_2, t_3, t_4 are the coordinates of the four distinct points P_1, P_2, P_3, P_4 of the line, then $\{t_1, t_2; t_3, t_4\} = -1$. So the four points can be permuted in every way by a projectivity. Thus $PGL(2,3) \cong S_4$ and $p(2,3) = 24$. Hence $PGL(2,3)$ acts sharply quadruply transitively on $PG(1,3)$. The four points form a superharmonic tetrad.

(iii) *PG*(1,4). $\gamma^+ = \{\infty, 0, 1, \omega, \omega^2\}$. Let the points be denoted by $P(t)$ for $t \in \gamma^+$. The cross-ratio $\{t_1, t_2; t_3, t_4\}$is ω or ω^2. Any permutation σ of the points can be performed by a collineation. If $P_1\sigma = P_2$, $P_2\sigma = P_2$, $P_3\sigma = P_3$, then either $P_4\sigma = P_4$ and $P_5\sigma = P_5$ in which case σ is the identity projectivity, or $P_4\sigma = P_5$ and $P_5\sigma = P_4$ in which case σ is a collineation. For example, if $P(\infty)\sigma = P(\infty)$, $P(0)\sigma = P(0)$, $P(1)\sigma = P(1)$, $P(\omega)\sigma = P(\omega^2)$, and $P(\omega^2)\sigma = P(\omega)$, then σ is an automorphism. Thus, $P\Gamma L(2,4) \cong S_5$ and since $PGL(2,4)$ is a subgroup of index two, $PGL(2,4) \cong A_5$.

(iv) *PG*(1,5). The six points have coordinates ∞, 0, 1, -1, 2, -2. So the possible values for the cross-ratio of four distinct points are -1, 2, -2 = 1/2. So every tetrad is harmonic. If P_1, P_2 are any two points and the other four satisfy $H(P_3, P_4; P_5, P_6)$, then also $H(P_1, P_2; P_3, P_4)$, and $H(P_1, P_2; P_5, P_6)$. For the harmonic conjugate of P_1 with respect to P_3, P_4 cannot be P_5 or P_6, since each of these has the other as harmonic conjugate. So $H(P_1, P_2; P_3, P_4)$ and similarly $H(P_1, P_2; P_5, P_6)$. Thus each pair of points determines a unique partition of the six points into three pairs, any two of which harmonically separate one another. Let us call such a partition of the six points a *harmonic syntheme*. The six points can be partitioned into three pairs in 15 ways, one-third of which are harmonic synthemes. The coordinates of the five harmonic synthemes are

$$\begin{array}{lll} \infty,\ 0; & 1,\ -1; & 2,\ -2 \\ \infty,\ 1; & 2,\ 0; & -2,\ -1 \\ \infty,\ 2; & -2,\ 1; & -1,\ 0 \\ \infty,\ -2; & -1,\ 2; & 0,\ 1 \\ \infty,\ -1; & 0,\ -2; & 1,\ 2. \end{array}$$

Any of these five synthemes can be transformed into another by a projectivity. For, if we have the harmonic synthemes $(P_1, P_2; P_3, P_4; P_5, P_6)$ and $(Q_1, Q_2; Q_3, Q_4; Q_5, Q_6)$, there is a projectivity $\mathfrak{T}$ such that $P_1\mathfrak{T} = Q_1$, $P_2\mathfrak{T} = Q_2$ and $P_3\mathfrak{T} = Q_3$. Then, by the harmonic property $P_4\mathfrak{T} = Q_4$ and so $\{P_5, P_6\}\mathfrak{T} = \{Q_5, Q_6\}$. So $PGL(2,5)$ acts transitively on the set of five harmonic synthemes. Since $p(2,5) = 120$, $PGL(2,5) \cong \mathbf{S}_5$.

(v) $PG(1,7)$. If we take any harmonic tetrad on the line, say with coordinates ∞, 0, 1, -1, then the remaining four points also form a harmonic tetrad, since $\{2, 3; -2, -3\} = -1$. The 42 harmonic tetrads thus fall into 21 pairs, where each pair provides a partition of the eight points.

The possible values for the cross-ratio ratio of four points are -1, 2, $-3 = \frac{1}{2}$, -2, 3. The last two values are the roots of $\lambda^2 - \lambda + 1 = 0$. So every tetrad is either harmonic or equianharmonic. As there are 70 tetrads on the line, 28 are equianharmonic. Since the harmonic tetrads partition the line in pairs, so also do the equianharmonic ones. Therefore the 28 equianharmonic tetrads form 14 partitions of the line.

Although $PGL(2,7)$ is transitive on the 14 pairs of disjoint equianharmonic tetrads, $PSL(2,7)$ is not. Any equianharmonic tetrad can be transformed to its complement by an element of $PSL(2,7)$; for example, $x \to -(x - 2)/(x + 1)$ transforms $(\infty, 0, 1, -2)$ to $(-1, 2, -3, 3)$. So it suffices to consider the equianharmonic tetrads containing a fixed point. The 14 such tetrads containing ∞ are given in Table 6.5. To transform $\{\infty, 0, 1, -2\}$ to $\{\infty, 0, 1, 3\}$, we may fix ∞. Then the three projectivities possible are $x \to 3x$, $-x + 1$,

$-2x + 3$, none of which is in $PSL(2,7)$. Hence the 14 pairs of disjoint equianharmonic tetrads fall into two orbits of seven under $PSL(2,7)$ as above, where the tetrad of the pair containing ∞ is given. We see immediately that the seven pairs of an orbit form a $PG(2,2)$ by taking the pairs as points and as lines those sets of three pairs where two elements of $GF(7)^+$ occur in the same tetrad of a pair. More simply, let one of the two elements be ∞. Then the dual plane is given by taking as points of $PG(2,2)$ the points other than ∞ on $PG(1,7)$, and as lines the trios other than ∞ in either of the columns in Table 6.5. Considering the first column, we have the $PG(2,2)$ of Fig. 4.

TABLE 6.5

∞,	0,	1,	-2	∞,	0,	1,	3
∞,	0,	-1,	-3	∞,	0,	-1,	2
∞,	0,	2,	3	∞,	0,	-2,	-3
∞,	1,	-1,	2	∞,	1,	-1,	-2
∞,	1,	3,	-3	∞,	1,	2,	-3
∞,	-1,	-2,	3	∞,	-1,	3,	-3
∞,	2,	-2,	-3	∞,	2,	-2,	3

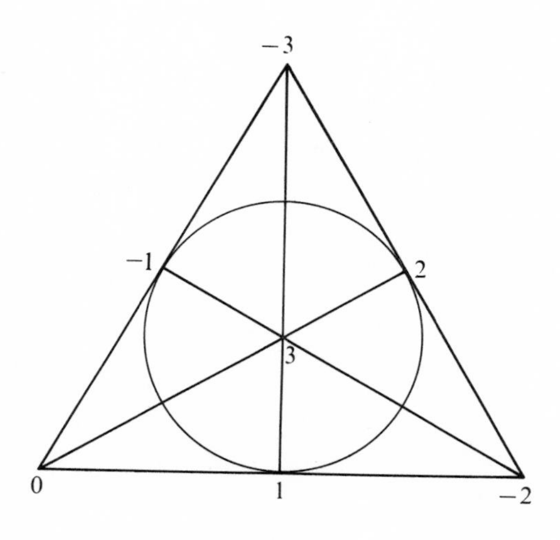

FIG. 4

Hence $PSL(2,7) \cong PGL(3,2) = PSL(3,2)$.

(vi) $PG(1,8)$. No tetrad can be harmonic, since $p = 2$, or equianharmonic, since $x^2 + x + 1$ is irreducible. So each tetrad admits as cross-ratios the six elements of $GF(8)\backslash\{0,1\}$. Hence every two tetrads are projectively equivalent, as are every two quintuplets.

If we take a tetrad and its associated point as in lemma 6.3.3, the residual tetrad on the line has the same associated point. For, suppose the four points have coordinates ∞, 0, 1, t where $t^7 = 1$ and $t \neq 1$, the associated point has coordinate $\sqrt{t} = t^4$. The remaining four points have coordinates t^2, t^3, t^5, t^6 and their associated point has coordinate r, where

$$r^2 = (t^{10} + t^{11} + t^{13} + t^{14})/(t^2 + t^3 + t^5 + t^6) = t^8 .$$

So $r = t^4$. Thus the $c(9,4) = 126$ tetrads on the line fall into 63 disjoint pairs.

Suppose S_5 is the associated point of $\{T_1, T_2, T_3, T_4\}$ and S_1, S_2, S_3, S_4 are the remaining points, then from above S_5 is the associated point of $\{S_1, S_2, S_3, S_4\}$. Also the other four tetrads taken from $\{S_1, S_2, S_3, S_4, S_5\}$ all have a different T_i as their associated point. For, $\{S_2, S_3, S_4, S_5\}$ cannot have S_1 as its associated point, as $\{T_1, T_2, T_3, T_4\}$ would also. So $\{S_2, S_3, S_4, S_5\}$ has T_1, say, as its associated point. As an involution is determined by its fixed point and another pair of corresponding points, it is not possible that $\{S_1, S_3, S_4, S_5\}$, say, also has T_1 as its associated point. Hence the assertion is proved.

Now consider any triad $\{P_1, P_2, P_3\}$ and let $\{Q_i | i \in \mathrm{N}_6\}$ be the residual set of points. If $\{P_1,P_2,P_3,Q_1\}$ has associated point Q_2 and $\{P_1,P_2,P_3,Q_2\}$ has associated point Q_3, then $\{P_1,P_2,P_3,Q_3\}$ has associated point Q_1. For, if the coordinates of P_1, P_2, P_3 are ∞, 0, 1 and Q_1 has coordinate t, where $t^7 = 1$ and $t \neq 1$, then Q_2 has coordinate $\sqrt{t} = t^4$, Q_3 has coordinate $\sqrt{(t^4)} = t^2$ and the associated point of $\{P_1, P_2, P_3, Q_3\}$ has coordinate $\sqrt{(t^2)} = t$ and so is Q_1. Similarly, if $\{Q_4, Q_5, Q_6\} = \{R_1, R_2, R_3\}$, the R_i can be chosen so that $\{P_1, P_2, P_3, R_1\}$,

$\{P_1, P_2, P_3, R_2\}$, $\{P_1, P_2, P_3, R_3\}$ have R_2, R_3, R_1 as their respective associated points. However, as the residual quintuplets to $\{R_1, R_2, R_3, Q_1\}$, $\{R_1, R_2, R_3, Q_2\}$, $\{R_1, R_2, R_3, Q_3\}$ are $\{P_1, P_2, P_3, Q_2, Q_3\}$, $\{P_1, P_2, P_3, Q_3, Q_1\}$, $\{P_1, P_2, P_3, Q_1, Q_2\}$, these tetrads have as their respective associated points Q_3, Q_1, Q_2. Also the associated point of $\{R_1, R_2, R_3, P_1\}$ cannot be Q_1, Q_2, or Q_3; for if it were Q_1, say, then (Q_2, Q_3) and (P_2, P_3) would be pairs in an involution with fixed point Q_1. However, from the initial position, (P_1, Q_2) and (P_2, P_3) are pairs in an involution with fixed point Q_1. As an involution is determined by a pair and a fixed point, this cannot be. So $\{R_1, R_2, R_3, P_1\}$, $\{R_1, R_2, R_3, P_2\}$, $\{R_1, R_2, R_3, P_3\}$ have as their respective associated points P_2, P_3, P_1 or P_3, P_1, P_2. Thus the three triads are symmetrically related and will be named a triple. Therefore, the $c(9,3) = 84$ triads of points fall into 28 triples, each of which partitions the line and any two triads of a triple are related as above.

(vii) $PG(1,9)$. From §6.1, the condition that $H(P_1, P_2; P_3, P_4)$ on $PG(1,q)$ with $p \neq 2$ is that the coordinates t_i of P_i satisfy

$$(t_1 + t_2)(t_3 = t_4) = 2(t_1 t_2 + t_3 t_4),$$

which becomes, when $p = 3$,

$$\Sigma'' t_i t_j = 0.$$

Thus any four points on $PG(1,9)$ whose coordinates are roots of a quartic in x with coefficient of x^2 zero forms a harmonic tetrad.

Given a harmonic tetrad $\{P_1, P_2, P_3, P_4\}$, then, by lemma 6.3.4, $\{P_1, P_2\}$ and $\{P_3, P_4\}$ are pairs in hyperbolic involution with fixed points Q_1, R_1, which harmonically separate each pair. Likewise $\{P_1, P_3\}$, $\{P_2, P_4\}$ determine an involution with fixed points Q_2, R_2 and $\{P_1, P_4\}$, $\{P_2, P_3\}$ an involution with fixed points Q_3, R_3. The points Q_i and R_i are all distinct. For, if P_1, P_2, P_3, P_4 have respective coordinates $\sigma, -\sigma, \sigma^3, -\sigma^3$ then the coordinates of the respective pairs $\{Q_1, R_1\}$, $\{Q_2, R_2\}$, $\{Q_3, R_3\}$ are $\{\infty, 0\}$,

$\{1, -1\}$, $\{\sigma^2, -\sigma^2\}$. We also note that $H(Q_i, R_i; Q_j, R_j)$ for $i \neq j$.

The number of ways that the line can be partitioned in this way is the same as the number of harmonic tetrads, namely 30. As there are 45 pairs of points, each pair occurs as a pair in exactly two of the sextuples residual to a harmonic tetrad. The 30 harmonic tetrads coincide in fact with the 30 Hermitian varieties, since each Hermitian variety is given by a form $H = a\,\bar{x}_0\,x_0 + b\,\bar{x}_0\,x_1 + \bar{b}\,\bar{x}_1\,x_0 + c\,\bar{x}_1\,x_1$, where $\bar{x} = x^3$ and $a, c \in GF(3)$: so $H = a\,x_0^4 + bx_0^3\,x_1 + b^3\,x_0\,x_1^3 + c\,x_1^4$. Since the variety is non-singular, $ac - b\bar{b} \neq 0$ and so is 1 or -1. The harmonic tetrads are then *positive* or *negative* according as the determinant of the corresponding Hermitian form is 1 or -1.

Each point occurs in 12 harmonic tetrads. Each pair of points occurs in four harmonic tetrads, since the involution with the pair as fixed points has four corresponding pairs on the line. It is possible to find a quintuple of harmonic tetrads so that each point lies in two tetrads and no pair of points occurs more than once. If we take a harmonic tetrad $\{P_1, P_2, P_3, P_4\}$ as above, then the number of tetrads containing $\{P_1, P_2\}$,$\{P_1, P_3\}$ or $\{P_1, P_4\}$ is ten. So there remain two tetrads containing P_1. Now, such a set cannot be $\{P_1, Q_1, R_1, R_2\}$, for the four pairs forming harmonic tetrads with $\{Q_1, R_1\}$ are $\{P_1, P_2\}$, $\{P_3, P_4\}$, $\{Q_2, R_2\}$, $\{Q_3, R_3\}$. Since Q_i and R_i have not been distinguished, they may be chosen so that the two tetrads required are $\{P_1, Q_1, Q_2, Q_3\}$ and $\{P_1, R_1, R_2, R_3\}$. If we select these two tetrads in our quintuple, the other three are determined. The quintuple is $\{P_1, Q_1, Q_2, Q_3\}$, $\{P_1, R_1, R_2, R_3\}$, $\{P_3, P_4, Q_1, R_1\}$, $\{P_2, P_4, Q_2, R_2\}$, $\{P_2, P_3, Q_3, R_3\}$. Twelve such quintuples exist and each harmonic tetrad lies in two of them. In fact, the tetrads in a quintuple are all positive or all negative. So, there are six *positive quintuples* and six *negative quintuples* (see Table 6.6). $PGL(2,9)$ acts as a transitive group on the 12 quintuples; $p(2,9) = 6!$ but $PGL(2,9)$ does not act as a group of degree six. However, let us consider $PSL(2,9)$, the subgroup of $PGL(2,9)$ of index two consisting of those elements of $PGL(2,9)$ represented by matrices with square determinant. An element of $PSL(2,9)$ acting on a Hermitian form leaves its determinant unaltered. So $PSL(2,9)$ fixes both the set of positive harmonic tetrads and the set of negative tetrads. So $PSL(2,9)$ acts transitively on the six positive quintuples and the six negative quintuples. As it has order 360, it is isomorphic to $\mathbf{A}_6$. Thus $PGL(2,9)$ has a subgroup of index two isomorphic to $\mathbf{A}_6$, but it is not isomorphic to $\mathbf{S}_6$. However, the automorphism $t \to t^3$ of $PG(1,9)$ fixes both the set of positive tetrads and the set of negative tetrads. Hence $P\Gamma SL(2,9) = \left\{ t \to \dfrac{a\tilde{t}+b}{c\tilde{t}+d} \;\middle|\; ad - bc = 1,\ \tilde{t} = t \text{ or } t^3 \right\}$ is isomorphic to $\mathbf{S}_6$. See Appendix I.

TABLE 6.6

The twelve quintuples of harmonic tetrads

Negative				
$\infty,\sigma,\sigma^2,-\sigma^3$;	$\infty,-\sigma,-\sigma^2,\sigma^3$;	$0,1,\sigma,\sigma^3$;	$0,-1,-\sigma,-\sigma^3$;	$1,-1,\sigma^2,-\sigma^2$;
$\infty,0,1,-1$;	$\infty,-\sigma,-\sigma^2,\sigma^3$;	$0,\sigma,-\sigma^2,\sigma^3$;	$1,\sigma,-\sigma,\sigma^2$;	$-1,\sigma^2,\sigma^3,-\sigma^3$;
$\infty,0,1,-1$;	$\infty,\sigma,\sigma^2,-\sigma^3$;	$0,-\sigma,\sigma^2,\sigma^3$;	$1,-\sigma^2,\sigma^3,-\sigma^3$;	$-1,\sigma,-\sigma,-\sigma^2$;
$\infty,0,\sigma^2,-\sigma^2$;	$\infty,1,-\sigma,-\sigma^3$;	$0,1,\sigma,\sigma^3$;	$-1,\sigma,-\sigma,-\sigma^2$;	$-1,\sigma^2,\sigma^3,-\sigma^3$;
$\infty,0,\sigma^2,-\sigma^2$;	$\infty,-1,\sigma,\sigma^3$;	$0,-1,-\sigma,-\sigma^3$;	$1,\sigma,-\sigma,\sigma^2$;	$1,-\sigma^2,\sigma^3,-\sigma^3$;
$\infty,1,-\sigma,-\sigma^3$;	$\infty,-1,\sigma,\sigma^3$;	$0,\sigma,-\sigma^2,-\sigma^3$;	$0,-\sigma,\sigma^2,\sigma^3$;	$1,-1,\sigma^2,-\sigma^2$.
Positive				
$\infty,0,\sigma,-\sigma$;	$\infty,1,\sigma^2,\sigma^3$;	$0,-1,-\sigma^2,\sigma^3$;	$1,-1,\sigma,-\sigma^3$;	$-\sigma,\sigma^2,-\sigma^2,-\sigma^3$;
$\infty,0,\sigma,-\sigma$;	$\infty,-1,-\sigma^2,-\sigma^3$;	$0,1,\sigma^2,-\sigma^3$;	$1,-1,-\sigma,\sigma^3$;	$\sigma,\sigma^2,-\sigma^2,\sigma^3$;
$\infty,1,\sigma^2,\sigma^3$;	$\infty,-1,-\sigma^2,-\sigma^3$;	$0,1,-\sigma,-\sigma^2$;	$0,-1,\sigma,\sigma^2$;	$\sigma,-\sigma,\sigma^3,-\sigma^3$;
$\infty,0,\sigma^3,-\sigma^3$;	$\infty,1,\sigma,-\sigma^2$;	$0,-1,\sigma,\sigma^2$;	$1,-1,-\sigma,\sigma^3$;	$-\sigma,\sigma^2,-\sigma^2,-\sigma^3$;
$\infty,0,\sigma^3,-\sigma^3$;	$\infty,-1,-\sigma,\sigma^2$;	$0,1,-\sigma,-\sigma^2$;	$1,-1,\sigma,-\sigma^3$;	$\sigma,\sigma^2,-\sigma^2,\sigma^3$;
$\infty,1,\sigma,-\sigma^2$;	$\infty,-1,-\sigma,\sigma^2$;	$0,1,\sigma^2,-\sigma^3$;	$0,-1,-\sigma^2,\sigma^3$;	$\sigma,-\sigma,\sigma^3,-\sigma^3$.

$\sigma^2 - \sigma - 1 = \sigma^3 - \sigma^2 - \sigma = \sigma^3 + \sigma^2 + 1 = \sigma^3 + \sigma - 1 = 1 + 1 + 1 = 0.$

6.5. Notes and references

§6.3. Lemma 6.3.3 comes from Segre [1956].

§6.4. The proof of the isomorphism of A_6 and *PSL*(2,9) comes from Edge [1955c] as does Table 6.6. See also Dieudonné [1954].

7
FIRST PROPERTIES OF $PG(2,q)$

7.1. Preliminaries

The plane $PG(2,q)$ contains $q^2 + q + 1$ points, $q^2 + q + 1$ lines, $q + 1$ points on every line and $q + 1$ lines through every point. Throughout $\mathbf{U}_0, \mathbf{U}_1, \mathbf{U}_2, \mathbf{U}$ denote the triangle of reference and unit point with $\mathbf{u}_0, \mathbf{u}_1, \mathbf{u}_2, \mathbf{u}$ their respective duals.

Theorem 3.2.1 for the case $n = 2$ is the following.

If $\mathscr{L}$ is a subset of $PG(2,q)$ such that every line of $PG(2,q)$ contains a point of $\mathscr{L}$, then $|\mathscr{L}| \geqslant q+1$ with equality if and only if $\mathscr{L}$ is a line. Dually, if $\mathscr{F}$ is a subset of the lines of $PG(2,q)$ such that every point of $PG(2,q)$ lies on a line of $\mathscr{F}$, then $|\mathscr{F}| \geqslant q + 1$ with equality if and only if $\mathscr{F}$ is a pencil, the set of lines through a point.

In §2.6, n-stigms with vertices forming an n-arc and n-grams with sides forming an n-side were defined. Let $l(n,q)$ be the number of points on the sides of an n-stigm, which dually is the number of lines through the vertices of an n-gram: write $l^*(n,q) = q^2 + q + 1 - l(n,q)$.

$l(n,q)$ and so $l^*(n,q)$ are determined for $n \leqslant 5$. The values are given in Table 7.1.

TABLE 7.1

n	2	3	4	5
$l(n,q)$	$q + 1$	$3q$	$6q - 5$	$10q - 20$
$l^*(n,q)$	q^2	$(q-1)^2$	$(q-2)(q-3)$	$(q-4)(q-5) + 1$

Let $L(n,q)$ denote the number of n-stigms in $PG(2,q)$. Then

$$L(2,q) = q(q + 1)(q^2 + q + 1)/2! \;;$$
$$L(3,q) = L(2,q)l^*(2,q)/3 = q^3(q + 1)(q^2 + q + 1)/3! \;;$$
$$L(4,q) = L(3,q)l^*(3,q)/4 = q^3(q^2 - 1)(q^3 - 1)/4! \;;$$
$$L(5,q) = L(4,q)l^*(4,q)/5 = q^3(q^2 - 1)(q^3 - 1)(q - 2)(q - 3)/5! \;;$$
$$L(6,q) = L(5,q)l^*(5,q)/6 = q^3(q^2-1)(q^3-1)(q-2)(q-3)(q^2-9q+21)/6! \;.$$

$L(n,q)$ is also the number of n-arcs, n-grams and n-sides in $PG(2,q)$.

The *diagonal points* of an n-stigm are the intersections of two sides which do not pass through the same vertex. A tetrastigm has three diagonal points and a pentastigm has 15 diagonal points.

7.1.1. LEMMA. *The three diagonal points of a tetrastigm in $PG(2,q)$ are collinear if and only if q is even.*

Proof. Let $\mathbf{U}_0,\mathbf{U}_1,\mathbf{U}_2,\mathbf{U}$ be the vertices of the tetrastigm. Then the diagonal points are

$$D_0 = \mathbf{U}\mathbf{U}_0 \cap \mathbf{U}_1\mathbf{U}_2 = \mathbf{P}(0,1,1),$$
$$D_1 = \mathbf{U}\mathbf{U}_1 \cap \mathbf{U}_0\mathbf{U}_2 = \mathbf{P}(1,0,1),$$
$$D_2 = \mathbf{U}\mathbf{U}_2 \cap \mathbf{U}_0\mathbf{U}_1 = \mathbf{P}(1,1,0).$$

So D_0, D_1, D_2 are collinear if and only if $2 = 0$; that is, q is even. □

A pentastigm contains five tetrastigms. We now consider the collinearity of diagonal points of a pentastigm, where each point is the diagonal point of a different tetrastigm.

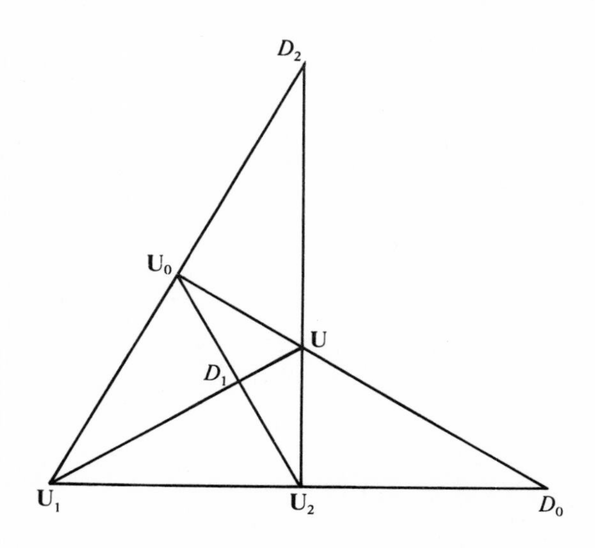

FIG. 5

7.1.2. LEMMA. (i) *If five diagonal points of a pentastigm in* $PG(2,q)$ *are collinear, then* $x^2 = x + 1$ *has a solution in* $GF(q)$.

(ii) *If two such collinearities of five diagonal points occur, then all six possible collinearities occur and* $q = 4^m$.

Proof. Let the vertices of the pentastigm $\mathscr{P}$ be P_0, P_1, P_2, P_3, P_4 with $P_0 = \mathbf{U}_0$, $P_1 = \mathbf{U}_1$, $P_2 = \mathbf{U}_2$, $P_3 = \mathbf{U}$ and $P_4 = \mathbf{P}(a_0, a_1, a_2)$. If we write $ij \cdot kl$ for $P_iP_j \cap P_kP_l$, then the diagonal points are as follows:

$01 \cdot 23 = \mathbf{P}(1,1,0)$; $\quad 01 \cdot 24 = \mathbf{P}(a_0, a_1, 0)$, $\quad 01 \cdot 34 = \mathbf{P}(a_0 - a_2, a_1 - a_2, 0)$,
$02 \cdot 13 = \mathbf{P}(1,0,1)$; $\quad 02 \cdot 14 = \mathbf{P}(a_0, 0, a_2)$, $\quad 03 \cdot 14 = \mathbf{P}(a_0, a_2, a_2)$,
$03 \cdot 12 = \mathbf{P}(0,1,1)$; $\quad 04 \cdot 12 = \mathbf{P}(0, a_1, a_2)$, $\quad 04 \cdot 13 = \mathbf{P}(a_2, a_1, a_2)$,
$02 \cdot 34 = \mathbf{P}(a_0 - a_1, 0, a_2 - a_1)$, $\quad 12 \cdot 34 = \mathbf{P}(0, a_1 - a_0, a_2 - a_0)$,
$03 \cdot 24 = \mathbf{P}(a_0, a_1, a_1)$, $\quad 13 \cdot 24 = \mathbf{P}(a_0, a_1, a_0)$,
$04 \cdot 23 = \mathbf{P}(a_1, a_1, a_2)$, $\quad 14 \cdot 23 = \mathbf{P}(a_0, a_0, a_2)$.

The possible collinearities among five diagonal points, which are in fact the ways of separating the ten sides of $\mathscr{P}$ into five pairs so that any pair do not meet at a vertex are as follows:

I. $01 \cdot 23$; $02 \cdot 14$; $04 \cdot 13$; $03 \cdot 24$; $12 \cdot 34$.
II. $01 \cdot 23$; $04 \cdot 12$; $03 \cdot 14$; $02 \cdot 34$; $13 \cdot 24$.
III. $02 \cdot 13$; $01 \cdot 24$; $03 \cdot 14$; $04 \cdot 23$; $12 \cdot 34$.
IV. $02 \cdot 13$; $04 \cdot 12$; $01 \cdot 34$; $03 \cdot 24$; $14 \cdot 23$.
V. $03 \cdot 12$; $01 \cdot 24$; $04 \cdot 13$; $02 \cdot 34$; $14 \cdot 23$.
VI. $03 \cdot 12$; $02 \cdot 14$; $01 \cdot 34$; $04 \cdot 23$; $13 \cdot 24$.

Since $\mathscr{P}$ is a pentastigm, P_4 cannot be collinear with any pair of other vertices; so

$$a_0\, a_1\, a_2\, (a_0 - a_1)(a_0 - a_2)(a_1 - a_2) \neq 0.$$

The conditions that five diagonal points are collinear are, in the respective cases:

I: $a_2 = a_0 + a_1$, $a_0^2 - a_1^2 = a_0a_1$; II: $a_2 = a_0 + a_1$, $a_0^2 - a_1^2 = -a_0a_1$;
III: $a_1 = a_0 + a_2$, $a_0^2 - a_2^2 = a_0a_2$; IV: $a_1 = a_0 + a_2$, $a_0^2 - a_2^2 = -a_0a_2$;
V: $a_0 = a_1 + a_2$, $a_1^2 - a_2^2 = a_1a_2$; VI: $a_0 = a_1 + a_2$, $a_1^2 - a_2^2 = -a_1a_2$.

If only one of these conditions is satisfied, we require the equation $x^2 = x + 1$ to have a root in $GF(q)$. If two conditions are satisfied, the field has characteristic two, $x^2 = x + 1$ has a root ω in $GF(q)$ and so $GF(q)$ is an extension of $GF(4)$: also, $P_4 = \mathbf{P}(1,\omega,\omega^2)$ or $\mathbf{P}(1,\omega^2,\omega)$ and all six conditions are satisfied.□

7.2. Conics

Let $Q(2,q)$ be the set of quadrics in $PG(2,q)$; that is, the varieties $\mathbf{V}(F)$, where

$$F = a_{00}x_0^2 + a_{11}x_1^2 + a_{22}x_2^2 + a_{01}x_0x_1 + a_{02}x_0x_2 + a_{12}x_1x_2 :$$

$|Q(2,q)| = \theta(5,q) = (q^6 - 1)/(q - 1)$. If $\mathbf{V}(F)$ is non-singular, then the quadric is a conic. If $\mathbf{V}(F)$ is singular, then F can be reduced to a form in one or two variables and the various types of forms correspond to the quadrics in $Q(1,q)$, §6.2.

TABLE 7.2

	Canonical F	Description of quadric	Number of elements in orbit
(i) singular	(a) x_0^2	*repeated line*	$q^2 + q + 1$
	(b) x_0x_1	*pair of distinct lines*	$q(q+1)(q^2+q+1)/2$
	(c) $x_0^2+ax_0x_1+bx_1^2$ *irreducible*	*a point* (*a pair of conjugate lines in* $PG(2,q^2)$ *with just their meet in* $PG(2,q)$)	$q(q-1)(q^2+q+1)/2$
(ii) non-singular	$x_0^2+x_1x_2$	*conic, consisting of* q+1 *points no three collinear*	$q^5 - q^2$

7.2.1. THEOREM. *The* $\theta(5,q)$ *elements of* $Q(2,q)$ *fall into four orbits under* $PGL(3,q)$ *as in Table* 7.2.

Proof. The canonical forms for the singular F follow from §6.2 and for non-singular F from theorem 5.2.4. The number of conics is calculated by taking the difference of $\theta(5,q)$ and the sum of the other types of quadrics. That a conic $\mathscr{C}$ consists of $q+1$ points follows since $\mathscr{C} = \mathbf{V}(x_0^2 + x_1x_2)$; so $\mathscr{C} = \{\mathbf{P}(t, -1, t^2) \mid t \in \gamma^+\}$. Then, since a line cannot meet $\mathscr{C}$ in more than two points, no three are collinear. See also theorem 5.2.6.□

COROLLARY. *In* $PG(2,q)$ *with* $q \geqslant 4$ *there is a unique conic through a* 5-*arc*.

Proof. Any quadric through a 5-arc $\mathscr{K}$ is non-singular by the theorem. There exists at least one conic through $\mathscr{K}$, since only five linear conditions are imposed on the coefficients a_{ij} of $\mathbf{V}(\Sigma' a_{ij}x_ix_j)$ to contain $\mathscr{K}$. If $\mathbf{V}(x_0{}^2 + x_1x_2)$ and $\mathbf{V}(\Sigma' a_{ij}x_ix_j)$ both contain $\mathscr{K}$, then the common points are $\mathbf{P}(t, -1, t^2)$, where t satisfies an equation of degree at most four. So these two conics are the same.□

That a conic $\mathscr{C}$ consists of $q+1$ points can be shown without resorting to the canonical form.

7.2.2. LEMMA. *If a conic contains one point, it contains exactly* $q+1$.

Proof. Suppose $\mathscr{C} = \mathbf{V}(F)$ and $\mathbf{P}(A)$ is on $\mathscr{C}$. Then $\mathbf{P}(A + tB)$ lies on $\mathscr{C}$ if

$$F(A) + tG(A,B) + t^2F(B) = 0,$$

similarly to (2.11). Since $\mathbf{P}(A)$ is on $\mathscr{C}$, so $F(A) = 0$. The one line meeting $\mathscr{C}$ in $\mathbf{P}(A)$ only is therefore $\mathbf{V}(G(A,X))$, namely the tangent, §2.6(xiii) and (xiv). Since $\mathscr{C}$ is non-singular, F is irreducible and so no line meets $\mathscr{C}$ in more than two points. So the $q+1$ lines of the plane through $\mathbf{P}(A)$ comprise the tangent and q bisecants. Hence $|\mathscr{C}| = q+1$.□

7.2.3. LEMMA. *Every conic in $PG(2,q)$ is a $(q+1)$-arc.*

Proof. From theorem 7.2.1, the number of conics is $q^5 - q^2$: this calculation was independent of the canonical form for a conic.

Let us now count the number N of conics which are $(q+1)$-arcs, that is which consist of $q+1$ points no three collinear. Since a conic is determined by the ratios of the coefficients $(a_{00}, a_{11}, a_{22}, a_{01}, a_{02}, a_{12})$, it is determined by five of its points. So, from §7.1,

$$\begin{aligned} N &= L(5,q)/\mathbf{c}(q+1,5) \\ &= (q^2+q+1)(q^2+q)q^2(q-1)^2(q-2)(q-3)/[(q+1)q(q-1)(q-2)(q-3)] \\ &= (q^2+q+1)q^2(q-1) \\ &= q^5-q^2. \end{aligned}$$

Since this is the number of conics, every conic is a $(q+1)$-arc.□

COROLLARY 1. *A conic in $PG(2,q)$ has the following canonical forms:*

(i) $\mathbf{V}(x_0^2+x_1x_2)$ *all* q;
(ii) $\mathbf{V}(x_0^2-x_1x_2)$ *all* q;
(iii) $\mathbf{V}(a_0x_0^2+a_1x_1^2+a_2x_2^2)$ $a_0a_1a_2 \neq 0$, q *odd*;
(iv) $\mathbf{V}(x_0^2+x_1^2+x_2^2)$ q *odd*.

Proof. It is only necessary to prove (iii). If $a_0x_0^2 + a_1x_1^2 + a_2x_2^2 = (b_0x_0 + b_1x_1 + b_2x_2)(c_0x_0 + c_1x_1 + c_2x_2)$ over some extension of γ, then $b_0b_1b_2c_0c_1c_2 \neq 0$ and $b_0c_1 = -b_1c_0$, $b_1c_2 = -b_2c_1$, $b_2c_0 = -b_0c_2$. So $b_0b_1b_2c_0c_1c_2 = -b_0b_1b_2c_0c_1c_2$, giving a contradiction since q is odd. So the form is irreducible over $GF(q^2)$ and defines a conic.□

COROLLARY 2. *The number of 6-arcs or hexastigms in $PG(2,q)$ whose vertices do not lie on a conic is*

$$q^3(q+1)(q-1)^2(q-2)(q-3)(q-5)^2(q^2+q+1)/6!\,.$$

Proof. The required number is

$$L(5,q)\{\ell^*(5,q) - (q-4)\}/6.\square$$

COROLLARY 3. *In* $PG(2,5)$ *a conic is a* 6*-arc and conversely.*□

COROLLARY 4. *In* $PG(2,q)$ *for* q *even, the* $q+1$ *tangents to a conic are concurrent.*

Proof. Let $\mathscr{C} = \mathbf{V}(x_0^2 + x_1x_2)$. The tangent to $\mathscr{C}$ at $\mathbf{P}(Y)$ with $Y = (y_0, y_1, y_2)$ is

$$\mathbf{V}(x_1y_2 + x_2y_1) = \pi(0,\ y_2,\ y_1).$$

Hence every tangent contains $\mathbf{U}_0$.□

The point of intersection of the tangents to a conic in $PG(2,q)$ with q even is the *nucleus*, as in §5.3.

COROLLARY 5. *In* $PG(2,q)$ *for* q *even, the nucleus of the conic*

$$\mathbf{V}(a_{00}x_0^2+a_{11}x_1^2+a_{22}x_2^2+a_{01}x_0x_1+a_{02}x_0x_2+a_{12}x_1x_2)$$

is the point $\mathbf{P}(a_{12}, a_{02}, a_{01})$.□

COROLLARY 6.
$$|PGO(3,q)| = p(3,q)/(q^5 - q^2) = q(q^2-1) = p(2,q).\square$$

COROLLARY 7. $PGO(3,q) \cong PGL(2,q)$.

Proof. Let $\mathscr{C}$ be the conic $\mathbf{V}(x_1^2 - x_0x_2)$: then $\mathscr{C} = \{\mathbf{P}(t^2,t,1) \mid t \in \gamma^+\}$. If t is the non-homogeneous coordinate of a point on $PG(1,q)$, then the element $\mathfrak{T} = \mathbf{M}(T)$ of $PGL(2,q)$ given by $\mathbf{P}(t,1) \to \mathbf{P}(t,1)\mathfrak{T}$, where

$$T = \begin{bmatrix} a & b \\ c & d \end{bmatrix}$$

can also be written $t \to (at + c)/(bt + d)$. It has the following effect on $\mathscr{C}$:

$$\begin{aligned} \mathbf{P}(t^2, t, 1) &\to \mathbf{P}((at + c)^2, (at + c)(bt + d), (bt + d)^2) \\ &= \mathbf{P}(t^2, t, 1)\,\mathfrak{T}', \end{aligned}$$

where $\mathfrak{T}' = \mathbf{M}(T')$ and

$$T' = \begin{bmatrix} a^2 & ab & b^2 \\ 2ac & ad + bc & 2bd \\ c^2 & cd & d^2 \end{bmatrix}.$$

So $\mathfrak{T}$ induces an element $\mathfrak{T}'$ of $PGO(3,q)$. Thus ϕ: $PGL(2,q) \to PGO(3,q)$ given by $\mathfrak{T}\phi = \mathfrak{T}'$ is a bijection. If $\mathbf{P}(t,1) \to \mathbf{P}(t,1)\mathfrak{T}_1\mathfrak{T}_2$, then on $\mathscr{C}$ we have $\mathbf{P}(t^2, t, 1) \to P(t^2, t, 1)\mathfrak{T}'_1\mathfrak{T}'_2$. So ϕ is an isomorphism.□

COROLLARY 8. *$PGO(3,q)$ acts triply transitively on $\mathscr{P}_2$.*

Proof. This follows from corollary 7 and the result of §6.3 that $PGL(2,q)$ acts similarly on $PG(1,q)$.□

For later reference it is useful to have the conditions for a plane quadric to be singular.

7.2.4. THEOREM. *If $\mathscr{F} = \mathbf{V}(F)$ where $F = \Sigma' a_{ij} x_i x_j$, then $\mathscr{F}$ is singular if and only if $\delta = 0$, where*

$$\delta = 4\, a_{00}\, a_{11}\, a_{22} + a_{01}\, a_{02}\, a_{12} - a_{00}\, a_{12}^2 - a_{11}\, a_{02}^2 - a_{22}\, a_{01}^2 .$$

Proof. $\mathscr{F}$ is singular at $Q = \mathbf{P}(y_0, y_1, y_2)$ if

$$\frac{\partial F}{\partial x_0} = \frac{\partial F}{\partial x_1} = \frac{\partial F}{\partial x_2} = 0 \text{ at } Q \text{ and } F(y_0, y_1, y_2) = 0.$$

Then

$$\begin{aligned} 2a_{00}\,y_0 + a_{01}\,y_1 + a_{02}\,y_2 &= 0, \\ a_{01}\,y_0 + 2a_{11}y_1 + a_{12}\,y_2 &= 0, \\ a_{02}\,y_0 + a_{12}\,y_1 + 2a_{22}\,y_2 &= 0. \end{aligned} \tag{7.1}$$

When $p > 2$, Q exists if and only if $\Delta = 0$, where

$$\Delta = \begin{bmatrix} 2a_{00} & a_{01} & a_{01} \\ a_{01} & 2a_{11} & a_{12} \\ a_{02} & a_{12} & 2a_{22} \end{bmatrix}.$$

But $\Delta = 2\delta$; so $\delta = 0$ is the required condition.

By Euler's theorem, $2F = x_0\dfrac{\partial F}{\partial x_0} + x_1\dfrac{\partial F}{\partial x_1} + x_2\dfrac{\partial F}{\partial x_2}$.

So, when $p = 2$, a point for which $\dfrac{\partial F}{\partial x_0} = \dfrac{\partial F}{\partial x_1} = \dfrac{\partial F}{\partial x_2} = 0$

does not necessarily lie on $\mathscr{F}$. However, when $p = 2$, a solution to (7.1) is given by $Q = \mathbf{P}(a_{12}, a_{02}, a_{01})$, which lies on $\mathscr{F}$ if and only if

$$a_{00}\,a_{12}^2 + a_{11}\,a_{02}^2 + a_{22}\,a_{01}^2 + a_{01}\,a_{02}\,a_{12} = 0;$$

that is, $\delta = 0$.□

COROLLARY. *If q is even and $\mathscr{F}$ is a conic, then the nucleus is* $\mathbf{P}(a_{12}, a_{02}, a_{01})$.

Proof. The nucleus of $\mathscr{F}$ is the point Q lying on all the tangents to $\mathscr{F}$ and so satisfies (7.1).□

7.2.5. THEOREM. *The line $\mathscr{G} = \mathbf{V}(G)$, $G = \Sigma\, b_i x_i$, is a component of the quadric $\mathscr{F} = \mathbf{V}(F)$, $F = \Sigma'\, a_{ij}x_ix_j$, if and only if*

$$\begin{aligned} a_{00}\,b_1^2 + a_{11}\,b_0^2 &= a_{01}\,b_0\,b_1, \\ a_{00}\,b_2^2 + a_{22}\,b_0^2 &= a_{02}\,b_0\,b_2, \\ a_{11}\,b_2^2 + a_{22}\,b_1^2 &= a_{12}\,b_1\,b_2. \end{aligned}$$

Proof. $\mathscr{G}$ is a component of $\mathscr{F}$ if and only if $\mathscr{F}$ contains any three points of $\mathscr{G}$. The selection of the three points $\mathbf{P}(b_1, -b_0, 0)$, $\mathbf{P}(-b_2, 0, b_0)$, $\mathbf{P}(0, b_2, -b_1)$ gives the three equations required.□

7.3. Hermitian curves

Let $U(2,q)$ be the set of Hermitian varieties in $PG(2,q)$ with q square. Thus, from §5.1, $U(2,q) = \{\mathbf{V}(\bar{X}HX^*)\,|\,\bar{H} = H^*\}$, where $H = (h_{ij})$, $\bar{H} = (\bar{h}_{ij})$, $H^* = (h_{ji})$, $\bar{x} = x^{\sqrt{q}}$ for any x in $GF(q)$. If we write out a Hermitian form in detail,

$$\begin{aligned}\bar{X}HX^* = {} & h_{00}\bar{x}_0x_0 + h_{11}\bar{x}_1x_1 + h_{22}\bar{x}_2x_2 + h_{01}\bar{x}_0x_1 + \bar{h}_{01}x_0\bar{x}_1 \\ & + h_{02}\bar{x}_0x_2 + \bar{h}_{02}x_0\bar{x}_2 + h_{12}\bar{x}_1x_2 + \bar{h}_{12}x_1\bar{x}_2.\end{aligned}$$

Since h_{00}, h_{11}, h_{22} are in $GF(\sqrt{q})$,

$$\begin{aligned}|U(2,q)| &= [(\sqrt{q})^3q^3 - 1]/(\sqrt{q} - 1) \\ &= (q + \sqrt{q} + 1)(q^3 + q\sqrt{q} + 1).\end{aligned}$$

7.3.1. THEOREM. *$U(2,q)$ falls into three orbits under $PGL(3,q)$ as in Table 7.3.*

TABLE 7.3

	Canonical Hermitian form	Description	Number of elements in orbit
(i) singular	(a) $\bar{x}_0x_0$	*line repeated $\sqrt{q}+1$ times*	q^2+q+1
	(b) $\bar{x}_0x_0+\bar{x}_1x_1$	*joins of $\sqrt{q}+1$ points on $\mathbf{u}_2$ to $\mathbf{U}_2$*	$\sqrt{q}(q+1)(q^2+q+1)$
(ii) non-singular	$\bar{x}_0x_0+\bar{x}_1x_1+\bar{x}_2x_2$	*Hermitian curve $\mathscr{U}_2$ of $q\sqrt{q}+1$ points*	$q\sqrt{q}(\sqrt{q}-1)(q+1)(q+\sqrt{q}+1)$

Proof. The canonical forms follow from theorem 5.1.2, corollary 1. So type (i)(a) consists of a single line and there are $q^2 + q + 1$ of these. Type (i)(b) consists of the joins of a point to the points of a Hermitian variety $\mathscr{U}_1$ on a line. By §6.2, there are $\sqrt{q}(q+1)$ non-singular Hermitian varieties on a line and so

$\sqrt{q}(q+1)(q^2+q+1)$ of type (i)(b). By subtraction from $|U(2,q)|$, we obtain the number of Hermitian curves in $PG(2,q)$. The number $\mu(2,q)$ of points on $\mathcal{U}_2$ was given in theorem 5.1.6.□

$\mu(2,q)$ is obtained independently below.

COROLLARY. $|PGU(3,q)| = p(3,q)/[q\sqrt{q}(\sqrt{q}-1)(q+1)(q+\sqrt{q}+1)]$
$= q\sqrt{q}(q-1)(q\sqrt{q}+1)$.□

7.3.2. LEMMA. (i) *The* q^2+q+1 *lines of* $PG(2,q)$ *consist of* $q\sqrt{q}+1$ *tangents to* $\mathcal{U}_2$*, each with* $(\sqrt{q}+1)$*-point contact, and* $q(q-\sqrt{q}+1)$ *lines, each meeting* $\mathcal{U}_2$ *in a* $\mathcal{U}_1$ *comprising* $\sqrt{q}+1$ *distinct points.*

(ii) *Through each point of* $\mathcal{U}_2$*, there is one tangent and* q *lines meeting* $\mathcal{U}_2$ *in a* $\mathcal{U}_1$*. Through each point not on* $\mathcal{U}_2$*, there are* $\sqrt{q}+1$ *tangents and* $q-\sqrt{q}$ *lines meeting* $\mathcal{U}_2$ *in a* $\mathcal{U}_1$*.*

Proof. Since a Hermitian form remains Hermitian under any linear transformation of the variables, the intersection of $\mathcal{U}_n$ with any subspace of $PG(n,q)$ is still a Hermitian variety. In particular, any line of $PG(2,q)$ meets $\mathcal{U}_2$ in either a single point or a $\mathcal{U}_1$ comprising $\sqrt{q}+1$ points §6.2.

Let $\mathrm{P}(A)$ be a point of $\mathcal{U}_2 = \mathrm{V}(H)$ with $H = x_0\bar{x}_0 + x_1\bar{x}_1 + x_2\bar{x}_2$. Then

$$H(A+tX) = H(A) + t(\bar{a}_0x_0 + \bar{a}_1x_1 + \bar{a}_2x_2) + \bar{t}(a_0\bar{x}_0 + a_1\bar{x}_1 + a_2\bar{x}_2) + \bar{t}t\,H(X). \tag{7.2}$$

As $\mathrm{P}(A) \in \mathcal{U}_2$, so $H(A) = 0$. Thus the unique tangent at $\mathrm{P}(A)$ is $\ell = \mathrm{V}(\bar{a}_0x_0 + \bar{a}_1x_1 + \bar{a}_2x_2)$. Since $\bar{a}_0x_0 + \bar{a}_1x_1 + \bar{a}_2x_2 = 0$ implies $a_0\bar{x}_0 + a_1\bar{x}_1 + a_2\bar{x}_2 = 0$, so ℓ meets $\mathcal{U}_2$ only at $P(A)$. Thus the $q+1$ lines through $\mathrm{P}(A)$ consist of one tangent meeting $\mathcal{U}_2$ only at $\mathrm{P}(A)$ and q lines meeting $\mathcal{U}_2$ in a $\mathcal{U}_1$. Hence $\mu(2,q) = 1 + q\sqrt{q}$.

If we take $\mu(2,q)$ as known, and there are r lines through $\mathrm{P}(A)$ meeting $\mathcal{U}_2$ there only, and s lines meeting $\mathrm{P}(A)$ in a $\mathcal{U}_1$, then $r+s = q+1$ and $1 + s\sqrt{q} = q\sqrt{q}+1$, whence $r = 1$ and $s = q$. If through Q not on $\mathcal{U}_2$ there are r' tangents and s' lines meeting $\mathcal{U}_2$ in a $\mathcal{U}_1$, that is a $(\sqrt{q}+1)$-secant, then $r' + s' = q+1$ and $r + s'(\sqrt{q}+1) = q\sqrt{q}+1$, whence

$r' = \sqrt{q} + 1$ and $s' = q - \sqrt{q}$.□

COROLLARY. *In the polarity* $\mathfrak{U}_2$ *of* $\mathcal{U}_2$, *a point* P *of* $\mathcal{U}_2$ *is polar to the tangent at* P, *and a point* Q *off* $\mathcal{U}_2$ *is polar to the* $(\sqrt{q}+1)$-*secant joining the points of contact of the* $\sqrt{q}+1$ *tangents through* Q.□

Once the coordinate system in $PG(2,q)$ is chosen, there is a naturally embedded subplane $PG(2,\sqrt{q})$ consisting of those points whose coordinates have ratios all in $GF(\sqrt{q})$. $\mathcal{U}_2$ has $\sqrt{q}+1$ points in common with the subplane; namely, those whose coordinates satisfy $x_0^2 + x_1^2 + x_2^2 = 0$, with x_i in $GF(\sqrt{q})$. The $\sqrt{q}+1$ points form a conic or a line according as q is odd or even.

However, it was shown in theorem 4.3.6, corollary 3, that $PG(2,q)$ can be partitioned into $q - \sqrt{q} + 1$ subgeometries $PG(2,\sqrt{q})$. Each of these contains $\sqrt{q} + 1$ points of $\mathcal{U}_2$, confirming that $\mathcal{U}_2$ contains $(\sqrt{q}+1)(q-\sqrt{q}+1) = q\sqrt{q}+1$ points. For example, in $PG(2,4)$, $\mathcal{U}_2$ consists of nine points which may be divided into three lines $PG(1,2)$. In $PG(2,9)$, $\mathcal{U}_2$ consists of 28 points which can be divided into seven conics over $GF(3)$, each containing four points in a subgeometry $PG(2,3)$.

Now, let us consider the curve $\mathcal{U}_2'$, called the *sub-Hermitian* curve, in $PG(2,q^2)$ with q square given by the same form as $\mathcal{U}_2$, namely

$$\mathcal{U}_2' = V_{2,q^2}(x_0^{\sqrt{q}+1} + x_1^{\sqrt{q}+1} + x_2^{\sqrt{q}+1}).$$

That is, we wish to consider what happens to $\mathcal{U}_2$ when the field is extended from $GF(q)$ to $GF(q^2)$.

7.3.3. THEOREM. *The sub-Hermitian curve* $\mathcal{U}_2'$ *in* $PG(2,q^2)$ *with* q *square has all its points on the Hermitian curve* $\mathcal{U}_2$ *in the subgeometry* $PG(2,q)$.

Proof. For any point P in $PG(2,q^2)$, let $\tilde{P}$ be its conjugate. So if $P = \mathbf{P}(a_0, a_1, a_2)$, then $\tilde{P} = \mathbf{P}(a_0{}^q, a_1{}^q, a_2{}^q)$; thus $P = \tilde{P}$ if and only if P is in the subgeometry $PG(2,q)$. Suppose P is a point of $\mathcal{U}_2'$ and $P \neq \tilde{P}$; then $\tilde{P}$ is also on $\mathcal{U}_2'$.

But $P\tilde{P}$ is a line, the ratios of whose coordinates lie in $GF(q)$. So, by lemma 7.3.2, $P\tilde{P}$ meets $\mathcal{U}'_2$ in one point or $\sqrt{q}+1$ points all of which lie in $PG(2,q)$. Thus we have a contradiction and $P = \tilde{P}$. So the points of $\mathcal{U}'_2$ are just those $q\sqrt{q}+1$ points in $PG(2,q)$ on $\mathcal{U}_2$.□

COROLLARY. *The $q^4 + q^2 + 1$ lines of $PG(2,q^2)$ with q square are of four types:*

(i) *$q\sqrt{q}+1$ tangents, with $(\sqrt{q}+1)$-point contact with $\mathcal{U}'_2$;*
(ii) *$q(q-\sqrt{q}+1)$ lines meeting $\mathcal{U}'_2$ in $\sqrt{q}+1$ distinct points;*
(iii) *$q(q-1)(q\sqrt{q}+1)$ lines meeting $\mathcal{U}'_2$ in just one point;*
(iv) *$q^2(q-1)(q-\sqrt{q}+1)$ lines skew to $\mathcal{U}'_2$.*□

7.4. $PGL(3,q)$

A projectivity $\mathfrak{T} = \mathrm{M}(T)$ in $PG(2,q)$ is given by the equation

$$tY = XT ,$$

where $Y = (y_0, y_1, y_2)$, $X = (x_0, x_1, x_2)$, $T = (t_{ij})$ and $t \in \gamma$. From the fundamental theorem, §2.1(ii), $\mathfrak{T}$ is determined uniquely if the vertices P_i, $i \in \mathrm{N}_4$, of a tetrastigm and their images $P_i\mathfrak{T}$, also the vertices of a tetrastigm, are given. $\mathfrak{T}$ is in fact determined by eight independent conditions. This can be seen in two ways. The matrix T has nine entries but $\mathfrak{T}$ is determined by their ratios and so by eight conditions; alternatively, given the four points P_i, each image $P_i\mathfrak{T}$ is determined by two conditions and so $\mathfrak{T}$ is determined by $4 \times 2 = 8$ conditions.

The order $p(3,q) = q^3(q^3-1)(q^2-1)$. The first target is to calculate $e(3,q)$, the number of conjugacy classes of $PGL(3,q)$.

As in the case of $PGL(2,q)$, we will firstly examine the general linear group and find the number $E(3,q)$ of conjugacy classes in $GL(3,q)$. For each canonical matrix T, we list the elementary divisors (powers of irreducible polynomials by definition) of $xI - T$, any special conditions on the divisors, and the number of distinct canonical matrices of this type (Table 7.4.).

TABLE 7.4

Type	Elementary divisors	Conditions	Number
I	$x - t_0,\ x - t_1,\ x - t_2$	$t_i \in \gamma_0$	$\mathbf{h}(q - 1,3)$
II	$(x - t_0)^2,\ x - t_1$	$t_i \in \gamma_0$	$(q - 1)^2$
III	$(x - t_0)^3$	$t_0 \in \gamma_0$	$q - 1$
IV	$x^2 - a_1x - a_0,\ x - t$	$t \in \gamma_0$	$(q - 1)\ N(2,q)$
V	$x^3 - a_2x^2 - a_1x - a_0$		$N(3,q)$

Now, $\mathbf{h}(q - 1,3) = q(q - 1)(q + 1)/6$, $N(2,q) = q(q - 1)/2$, $N(3,q) = q(q - 1)(q + 1)/3$, §1.6(vii). So

$$E(3,q) = (q^3 - q)/6 + (q - 1)^2 + (q - 1) + q(q - 1)^2/2 + (q^3 - q)/3$$
$$= q^3 - q.$$

The elementary divisors of $xI - T$ where T is the matrix of a projectivity $\mathfrak{T}$ can then be listed as follows with the possible symbols for $\mathfrak{T}$.

$\mathscr{P}(1)$	$x - 1,\ x - t_0,\ x - t_1$	[(1)1,(1)1,(1)1], [(1,1)1,(1)1], [(1,1,1)1]
$\mathscr{P}(2)$	$(x - 1)^2,\ x - t$	[(2)1,(1)1], [(2,1)1]
$\mathscr{P}(3)$	$(x - 1)^3$	[(3)1]
$\mathscr{P}(4)$	$x^2 - a_1x - a_0,\ x - 1$	[(1)2, (1)1]
$\mathscr{P}(5)$	$x^3 - b_2x^2 - b_1x - b_0$	[(1)3]

For the purposes of enumeration, some of the $\mathscr{P}(i)$ must be split into subtypes; also the cases $(q - 1,3) = 1$ and $(q - 1,3) = 3$ must be distinguished. Firstly these will be listed in Table 7.5; then the properties of the different types of projectivity will be discussed. From Table 7.5, we see that, for $(q - 1,3) = 1$, $e(3,q) = q^2 + q$; and, for $(q - 1,3) = 3$, $e(3,q) = q^2 + q + 2$.

7.4.1. THEOREM. *The number of conjugacy classes in $GL(3,q)$ and $PGL(3,q)$ are respectively*

TABLE 7.5

Type	Elementary divisors	$(q-1,3)=1$ Conditions	Number	$(q-1,3)=3$ Conditions	Number	Symbol
$\mathscr{P}(1,1)$	$x-1, x-t_0, x-t_1$	$t_i \in \gamma_{01}$, $t_0 \neq t_1$	$\mathbf{c}(q-2,2)/3$	$t_i \in \gamma_{01} \setminus \{\omega, \omega^2\}$	$[\mathbf{c}(q-2,2)-1]/3$	[(1)1,(1)1,(1)1]
$\mathscr{P}(1,1a)$	$x-1, x-\omega, x-\omega^2$		—	$1+\omega+\omega^2=0$	1	[(1)1,(1)1,(1)1]
$\mathscr{P}(1,2)$	$x-1, x-1, x-t$	$t \in \gamma_{01}$	$q-2$	$t \in \gamma_{01}$	$q-2$	[(1,1)1,(1)1]
$\mathscr{P}(1,3)$	$x-1, x-1, x-1$		1		1	[(1,1,1)1]
$\mathscr{P}(2,1)$	$(x-1)^2, x-t$	$t \in \gamma_{01}$	$q-2$	$t \in \gamma_{01}$	$q-2$	[(2)1,(1)1]
$\mathscr{P}(2,2)$	$(x-1)^2, x-1$		1		1	[(2,1)1]
$\mathscr{P}(3)$	$(x-1)^3$		1		1	[(3)1]
$\mathscr{P}(4)$	$x^2-a_1x-a_0, x-1$		$N(2,q)$		$N(2,q)$	[(1)2,(1)1]
$\mathscr{P}(5)$	$x^3-b_2x^2-b_1x-b_0$		$N(3,q)/(q-1)$	b_1, b_2 *not both zero*	$[N(3,q)-2(q-1)/3]/(q-1)$	[(1)3]
$\mathscr{P}(5a)$	x^3-b_0		—	$b_0 \in \gamma_0$	2	[(1)3]

(i) $E(3,q) = q^3 - q$; (ii) $e(3,q) = q^2 + q - 1 + (q-1,3)$.□ This agrees with (2.6), §2.3.

The $q^2 + q - 1 + (q - 1,3)$ distinct types of projectivity have been divided into $7 + (q - 1,3)$ distinct classes. For each class we give the canonical matrix T of the projectivity $\mathfrak{T}$ and consider some properties of $\mathfrak{T}$: write $\tau = o(\mathfrak{T})$. If $a_1,\ldots,a_k \in \gamma$, write $|a_1,\ldots,a_k|$ for the smallest positive integer n such that $a_1^n = a_2^n = \ldots = a_k^n$.

$\mathscr{P}(1,1)$. $T = \text{diag}(1,t_0,t_1)$, $t_0,t_1 \in \gamma_{01}$, $t_0 \neq t_1$.

(i) $\tau = [o(t_0), o(t_1)]$ where $[m,n]$ denotes the least common multiple of m and n: $\tau | (q - 1)$.

(ii) Fixed points: $\mathbf{U}_0,\mathbf{U}_1,\mathbf{U}_2$.

(iii) Fixed lines: $\mathbf{u}_0,\mathbf{u}_1,\mathbf{u}_2$.

(iv) Orbits: the $(q - 1)^2$ points apart from $\mathbf{u}_0,\mathbf{u}_1,\mathbf{u}_2$ lie in $(q - 1)^2/\tau$ orbits of τ points each. On $\mathbf{u}_0$, the $q - 1$ points other than $\mathbf{U}_1,\mathbf{U}_2$ lie in $(q - 1)/|t_0,t_1|$ orbits of $|t_0,t_1|$ points each. Similarly for $\mathbf{u}_1$ and $\mathbf{u}_2$.

(v) Fixed curves: $\mathbf{V}(k_0x_0^n + k_1x_1^n + k_2x_2^n)$ for $n = k\tau$ with k in $\mathbf{N}$. Each curve contains $r\tau$ points for some r in $\mathbf{N}$ and each orbit lies on at least one of these curves.

(vi) Name: *standard projectivity*.

$\mathscr{P}(1,1a)$. $T = \text{diag}(1,\omega,\omega^2)$, $\omega^2 + \omega + 1 = 0$, $(q-1,3) = 3$.

(i) $\tau = 3$, (ii) - (v) are as in $\mathscr{P}(1,1)$ with $\tau = 3$.

$\mathscr{P}(1,2)$. $T = \text{diag}(1,1,t)$, $t \in \gamma_{01}$.

(i) $\tau = o(t)$, $\tau | (q-1)$.

(ii) Fixed points: $\mathbf{U}_2$ and every point on $\mathbf{u}_2$.

(iii) Fixed lines: $\mathbf{u}_2$ and every line through $\mathbf{U}_2$.

(iv) Orbits: apart from $\mathbf{U}_2$ and $\mathbf{u}_2$, the $q^2 - 1$ points form $(q^2 - 1)/\tau$ orbits of τ points each.

(v) Each orbit lies on a line $\mathbf{V}(k_0x_0 + k_1x_1)$.

(vi) Name: *homology*. $\mathbf{U}_2$ and $\mathbf{u}_2$ are respectively the *centre* and *axis* of the homology.

(vii) If $t = -1$, then $\tau = 2$. If P is not a fixed point and $P\mathbf{U}_2 \cap \mathbf{u}_2 = Q$, then $P\mathfrak{T}$ is the harmonic conjugate of P with respect to $\mathbf{U}_2$ and Q. $\mathfrak{T}$ is called a *harmonic homology*.

$\mathscr{P}(1,3)$. $T = \text{diag}(1,1,1)$.

$\tau = 1$, every point and every line is fixed, and $\mathfrak{T}$ is the *identity*.

$\mathscr{P}(2,1)$. $T = \begin{bmatrix} 1 & 1 & 0 \\ 0 & 1 & 0 \\ 0 & 0 & t \end{bmatrix}$ $t \in \gamma_{01}$

(i) $\tau = pd$, where $d = o(t)$ and $d|(q-1)$.

(ii) Fixed points: $\mathbf{U}_1, \mathbf{U}_2$.

(iii) Fixed lines: $\mathbf{u}_0, \mathbf{u}_2$.

(iv) Orbits: $T^p = \mathrm{diag}(1,1,t^p)$. So $\mathfrak{T}^p$ has $\mathbf{U}_2$ and every point on $\mathbf{u}_2$ as fixed points.

$$T^d = \begin{bmatrix} 1 & d & 0 \\ 0 & 1 & 0 \\ 0 & 0 & 1 \end{bmatrix} \sim \begin{bmatrix} 1 & 1 & 0 \\ 0 & 1 & 0 \\ 0 & 0 & 1 \end{bmatrix}$$

So $\mathfrak{T}^d$ has every point on $\mathbf{u}_2$ fixed. Thus, on $\mathbf{u}_2$, the points apart from $\mathbf{U}_1$ fall into p^{h-1} orbits of p points each ($q = p^h$); on $\mathbf{u}_0$, the $q-1$ points apart from $\mathbf{U}_1$ and $\mathbf{U}_2$ fall into $(q-1)/d$ orbits of d points each. The remaining $q^2 - q$ points fall into $p^{h-1}(q-1)/d$ orbits of τ points each.

(v) Fixed quadrics: $p \neq 2$, only $\mathbf{V}(x_0{}^2)$, $\mathbf{V}(x_2{}^2)$, $\mathbf{V}(x_0x_2)$; $p=2$, these three and $\mathbf{V}(bx_0{}^2 + cx_1(x_0 + x_1))$ for b, c in γ.

(vi) Fixed curves: $\mathbf{V}(bx_0{}^p + cx_1(x_1{}^{p-1} - x_0{}^{p-1}))$.

$\mathscr{P}(2,2)$.

$$T = \begin{bmatrix} 1 & 1 & 0 \\ 0 & 1 & 0 \\ 0 & 0 & 1 \end{bmatrix}$$

(i) $\tau = p$.

(ii) Fixed points: all points on $\mathbf{u}_2$.

(iii) Fixed lines: all lines through $\mathbf{U}_1$.

(iv) Fixed quadrics: $p \neq 2$, pairs of fixed lines;

$$p = 2,\ \mathbf{V}(a_0x_0^2 + a_1x_1^2 + a_2x_2^2 + a_{02}x_0x_2 + a_1x_0x_1).$$

(v) Orbits: on each of the q lines $\mathbf{V}(bx_0 + x_2)$, $b \in \gamma$, through $\mathbf{U}_1$, the q points apart from $\mathbf{U}_1$ fall into p^{h-1} orbits of p points each. For $p = 2$, $\mathfrak{T}$ induces an involution on each of these lines with $\mathbf{U}_1$ as fixed point.

(vi) Name: *elation*.

$\mathscr{P}(3)$.

$$T = \begin{bmatrix} 1 & 1 & 0 \\ 0 & 1 & 1 \\ 0 & 0 & 1 \end{bmatrix}$$

(i) $\tau = p(p,2)$; that is, $\tau = 4$ if $p = 2$ and $\tau = p$ if $p > 2$.

(ii) Fixed point: $\mathbf{U}_2$.

(iii) Fixed line: $\mathbf{u}_0$.

(iv) Fixed quadrics: $\mathbf{V}(bx_0^2 + c(x_1^2 - x_0x_1 - 2x_0x_2))$, for b,c in γ.

For $p > 2$, the pencil of quadrics consists of the repeated line $\mathbf{u}_0$ and q conics, each of which touches $\mathbf{u}_0$ at $\mathbf{U}_2$ and any two of which have quadruple intersection at $\mathbf{U}_2$. Apart from the point $\mathbf{U}_2$, each member of the pencil contains p^{h-1} orbits of p points each.

For $p = 2$, the pencil of quadrics consists of the repeated line $\mathbf{u}_0$, the single point $\mathbf{U}_2$ taken $q/2$ times, and $q/2$ pairs of lines through $\mathbf{U}_2$, the lines of each pair being transposed by $\mathfrak{T}$. On $\mathbf{u}_0$, $\mathfrak{T}$ induces an involution with $\mathbf{U}_2$ as fixed point. The remaining q^2 points fall into $q^2/4$ orbits of four points each with the following property. Suppose $\{l_1,l_2\}$ is one of the fixed line pairs through $\mathbf{U}_2$ and $P_1 \in l_1$ but $P_1 \neq \mathbf{U}_2$. Then $P_1\mathfrak{T} = P_2 \in l_2$, $P_2\mathfrak{T} = P_3 \in l_1$, $P_3\mathfrak{T} = P_4 \in l_2$ and $P_4\mathfrak{T} = P_1$. Let $D_0 = P_1P_2 \cap P_3P_4$ and $D_1 = P_1P_4 \cap P_2P_3$; then $D_0\mathfrak{T} = D_1$ and $D_1\mathfrak{T} = D_0$. So D_0D_1 is fixed by $\mathfrak{T}$ and so is $\mathbf{u}_0$. This incidentally shows that the three diagonal points $D_0,D_1,\mathbf{U}_2$ of the tetrastigm with vertices P_1, P_2, P_3, P_4 are collinear.

$\mathscr{P}(4)$.

$$T = \begin{bmatrix} 0 & 1 & 0 \\ a_0 & a_1 & 0 \\ 0 & 0 & 1 \end{bmatrix} \quad \begin{array}{l} x^2-a_1x-a_0 \text{ irreducible;} \\ \text{over } GF(q^2), x^2-a_1x-a_0=(x-\alpha)(x-\alpha^q); \end{array}$$

here m,n are respectively the smallest integers such that $\alpha^m \in GF(q)$ and $\alpha^n = 1$.

(i) $\tau = n$, $n|(q^2 - 1)$, $m|(q+1)$, $m|n$, $n|[m(q-1)]$; if $a_1 = 0$, then $m = 2$ and $\tau = 2o(a_0)$.

(ii) Fixed point: U_2.

(iii) Fixed line: $\mathbf{u}_2$.

(iv) Orbits: $T^m = \text{diag}(\alpha^m, \alpha^m, 1)$. So $\mathfrak{T}^m$ has U_2 and all points of $\mathbf{u}_2$ as fixed points. Therefore, under $\mathfrak{T}$, the $q + 1$ points on $\mathbf{u}_2$ fall into $(q + 1)/m$ orbits of m points each. The remaining $q^2 - 1$ points fall into $(q^2 - 1)/n$ orbits of n points each.

(v) Fixed conics: When $a_0 = -1$, $\mathfrak{T}$ fixes the pencil of quadrics

$$\mathbf{V}(b(x_0{}^2 + a_1 x_0 x_1 + x_1{}^2) + c\, x_2^2).$$

Apart from the point U_2 and the repeated line $\mathbf{u}_2$, the other q-1 members of the pencil are conics $\mathscr{C}_1, \mathscr{C}_2, \ldots, \mathscr{C}_{q-1}$. Any two $\mathscr{C}_i$ are disjoint. For q odd, U_2 is the pole of $\mathbf{u}_2$ with respect to each $\mathscr{C}_i$; for q even, U_2 is the nucleus of each $\mathscr{C}_i$. Since $a_0 = -1$, $\alpha^{q+1} = 1$; so $\tau | q+1$. Thus each $\mathscr{C}_i$ contains $(q+1)/\tau$ orbits of τ points each. In the case that q+1 is prime (which occurs for $q = 2^{2^k}$ with $k = 0,1,2,3,4$), τ is necessarily q+1.

$\mathscr{P}(5)$.

$$T = \begin{bmatrix} 0 & 1 & 0 \\ 0 & 0 & 1 \\ b_0 & b_1 & b_2 \end{bmatrix}$$

$F(x) = x^3 - b_2 x^2 - b_1 x - b_0$ is irreducible over $GF(q)$; over $GF(q^3)$, $F(x) = (x-\alpha)(x-\alpha^q)(x-\alpha^{q^2})$.

Let m be the smallest positive integer such that $\alpha^m \in GF(q)$.

(i) $\tau = m$, $\tau | (q^2 + q + 1)$.

(ii) There are no fixed points or lines.

(iii) Orbits: there are $(q^2 + q + 1)/m$ orbits of m points each. When $m = q^2 + q + 1$, that is when α is a 3-subprimitive root of $GF(q^3)$, §1.6(v), then $\mathfrak{T}$ is cyclic, theorem 4.2.1; so there is just one orbit. When $q^2 + q + 1$ is prime (for example, $q = 2,3,5,8$), then $\mathfrak{T}$ is necessarily cyclic.

$\mathscr{P}(5a)$.

$$T = \begin{bmatrix} 0 & 1 & 0 \\ 0 & 0 & 1 \\ b_0 & 0 & 0 \end{bmatrix} \quad \begin{matrix} x^3 - b_0 \text{ irreducible} \\ (q-1,3) = 3. \end{matrix}$$

(i) $\tau = 3$.

(ii) No fixed points or lines.

(iii) $(q^2 + q + 1)/3$ orbits of three points each.

A list of all the types of projectivities leaving a conic fixed can be compiled using the isomorphism of $PGO(3,q)$ with $PGL(2,q)$, lemma 7.2.3 corollary 7, and the classification of $PGL(2,q)$, §6.3. An element $\mathfrak{T} = \mathrm{M}(T)$ of $PGL(2,q)$ induces an element $\mathfrak{T}' = \mathrm{M}(T')$ of $PGO(3,q)$, lemma 7.2.3, corollary 7, where

$$T = \begin{bmatrix} a & b \\ c & d \end{bmatrix} \qquad T' = \begin{bmatrix} a^2 & ab & b^2 \\ 2ac & ad+bc & 2bd \\ c^2 & cd & d^2 \end{bmatrix}$$

Let us take the three types of $\mathfrak{T}$ as in Table 6.2.

I. $T = \mathrm{diag}(1,d) \to T' = \mathrm{diag}(1,d,d^2)$

II. $T = \begin{bmatrix} 1 & 1 \\ 0 & 1 \end{bmatrix} \to T' = \begin{bmatrix} 1 & 1 & 1 \\ 0 & 1 & 2 \\ 0 & 0 & 1 \end{bmatrix}$

T' has canonical form S' where, if $p \neq 2$,

$$S' = \begin{bmatrix} 1 & 1 & 0 \\ 0 & 1 & 1 \\ 0 & 0 & 1 \end{bmatrix} \text{ ; if } p = 2,\ S' = \begin{bmatrix} 1 & 1 & 0 \\ 0 & 1 & 0 \\ 0 & 0 & 1 \end{bmatrix}$$

III. $T = \begin{bmatrix} 0 & 1 \\ c & d \end{bmatrix} \to T' = \begin{bmatrix} 0 & 0 & 1 \\ 0 & c & 2d \\ c^2 & cd & d^2 \end{bmatrix}$

Then the canonical form S' is as follows:

(i) $d \neq 0$, $S' = \begin{bmatrix} 0 & 1 & 0 \\ -1 & a_1 & 0 \\ 0 & 0 & 1 \end{bmatrix}$ $\quad a_1 = 2 + d^2/c$.

(ii) $d = 0$, $S' = \text{diag}(1,1,-1)$, which is the same as type I with $d = -1$.

7.5. Perspectivities

Although projectivities on a line have already been investigated, there are some important classical results, which we would be loath to omit, on projectivities from one line to another when both lines lie in the same plane.

Let Π_1, Π_1' be any two lines in $PG(2,q)$ and let P be any point in the plane not on Π_1 or Π_1'. Define the function $\mathfrak{T}:\Pi_1 \to \Pi_1'$ by, if $A \in \Pi_1$, then $A' = A\mathfrak{T} = PA \cap \Pi_1'$: $\mathfrak{T}$ is called a *perspectivity* and P is its *centre*.

7.5.1. LEMMA. *A perspectivity is a projectivity.*

Proof. Let $\Pi_1 = \mathbf{u}_0$, let $\Pi_1' = \mathbf{u}_1$ and let $P = \mathbf{U}$. If $A = \mathbf{P}(0,1,t)$ and $A' = \mathbf{P}(1,0,t')$, then $t' = 1-t$. So $\mathfrak{T}$ is a projectivity (Fig. 6).□

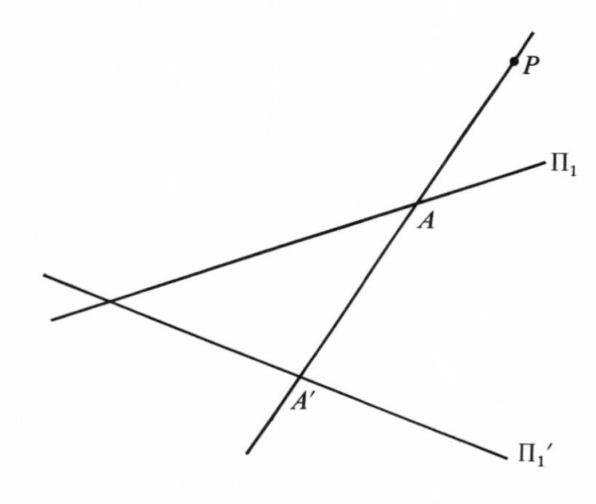

FIG. 6

If $A_i\mathfrak{T} = B_i$ in the above perspectivity for i in $\mathbf{N}_r$, write

$$(A_1, A_2, \ldots, A_r) \overset{P}{\barwedge} (B_1, B_2, \ldots, B_r).$$

If there is a projectivity $\mathfrak{T}\colon \Pi_1 \to \Pi_1'$ for which $A_i\mathfrak{T} = B_i$, i in N_r, write

$$(A_1, A_2, \ldots, A_r) \barwedge (B_1, B_2, \ldots, B_r).$$

7.5.2. LEMMA. *If $\mathfrak{S}\colon \Pi_1 \to \Pi_1'$ is a projectivity such that $Q\mathfrak{S} = Q$, where $Q = \Pi_1 \cap \Pi_1'$ then $\mathfrak{S}$ is a perspectivity.*

Proof. Suppose $A_1\mathfrak{S} = B_1$ and $A_2\mathfrak{S} = B_2$, where neither A_1 nor A_2 is Q. Let $P = A_1B_1 \cap A_2B_2$. So $QA_1A_2 \overset{P}{\barwedge} QB_1B_2$. As a projectivity is determined by the images of three points, this perspectivity is the projectivity $\mathfrak{S}$ (Fig. 7).□

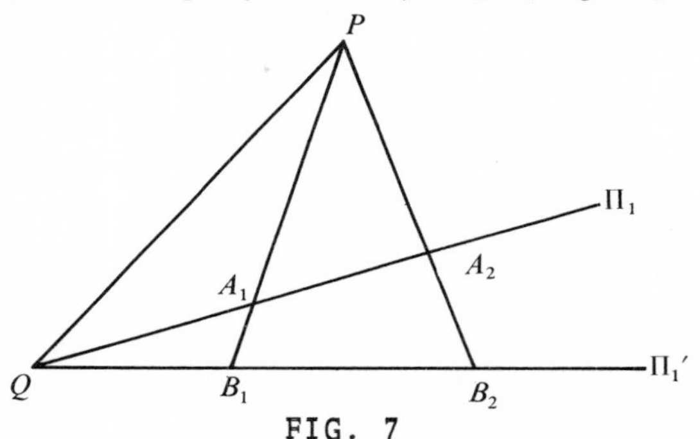

FIG. 7

In §6.1, properties of harmonic tetrads on a line were considered. When the line is imbedded in the plane and the points P_1, P_2, P_3 on the line given, then P_4 can be constructed so that $H(P_1, P_2; P_3, P_4)$.

Given the line $P_1P_2P_3$ take any other line $C\,A\,P_1$. Let $B = AP_3 \cap CP_2$, $D = AP_2 \cap BP_1$ and $P_4 = CD \cap P_1P_2$ (Fig. 8).

7.5.3. THEOREM. $H(P_1, P_2; P_3, P_4)$.

Proof. Let $E = AB \cap CD$. Then

$$(P_1, P_2, P_3, P_4) \overset{C}{\barwedge} (A, B, P_3, E) \overset{D}{\barwedge} (P_2, P_1, P_3, P_4).$$

So $(P_1, P_2, P_3, P_4) \barwedge (P_2, P_1, P_3, P_4)$. If λ is the cross-ratio

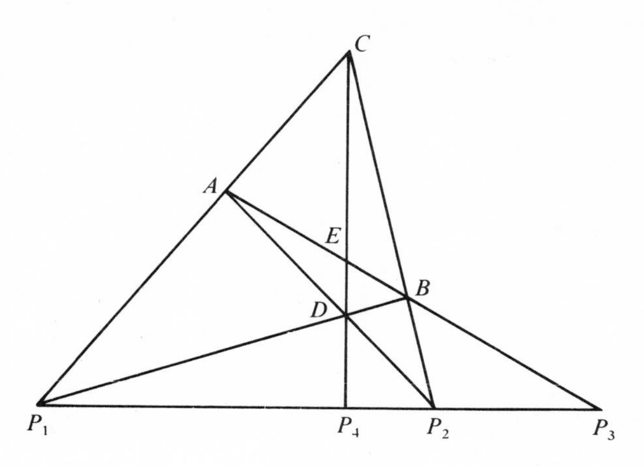

FIG. 8

$\{P_1,P_2;P_3,P_4\}$, then $\{P_2,P_1;P_3,P_4\} = 1/\lambda$. So $\lambda^2 = 1$ and $\lambda = -1$. So $H(P_1,P_2;P_3,P_4)$.

Alternatively, the coordinate system can be chosen so that $P_1 = \mathbf{U}_0$, $P_2 = \mathbf{U}_1$, $P_3 = \mathbf{P}(1,1,0)$, $C = \mathbf{U}_2$ and $A = \mathbf{P}(1,0,1)$. Then $B = \mathbf{P}(0,-1,1)$, $D = \mathbf{P}(1,-1,1)$ and $P_4 = \mathbf{P}(1,-1,0)$. So, again, $H(P_1,P_2;P_3,P_4)$.□

We note that, when $p = 2$, the construction fails, as $P_4 = P_3$.

7.5.4. THEOREM. (*Pappus*) *Given two lines* Π_1, Π_1' *in* $PG(2,q)$ *such that* A, B, $C \in \Pi_1$ *and* A', B', $C' \in \Pi_1'$, *then* $L = BC' \cap CB'$, $M = CA' \cap AC'$ *and* $N = AB' \cap BA'$ *are collinear* (Fig. 9).

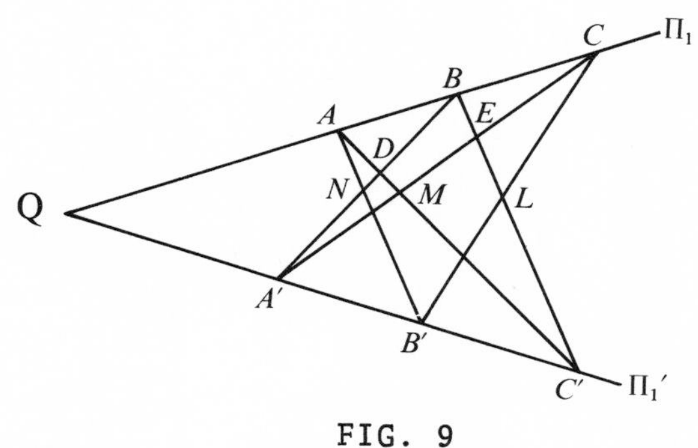

FIG. 9

Proof. Let $D = AC' \cap BA'$, $E = BC' \cap CA'$ and $Q = \Pi_1 \cap \Pi_1'$. Then

$$(A',N,D,B) \overset{A}{\underset{\wedge}{=}} (A',B',C',Q) \overset{C}{\underset{\wedge}{=}} (E,L,C',B).$$

So B is a fixed point in the projectivity $(A',N,D) \barwedge (E,L,C')$. By lemma 7.5.2, this is a perspectivity with centre $A'E \cap DC' = M$. So L, M, and N are collinear.□

7.5.5. THEOREM. (*Desargues*) *If two triangles ABC and A'B'C' are in perspective from R, then the corresponding joins of their sides are collinear.*

Proof.

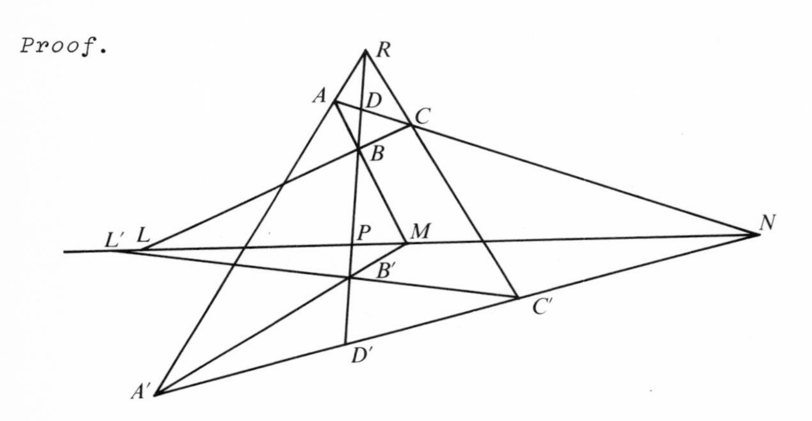

FIG. 10

$M = AB \cap A'B'$, $N = AC \cap A'C'$, $L = BC \cap MN$ and $L' = B'C' \cap MN$. It must be shown that $L = L'$. Only the case that all ten lines drawn and all the points named are distinct will be considered (see Fig. 10).

Let $D = BB' \cap AC$, $D' = BB' \cap A'C'$ and $P = BB' \cap MN$. Then

$$(P,M,N,L) \overset{B}{\underset{\wedge}{=}} (D,A,N,C) \overset{R}{\underset{\wedge}{=}} (D',A',N,C') \overset{B'}{\underset{\wedge}{=}} (P,M,N,L').$$

So $(P,M,N,L) \barwedge (P,M,N,L')$ and this projectivity must be the identity. Hence $L = L'$.□

The two configurations in the theorems of Pappus and Desargues give examples of tactical configurations, §2.2. If $\mathscr{S} = \{A,B,C,A',B',C',L,M,N\}$, then in theorem 7.5.4, these nine points and the nine lines through them form a $(9_3,9_3)$ configuration. Similarly, using the same notation in theorem 7.5.5, the ten points $\mathscr{S} \cup \{R\}$ and the ten lines through them form a $(10_3, 10_3)$ configuration.

Finally, a method is given for constructing the image P' of a point P in a projectivity $\mathfrak{T}: \Pi_1 \to \Pi_1'$ for which $A' = A\mathfrak{T}$, $B' = B\mathfrak{T}$ and $C' = C\mathfrak{T}$.

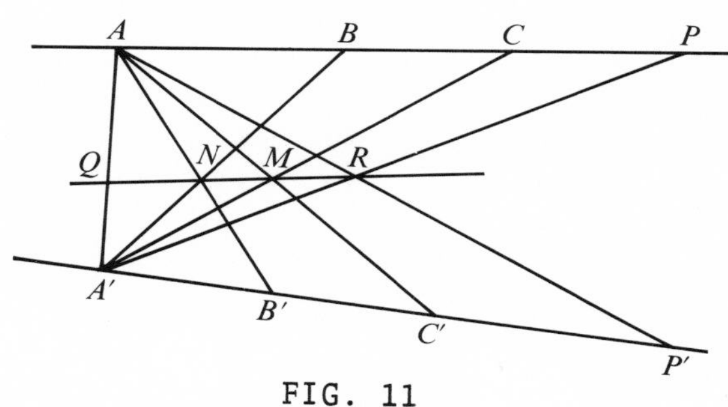

FIG. 11

Let $M = AC' \cap CA'$, $N = AB' \cap BA'$ and $R = MN \cap A'P$. Then $P' = AR \cap \Pi_1'$. For, if $Q = MN \cap AA'$,

$$(A,B,C,P) \overset{A'}{\underset{\wedge}{=}} (Q,N,M,R) \overset{A}{\underset{\wedge}{=}} (A',B',C',P').$$

Hence $(A,B,C,P) \barwedge (A',B',C',P')$ and $P' = P\mathfrak{T}$. By the theorem of Pappus, MN contains all points $PS' \cap SP'$ where $S' = S\mathfrak{T}$. The line is called the *cross-axis* of the projectivity.

7.6. Notes and references

§7.1. For another approach to lemma 7.1.1, see D'Orgeval [1975a]. See also Pickert [1952]. Lemma 7.1.2 comes from Segre [1959a].

§7.2. Properties of a conic are further developed in Chapter 8. For the classification of pencils of plane quadrics, see Campbell [1927b], Dickson [1908], [1915h]. For nets of plane quadrics, see Campbell [1928b] for q even and Wilson [1914] for q odd; for comparison, see Campbell [1928e] for the real case and Jordan [1906] for the complex case, as well as C.T.C. Wall [1977]. For other families of plane quadrics, see Campbell [1927a].

§7.3. For properties of Hermitian curves, see Segre [1965a], [1967], as well as Bose [1959]. For the idea of a sub-Hermitian curve, see Hirschfeld [1967c].

§7.4. For the classification of projectivities, see Mitchell [1911] for q odd and Hartley [1926] for q even. See also Dickson [1901], Hughes and Piper [1973], and assorted papers in the Bibliography on groups acting on projective planes.

§7.5. For properties of the Pappus configuration, see Coxeter [1950], [1976], Mielants [1971], Thas [1971b]. For the Desargues configuration, see Coxeter [1950], [1975b], Ostrom [1955], Ramamurti [1933], Thas and Puystjens [1970].

8
OVALS

8.1. *k-arcs*

As in §3.3, a *k-arc* in $PG(2,q)$ is a set of k points no three of which are collinear. A k-arc $\mathcal{K}$ is *complete* if it is not contained in a $(k+1)$-arc. The maximum number of points that a k-arc can have is $m(2,q)$, and a k-arc with this number of points is an *oval*.

In Chapters 8, 9, and 10, plane k-arcs will be examined in some detail. The main point of interest is how closely they resemble conics. $PG(2,q)$ will mostly be denoted by Π.

Firstly, we will determine $m(2,q)$. For a k-arc $\mathcal{K}$, each line of Π is a 2-secant, a 1-secant, or a 0-secant. As in §3.3, a 2-secant is called a *bisecant* and a 1-secant is called a *unisecant*. Here we will call a 0-secant an *external line*. Thus $\mathcal{K}$ is complete if every point of Π lies on some bisecant of $\mathcal{K}$.

Let P be a point of the k-arc $\mathcal{K}$ and let $t(P)$ be the number of unisecants through P. Also let τ_i be the number of i-secants of $\mathcal{K}$ in the plane; that is, τ_2 is the number of bisecants, τ_2 the number of unisecants, and τ_0 the number of external lines.

8.1.1. LEMMA. (i) $t(P) = q + 2 - k = t$.

(ii) $\tau_2 = k(k-1)/2$, $\tau_1 = kt$, $\tau_0 = q(q-1)/2 + t(t-1)/2$.

Proof. There are $k-1$ bisecants of $\mathcal{K}$ through P and $q+1$ lines altogether; hence $t(P) = (q+1) - (k-1) = t$. Hence there are $c(k, 2)$ bisecants, kt unisecants, and $q^2 + q + 1 - k(k-1)/2 - kt = q(q-1)/2 + t(t-1)/2$ external lines.□

COROLLARY 1. *For a $(q+1)$-arc, $t = 1$, $\tau_2 = q(q+1)/2$, $\tau_1 = q + 1$, $\tau_0 = q(q-1)/2$.*□

COROLLARY 2. *Since a conic is a $(q+1)$-arc, its*

unisecants are its tangents. □

Let Q be a point of Π not on the k-arc $\mathscr{K}$. Let $\sigma_i(Q)$ be the number of i-secants through Q. The number of bisecants $\sigma_2(Q)$ is called the *index of* Q *with respect to* $\mathscr{K}$ and the number of unisecants $\sigma_1(Q)$ is called the *grade of* Q *with respect to* $\mathscr{K}$.

8.1.2. LEMMA. *For any point* Q *in* $\Pi\backslash\mathscr{K}$,

$$\sigma_1(Q) + 2\sigma_2(Q) = k. \square$$

8.1.3. THEOREM. $m(2,q) = \begin{cases} q+2 \text{ for } q \text{ even} \\ q+1 \text{ for } q \text{ odd.} \end{cases}$

Proof. Since $t(P) = q + 2 - k = t \geq 0$ for any k-arc $\mathscr{K}$, so $k \leq q + 2$. When q is even, the unisecants to a conic $\mathscr{C}$ are concurrent at the nucleus N, lemma 7.2.3, corollary 4. So $\mathscr{C} \cup \{N\}$ is a $(q+2)$-arc.

Suppose there exists a $(q+2)$-arc $\mathscr{K}$ for q odd. Then $t = 0$ and so $\sigma_1(Q) = 0$ for any Q in $\Pi\backslash\mathscr{K}$. Thus $2\sigma_2(Q) = q + 2$, which is impossible as q is odd. So $m(2,q) \leq q + 1$. But a conic is always a $(q+1)$-arc. Hence $m(2,q) = q + 1$ for q odd. □

This theorem motivates the problems:

I. To characterize ovals.

II. To find all complete arcs.

The relation between $(q+1)$-arcs and ovals is clarified by the following.

8.1.4. LEMMA. *The* $q+1$ *unisecants to a* $(q+1)$*-arc* $\mathscr{K}$ *in* $PG(2,q)$ *with* q *even are concurrent.*

Proof. For any point Q not in $\mathscr{K}$, $\sigma_1(Q) + 2\sigma_2(Q) = q + 1$. So $\sigma_1(Q) \geq 1$. Hence through every point on a bisecant of $\mathscr{K}$ there passes at least one unisecant. By lemma 8.1.1, $\mathscr{K}$ has $q+1$ unisecants. So through every point on a bisecant of $\mathscr{K}$ there passes exactly one unisecant. Thus two unisecants cannot intersect on a bisecant. So all unisecants are concurrent. □

As for a conic, the point of concurrency of the

unisecants to a $(q+1)$-arc is called the *nucleus*.

COROLLARY. *For q even, a $(q+1)$-arc can be uniquely completed to an oval by adding the nucleus.*□

Another small result on ovals when q is even is the following.

8.1.5. LEMMA. *If two ovals have more than half their points in common when q is even, they coincide.*

Proof. Suppose the ovals $\mathscr{K}$ and $\mathscr{K}'$ have $(q+2)/2 + n$ points in common, with $n > 0$. Take P in $\mathscr{K}'\backslash\mathscr{K}$. As $\mathscr{K}$ has no unisecants any line through P and a point of $\mathscr{K}\backslash\mathscr{K}'$ meets $\mathscr{K}$ again in a point of $\mathscr{K}\backslash\mathscr{K}'$. So $\mathscr{K}$ has $2[(q+2)/2+n] = (q+2) + 2n$ points, contradicting theorem 8.1.3. So there is no such P and $\mathscr{K}' = \mathscr{K}$.□

For q even, we have the example of an oval as a conic plus its nucleus. Then problem I above asks if every oval is of this type. Except for small values of q, the answer is no. This will be shown later in §8.4.

8.2. Ovals for q odd

For q odd, ovals can be classified and this we proceed to do.

8.2.1. LEMMA. *In $\Pi = PG(2,q)$ for q odd, every point off the oval $\mathscr{K}$ lies on exactly two or no unisecants of $\mathscr{K}$.*

Proof. Since $\mathscr{K}$ has $q+1$ points, there is exactly one unisecant through each point of $\mathscr{K}$. Let l be the unisecant at P to $\mathscr{K}$ and let $Q \in l\backslash\{P\}$. Then $\sigma_1(Q) + 2\sigma_2(Q) = q+1$. Since q is odd, $\sigma_1(Q)$ is even. As l itself is a unisecant through Q, $\sigma_1(Q) \geq 2$. Since this is true for each of the q points Q in $l\backslash\{P\}$ and since $\mathscr{K}$ has exactly $q+1$ unisecants, $\sigma_1(Q) = 2$ for all such points Q. That is, through every point of $l\backslash\{P\}$, there is exactly one other unisecant.□

A point of Π is called *external* or *internal* to the oval $\mathscr{K}$ according as it lies on two or no unisecants of $\mathscr{K}$. Hence, with respect to $\mathscr{K}$, the $q^2 + q + 1$ points of Π are partitioned into three classes: $q+1$ points on $\mathscr{K}$, $q(q+1)/2$ external points and $q(q-1)/2$ internal points. $\mathscr{K}$ also exhibits a dual property. The $q^2 + q + 1$ lines of Π are partitioned into three classes with respect to $\mathscr{K}$: $q+1$ unisecants, $q(q+1)/2$ bisecants, $q(q-1)/2$ external lines. Further we have Table 8.1, where each entry gives the number of points of column type on a line of row type.

TABLE 8.1

	Point of $\mathscr{K}$	External point	Internal point
Unisecant	1	q	0
Bisecant	2	$(q-1)/2$	$(q-1)/2$
External line	0	$(q+1)/2$	$(q+1)/2$

Dually, we have Table 8.2 in which each entry gives the number of lines of column type through a point of row type.

TABLE 8.2

	Unisecant	Bisecant	External line
Point of $\mathscr{K}$	1	q	0
External point	2	$(q-1)/2$	$(q-1)/2$
Internal point	0	$(q+1)/2$	$(q+1)/2$

For any k-arc $\mathscr{K}$ with $3 \leqslant k \leqslant q+1$, choose three of its points as the triangle of reference $\mathbf{U}_0\ \mathbf{U}_1\ \mathbf{U}_2$ of the coordinate system. A unisecant to $\mathscr{K}$ through $\mathbf{U}_0$, $\mathbf{U}_1$ or $\mathbf{U}_2$ will have the respective form

$$\mathbf{V}(x_1 - dx_2),\ \mathbf{V}(x_2 - dx_0),\ \text{or}\ \mathbf{V}(x_0 - dx_1),\ \text{with}\ d \neq 0.$$

As in §2.8, d is called the *coordinate* of such a line. Suppose the $t = q+2-k$ unisecants to $\mathscr{K}$ at each of $\mathbf{U}_0$, $\mathbf{U}_1$, $\mathbf{U}_2$ are

$$\mathbf{V}(x_1 - a_i x_2),\ \mathbf{V}(x_2 - b_i x_0),\ \mathbf{V}(x_0 - c_i x_1),\ i \in \mathbf{N}_t.$$

8.2.2. LEMMA. *The coordinates a_i, b_i, c_i of the unisecants at*

U_0, U_1, U_2 *to a k-arc* $\mathcal{K}$ *through these points satisfy* $\prod_{i=1}^{t}(a_i\, b_i\, c_i) = -1$.

Proof. If $P = \mathbf{P}(d_0, d_1, d_2)$ is any point of $\mathcal{K}$ other than U_0, U_1, U_2, then the lines PU_0, PU_1, PU_2 are respectively

$$\mathbf{V}(x_1 - ex_2),\ \mathbf{V}(x_2 - fx_0),\ \mathbf{V}(x_0 - gx_1)$$

where

$$e = d_1/d_2,\ f = d_2/d_0,\ g = d_0/d_1;$$

and so

$$e\,f\,g = 1. \tag{8.1}$$

Through U_0 there are q-1 lines other than U_0U_1 and U_0U_2; they consist of t unisecants $\mathbf{V}(x_1 - a_i x_2)$, i in N_t, and k-3 bisecants $\mathbf{V}(x_1 - e_j x_2)$, j in N_{k-3}. Since the product of the non-zero elements of γ is -1, we have $\Pi a_i\ \Pi e_j = -1$. Similarly, for the q-1 lines through U_1 and U_2 other than the sides of the triangle of reference,

$$\Pi b_i\ \Pi f_j = -1 \text{ and } \Pi c_i\ \Pi g_j = -1.$$

Hence

$$\Pi(a_i\, b_i\, c_i)\ \Pi(e_j\, f_j\, g_j) = -1.$$

But $e_j\, f_j\, g_j = 1$ as in (8.1) for each j. So $\Pi(a_i\, b_i\, c_i) = -1$.□

COROLLARY. *If the unisecants to a* $(q+1)$*-arc* $\mathcal{K}$ *at the points* U_0, U_1, U_2 *of* $\mathcal{K}$ *are respectively* $\mathbf{V}(x_1 - ax_2)$, $\mathbf{V}(x_2 - bx_0)$, $\mathbf{V}(x_0 - cx_1)$, *then* $abc = -1$.□

8.2.3. LEMMA. *The triangles formed by three points of an* $(q+1)$*-arc* $\mathcal{K}$ *and the unisecants at these points are in perspective.*

Proof. Choose the three points as $\mathbf{U}_0$, $\mathbf{U}_1$, $\mathbf{U}_2$; then the unisecants are $\mathbf{V}(x_1 - ax_2)$, $\mathbf{V}(x_2 - bx_0)$, $\mathbf{V}(x_0 - cx_1)$, with $abc = -1$. The vertices V_0, V_1, V_2 of the triangle formed by the unisecants are $\mathbf{P}(c, 1, bc)$, $\mathbf{P}(ca, a, 1)$, $\mathbf{P}(1, ab, b)$. Hence the lines $\mathbf{U}_0V_0$, $\mathbf{U}_1V_1$, $\mathbf{U}_2V_2$ are respectively $\mathbf{V}(x_2 - bcx_1)$, $\mathbf{V}(x_0 - cax_2)$, $\mathbf{V}(x_1 - abx_0)$. These three lines are concurrent if and only if $a^2b^2c^2 = 1$, which is satisfied since $abc = -1$.□

8.2.4. THEOREM. *In $PG(2,q)$ with q odd, every oval is a conic.*

Proof. Let $\mathscr{K}$ be the oval and choose any three of its points as $\mathbf{U}_0$, $\mathbf{U}_1$, $\mathbf{U}_2$. Also let the unit point $\mathbf{U}$ be the point of perspectivity of the triangle $\mathbf{U}_0\mathbf{U}_1\mathbf{U}_2$ and its circumscribed triangle as in the previous lemma. But, by the previous lemma, $\mathbf{U}$ is the intersection of $\mathbf{V}(x_0 - cax_2)$ and $\mathbf{V}(x_1 - abx_0)$, which is $\mathbf{P}(ca, -a, 1)$. So $a = b = c = -1$. Thus the unisecants at $\mathbf{U}_0$, $\mathbf{U}_1$, $\mathbf{U}_2$ to $\mathscr{K}$ are $\mathbf{V}(x_1 + x_2)$, $\mathbf{V}(x_2 + x_0)$, $\mathbf{V}(x_0 + x_1)$. Let $P = \mathbf{P}(y_0, y_1, y_2)$ be any other point of $\mathscr{K}$ with $l = \mathbf{V}(k_0x_0 + k_1x_1 + k_2x_2)$ the unisecant to $\mathscr{K}$ at P. If lemma 8.2.3 is applied to the triangle $P\mathbf{U}_1\mathbf{U}_2$, we obtain

$$(k_0 - k_1 - k_2)\,[k_1(y_0 + y_1) - k_2(y_0 + y_2)] = 0.$$

However, by lemma 8.2.1, the line l cannot pass through the intersection of the unisecants at $\mathbf{U}_1$ and $\mathbf{U}_2$; this point is $\mathbf{P}(1, -1, -1)$ and so $k_0 - k_1 - k_2 \neq 0$. Thus

$$k_1(y_0 + y_1) = k_2(y_0 + y_2);$$

and similarly from the triangles $P\mathbf{U}_2\mathbf{U}_0$ and $P\mathbf{U}_0\mathbf{U}_1$,

$$k_2(y_1 + y_2) = k_0(y_1 + y_0),$$
$$k_0(y_2 + y_0) = k_1(y_2 + y_1).$$

Hence

$$k_0 : k_1 : k_2 = y_1 + y_2 : y_2 + y_0 : y_0 + y_1.$$

But $k_0 y_0 + k_1 y_1 + k_2 y_2 = 0$. So $y_1 y_2 + y_2 y_0 + y_0 y_1 = 0$. Hence all points of $\mathscr{K}$ lie on the conic

$$\mathbf{V}(x_1x_2 + x_2x_0 + x_0x_1).$$

As this conic consists of $q+1$ points, it coincides with $\mathscr{K}$. □

The above method of proof is typical in the subject. A known property of conics is proved for ovals and this property is used to show that an oval is a conic. In other words, a property which characterizes conics among algebraic curves also characterizes ovals among k-arcs.

8.3. Polarities

Hermitian polarities, when q is square, have already been described in §7.3. It remains to consider ordinary polarities for q odd and pseudo polarities for q even.

From §2.1(v), an ordinary polarity in $PG(2,q)$ for q odd has a conic $\mathscr{C} = \mathbf{V}(XTX^*)$ as its set of self-conjugate points. Two points $\mathbf{P}(A)$ and $\mathbf{P}(B)$ are conjugate if $ATB^* = 0$. When $\mathscr{C} = \mathscr{P}_2 = \mathbf{V}(x_0^2 + x_1x_2)$, then

$$ATB^* = 2a_0b_0 + a_1b_2 + a_2b_1.$$

Several properties of the polarity are similar to those in the real plane $PG(2,\mathbf{R})$.

8.3.1. LEMMA. *Let $\mathscr{C} = \mathbf{V}(XTX^*)$ be a conic in $PG(2,q)$, q odd, and let P be a point not on $\mathscr{C}$. Then the set of points Q such that*

(i) *PQ is a bisecant of $\mathscr{C}$,*

(ii) *if $PQ \cap \mathscr{C} = \{R, S\}$ then $H(P,Q;\ R,S)$,*

lies on the polar of P with respect to $\mathscr{C}$.

Proof. Let $P = \mathbf{P}(A)$ and $Q = \mathbf{P}(B)$. Then $P_t = \mathbf{P}(A + tB)$ lies on $\mathscr{C}$ if

$$(A + tB)\ T(A + tB)^* = 0;$$

that is

$$ATA^* + t(ATB^* + BTA^*) + t^2BTB^* = 0. \qquad (8.2)$$

As $T = T^*$, so $BTA^* = (BTA^*)^* = AT^*B^* = ATB^*$. Therefore, (8.2) becomes

$$ATA^* + 2tATB^* + t^2BTB^* = 0. \qquad (8.3)$$

By condition (ii), eqn (8.3) has two solutions $t = r$ and $t = s$, so that $P_r = R$ and $P_s = S$. Then, since $P_0 = P$, and $P_\infty = Q$, we have that $H(P,Q;\ R,S)$ if and only if $r + s = 0$, §6.1; that is, $ATB^* = 0$. Hence Q lies on the polar $\mathbf{V}(ATX^*)$ of P.□

For q odd, let P be any point of $\Pi = PG(2,q)$ and l its polar with respect to $\mathscr{C}$.

8.3.2. THEOREM. *The polar l of P is the tangent to $\mathscr{C}$ at P, an external line, or a bisecant according as P is a point of $\mathscr{C}$, an internal point, or an external point.*

Proof. From the equation of l, it is the tangent at P when P is on $\mathscr{C}$. If P is an external point of $\mathscr{C}$, there are two tangents to $\mathscr{C}$ through P with points of contact P_1 and P_2; then l is the bisecant P_1P_2. From §8.2, $\mathscr{C}$ has q+1 points and tangents, $q(q+1)/2$ external points and bisecants, $q(q-1)/2$ internal points and external lines. Since the polarity is a bijection of Π with its dual, and since the points of $\mathscr{C}$ and its tangents, the external points and the bisecants are paired in the polarity, so are the internal points and the external lines. Thus the polar of each internal point is an external line.

The last statement may also be shown directly as follows. Let A_1A_2 and A_3A_4 be any two bisecants of $\mathscr{C}$ through P and let the diagonal triangle of the tetrastigm with vertices A_1, A_2, A_3, A_4 be PQR. Then l is QR. For, if A_1A_2 meets QR in S_1, then $H(A_1,\ A_2;\ P,\ S_1)$ and so S_1 is on l; similarly, if A_3A_4 meets QR in S_2, then $H(A_3,\ A_4;\ P,\ S_2)$ and S_2 is on l. Equally well, if l_1, l_2, l_3, l_4 are the tangents to $\mathscr{C}$ at

$\mathscr{C}$ at A_1, A_2, A_3, A_4 and $P_1 = l_1 \cap l_2$ and $P_2 = l_3 \cap l_4$, then $l = P_1P_2$. Now, the line QR is the same for any two bisecants of $\mathscr{C}$ through P. So QR meets each bisecant of $\mathscr{C}$ through P in a point not on $\mathscr{C}$, and so QR cannot contain any point of $\mathscr{C}$. Thus QR is an external line (see Fig. 12).□

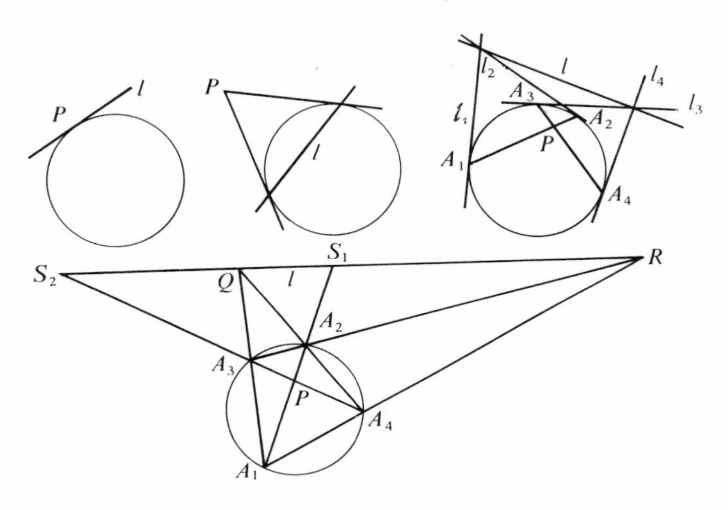

FIG. 12

The concept of internal and external point can also be represented analytically. Let $\mathscr{C} = \mathbf{V}(x_1^2 - x_0x_2)$.

8.3.3. THEOREM. *The point* $\mathbf{P}(Y)$ *with* $Y = (y_0, y_1, y_2)$ *is on* $\mathscr{C}$, *external to* $\mathscr{C}$, *or internal to* $\mathscr{C}$, *according as* $y_1^2 - y_0 y_2$ *is zero, a non-zero square, or a non-square in* γ.

Proof. The polar of $\mathbf{P}(Y)$ is the line $l = \mathbf{V}(x_0y_2 - 2x_1y_1 + x_2y_0)$. So l meets $\mathscr{C}$ at points $\mathbf{P}(s^2, s, 1)$, where s satisfies $s^2y_2 - 2sy_1 + y_0 = 0$. Hence l is a tangent, a bisecant, or an external line, according as $y_1^2 - y_0y_2$ is zero, a non-zero square, or a non-square in γ. The theorem then follows from the previous theorem.□

A *self-polar* triangle of $\mathscr{C}$ is a triangle each of whose vertices has the opposite side as polar. We can now ask how many self-polar triangles there are and what form they take.

The polar of an external point P is a bisecant l, whose points of contact with $\mathscr{C}$ can be taken as $Q = \mathbf{P}(s^2, s, 1)$ and

$R = \mathbf{P}(t^2, t, 1)$. So conjugate pairs on l are pairs which harmonically separate Q and R, namely

$$P_1 = \mathbf{P}(s^2 + \lambda t^2, s + \lambda t, 1 + \lambda)$$

and

$$P_2 = \mathbf{P}(s^2 - \lambda t^2, s - \lambda t, 1 - \lambda).$$

For P_1, $x_1^2 - x_0x_2 = (s + \lambda t)^2 - (s^2 + \lambda t^2)(1 + \lambda) = -\lambda(s - t)^2$; and so, for P_2, $x_1^2 - x_0x_2 = \lambda(s - t)^2$. Thus we must separate the cases in which -1 is a square or a non-square in γ; that is, $q \equiv 1 \pmod 4$ or $q \equiv -1 \pmod 4$, §1.5(viii), (ix).

Let I represent internal points of $\mathscr{C}$ and E external points of $\mathscr{C}$.

8.3.4. THEOREM. *For a conic $\mathscr{C}$ in $PG(2,q)$, q odd, the following are the types and numbers of self-polar triangles.*

(i) $q \equiv 1 \pmod 4$ (a) EEE $q(q^2 - 1)/24$;
(b) EII $q(q^2 - 1)/8$;

(ii) $q \equiv -1 \pmod 4$ (a) III $q(q^2 - 1)/24$;
(b) EEI $q(q^2 - 1)/8$.

Proof. If P is an external point of $\mathscr{C}$ and P_1, P_2 are two conjugate points on the polar l of P, then if $q \equiv 1 \pmod 4$, the pair $\{P_1, P_2\}$ is of type EE or II, and if $q \equiv -1 \pmod 4$ the pair $\{P_1, P_2\}$ is of type EI. Hence, if $q \equiv 1 \pmod 4$, l contains $(q-1)/4$ conjugate pairs EE and $(q-1)/4$ conjugate pairs II: if $q \equiv -1 \pmod 4$, l contains $(q-1)/2$ conjugate pairs EI.

Now let P be an internal point of $\mathscr{C}$ and P_1, P_2 two conjugate points on its polar l. Firstly let $q \equiv 1 \pmod 4$ and suppose $\{P_1, P_2\}$ is of type EE; then on the polar PP_2 of the external point P_1, the conjugate pair $\{P, P_2\}$ is of type EI, contradicting the above. Hence each external point on l has a conjugate point on l that is internal to $\mathscr{C}$. Similarly, when $q \equiv -1 \pmod 4$, suppose that $\{P_1, P_2\}$ is of type EI; then an external point would have a conjugate pair II on its polar.

So there is no conjugate pair of type EI on l. Therefore, if $q \equiv 1 \pmod 4$, l contains $(q+1)/2$ conjugate pairs EI; if $q \equiv -1 \pmod 4$, l contains $(q+1)/4$ conjugate pairs EE and $(q+1)/4$ conjugate pairs II.

If $q \equiv 1 \pmod 4$, the number of self-polar triangles of type EEE is $[q(q+1)/2][(q-1)/4]/3 = q(q^2 - 1)/24$ and of type EII is $[q(q+1)/2][(q-1)/4] = q(q^2 - 1)/8$. If $q \equiv -1 \pmod 4$, the number of triangles of type III is $[q(q-1)/2][(q+1)/4]/3 = q(q^2 - 1)/24$ and of type EEI is $[q(q-1)/2][(q+1)/4] = q(q^2 - 1)/8$.□

For q even, the bilinear form of a pseudo polarity is the same as that for an ordinary polarity for q odd. The points $\mathbf{P}(A)$ and $\mathbf{P}(B)$ are conjugate under the pseudo polarity $\mathfrak{T}$ if $ATB^* = 0$, where $T = T^*$. If $T = (t_{ij})$, then, with $A = (a_0, a_1, a_2)$ and $B = (b_0, b_1, b_2)$,

$$ATB^* = t_{00}a_0b_0 + t_{11}a_1b_1 + t_{22}a_2b_2 + t_{01}(a_0b_1 + a_1b_0) + t_{02}(a_0b_2 + a_2b_0)$$
$$+ t_{12}(a_1b_2 + a_2b_1).$$

The set of self-conjugate points is the line $l_0 = \pi(\surd t_{00}, \surd t_{11}, \surd t_{22})$, and the pole of l_0 is the point $P_0 = \mathbf{P}((\surd t_{00}, \surd t_{11}, \surd t_{22})T^{-1})$, §2.1(v), which is not on l. The polar of a point P on l_0 is therefore PP_0.

In the canonical form of theorem 5.3.3(i),

$$ATB^* = a_0b_0 + a_1b_2 + a_2b_1. \tag{8.4}$$

In this case, $l_0 = \mathbf{u}_0$ and $P_0 = \mathbf{U}_0$.

8.3.5. LEMMA. *For a pseudo polarity in* $PG(2,q)$, *q even, the number of self-polar triangles is* $q(q^2 - 1)/6$.

Proof. A vertex can be any point P other than P_0 or a point of l_0. The next vertex can be any point of $l \setminus l_0$, where l is the polar of P. The third vertex is then determined. Hence the number is $(q^2 - 1)\cdot q/(3\cdot 2) = q(q^2 - 1)/6$.□

8.3.6. LEMMA. $PGPs(3,q) \cong PGSp(2,q)$
$\cong PGL(2,q)$.

Proof. Let the pseudo polarity $\mathfrak{T}$ be given by eqn (8.4). The projectivities commuting with $\mathfrak{T}$ fix $\mathbf{U}_0$, $\mathbf{u}_0$ and the null polarity on $\mathbf{u}_0$ given by $x_1y_2 + x_2y_1 = 0$. So we have the first isomorphism; for the second, see lemma 6.2.2.(i).□

8.4. Ovals for q even

In $PG(2,q)$, q even, there exist ovals comprising the points of conic and its nucleus, theorem 8.1.3: this type of oval is called a *regular* oval.

8.4.1. LEMMA. *For $q = 2$ and 4, every oval in $PG(2,q)$ is regular.*

Proof. In $PG(2,2)$, an oval consists of four points, any three of which form a conic. So, not only is an oval regular, but it consists of a conic plus its nucleus in four ways.

Similarly in $PG(2,4)$, an oval consists of six points, any five of which form a conic. So every oval is regular, and consists of a conic plus its nucleus in six ways.□

To investigate ovals further, it is necessary to find a canonical form for them.

As in §1.3(v), let $\mathscr{P}(q;x)$ denote the set of permutation polynomials over γ in the indeterminate x.

8.4.2. THEOREM. *An oval $\mathcal{O}$ in $PG(2,q)$, q even and $q > 2$, can be written as*

$$\mathscr{D}(F) = \{P(t) = \mathbf{P}(F(t),\ t,\ 1) \mid t \in \gamma^+\} \cup \{\mathbf{U}_1\},$$

such that

(i) $F \in \mathscr{P}(q;\ x)$, $F(0) = 0$, $F(1) = 1$;

(ii) *for each s in γ,*

$$F_s \in \mathscr{P}(q;\ x) \text{ and } F_s(0) = 0,$$

where $$F_s(x) = [F(x+s) + F(s)]/x.$$

Proof. Let $\mathcal{K}$ be a $(q+1)$-arc. Choose $\mathbf{U}_1$ as the nucleus of $\mathcal{K}$ and let $\mathbf{U}_0$, $\mathbf{U}_2$, and $\mathbf{U}$ be points of $\mathcal{K}$. Then the oval $\mathcal{O} = \mathcal{K} \cup \{\mathbf{U}_1\}$.

As $\mathcal{O}$ contains $\mathbf{U}_0$ and $\mathbf{U}_1$ on $\mathbf{u}_2$, it has no other points on this line. Hence $\mathcal{O} \setminus \{\mathbf{U}_0, \mathbf{U}_1\} = \{\mathbf{P}(s_i, t_i, 1) \mid i \in \mathbf{N}_q\}$. Since each line through $\mathbf{U}_0$ contains exactly one other point of $\mathcal{O}$, $t_i \neq t_j$ for $i \neq j$. Similarly, since each line through $\mathbf{U}_1$ contains exactly one other point of $\mathcal{O}$, $s_i \neq s_j$ for $i \neq j$. Hence $\{s_i \mid i \in \mathbf{N}_q\} = \{t_i \mid i \in \mathbf{N}_q\} = \gamma$ and there exists a unique F in $\mathcal{P}(q; x)$ such that $F(t_i) = s_i$, all i. Hence $\mathcal{O} \setminus \{\mathbf{U}_0, \mathbf{U}_1\} = \mathcal{K} \setminus \{\mathbf{U}_0\} = \{\mathbf{P}(F(t), t, 1) \mid t \in \gamma\}$. Now, deg $F > 1$, as otherwise the points of $\mathcal{K} \setminus \{\mathbf{U}_0\}$ would lie on a line, which is impossible for $q > 2$. Therefore we may write $\mathcal{O} = \{\mathbf{P}(F(t), t, 1) \mid t \in \gamma^+\}$, where $t = \infty$ parametrizes $\mathbf{U}_0$. Since $\mathbf{U}_2$ and $\mathbf{U}$ lie on $\mathcal{K}$, we have $F(0) = 0$ and $F(1) = 1$. Hence $\mathcal{O}$ takes the form $\mathcal{D}(F)$ with condition (i) satisfied.

It remains to show that condition (ii) is equivalent to no three points of $\mathcal{D}(F) \setminus \{\mathbf{U}_0, \mathbf{U}_1\}$ being collinear, This is true if and only if

$$\begin{vmatrix} F(t_1) & t_1 & 1 \\ F(t_2) & t_2 & 1 \\ F(t_3) & t_3 & 1 \end{vmatrix} \neq 0,$$

for all distinct t_1, t_2, t_3 in γ; that is,

$$\frac{F(t_1) + F(t_2)}{t_1 + t_2} \neq \frac{F(t_1) + F(t_3)}{t_1 + t_3} .$$

Equivalently, for each s in γ, $[F(t) + F(s)]/(t + s)$ takes a different value in γ_0 for each t in $\gamma \setminus \{s\}$; we could not have $[F(t) + F(s)]/(t + s) = 0$, for then $F(t) = F(s)$ and so $t = s$. So the substitution of $x + s$ for t means that $F_s(x) = [F(x + s) + F(s)]/x$ takes a different value in γ_0 for each x in γ; that is, for each s in γ, the polynomial $F_s(x)$ defines a permutation of γ_0. However, the degree of F_s is less than q-1. So, by §1.3(iv), $F_s(0) = 0$ and $F_s \in \mathcal{P}(q; x)$. Therefore

(ii) is the condition that no three points of $\mathscr{D}(F) \setminus \{U_0, U_1\}$ are collinear.□

When, in the rest of this chapter, we refer to the oval $\mathscr{D}(F)$, then this means $\mathscr{D}(F)$ with conditions (i) and (ii) of the theorem satisfied.

COROLLARY 1. *If $\mathscr{D}(F)$ with $F(x) = \sum_{i=1}^{q-2} a_i x^i$ is an oval in $PG(2,q)$ with q even and $q > 2$, then $F(x) = \sum_{j=1}^{(q-2)/2} a_{2j} x^{2j}$.*

Proof. Since $F_s(x) = [F(x+s) + F(s)]/x$, so $F_s(0) = a_1 + a_3 s^2 + a_5 s^4 + \ldots + a_{q-3} s^{q-4}$. Since $F_s(0) = 0$ for all s in γ, we have $a_1 = a_3 = a_5 = \ldots = a_{q-3} = 0$.□

When F is a monomial, the conditions of the theorem can be simplified. If $F(x) = x^k$, write $\mathscr{D}(k)$ for $\mathscr{D}(F)$.

COROLLARY 2. *In $PG(2,q)$, with q even and $q > 2$, $\mathscr{D}(k)$ is an oval if and only if*

(i) $(k, q-1) = 1$;

(ii) $(k-1, q-1) = 1$;

(iii) $[(x+1)^k + 1]/x \in \mathscr{P}(q;\, x)$.

Proof. $x^m \in \mathscr{P}(q;\, x)$ if and only if $x^m = c$ has a unique solution in γ for each c in γ, which occurs if and only if $(m, q-1) = 1$, §1.5(iv). So condition (i) of the theorem becomes condition (i) here. Similarly, condition (ii) of the theorem for $s = 0$ becomes (ii) here. For $s \neq 0$,

$$F_s(x) = [(x+s)^k + s^k]/x = s^{k-1}[(x/s + 1)^k + 1]/(x/s),$$

which is in $\mathscr{P}(q;\, x)$ if and only if $[(x+1)^k + 1]/x$ is.□

COROLLARY 3. *In $PG(2,2^h)$ with $h > 1$,* $\mathscr{D}(2^n)$ is an oval if *and only if $(n,h) = 1$.*

Proof. If $k = 2^n$, then $[(x+1)^k + 1]/x = x^{k-1}$, so, in corollary 2, (iii) is true if and only if (ii) is. Now, $(2^n, 2^h-1) = 1$; so (i) is satisfied. Also $(2^n-1, 2^h-1) = 2^{(n,h)}-1$. Therefore (ii) is satisfied if and only if $(n,h) = 1$.□

COROLLARY 4. *In $PG(2,2^h)$ with $h > 1$, $\mathscr{D}(2^n)$ is a regular oval if and only if $n = 1$ or $n = h-1$.*

Proof. $\mathscr{D}(2)$ contains the conic $\mathbf{V}(x_1^2 - x_0x_2)$, $\mathscr{D}(2^{h-1})$ the conic $\mathbf{V}(x_2^2 - x_0x_1)$. If $\mathscr{D}(2^n)$ with $1 < n < h - 1$ contains a conic $\mathscr{C} = \mathbf{V}(\Sigma' a_{ij}x_ix_j)$, then $\mathscr{C}$ has at least q points in common with $\mathscr{D}(2^n) \setminus \{\mathbf{U}_1\}$. With $k = 2^n$, these solutions are given by $a_{00} + a_{11}\, t^2 + a_{22}\, t^{2k} + a_{01}\, t + a_{02}\, t^k + a_{12}\, t^{k+1} = 0$. As $1 < n < h - 1$, so $2 < k < k+1 < 2k < q/2$; so there cannot be q solutions. □

COROLLARY 5. *In $PG(2,2^h)$, irregular ovals exist for $h = 5$ and $h \geq 7$.*

Proof. $\phi(h)$, the Euler function, is the number of integers n with $1 \leq n \leq h$ such that $(n, h) = 1$. Since both $(1, h) = 1$ and $(h-1, h) = 1$, corollary 4 tells us that irregular ovals exist if $\phi(h) > 2$. If $p_1, \ldots, p_r$ are the primes dividing h, then $\phi(h) = h(1-1/p_1) \ldots (1-1/p_r)$. For small values, we have

$h =$	1	2	3	4	5	6	7	8	9
$\phi(h) =$	1	1	2	2	4	2	6	4	6.

For $h = 5$ and $h \geq 7$, $\phi(h) > 2$. □

This still leaves the cases $h = 3$, 4 and 6. For $h = 3$, we will see below that every oval is regular. For $h = 4$, all the ovals can be found using a computer and, for example, $\mathscr{D}(F)$ is an irregular oval with

$$F(x) = (\eta^2 x^7 + \eta^{12}x^6 + \eta^6 x^5 + \eta^9 x^4 + \eta^5 x^3 + \eta^5 x^2 + \eta^6 x)^2,$$

where η is a primitive root of $GF(16)$ and $\eta^4 = \eta + 1$, as in §1.7 and Appendix I. In fact, every irregular oval in $PG(2,16)$ is projectively equivalent to this one. It has a projective group of order 36 transitive on its 18 points, and a collineation group of order 144.

COROLLARY 6. (i) *The projective group G of a regular oval $\mathcal{O}$ in $PG(2,q)$ is isomorphic to*

(a) $PGL(2,q)$ *for* $q > 4$;

(b) $\mathbf{A}_6$ *for* $q = 4$;

(c) $\mathbf{S}_4$ *for* $q = 2$.

(ii) *The projective group of an irregular oval $\mathcal{D}(2^n)$ is isomorphic to* $\mathbf{Z}_q\mathbf{Z}_{q-1}$.

Proof. (i) If $\mathcal{O}$ is regular and $q > 4$, then it contains one conic $\mathcal{C}$. So G may be regarded either as the group fixing $\mathcal{C}$ or as the group fixing the pencil of lines through the nucleus of $\mathcal{C}$. For $q=4$ and $q=2$, the group G is transitive on the points of $\mathcal{O}$: see §14.3 and §14.1 for the respective details. With $\mathcal{O} = \mathcal{D}(2)$, each transformation $t \to (at+b)/(ct+d)$, $ad - bc \neq 0$, fixes $\mathcal{O}$.

(ii) When $\mathcal{D}(2^n)$ is an irregular oval, $G = \{t \to ct + d \mid c \in \gamma_0,\ d \in \gamma\}$. Then $G_1 = \{t \to ct \mid c \in \gamma_0\} \cong \mathbf{Z}_{q-1}$ and $G_2 = \{t \to t+d \mid d \in \gamma\} \cong \mathbf{Z}_q$. Hence $G_2 \lhd G$ and $G = G_1G_2$. □

From the example of an irregular oval for $GF(16)$, it would seem that the characterization of ovals is not a simple problem. To restrict the problem, we consider ovals $\mathcal{D}(F)$ firstly when $\mathcal{D}(F) = \mathcal{D}(k)$ for some integer k and then when F has low degree. If $\mathcal{D}(k)$ and $\mathcal{D}(m)$ are projectively equivalent, write $\mathcal{D}(k) \sim \mathcal{D}(m)$.

8.4.3. THEOREM. *If $\mathcal{D}(k)$ is an oval in $PG(2,q)$ with q even, then*

$$\mathcal{D}(k) \sim \mathcal{D}(k_1) \sim \mathcal{D}(k_2) \sim \mathcal{D}(k_3),$$

where k_1, k_2, k_3 are defined by

$$k\,k_1 \equiv 1 \pmod{q-1} \text{ and } 1 < k_1 < q-1\,,$$

$$(k-1)(k_2-1) \equiv 1 \pmod{q-1} \text{ and } 1 < k_2 < q-1,$$

$$k + k_3 = q.$$

Proof. By theorem 8.4.2, corollary 2, we have

$(k, q-1) = 1$ and $(k-1, q-1) = 1$; so the integers k_1, k_2, and k_3 exist.

$\mathscr{D}(k) = \{\mathbf{P}(t^k, t, 1)\} \cup \{\mathbf{U}_1\}$. The substitution $t = u^{k_1}$ makes $\mathscr{D}(k) = \{\mathbf{P}(u, u^{k_1}, 1)\} \cup \{\mathbf{U}_0\} \sim (k_1)$. The substitution $t = s^{-1}$ makes $\mathscr{D}(k) = \{\mathbf{P}(1, s^{k-1}, s^k)\} \cup \{\mathbf{U}_1\}$. The further substitution $s = v^{k_2-1}$ makes $\mathscr{D}(k) = \{\mathbf{P}(1, v, v^{k_2})\} \cup \{\mathbf{U}_1\} \sim$ $\sim \mathscr{D}(k_2)$. The substitution $t = s^{-1}$ above also makes $\mathscr{D}(k) =$ $\{\mathbf{P}(s^{-k}, s^{-1}, 1)\} \cup \{\mathbf{U}_1\} = \{\mathbf{P}(s^{q-k}, 1, s)\} \cup \{\mathbf{U}_2\} \sim \mathscr{D}(q-k) =$ $\mathscr{D}(k_3)$.□

COROLLARY 1. *If $\mathscr{D}(k)$ is an oval in $PG(2,q)$, it is regular if and only if $k = 2$, $q/2$ or $q-2$.*

Proof. When $k = 2$, then $k_1 = q/2$ and $k_3 = q-2$. So $\mathscr{D}(2)$, $\mathscr{D}(q/2)$ and $\mathscr{D}(q-2)$ are all regular and contain the respective conics $\mathbf{V}(x_1^2 - x_0x_2)$, $\mathbf{V}(x_0^2 - x_1x_2)$, $\mathbf{V}(x_2^2 - x_0x_1)$. As in theorem 8.4.2, corollary 4, if $\mathscr{D}(k)$ with $2 < k < q/2$ or $q/2 < k < q-2$ contains a conic $\mathscr{C} = \mathbf{V}(\Sigma' a_{ij}x_ix_j)$, then $\mathscr{C}$ has at least q points in common with $\mathscr{D}(k) \setminus \{\mathbf{U}_1\}$. These points are given by the solutions of

$$a_{00} + a_{11}t^2 + a_{22}t^{2k} + a_{01}t + a_{02}t^k + a_{12}t^{k+1} = 0, \qquad (8.5)$$

of which one may be $t = \infty$. If $2 < k < q/2$, then $2 < k < k+1 <$ $< 2k \leqslant q-2$ since k is even, theorem 8.4.2, corollary 1; so there cannot be q solutions. If $q/2 < k < q-2$, then $q \geqslant 16$ and, since k is even, $q/2 + 2 \leqslant k \leqslant q-4$; then no two terms in (8.5) have the same power modulo $q-1$, and the highest power possible is $q-3$. So again there cannot be q solutions.□

COROLLARY 2. *In $PG(2,8)$, $\mathscr{D}(2)$, $\mathscr{D}(4)$, and $\mathscr{D}(6)$ are all regular ovals.*□

Using theorem 8.4.3, we can analyse the ovals $\mathscr{D}(k)$ in $PG(2,q)$ for $q \leqslant 256$ without much difficulty. This is done in §14.8. For ovals $\mathscr{D}(F)$ with deg F small, we can say the following.

8.4.4. THEOREM. *In $PG(2,q)$ with $q = 2^h$,*

(i) *if* $\deg F = 2$, *then* $\mathscr{D}(F)$ *is an oval if and only if* $\mathscr{D}(F) = \mathscr{D}(2)$;

(ii) *if* $\deg F = 4$, *then* $\mathscr{D}(F)$ *is an oval if and only if* h *is odd and* $\mathscr{D}(F) = \mathscr{D}(4)$;

(iii) *if* $\deg F = 6$, *then* $\mathscr{D}(F)$ *is an oval if and only if* h *is odd and* $F(x) = (x^6 + \lambda x^4 + \lambda^2 x^2)/(1 + \lambda + \lambda^2)$, *some* λ *in* γ: *in this case* $\mathscr{D}(F) \sim \mathscr{D}(6)$.

Proof. By theorem 8.4.2, corollary 1, F has only even terms and no constant term.

(i) $F(x) = a_2 x^2$. As $F(1) = 1$, so $a_2 = 1$, $F(x) = x^2$, and $\mathscr{D}(F) = \mathscr{D}(2)$.

(ii) $F(x) = a_2 x^2 + a_4 x^4$ with $a_4 \neq 0$. As $F(t) = 0$ implies $t = 0$, so $a_2 = 0$; as $F(1) = 1$, so $a_4 = 1$. Hence $F(x) = x^4$ and $\mathscr{D}(F) = \mathscr{D}(4)$.

(iii) $F(x) = a_2 x^2 + a_4 x^4 + a_6 x^6$ with $a_6 \neq 0$. Then $F \in \mathscr{P}(q; x)$ if $F(t) = F(s)$ has no solutions other than $t = s$. If $F(t) + F(s) = 0$, then

$$a_2 + a_4(S + T) + a_6(S^2 + ST + T^2) = 0 ,$$

where $T = t^2$ and $S = s^2$. So, for F to be in $\mathscr{P}(q; x)$, the plane quadric $\mathscr{C}$ with equation

$$a_2 R^2 + a_4 R(S + T) + a_6(S^2 + ST + T^2) = 0$$

has points only for $S = T$ or $R = 0$. As $a_6 \neq 0$, there is a term in ST; so $\mathscr{C}$ is not a repeated line. The equation can also be written as

$$(S + T)(a_4 R + a_6 S) + (\sqrt{a_2} R + \sqrt{a_6} T)^2 = 0.$$

$\mathscr{C}$ is a conic if $S + T$, $a_4 R + a_6 S$ and $\sqrt{a_2} R + \sqrt{a_6} T$ are linearly independent, and this occurs if and only if $a_2 a_6 + a_4^2 \neq 0$. In this case $\mathscr{C}$ has one point with $R = 0$ and at most two points with $S = T$. Since $a_6 \neq 0$, $q \geq 8$ and so there is a point on $\mathscr{C}$ with $R = 1$ and $S \neq T$. Then $F \in \mathscr{P}(q; x)$. If $a_2 a_6 + a_4^2 = 0$, then the equation becomes

$$(\sqrt{a_2}\,R + \sqrt{a_6}\,T)^2 + (\sqrt{a_2}\,R + \sqrt{a_6}\,T)(\sqrt{a_2}\,R + \sqrt{a_6}\,S) + (\sqrt{a_2}\,R + \sqrt{a_6}\,S)^2 = 0.$$

Hence, if h is even and so $x^2 + x + 1 = 0$ has two solutions, $\mathscr{C}$ is a pair of lines. As they are not given by $R(S+T) = 0$, $\mathscr{C}$ contains a point with $R = 1$, $S \neq T$. So, again, $F \notin \mathscr{P}(q;\ x)$. If h is odd and so $x^2 + x + 1$ is irreducible, $\mathscr{C}$ consists of the single point $\mathbf{P}(\sqrt{a_6},\ \sqrt{a_2},\ \sqrt{a_2})$ for which $S = T$. Hence, $F \in \mathscr{P}(q;\ x)$. So $a_2 a_6 = a_4^2$ and, since $F(1) = 1$, $a_2 + a_4 + a_6 = 1$. Since $a_6 \neq 0$, we have the solution

$$a_2 = \lambda^2/(1 + \lambda + \lambda^2),\quad a_4 = \lambda/(1 + \lambda + \lambda^2),\quad a_6 = 1/(1 + \lambda + \lambda^2)$$

and F has the required form. If we now apply the projectivity $x_2 \to x_2$, $x_1 \to \sqrt{\lambda}x_2 + x_1$, $x_0 \to \lambda^3 x_2 + (1 + \lambda + \lambda^2)x_0$, then

$$\begin{aligned}
\mathscr{D}(F) &= \{\mathbf{P}((t^6 + \lambda t^4 + \lambda^2 t^2)/(1 + \lambda + \lambda^2),\ t,\ 1)\} \cup \{\mathbf{U}_1\} \\
&\sim \{\mathbf{P}(t^6 + \lambda t^4 + \lambda^2 t^2 + \lambda^3,\ t + \sqrt{\lambda},\ 1)\} \cup \{\mathbf{U}_1\} \\
&= \{\mathbf{P}((t + \sqrt{\lambda})^6,\ t + \sqrt{\lambda},\ 1)\} \cup \{\mathbf{U}_1\} \\
&= \{\mathbf{P}(t^6,\ t,\ 1)\} \cup \{\mathbf{U}_1\} \\
&= \mathscr{D}(6).
\end{aligned}$$

Now it must be shown that $\mathscr{D}(6)$ is an oval if and only if h is odd. Firstly, if h is odd then $(6,\ 2^h - 1) = (5,\ 2^h - 1) = 1$. If h is even, then $(6,\ 2^h - 1) = 3$.

It remains to show that $[(x + 1)^6 + 1]/x \in \mathscr{P}(q;\ x)$. Now, $[(x + 1)^6 + 1]/x = x^5 + x^3 + x$. So, consider

$$t^5 + t^3 + t = s^5 + s^3 + s. \tag{8.6}$$

Division of (8.6) by $s + t$ gives

$$1 + s^2 + st + t^2 + s^4 + s^3 t + s^2 t^2 + st^3 + t^4 = 0.$$

Substituting $S = s^2 + t^2$ and $T = st$, we have

$$S^2 + S(T + 1) + T^2 + T + 1 = 0. \tag{8.7}$$

If there is a solution in s and t of (8.6) with $s \neq t$, there is a solution in S and T of (8.7). However, (8.7) becomes

$$(S + \omega T + \omega^2)(S + \omega^2 T + \omega) = 0$$

with $\omega^2 + \omega + 1 = 0$. This occurs for h even but not for h odd, §1.5(xi), (xii). So, for h odd, $x^5 + x^3 + x \in \mathscr{P}(q;\ x)$ and $\mathscr{D}(6)$ is an oval.□

COROLLARY 1. *In PG(2,8), every oval is regular.*

Proof. If $\mathscr{D}(F)$ is an oval, deg $F = 2$, 4, or 6, theorem 8.4.2, corollary 1. Then, by the theorem $\mathscr{D}(F) = \mathscr{D}(2)$, $\mathscr{D}(F) = \mathscr{D}(4)$, or $\mathscr{D}(F) \sim \mathscr{D}(6)$, all of which are regular ovals, theorem 8.4.3, corollary 2.□

This corollary is proved by a simpler method in theorem 9.2.3.

8.5. Translation ovals

In $PG(2,q)$ for q even, there is one type of oval that can be characterized. If $\mathscr{D}(F)$ is an oval such that F induces an endomorphism of the additive group of γ, that is, $F(s + t) = F(s) + F(t)$ for all s,t in γ, then $\mathscr{D}(F)$ is called a *translation* oval. This is because $\mathscr{D}(F)$ remains fixed under the elation $\mathfrak{T}_c$, for $c \in \gamma$, given by $x_2 \to x_2$, $x_1 \to x_1 + cx_2$, $x_0 \to x_0 + F(c)x_2$: in the affine plane whose line at infinity is $\mathbf{u}_2$, $\mathfrak{T}_c$ is a translation. $\mathscr{D}(2^n)$ with $(n,h) = 1$ is a translation oval. In fact we wish to show that every translation oval is of this form.

From theorem 8.4.2 and the definition, $\mathscr{D}(F)$ is a translation oval if and only if

I. $F(s + t) = F(s) + F(t)$ for all s, t in γ;
II. $F(x) \in \mathscr{P}(q;\ x)$;
III. $F(x)/x \in \mathscr{P}(q;\ x)$.

8.5.1. LEMMA. *Every endomorphism of $GF(q)$, $q = p^h$, as an additive group is given by a polynomial of the form*

$$F(x) = a_0 x + a_1 x^p + \ldots + a_{h-1} x^{p^{h-1}} \tag{8.8}$$

Proof. $GF(q)$ is a vector space over $GF(p)$. So, let it have a basis $\{x_1, \ldots, x_h\}$. Then an endomorphism of $GF(q)$ is determined by the images of all the elements x_i. As each x_i can have any element of $GF(q)$ as its image, there are q^h endomorphisms of $GF(q)$. However, each polynomial of the form (8.8) induces a distinct endomorphism of $GF(q)$. Since there are q^h such polynomials, each endomorphism of $GF(q)$ is given by such a polynomial.□

8.5.2. LEMMA. *If* $a_m a_n \neq 0$ *and* $m < n < h$, *then* $a_m x^{2^m-1} + a_n x^{2^n-1}$ *is not in* $\mathscr{P}(2^h; x)$.

Proof. By Dickson's criterion, §1.3(vii), it suffices to show that there exists an odd integer $r < 2^h - 2$ such that $(a_m x^{2^m-1} + a_n x^{2^n-1})^r$ modulo $x^{2^h}-x$ contains a term in x^{2^h-1}. The power of the general term in this expression expanded is

$$r(2^m - 1) + k(2^n - 2^m).$$

Let $r = (2^h - 1) - z(2^n - 2^m)$. Then, since we require that

$$r(2^m - 1) + k(2^n - 2^m) \equiv 0 \pmod{2^h - 1},$$

so

$$k(2^n - 2^m) - z(2^m - 1)(2^n - 2^m) \equiv 0.$$

Now, $k = z(2^m - 1)$ is a solution of this equation. Let $d = 2^{(n-m,h)}-1$ and let $R = (2^h - 1)/d$. Then, as $(2^n - 2^m, 2^h - 1) = d$, there are d solutions given by

$$k \equiv z(2^m - 1) + RN, \quad \text{for} \quad N = 0,1,2,\ldots, d-1.$$

We require z such that there is a unique k with $0 < k < r$. In particular, $r = (2^h - 1) - z(2^n - 2^m)$ and $k = z(2^m - 1)$ fulfil our requirements if $k < r < R$.

Put $z = 2^{h-n}$. Then $k = 2^{h-n+m} - 2^{h-n} < 2^{h-n+m} - 1 = r \leqslant$ $\leqslant (2^{h-n+m} - 1)(2^{n-m} - 1)/d < (2^h - 1)/d = R$. So $r = 2^{h-n+m} - 1$

and $k = 2^{h-n+m} - 2^{h-n}$. Then $(1+x)^r = \sum_{i=0}^{r} x^i$; therefore the coefficient of x^k is 1.

Thus it has been shown that, if $a_m a_n \neq 0$, $(a_m x^{2^m-1} + a_n x^{2^n-1})^{2^{h-n+m}-1}$ has exactly one term in x^{2^h-1}. Therefore $a_m x^{2^m-1} + a_n x^{2^n-1}$ is not in $\mathscr{P}(2^h;x)$.□

8.5.3. LEMMA. *If* $a_m a_n \neq 0$ *and* $m < n < h$, *then* $a_m x^{2^m-1} + a_{m+1}x^{2^{m+1}-1} + \ldots + a_n x^{2^n-1}$ *is not in* $\mathscr{P}(2^h;x)$.

Proof. The previous lemma suggests the proof. We again use Dickson's criterion and the same r, namely $r = 2^{h-n+m}-1$, to show that $(a_m x^{2^m-1} + \ldots + a_n x^{2^n-1})^r$ modulo $x^{2^h}-x$ always contains a term in x^{2^h-1}.

Intrinsic to the previous was the identity

$$(2^{h-n+m} - 1)(2^m - 1) + (2^{h-n+m} - 2^{h-n})(2^n - 2^m) = (2^m - 1)(2^h - 1)$$

or

$$(2^{h-n} - 1)2^m + (2^{h-n+m} - 2^{h-n})2^n - 2^{h-n+m} - 1 = (2^m - 1)(2^h - 1).$$

It suffices to take $a_i = 1$, all i. Then

$$(x^{2^m-1} + x^{2^{m+1}-1} + \ldots + x^{2^n-1})^{2^{h-n+m}-1}$$
$$= (x^{2^m} + x^{2^{m+1}} + \ldots + x^{2^n})^{1+2^1+2^2+\ldots+2^{h-n+m-1}} / x^{2^{h-n+m}-1}$$
$$= (x^{2^m} + \ldots + x^{2^n})(x^{2^{m+1}} + \ldots + x^{2^{n+1}}) \ldots (x^{2^{h-n+2m-1}} + \ldots + x^{2^{h+m-1}}) / x^{2^{h-n+m}-1}$$
$$= \Sigma\, x^{2^{m_0}+2^{m_1}+\ldots+2^{m_s}-2^{h-n+m}+1},$$

where $s = h-n+m-1$ and $m + i \leqslant m_i \leqslant n + i$ for i in $\overline{\mathbf{N}}_s$. We require solutions for

$$2^{m_0} + 2^{m_1} + \ldots + 2^{m_s} \equiv 2^{h-n+m} - 1 \pmod{2^h - 1}. \quad (8.9)$$

From the previous lemma (or by the initial identity), there is a solution

$$m_i = m + i \text{ for } i = 0,1,\ldots, h-n-1;$$
$$m_i = n + i \text{ for } i = h-n, h-n+1, \ldots, s.$$

It must be shown that this is the only solution.

Put $m_i = m + r_i$; then $i \leqslant r_i \leqslant n - m + i$ for i in $\bar{\mathrm{N}}_s$. Equation (8.9) now becomes

$$2^m(2^{r_0} + \ldots + 2^{r_s}) \equiv 2^{h-n+m} - 1 (\mathrm{mod}\ 2^h - 1).$$

Since $(2^m, 2^h-1) = 1$ and $2^{h-m}\cdot 2^m - (2^h - 1) = 1$

$$2^{r_0} + 2^{r_1} + \ldots + 2^{r_s} \equiv 2^{h-m}(2^{h-n+m} - 1) \equiv 2^{2h-n} - 2^{h-m}.$$

As $r_i \geqslant i$, so $\Sigma 2^{r_i} \geqslant 1 + 2 + \ldots + 2^s = 2^{h-n+m} - 1$. As $r_i \leqslant n - m + i$, so $\Sigma 2^{r_i} \leqslant 2^{n-m}(2^{h-n+m} - 1) = 2^h - 2^{n-m}$. However, $(2^h - 2^{n-m}) - (2^{h-n+m} - 1) < 2^h - 1$. Therefore $\Sigma 2^{r_i}$ takes a definite value such that $2^{h-n+m} - 1 \leqslant \Sigma 2^{r_i} \leqslant 2^h - 2^{n-m}$. In fact $2^{2h-n} - 2^{h-m} - (2^{h-m} - 1)(2^h - 1) = 2^h - 2^{h-m} + 2^{h-n} - 1$, which lies in the required range. Thus

$$\Sigma 2^{r_i} = 2^h - 2^{h-m} + 2^{h-n} - 1 = 1 + 2 + \ldots + 2^{h-n-1} + 2^{h-m} + \ldots + 2^{h-1}.$$

Written in the binary scale, the number on the right has exactly exactly $h-n+m$ unit digits, which is the number of summands on the left. As $i \leqslant r_i \leqslant n - m + i$, the unique solution is

$$r_i = i \quad \text{for} \quad i = 0,1,\ldots, h-n-1$$

and

$$r_i = n - m + i \quad \text{for} \quad i = h-n, h-n+1, \ldots, h-n+m-1.$$

So there is always a term in x^{2^h-1} in the expansion of $(a_m x^{2^m-1} + \ldots + a_n x^{2^n-1})^{2^{h-n+m}-1}$ provided $a_m a_n \neq 0$.□

8.5.4. THEOREM. *In* $PG(2,2^h)$, $\mathscr{D}(F)$ *is a translation oval if* *and only if* $\mathscr{D}(F) = \mathscr{D}(2^n)$ *with* $(n,h) = 1$.

Proof. If $\mathscr{D}(F) = \mathscr{D}(2^n)$ with $(n,h) = 1$ then by theorem 8.4.2, corollary 3, $\mathscr{D}(F)$ is an oval. Since $(x + y)^{2^n} = x^{2^n} + y^{2^n}$, $\mathscr{D}(F)$ is a translation oval.

Conversely, if $\mathscr{D}(F)$ is a translation oval, then by condition I and lemma 8.5.1,

$$F(x) = a_0 x + a_1 x^2 + \ldots + a_{h-1} x^{2^{h-1}} .$$

By theorem 8.4.2, corollary 1, $a_0 = 0$. By condition III and lemma 8.5.3, $F(x) = a_n x^{2^n}$ for some n in $0 < n < h$. Since $F(1) = 1$ so $a_n = 1$. Finally, again by theorem 8.4.2, corollary 3, for $\mathscr{D}(2^n)$ to be an oval, it is necessary that $(n,h) = 1$. So $\mathscr{D}(F) = \mathscr{D}(2^n)$ with $(n,h) = 1$.□

8.6. q-arcs for q odd

In Chapter 10, we shall consider the problem of how big a k-arc must be in order that it is necessarily contained in an oval. In this section, we shall show that, for q odd, a q-arc in $PG(2,q)$ is contained in a conic; further, if $q \neq 3$, the conic is unique.

Throughout this section, the space will be $PG(2,q)$ with q odd.

8.6.1. LEMMA. *Let* $\mathscr{K}$ *be a* q*-arc. Through each point* P *of* $\mathscr{K}$ *there are exactly two unisecants and through each point* Q *off* $\mathscr{K}$ *there is an odd number of unisecants.*

Proof. This follows directly from lemmas 8.1.1 and 8.1.2.□

As in §8.1, let $\sigma_1(Q)$ be the number of unisecants through a point Q off $\mathscr{K}$.

8.6.2. LEMMA. *Let* $\mathscr{K}$ *be a* q*-arc with* $q > 3$. *Then, for some point* Q *off* $\mathscr{K}$, $\sigma_1(Q) \geq 5$.

Proof. Assume the result is false. Then, as $\sigma_1(Q)$ is odd, so $\sigma_1(Q) = 1$ or 3 for each point Q off $\mathscr{K}$.

Let P be a point of $\mathscr{K}$ and l a unisecant through P. By lemma 8.6.1, $\mathscr{K}$ has $2q$ unisecants. Hence the $2(q-1)$ unisecants not through P meet l in the q points other than P. Let t_i be the number of these points for which $\sigma_1(Q) = i$. Then $t_1 + t_3 = q$, $2t_3 = 2(q-1)$, whence $t_1 = 1$, $t_3 = q-1$. So the points of the plane are partitioned into q points on $\mathscr{K}$, $2q$ points on exactly one unisecant and $q^2 + q + 1 - 2q - q = (q - 1)^2$ points on three unisecants. If we count the number of pairs (Q, s) where Q is a point off and s a unisecant through Q, then

$$3(q - 1)^2 = 2q(q - 1).$$

So $(q - 1)(q - 3) = 0$, whence $q = 1$ or 3: a contradiction.□

Let $\mathscr{K} = \{\mathbf{U}_0, \mathbf{U}_1, \mathbf{U}_2, \mathbf{U}, V, W\}$ be a 6-arc. Then, as no coordinate of V or W can be zero, we may write

$$V = \mathbf{P}(v_0^{-1}, v_1^{-1}, v_2^{-1}), \ W = \mathbf{P}(w_0^{-1}, w_1^{-1}, w_2^{-1}).$$

Then the conditions that $\mathscr{K}$ is a 6-arc are

$$\prod_{i \neq j} (v_i - v_j)(w_i - w_j)(v_i/v_j - w_i/w_j) \neq 0,$$

$$\begin{vmatrix} 1 & 1 & 1 \\ v_0^{-1} & v_1^{-1} & v_2^{-1} \\ w_0^{-1} & w_1^{-1} & w_2^{-1} \end{vmatrix} \neq 0 .$$

Let $\mathfrak{T}_1$ and $\mathfrak{T}_2$ be projectivities leaving each of $\mathbf{U}_0, \mathbf{U}_1, \mathbf{U}_2$ fixed and such that

$$\mathbf{U}\mathfrak{T}_1 = V, \quad W\mathfrak{T}_2 = V. \tag{8.10}$$

Let $V\mathfrak{T}_1 = V_1$ and $V\mathfrak{T}_2 = V_2$.

8.6.3. LEMMA. *The 6-arc $\mathscr{K}$ lies on a conic if and only if $VV_1 = VV_2$, in which case VV_1 is the tangent to the conic at V.*

Proof. Suppose $\mathscr{K}$ lies on a conic $\mathscr{C}$. Then

$$= V(c_0x_1x_2 + c_1x_0x_2 + c_2x_0x_1),$$

where

$$\begin{aligned} c_0 + c_1 + c_2 &= 0, \\ c_0v_0 + c_1v_1 + c_2v_2 &= 0, \\ c_0w_0 + c_1w_1 + c_2w_2 &= 0. \end{aligned} \tag{8.11}$$

Hence, if

$$D = \begin{vmatrix} 1 & 1 & 1 \\ v_0 & v_1 & v_2 \\ w_0 & w_1 & w_2 \end{vmatrix},$$

then $D = 0$ is the condition that $\mathscr{K}$ lies on $\mathscr{C}$.

The projectivities $\mathfrak{T}_i$ are given by $\mathbf{P}(X') = \mathbf{P}(X)\mathfrak{T}_i$ or $\rho X' = XT_i$, where

$$T_1 = \operatorname{diag}(v_0^{-1}, \quad v_1^{-1}, \quad v_2^{-1}),$$

$$T_2 = \operatorname{diag}(w_0v_0^{-1}, w_1v_1^{-1}, w_2v_2^{-1}).$$

So $V_1 = \mathbf{P}(v_0^{-2}, v_1^{-2}, v_2^{-2})$, $V_2 = \mathbf{P}(w_0v_0^{-2}, w_1v_1^{-2}, w_2v_2^{-2})$.

Then

$$\begin{vmatrix} v_0^{-1} & v_1^{-1} & v_2^{-1} \\ v_0^{-2} & v_1^{-2} & v_2^{-2} \\ w_0v_0^{-2} & w_1v_1^{-2} & w_2v_2^{-2} \end{vmatrix} = -(v_0v_1v_2)^{-2}\, D.$$

So $D = 0$ is also the condition that V, V_1, and V_2 are collinear.

The tangent l at V to $\mathscr{C}$ is

$$l = \pi(c_2v_1^{-1} + c_1v_2^{-1},\ c_2v_0^{-1} + c_0v_2^{-1},\ c_1v_0^{-1} + c_0v_1^{-1}).$$

From (8.11),

$$c_0 : c_1 : c_2 = v_2 - v_1 : v_0 - v_2 : v_1 - v_0.$$

Hence

$$l = \pi(v_0^2(v_1 - v_2), v_1^2(v_2 - v_0), v_2^2(v_0 - v_1)) = VV_1. \square$$

Let (A_0, B_0), (A_1, B_1), (A_2, B_2) be pairs of points on the respective sides of the triangle $\mathbf{U}_0\mathbf{U}_1\mathbf{U}_2$ but distinct from the vertices; that is,

$$\begin{aligned} A_0 &= \mathbf{P}(0, a_0, 1), & B_0 &= \mathbf{P}(0, b_0, 1), \\ A_1 &= \mathbf{P}(1, 0, a_1), & B_1 &= \mathbf{P}(1, 0, b_1), \\ A_2 &= \mathbf{P}(a_2, 1, 0), & B_2 &= \mathbf{P}(b_2, 1, 0). \end{aligned} \tag{8.12}$$

Also let A'_0, B'_0 be such that $H(\mathbf{U}_1, \mathbf{U}_2; A_0, A'_0)$, $H(\mathbf{U}_1, \mathbf{U}_2; B_0, B'_0)$ and similarly define A'_1, B'_1, A'_2, B'_2.

8.6.4. LEMMA. (i) *The six points A_0, A_1, A_2, B_0, B_1, B_2 lie on a conic if and only if $a_0a_1a_2b_0b_1b_2 = 1$.*

(ii) *If an odd number of the six points P in* (i) *are replaced by P', then the new set of six points lies on a conic if and only if $a_0a_1a_2b_0b_1b_2 = -1$.*

Proof. (i) The condition for the six points to lie on the conic $\mathscr{C} = \mathbf{V}(\Sigma' c_{ij} x_i x_j)$ is that the determinant of the matrix with rows $(x_0^2, x_1^2, x_2^2, x_1x_2, x_0x_2, x_0x_1)$, obtained by substituting for the six points, is zero; that is

$$(a_0 - b_0)(a_1 - b_1)(a_2 - b_2)(a_0a_1a_2b_0b_1b_2 - 1) = 0.$$

Since $A_i \neq B_i$, this becomes

$$a_0a_1a_2b_0b_1b_2 = 1. \tag{8.13}$$

(ii) Since $A'_0 = \mathbf{P}(0, -a_0, 1)$, the condition for the six points obtained by substituting an odd number of harmonic conjugates to lie on a conic is

$$a_0a_1a_2b_0b_1b_2 = -1.\square \qquad (8.14)$$

If the three pairs $(A_0, B_0), (A_1, B_1), (A_2, B_2)$ of (8.12) satisfy (8.14), they form a *quasi-conical set*.

8.6.5. LEMMA. *If B_0, B_1, B_2 are collinear, then the following are equivalent:*

(i) (A_0, B_0), (A_1, B_1), (A_2, B_2) form a quasi-conical set;

(ii) the triangle $A_0A_1A_2$ is in perspective with the triangle whose sides are A_0B_0, A_1B_1, A_2B_2;

(iii) there exists a conic tangent to A_0B_0, A_1B_1, A_2B_2 at A_0, A_1, A_2 respectively.

Proof. If B_0, B_1, B_2 are collinear, then $b_0b_1b_2 = -1$. So the condition for (i) is $a_0a_1a_2 = 1$. Since $A_iB_i = \mathbf{u}_i$, condition (ii) is that $A_0A_1A_2$ and $\mathbf{U}_0\mathbf{U}_1\mathbf{U}_2$ are in perspective. As $A_0\mathbf{U}_0 = \mathbf{V}(x_1 - a_0x_2)$, $A_1\mathbf{U}_1 = V(x_2 - a_1x_0)$, $A_2\mathbf{U}_2 = \mathbf{V}(x_0 - a_2x_1)$, so (ii) is satisfied when these three lines are concurrent; that is, $a_0a_1a_2 = 1$. Condition (iii) is equivalent to (ii) by lemma 8.2.3.$\square$

8.6.6. LEMMA. *If l is any line containing none of $\mathbf{U}_i$, A_i, B_i but containing a pair of points (A, B) which form a quasi-conical set with every two of the three pairs (A_0, B_0), (A_1, B_1), (A_2, B_2), then these three pairs themselves form a quasi-conical set if and only if the pair (A, B) is unique.*

Proof. We may take $l = \mathbf{u}$. Let $A = \mathbf{P}(\alpha_0, \alpha_1, \alpha_2)$, $B = \mathbf{P}(\beta_0, \beta_1, \beta_2)$, so that $\alpha_0 + \alpha_1 + \alpha_2 = \beta_0 + \beta_1 + \beta_2 = 0$. Let the projectivity $\mathfrak{T}$ be given by

$$x_0' = x_0, \; x_1' = x_1, \; x_2' = x_0 + x_1 + x_2.$$

Then, with $P' = P\mathfrak{T}$, we have

$$A'_0 = \mathbf{P}(0, a_0/(1 + a_0), 1), \quad B'_0 = \mathbf{P}(0, b_0/(1 + b_0), 1),$$
$$A'_1 = \mathbf{P}(1, 0, 1 + a_1), \quad B'_1 = \mathbf{P}(1, 0, 1 + b_1),$$
$$A' = \mathbf{P}(\alpha_0/\alpha_1, 1, 0), \quad B' = \mathbf{P}(\beta_0/\beta_1, 1, 0).$$

So (A, B), (A_0, B_0), (A_1, B_1) form a quasi-conical set if

$$a_0 b_0 (1 + a_1)(1 + b_1)\alpha_0\beta_0 = -(1 + a_0)(1 + b_0)\,\alpha_1\beta_1. \quad (8.15)$$

Similarly, (A, B) forms a quasi-conical set with (A_1, B_1), (A_2, B_2) and with (A_0, B_0), (A_2, B_2) if respectively

$$a_1 b_1\,(1 + a_2)(1 + b_2)\,\alpha_1\beta_1 = -(1 + a_1)(1 + b_1)\,\alpha_2\beta_2, \quad (8.16)$$

$$a_2 b_2\,(1 + a_0)(1 + b_0)\,\alpha_2\beta_2 = -(1 + a_2)(1 + b_2)\,\alpha_0\beta_0. \quad (8.17)$$

These three relations multiplied together give (8.14).

On the other hand, given (8.14), only two of (8.15), (8.16), (8.17) are independent. We may take $\alpha_2 = \beta_2 = 1$. Then (8.16), (8.17) give $\alpha_0\beta_0$ and $\alpha_1\beta_1$. However, as $\alpha_0 = -(\alpha_1 + 1)$ and $\beta_0 = -(\beta_1 + 1)$, (8.16) and (8.17) determine $\alpha_1 + \beta_1$ and $\alpha_1\beta_1$. So α_1, β_1 are roots of a quadratic. As α_1 and β_1 exist in γ, they are unique, as are α_0 and β_0.□

Dual to (8.12), we may consider the following lines through U_0, U_1, U_2:

$$\begin{aligned} l_0 &= \pi(0, 1, -a_0), & m_0 &= \pi(0, 1, -b_0), \\ l_1 &= \pi(-a_1, 0, 1), & m_1 &= \pi(-b_1, 0, 1), \\ l_2 &= \pi(1, -a_2, 0), & m_2 &= \pi(1, -b_2, 0). \end{aligned} \quad (8.18)$$

Then the three pairs (l_0, m_0), (l_1, m_1), (l_2, m_2) form a *quasi-conical* set if (8.14) is satisfied. The duals to lemma 8.6.5 and 8.6.6 will be used below.

8.6.7. LEMMA. *If $\mathcal{K}$ is a q-arc in PG(2, q), q odd, then the three pairs of unisecants at any three points of $\mathcal{K}$ form a quasi-conical set.*

Proof. Let U_0, U_1, U_2 be points of $\mathcal{K}$ and let the unisecants be (8.18). Then, by lemma 8.2.2, $a_0a_1a_2b_0b_1b_2 = -1$.□

If $\mathcal{C}$ is a conic and N one of its points, then $\mathcal{K} = \mathcal{C} \setminus \{N\}$ is a q-arc. At any point P of $\mathcal{K}$, one of the unisecants is the tangent to $\mathcal{K}$ at P and the other is PN.

Suppose conversely that the q-arc $\mathcal{K}$ has the property that one of the unisecants through each of its points passes through a fixed point N. Then $\mathscr{K} \cup \{N\}$ is a $(q+1)$-arc and therefore a conic. Our strategy is to find N.

8.6.8. LEMMA. *Let $\mathscr{K}$ be a q-arc such that through the point N there are four unisecants to $\mathscr{K}$ with points of contact P_0, P_1, P_2, P_3. Then there exists a conic $\mathscr{C}$ through these four points such that the tangent to $\mathscr{C}$ at P_i is the unisecant to $\mathscr{K}$ through P_i other than P_iN.*

Proof. Let l_i be the unisecant to $\mathscr{K}$ at P_i other than P_iN, $i = 0,1,2,3$. By lemma 8.6.7, the unisecants at the three points P_i, P_j, P_k form a quasi-conical set. As three of these unisecants are concurrent at N, the dual of lemma 8.6.5 gives that the triangle $P_iP_jP_k$ is in perspective with the triangle $l_il_jl_k$ and there is a conic touching l_i, l_j, l_k at P_i, P_j, P_k respectively.

Take $P_0 = \mathbf{U}_0$, $P_1 = \mathbf{U}_1$, $P_2 = \mathbf{U}_2$. Let the conic touching l_0, l_1, l_2 at $\mathbf{U}_0$, $\mathbf{U}_1$, $\mathbf{U}_2$ be

$$\mathscr{C} = \mathbf{V}(x_0x_1 + x_0x_2 + x_1x_2).$$

So $l_0 = \mathbf{V}(x_1 + x_2)$, $l_1 = \mathbf{V}(x_0 + x_2)$, $l_2 = \mathbf{V}(x_0 + x_1)$.
Let $P_3 = \mathbf{P}(a_0, a_1, a_2)$ and $l_3 = \pi(\alpha_0, \alpha_1, \alpha_2)$.
Since P_3 lies on neither l_i nor $\mathbf{U}_j\mathbf{U}_k$, $i,j,k = 0,1,2$, so

$$a_0a_1a_2\,(a_0 + a_1)(a_0 + a_2)(a_1 + a_2) \neq 0\,. \tag{8.19}$$

Consider $P_1P_2P_3$ and $l_1l_2l_3$. Let the vertices of the triangle $l_1l_2l_3$ be P_1', P_2', P_3'. Then

$$\begin{aligned}
P'_1 &= \mathbf{P}(\alpha_2, -\alpha_2, \alpha_1 - \alpha_0), & P_1P'_1 &= \pi(\alpha_0 - \alpha_1, 0, \alpha_2)\\
P'_2 &= \mathbf{P}(\alpha_1, \alpha_2 - \alpha_0, -\alpha_1), & P_2P'_2 &= \pi(\alpha_0 - \alpha_2, \alpha_1, 0)\\
P'_3 &= \mathbf{P}(-1, 1, 1), & P_3P'_3 &= \pi(a_1 - a_2, -a_0 - a_2, a_0 + a_1).
\end{aligned}$$

As the triangles are in perspective, so $D = 0$, where

$$D = \begin{vmatrix} \alpha_0 - \alpha_1 & 0 & \alpha_2 \\ \alpha_0 - \alpha_2 & \alpha_1 & 0 \\ a_1 - a_2 & -a_0 - a_2 & a_0 + a_1 \end{vmatrix}.$$

Thus $a_0a_1a_2D = \begin{vmatrix} a_0(\alpha_0 - \alpha_1) & 0 & a_2\alpha_2 \\ a_0(\alpha_0 - \alpha_2) & a_1\alpha_1 & 0 \\ a_0(a_1 - a_2) & -a_1(a_0 + a_2) & a_2(a_0 + a_1) \end{vmatrix}.$ (8.20)

Since P_3 lies on l_3, we have

$$a_0\alpha_0 + a_1\alpha_1 + a_2\alpha_2 = 0 \tag{8.21}$$

So, the addition of the other columns to the first in (8.20) gives

$$\begin{aligned} a_0D &= \begin{vmatrix} -\alpha_1(a_0 + a_1) & 0 & \alpha_2 \\ -\alpha_2(a_0 + a_2) & \alpha_1 & 0 \\ 0 & -a_0 - a_2 & a_0 + a_1 \end{vmatrix} \\ &= -\alpha_1{}^2(a_0 + a_1)^2 + \alpha_2{}^2(a_0 + a_2)^2. \end{aligned}$$

Thus

$$\alpha_1{}^2\ (a_0 + a_1)^2 = \alpha_2{}^2\ (a_0 + a_2)^2$$

and, from the remaining pairs of perspective triangles,

$$\alpha_2{}^2\ (a_1 + a_2)^2 = \alpha_0{}^2\ (a_1 + a_0)^2,$$

$$\alpha_0{}^2\ (a_2 + a_0)^2 = \alpha_1{}^2\ (a_2 + a_1)^2.$$

So, $\alpha_0 : \alpha_1 : \alpha_2 = \varepsilon_0(a_1 + a_2) : \varepsilon_1(a_0 + a_2) : \varepsilon_2(a_0 + a_1)$.

where $\varepsilon_i{}^2 = 1$, each i. Therefore, we may either take exactly two or all three of the ε_i to be 1. If, say $\varepsilon_0 = \varepsilon_1 = 1$, $\varepsilon_2 = -1$, then, by (8.21),

$$a_0(a_1 + a_2) + a_1(a_0 + a_2) - a_2(a_0 + a_1) = 0,$$

whence

$$2a_0\, a_1 = 0,$$

in contradiction to (8.19). So

$$l_3 = \pi(a_1 + a_2,\ a_0 + a_2,\ a_0 + a_1).$$

Hence l_3 is the polar of P_3 with respect to $\mathscr{C}$. As P_3 lies on l_3, so P_3 lies on $\mathscr{C}$ and l_3 is the tangent to $\mathscr{C}$ at P_3.□

We now prove the essential lemma. For a point V, let $\mathfrak{T} = \mathfrak{T}_{P_3,V}$ be the projectivity such that $P_0\mathfrak{T} = P_0$, $P_1\mathfrak{T} = P_1$, $P_2\mathfrak{T} = P_2$, $P_3\mathfrak{T} = V$. Write $V\mathfrak{T} = V'$.

Consider the case that $P_0 = \mathbf{U}_0$, $P_1 = \mathbf{U}_1$, $P_2 = \mathbf{U}_2$. Then, from (8.10), $\mathfrak{T}_{\mathbf{U},V} = \mathfrak{T}_1$ with $V' = V_1$, and $\mathfrak{T}_{W,V} = \mathfrak{T}_2$ with $V' = V_2$.

8.6.9. LEMMA. *Let $\mathscr{K}$ be a q-arc such that through the point N there are four unisecants with points of contact P_0, P_1, P_2, P_3. If V is any other point of $\mathscr{K}$ such that VV' does not contain N, then V lies on a fixed conic $\mathscr{C}$.*

Proof. We use almost the same notation as the previous lemma. Let $P_0 = \mathbf{U}_0$, $P_1 = \mathbf{U}_1$, $P_2 = \mathbf{U}_2$. However, in this case, let $P_3 = \mathbf{U}$. So $\mathfrak{T} = \mathfrak{T}_1$ and $V' = V_1$. Then by lemma 8.6.8, there exists a conic $\mathscr{C}$ through P_0, P_1, P_2, P_3 such that the unisecant through P_i other than P_iN is tangent to $\mathscr{C}$ at P_i. So

$$\mathscr{C} = \mathbf{V}(c_0x_1x_2 + c_1x_0x_2 + c_2x_0x_1)$$

where
$$c_0 + c_1 + c_2 = 0.$$

Let $V = \mathbf{P}(v_0^{-1},\ v_1^{-1},\ v_2^{-1})$ and $N = \mathbf{P}(n_0,\ n_1,\ n_2)$.

We now apply lemma 8.6.7 to each triad of points in the set $\{\mathbf{U}_0, \mathbf{U}_1, \mathbf{U}_2, \mathbf{U}, V\}$. The pairs of unisecants to $\mathscr{K}$ at $\mathbf{U}_0$, $\mathbf{U}_1$, $\mathbf{U}_2$ are respectively

$$\mathbf{V}((c_1x_2 + c_2x_1)(n_1x_2 - n_2x_1)), \qquad (8.22_0)$$

$$\mathbf{V}((c_2x_0 + c_0x_2)(n_2x_0 - n_0x_2)), \qquad (8.22_1)$$

$$\mathbf{V}((c_0x_1 + c_1x_0)(n_0x_1 - n_1x_0)). \qquad (8.22_2)$$

Let

$$\begin{aligned} r_0 &= (c_1v_1 + c_2v_2)(n_1v_1 - n_2v_2), \\ r_1 &= (c_2v_2 + c_0v_0)(n_2v_2 - n_0v_0), \\ r_2 &= (c_0v_0 + c_1v_1)(n_0v_0 - n_1v_1). \end{aligned} \qquad (8.23)$$

As V lies on none of the unisecants to $\mathscr{K}$ through $\mathbf{U}_0$, $\mathbf{U}_1$, $\mathbf{U}_2$, so $r_0r_1r_2 \neq 0$. The unisecants to $\mathscr{K}$ through $\mathbf{U}$ are

$$\mathbf{V}([(c_1 + c_2)x_0 + (c_2 + c_0)x_1 + (c_0 + c_1)x_2][(n_1 - n_2)x_0 + (n_2 - n_0)x_1 + (n_0 - n_1)x_2]). \qquad (8.24)$$

Since the unisecants to $\mathscr{K}$ through V contain none of $\mathbf{U}_0$, $\mathbf{U}_1$, $\mathbf{U}_2$, they are

$$\mathbf{V}([(v_1x_1 - v_0x_0) - \alpha(v_2x_2 - v_0x_0)][(v_1x_1 - v_0x_0) - \beta(v_2x_2 - v_0x_0)]), \qquad (8.25)$$

where $\alpha\beta(\alpha - \beta) \neq 0$.

With some manipulation the conditions that (8.25) forms a quasi-conical set with pairs of (8.22_i) are

$$\begin{aligned} \alpha\beta &= r_2/r_1, \\ (\alpha - 1)(\beta - 1)/(\alpha\beta) &= r_0/r_2, \\ 1/[(\alpha - 1)(\beta - 1)] &= r_1/r_0, \end{aligned}$$

whence

$$\alpha + \beta = (r_1 + r_2 - r_0)/r_1.$$

Some more manipulation gives the condition for (8.25) to form a quasi-conical set with (8.24) and (8.22_0); this is

$$F_0F_1 = F_2,$$

where

$$F_0 = c_0v_1v_2 + c_1v_0v_2 + c_2v_0v_1 ,$$
$$F_1 = n_0v_0(v_1 - v_2) + n_1v_1(v_2 - v_0) + n_2v_2(v_1 - v_0),$$
$$F_2 = r_0(v_0 - v_1)(v_0 - v_2) + r_1(v_1 - v_0)(v_1 - v_2) + r_2(v_2 - v_0)(v_2 - v_1).$$

Using the fact that $c_0 + c_1 + c_2 = 0$, we obtain that

$$F_0F_1 - F_2 = - 2G_0G_1 ,$$

where

$$G_0 = c_0v_0 + c_1v_1 + c_2v_2 ,$$
$$G_1 = n_0v_0^2(v_1 - v_2) + n_1v_1^2(v_2 - v_0) + n_2v_2^2(v_0 - v_1).$$

Hence $G_0 = 0$ or $G_1 = 0$. But $G_1 = 0$ is just the condition that the points $N = \mathrm{P}(n_0, n_1, n_2)$, $V = \mathrm{P}(v_0^{-1}, v_1^{-1}, v_2^{-1})$ and $V' = \mathrm{P}(v_0^{-2}, v_1^{-2}, v_2^{-2})$ are collinear. Therefore, if $G_1 \neq 0$, then $G_0 = 0$, which is exactly the condition that V lies on $\mathscr{C}$.□

8.6.10. THEOREM. *In $PG(2,q)$, q odd, a q-arc $\mathscr{K}$ lies on a conic $\mathscr{C}$; the number of such conics is one or four as $q \neq 3$ or $q = 3$.*

Proof. If $q = 3$, the bisecants of $\mathscr{K}$ contain nine points leaving four in the plane, any of which form a conic with $\mathscr{K}$.

Let $q > 3$. Then, by lemma 8.6.2, there is a point N through which there are at least five unisecants of $\mathscr{K}$. Let the points of contact be U_0, U_1, U_2, U, and W. By lemma 8.6.8, there is a conic $\mathscr{C}$ through U_0, U_1, U_2, and U such that $\mathscr{C}$ touches the unisecant through these points which does not contain N. Similarly, there is a conic $\mathscr{C}'$ through U_0, U_1, U_2, and W touching the appropriate unisecants. However, $\mathscr{C}' = \mathscr{C}$, since the conic is determined by passing through U_0, U_1, U_2 and having as tangents the unisecants to $\mathscr{K}$ other than U_iN.

Let V be a point of $\mathscr{K}$ and suppose that V is not on $\mathscr{C}$. In lemma 8.6.9, let $P_0 = \mathbf{U}_0$, $P_1 = \mathbf{U}_1$, $P_2 = \mathbf{U}_2$. Then the lemma says that, with $P_3 = \mathbf{U}$, the line VV_1 contains N, and, with $P_3 = W$, that VV_2 contains N. Therefore $VV_1 = VV_2$ and so, by lemma 8.6.3, V lies on $\mathscr{C}$: a contradiction. So V lies on $\mathscr{C}$ and $\mathscr{K} \subseteq \mathscr{C}$. In fact, $\mathscr{C} = \mathscr{K} \cup \{N\}$.□

8.7. *q*-arcs for *q* even

In this section, we wish to show a comparable result to the last section: in $PG(2,q)$ for q even, a q-arc is contained in an oval, which is unique for $q > 2$. The methods of §8.6 can be adapted to prove this, but we give an independent method here.

8.7.1. LEMMA. *In $PG(2,q)$, q even and $q > 2$, let l be a unisecant to the q-arc $\mathscr{K}$ and let $l \cap \mathscr{K} = \{P\}$. Then there exists a point Q on l for which $\sigma_1(Q) \geq 3$.*

Proof. If there is no such point Q, then the $2(q-1)$ unisecants through the points of $\mathscr{K} \setminus \{P\}$ all meet l in distinct points. Hence $q \geq 2(q-1)$ and so $q \leq 2$: a contradiction.□

8.7.2. THEOREM. *In $PG(2,q)$, q even, a q-arc $\mathscr{K}$ is contained in an oval; the number of such ovals is one or two as $q > 2$ or $q = 2$.*

Proof. Firstly, for $q = 2$, let $\mathscr{K} = \{P_1, P_2\}$. Then there are four points off the line P_1P_2. If we choose one of these, say P_3, then there is only one point on no side of the triangle with vertices P_1, P_2, P_3. Hence $\mathscr{K}$ is contained in $4/2 = 2$ ovals.

Now let $q > 2$. From lemma 8.7.1, let Q be a point off through which three unisecants l_0, l_1, l_2 pass. Choose a coordinate system such that $l_0 \cap \mathscr{K} = \{\mathbf{U}_0\}$, $l_1 \cap \mathscr{K} = \{\mathbf{U}_1\}$, $l_2 \cap \mathscr{K} = \{\mathbf{U}_2\}$. Then

$$l_0 = \mathbf{V}(x_1 + a_0x_2),\ l_1 = \mathbf{V}(x_2 + a_1x_0),\ l_2 = \mathbf{V}(x_0 + a_2x_1), \quad (8.20)$$

for some a_0, a_1, a_2 in γ_0 with

$$a_0\, a_1\, a_2 = 1. \tag{8.27}$$

Let the other unisecants at $\mathbf{U}_0$, $\mathbf{U}_1$, $\mathbf{U}_2$ be m_0, m_1, m_2 respectively. Then

$$m_0 = \mathbf{V}(x_1 + b_0 x_2),\ m_1 = \mathbf{V}(x_2 + b_1 x_0),\ m_2 = \mathbf{V}(x_0 + b_2 x_1).$$

By lemma 8.2.2,

$$a_0 a_1 a_2 b_0 b_1 b_2 = 1, \tag{8.28}$$

whence, from (8.27),

$$b_0 b_1 b_2 = 1.$$

So m_0, m_1, m_2 are concurrent at a point Q'. Let $\mathscr{K}' = \mathscr{K} \cup \{Q, Q'\}$. Then we show that $\mathscr{K}'$ is an oval.

Let P be a point of $\mathscr{K} \setminus \{\mathbf{U}_0, \mathbf{U}_1, \mathbf{U}_2\}$ and let l and m be the unisecants to $\mathscr{K}$ at P. It suffices to show that $\{l, m\} = \{PQ, PQ'\}$; for, then PQ and PQ' are unisecants to $\mathscr{K}$, whence $\mathscr{K}'$ is an oval.

Choose the coordinate system so that $P = \mathbf{U}$. Let

$$l = \pi(1, c, 1+c), \qquad m = \pi(1, d, 1+d),$$
$$\mathbf{U}Q = \pi(1, c', 1+c'), \qquad \mathbf{U}Q' = \pi(1, d', 1+d').$$

Now, we apply lemma 8.2.2 to the triangles $\mathbf{U}_0\mathbf{U}_1\mathbf{U}$ and $\mathbf{U}_0\mathbf{U}_2\mathbf{U}$. To do this, we use the projectivity $\mathfrak{T}$ which fixes $\mathbf{U}_0$ and $\mathbf{U}_1$, and interchanges $\mathbf{U}$ and $\mathbf{U}_2$. In coordinates, $\mathbf{P}(X') = \mathbf{P}(X)\mathfrak{T}$ where $x_0' = x_0 + x_2$, $x_1' = x_1 + x_2$, $x_2' = x_2$. Then

$$l_0\mathfrak{T} = \pi(0, 1, t_0), \qquad m_0\mathfrak{T} = \pi(0, 1, t_0'),$$
$$l_1\mathfrak{T} = \pi(t_1, 0, 1), \qquad m_1\mathfrak{T} = \pi(t_1', 0, 1),$$
$$l\mathfrak{T} = \pi(1, \text{[illegible]}\ 0), \qquad m\mathfrak{T} = \pi(1, d, 0),$$
$$(\mathbf{U}Q)\mathfrak{T} = \pi(1, c', 0), \qquad (\mathbf{U}Q')\mathfrak{T} = \pi(1, d', 0).$$

So, by lemma 8.2.2 as in (8.28),

$$t_0 t_1 c\, t'_0 t'_1 d = 1 \tag{8.29}$$

As l_0, l_1, and UQ are concurrent at Q, and as m_0, m_1, and UQ' are concurrent at Q', we have respectively

$$t_0 t_1 c' = 1, \qquad t'_0 t'_1 d' = 1. \tag{8.30}$$

From (8.29) and (8.30),

$$cd = c'd' . \tag{8.31}$$

Now, for UU_0U_2, we use the projectivity $\mathfrak{S}$ which fixes U_0 and U_2, and interchanges U and U_1. It is given by $x'_0 = x_0 + x_1$, $x'_1 = x_1$, $x'_2 = x_1 + x_2$. Then

$$l\mathfrak{S} = \pi(1,\ 0,\ 1+c), \qquad m\mathfrak{S} = \pi(1,\ 0,\ 1+d),$$

$$(UQ)\mathfrak{S} = \pi(1,\ 0,\ 1+c'), \qquad (UQ')\mathfrak{S} = \pi(1,\ 0,\ 1+d').$$

So, similarly to (8.31), we obtain

$$(1 + c)(1 + d) = (1 + c')(1 + d'), \tag{8.32}$$

whence, from (8.31)

$$c + d = c' + d'.$$

Hence $\{c, d\} = \{c', d'\}$ and so $\{l, m\} = \{PQ, PQ'\}$.

If $\mathcal{K}$ were contained in a second oval $\mathcal{K}'$, then, from lemma 8.1.5, $q \leqslant (q + 2)/2$ and $q \leqslant 2$.□

8.8. Notes and references

§8.1. The value of $m(2,q)$ is due to Bose [1947]. See also Qvist [1952].

§8.2. This is taken from Segre [1955a], [1960a]. An alternative proof of theorem 8.2.4 was given by Cossu [1960]; see also Barlotti [1965].

For q odd, the external lines of a conic were characterized by Segre and Korchmáros [1977] as the sets $\mathscr{L}$ of points such that every tangent or bisecant of the conic meets $\mathscr{L}$ in exactly one point. For q even, the external lines of a conic were characterized by Bruen and Thas [1975] as the sets $\mathscr{L}$ of $q+1$ points such that the join of any pair of points of $\mathscr{L}$ is an external line of the conic.

For an abstract account of ovals, see Buekenhout [1966a], [1966c], Korchmáros [1975/6]. See also Segre [1973].

§8.3. Theorem 8.3.3 is due to Qvist [1952]. Theorem 8.3.4 is taken from Edge [1956]. See also Raber [1975].

§8.4. This treatment of ovals is taken from Hirschfeld [1975], although the results of theorems 8.4.2 and its corollaries are due to Segre [1957a], [1962b]. See also Hirschfeld [1971], Segre and Bartocci [1971].

The first example of an irregular oval in $PG(2,16)$ was given by Lunelli and Sce [1958a]. That this example is unique was shown by Hall [1975], whence the oval given is derived. Hall [1975] also showed that the oval was fixed by a group of order 144. That this group is the full collineation group of the oval has been shown by S.E. Payne and J.E. Conklin.

§8.5. Theorem 8.5.4 is due to Payne [1971a], although this treatment follows Hirschfeld [1975].

§8.6. This section is due to Segre [1955b], [1960a], with a slight amendment by Büke [1974].

§8.7. Theorem 8.7.2 is due to Tallini [1957b], but the proof given here was communicated by J.A. Thas.

9
ARITHMETIC OF k-ARCS

9.1. The basic equations

In §8.1, the concepts of bisecant, unisecant, and external line were defined for any k-arc $\mathcal{K}$. When q is odd and $k = q + 1$, internal and external points were defined in §8.2. Now, for any k-arc $\mathcal{K}$ in $\Pi = PG(2,q)$ for any q, a point Q in $\Pi \setminus \mathcal{K}$ is an *internal* or *external* point of $\mathcal{K}$ according as it lies on at most one or at least two unisecants of $\mathcal{K}$. In §8.1, the index and grade of Q were defined as the respective number of bisecants and unisecants of $\mathcal{K}$ through Q.

For a k-arc $\mathcal{K}$, let c_i be the number of points of $\Pi \setminus \mathcal{K}$ of index exactly i, let t_i be the number of points of grade exactly i and let e_i be the number of points through which exactly i external lines of $\mathcal{K}$ pass.

Let $k' = [k/2]$, the integral part of $k/2$. Then $c_i = 0$ for $i > k'$. $\mathcal{K}$ is complete if and only if $c_0 = 0$. Let α be the smallest i for which $c_i \neq 0$; α is called the *index* of $\mathcal{K}$. Similarly, let β be the largest i for which $c_i \neq 0$; β is called the *rank* of $\mathcal{K}$. Then $0 \leqslant \alpha \leqslant \beta \leqslant k'$.

9.1.1. LEMMA. *The constants c_i of a k-arc $\mathcal{K}$ satisfy the following equations with the summation taken from α to β:*

$$\Sigma\, c_i = q^2 + q + 1 - k\,; \qquad (9.1_1)$$

$$\Sigma\, ic_i = k(k-1)(q-1)/2\,; \qquad (9.1_2)$$

$$\Sigma\, i(i-1)c_i/2 = k(k-1)(k-2)(k-3)/8. \qquad (9.1_3)$$

Proof. The equations express in different ways the cardinality of the following sets:

(i) $\{Q \mid Q \in \Pi\setminus\mathcal{K}\}$;

(ii) $\{(Q,l) \mid Q \in l\setminus\mathcal{K},\ l$ a bisecant of $\mathcal{K}\}$;

(iii) $\{(Q,\{l,l'\}) \mid Q \in (l \cap l')\setminus\mathcal{K}, l$ and l' bisecants of $\mathcal{K}\}$.□

For a bisecant l of $\mathcal{K}$, let l_i be the number of points on

l of index i.

9.1.2. LEMMA. *The constants l_i of a bisecant of a k-arc $\mathcal{K}$ satisfy the following equations with the summation taken from 1 to β:*

$$\Sigma\ l_i = q - 1; \tag{9.2$_1$}$$

$$\Sigma\ (i - 1)l_i = (k - 2)(k - 3)/2. \tag{9.2$_2$}$$

Proof. The equations give the cardinality of the following sers:

(i) $\{Q \mid Q \in l\backslash\mathcal{K}\}$;

(ii) $|(Q,l') \mid Q \in (l \cap l')\backslash\mathcal{K},\ l'$ a bisecant of $\mathcal{K}\}$. □

If the bisecants of $\mathcal{K}$ are $l(j)$, $j \in \mathbf{N}_n$ with $n = k(k - 1)/2$, then there is the following result.

9.1.3. LEMMA. *For a k-arc $\mathcal{K}$, the constants c_i and $l(j)_i$ are connected by the following equation with the summation taken over j from 1 to n:*

$$ic_i = \Sigma\ l(j)_i. \tag{9.3$_i$}$$

Proof. The equation gives the cardinality of $\{(Q,l) \mid Q \in l$, l a bisecant of $\mathcal{K}$, $\sigma_2(Q) = i\}$.□

If the equations (9.3) are summed from α to β and (9.2_1) is used, the equation (9.1_2) is obtained.

Instead of the constants c_i, the constants t_i may be used. Since, for Q in $\Pi\backslash\mathcal{K}$, $\sigma_1(Q) + 2\sigma_2(Q) = k$, we have $c_i = t_j$ where $2i + j = k$. So as i varies from α to β, j varies from $k - 2\alpha$ to $k - 2\beta$.

9.1.4. LEMMA. *The constants t_i satisfy the following equations with the summation taken from 0 to k or, less superfluously, from $k - 2\beta$ to $k - 2\alpha$:*

$$\Sigma\ t_i = q^2 + q + 1 - k; \tag{9.4$_1$}$$

$$\Sigma\, it_i = ktq; \tag{9.4$_2$}$$

$$\Sigma\, i(i-1)t_i/2 = t^2k(k-1)/2. \tag{9.4$_3$}$$

Proof. The three equations differently express the cardinality of the following sets:

(i) $\{Q \mid Q \in \Pi\backslash\mathscr{K}\}$;

(ii) $\{(Q,l) \mid Q \in l\backslash\mathscr{K},\ l$ unisecant to $\mathscr{K}\}$;

(iii) $\{(Q,\{l,l'\}) \mid Q \in (l \cap l')\backslash\mathscr{K},\ l$ and l' unisecants to $\mathscr{K}\}$.□

If k is even, then $t_{2i+1} = 0$, all i: if k is odd, then $t_{2i} = 0$, all i. So, to distinguish k even and odd, there is the following.

COROLLARY 1. *If k is even, the constants t_i of a k-arc $\mathscr{K}$ satisfy the following equations summed from $k/2 - \beta$ to $k/2 - \alpha$:*

$$\Sigma\, t_{2j} = q^2 + q + 1 - k; \tag{9.5$_1$}$$

$$\Sigma\, jt_{2j} = ktq/2; \tag{9.5$_2$}$$

$$\Sigma\, j(j-1)t_{2j} = kt(k-2)(t-1)/4. \square \tag{9.5$_3$}$$

COROLLARY 2. *If k is odd, the constants t_i satisfy the following equations summed from $(k-1)/2 - \beta$ to $(k-1)/2 - \alpha$:*

$$\Sigma\, t_{2j+1} = q^2 + q + 1 - k; \tag{9.6$_1$}$$

$$\Sigma\, jt_{2j+1} = (k-1)(qt-q+1)/2; \tag{9.6$_2$}$$

$$\Sigma\, j(j-1)t_{2j+1} = (k-1)(k-3)(t^2-3t+3)/4. \square \tag{9.6$_3$}$$

In their respective cases, (9.5) and (9.6) may be expressed in terms of the constants c_i by substituting $c_{k/2-j} = t_{2j}$ and $c_{(k-1)/2-j} = t_{2j+1}$. However, after manipulation, these only give the equations (9.1) and so no further information.

On a unisecant m to $\mathscr{K}$, let m_i be the number of points of grade i.

9.1.5. LEMMA. *For a unisecant m to $\mathscr{K}$, the integers m_i satisfy the following equations summed from 1 to k:*

$$\Sigma\, m_i = q; \tag{9.7$_1$}$$

$$\Sigma\, (i - 1)m_i = t(k - 1)\,. \tag{9.7$_2$}$$

Proof. The following sets are counted:

(i) $\{Q \mid Q \in m\backslash\mathscr{K}\}$;

(ii) $\{(Q,m') \mid Q \in (m \cap m')\backslash\mathscr{K},\ m'$ unisecant to $\mathscr{K}\}$.□

Let the unisecants to $\mathscr{K}$ be $m(j)$, $j \in \mathrm{N}_{tk}$.

9.1.6. LEMMA. *The constants t_i and $m(j)_i$ are connected by the following equation:*

$$it_i = \Sigma\, m(j)_i\,. \tag{9.8$_i$}$$

Proof. The set $\{(Q,m) \mid Q \in m,\ m$ unisecant, $\sigma_1(Q) = i\}$ is counted.□

For completeness, we include the equations for the constants e_i. If $Q \in \Pi\backslash\mathscr{K}$, then the *rank* of Q is the number of external lines of $\mathscr{K}$ through Q: so $e_i = |\{Q \mid \text{rank } Q = i\}|$.

9.1.7. LEMMA. *The constants e_i of a k-arc $\mathscr{K}$ satisfy the following equations summed from $q + 1 - k + \alpha$ to $q + 1 - k + \beta$:*

$$\Sigma\, e_i = q^2 + q + 1 - k\ ; \tag{9.9$_1$}$$

$$\Sigma\, ie_i = (q + 1)(q^2 - q + t^2 - t)/2; \tag{9.9$_2$}$$

$$\Sigma\, i(i - 1)e_i/2 = (q^2 - q + t^2 - t)(q^2 - q + t^2 - t - 2)/4. \tag{9.9$_3$}$$

Proof. The following sets are counted:

(i) $\{Q \mid Q \in \Pi\backslash\mathscr{K}\}$;

(ii) $\{(Q,l) \mid Q \in l, l$ an external line of $\mathscr{K}\}$,

(iii) $\{(Q,\{l,m\})\,|\,Q = l \cap m,\ l \text{ and } m \text{ external lines of } \mathscr{K}\}$.□

If an external line n has n_i points of rank i and the external lines are $n(j)$ for j in $\mathbf{N}_r$ with $r = q(q-1)/2 + t(t-1)/2$, then we have the following.

9.1.8. LEMMA. *For an external line n to the k-arc $\mathscr{K}$, the constants n_i satisfy*

$$\Sigma\, n_i = q + 1; \tag{9.10$_1$}$$

$$\Sigma\, (i-1)n_i = (q^2 - q + t^2 - t - 2)/2.\square \tag{9.10$_2$}$$

9.1.9. LEMMA. *The constants e_i and $n(j)_i$ satisfy*

$$ie_i = \Sigma\, n(j)_i.\square \tag{9.11$_i$}$$

Problem II in §8.1 was to find all complete arcs. We can make the following start.

9.1.10. THEOREM. *If the k-arc $\mathscr{K}$ in $PG(2,q)$ satisfies*

$$q \geqslant k(k-3)/2 + 2,$$

then $\mathscr{K}$ is incomplete.

Proof. Suppose l is a unisecant at P to $\mathscr{K}$. If the $(k-1)(k-2)/2$ bisecants of $\mathscr{K}\backslash\{P\}$ do not contain all the points of $l\backslash\{P\}$, then $\mathscr{K}$ is incomplete. So $q > (k-1)(k-2)/2$ implies that $\mathscr{K}$ is incomplete. But $q > (k-1)(k-2)/2$ is equivalent to $q \geqslant (k-1)(k-2)/2 + 1 = k(k-3)/2 + 2$.□

This result can be improved slightly when k is odd. From the equations (9.1$_i$)

$$\alpha \leqslant \Sigma\, ic_i/\Sigma\, c_i \leqslant \beta;$$

and

$$\alpha \leqslant \Sigma\, i^2c_i/\Sigma\, ic_i \leqslant \beta.$$

Further $\Sigma(i-\alpha)(i-\beta)c_i \leqslant 0$. Hence

$$\alpha[\beta \Sigma c_i - \Sigma ic_i] \leqslant \beta \Sigma ic_i - \Sigma i^2 c_i. \qquad (9.1_4)$$

As a curiosity, there is the following.

9.1.11. THEOREM. *For a k-arc $\mathcal{K}$, the rank and index are the same if and only if q is even, $\mathcal{K}$ is an oval and $\alpha = \beta = q/2 + 1$.*

Proof. If $\mathcal{K}$ is an oval and q is even, then $k = q + 2$. If $Q \notin \mathcal{K}$, then, as $\mathcal{K}$ has no unisecants, there are $q/2 + 1$ bisecants and $q/2 + 1$ external lines through Q. Hence $\alpha = \beta$.

If $\alpha = \beta$, then (9.1_1), (9.1_2), (9.1_3) become

$$c_\alpha = q^2 + q + 1 - k,$$
$$\alpha c_\alpha = k(k - 1)(q - 1)/2,$$
$$\alpha(\alpha - 1)c_\alpha = k(k - 1)(k - 2)(k - 3)/4.$$

By eliminating α and c_α, we obtain

$$(k - 2)(k - 3)/[2(q - 1)] + 1 = k(k - 1)(q - 1)/[2(q^2 + q + 1 - k)].$$

Thus

$$(q^2 + q + 1 - k)[(k - 2)(k - 3) + 2(q - 1)] = k(k - 1)(q - 1)^2.$$

Hence

$$(q - k + 2)[2q(q - k + 1) + (k - 1)(k - 2)] = 0.$$

The second factor is always positive and so $k = q + 2$. Hence $\mathcal{K}$ is an oval with q even and the result follows by the first part of the proof.□

The equation (9.1_4) allows a slight improvement in theorem 9.1.10.

9.1.12. THEOREM. *If k is odd and $q \geqslant k(k - 3)/2$, then the k-arc $\mathcal{K}$ is incomplete.*

Proof. From (9.1_4) and using (9.1_1), (9.1_2), (9.1_3), we have

$$\alpha \leqslant \frac{2(\beta - 1)(q - 1)\,k(k - 1) - k(k - 1)(k - 2)(k - 3)}{4\beta(q^2 + q + 1 - k) - 2(q - 1)\,k(k - 1)} = A.$$

Now,

$$A < 1 \iff \beta < \frac{k(k - 1)(k - 2)(k - 3)}{2(q - 1)\,k(k - 1) - 4(q^2 + q + 1 - k)} = B.$$

Hence $\beta < B \Rightarrow \alpha < 1 \Rightarrow \alpha = 0 \Rightarrow c_0 \neq 0 \Rightarrow \mathscr{K}$ is incomplete. Since $\beta \leqslant k' = [k/2]$, if $[k/2] < B$ then $\beta < B$. Now, $[k/2] < B \iff$

$$2[k/2]\{(q - 1)\,k(k - 1) - 2(q^2 + q + 1 - k)\} < k(k - 1)(k - 2)(k - 3). \tag{9.12}$$

If k is even, $2[k/2] = k$ and the same result as theorem 9.1.10 is obtained. If k is odd, $2[k/2] = k - 1$ and (9.12) becomes

$$(q - 1)\,k(k - 1) - 2(q^2 + q + 1 - k) < k(k - 2)(k - 3).$$

Since both sides are even, we may write

$$(q - 1)\,k(k - 1) - 2(q^2 + q - k) \leqslant k(k - 2)(k - 3).$$

Hence $$(q - k + 1)\,\{k(k - 3) - 2q\} \leqslant 0.$$

Therefore $q \geqslant k(k - 3)/2 \Rightarrow \mathscr{K}$ is incomplete.□

A k-arc $\mathscr{K}$ is *symmetric* if each bisecant l has the same set of constants $l_1, l_2, \ldots, l_{k'}$; thus, for each i, $l(j)_i = l_i$.

9.1.13. LEMMA. *If a k-arc $\mathscr{K}$ is complete and symmetric with k an odd prime, then k divides $q^2 + q + 1$.*

Proof. For a symmetric arc, eqn (9.3_i) becomes

$$ic_i = l_i\,k(k - 1)/2.$$

Since $k > i$, so $k | c_i$ for $i > 0$. From (9.1_1), we have $\sum_{i=1}^{k'} c_i = q^2 + q + 1 - k$, whence $k | (q^2 + q + 1)$.□

An example of a complete, symmetric arc is an oval in $PG(2,4)$, for which $c_0 = c_1 = c_2 = 0$, $c_3 = 15$ and for any bisecant l, $l_1 = 2$, $l_2 = 0$, $l_3 = 1$.

9.2. Small arcs

By the way of example, let us write down the values of the constants c_i of a k-arc $\mathscr{K}$ for $k = 4,5,6$, and 7.

$$k = 2,\ c_0 = q^2,\quad c_1 = q - 1;$$
$$k = 3,\ c_0 = (q-1)^2,\quad c_1 = 3(q-1);$$
$$k = 4,\ c_0 = (q-2)(q-3),\quad c_1 = 6(q-2),\quad c_2 = 3;$$
$$k = 5,\ c_0 = (q-4)(q-5)+1,\quad c_1 = 10(q-4),\quad c_2 = 15;$$
$$k = 6,\ c_0 = (q-7)^2 + 6 - c_3,\quad c_1 = 3\{5(q-7) + c_3\},\quad c_2 = 3\{15 - c_3\};$$
$$k = 7,\ c_0 = (q-10)^2 + 20 - c_3,\quad c_1 = 3\{7(q-11) + c_3\},\quad c_2 = 3\{35 - c_3\}.$$

9.2.1. LEMMA. (i) *A complete 4-arc exists only for $q = 2$ and 3.*

(ii) *A 5-arc is never complete.*

(iii) *A complete 6-arc is as in Table 9.1.*□

TABLE 9.1

q	c_0	c_1	c_2	c_3
4	0	0	0	15
5	0	0	15	10
7	0	18	27	6
8	0	36	24	7
9	0	60	15	10

From theorem 9.1.10, if $q \geq 11$, a 6-arc is incomplete. Complete 6-arcs exist in $PG(2,4)$ and $PG(2,5)$ since in these planes, they are ovals. Let us show that a complete 6-arc exists in $PG(2,7)$.

9.2.2. THEOREM. *Every 7-arc in $PG(2,7)$ lies on a conic.*

Proof. Suppose the 7-arc $\mathscr{K}$ in $PG(2,7)$ is complete.

Then $c_0 = 0$, $c_1 = 3$, $c_2 = 18$, $c_3 = 29$. Equations (9.2_1) and (9.2_2) become $l_1 + l_2 + l_3 = 6$, $l_2 + 2l_3 = 10$ giving either $l_1 = 0$, $l_2 = 2$, $l_3 = 4$ or $l_1 = 1$, $l_2 = 0$, $l_3 = 5$. Suppose among the 21 bisecants $l(j)$ there are M of the former type and N of the latter: $M + N = 21$. Equation (9.3_3) becomes

$$87 = \Sigma\, l(j)_3 = 4M + 5N.$$

Hence $M = 18$, $N = 3$. So consider a bisecant for which $l_3 = 5$. This means that the pentastigm formed by the five points of $\mathscr{K}$ other than those on l has five diagonal points collinear on l. This can only occur if $x^2 = x + 1$ has a solution in $GF(7)$, lemma 7.1.2. Since 5 is a non-square in $GF(7)$, the equation has no solutions. So the 7-arc is not complete and hence lies on an 8-arc, which is a conic.□

COROLLARY. *In* $PG(2,7)$,

(i) *every* 6-*arc not on a conic is complete;*

(ii) *there are* $2^5 \cdot 3 \cdot 7^3 \cdot 19$ *complete* 6-*arcs;*

(iii) *the only complete arcs are* 6-*arcs and conics.*

Proof. (ii) From lemma 7.2.3, corollary 2, the number of 6-arcs not on a conic is as given.□

The structure of a complete 6-arc in $PG(2,q)$ for q = 7,8, and 9 will be further examined in the sections devoted to these planes. In the meantime, elementary methods yield results for $PG(2,8)$ somewhat similar to $PG(2,7)$.

9.2.3. THEOREM. *In* $PG(2,8)$,

(i) *there are two ovals through every* 5-*arc;*

(ii) *every oval is regular.*

Proof. Part (ii) was already shown in theorem 8.4.4, corollary 1. We now give a much more direct method.

It must be shown that every 10-arc contains a conic. For a 5-arc in $PG(2,8)$, $c_0 = 13$, $c_1 = 40$, $c_2 = 15$. So there are 13 6-arcs containing a 5-arc. But, by lemma 8.1.5, any two 10-arcs can intersect in at most five points.

So through any 5-arc, there pass at most two 10-arcs. Hence, since the number of 5-arcs has been denoted by $L(5,8)$, the number of 10-arcs in the plane is at most $2L(5,8)/\mathbf{c}(10,5)$. However, the number of conics in the plane and hence the minimum number of 10-arcs is $L(5,8)/\mathbf{c}(9,5)$. But $\mathbf{c}(9,5) = \mathbf{c}(10,5)/2$: so the above numbers are equal. Hence there are two 10-arcs through a 5-arc and every 10-arc contains exactly one conic.□

Now, we can show for $PG(2,8)$ the result comparable to theorem 9.2.2, corollary (iii); that is, the complete arcs can be identified. Firstly, we give an extension of lemma 8.1.5.

9.2.4. LEMMA. *In* $\Pi = PG(2,q)$ *with* q *even,* $\mathscr{K}$ *is a* k*-arc with* $k > q/2 + 1$ *lying on a conic* $\mathscr{C}$ *with nucleus* N*, and* $\mathscr{K}' = (\mathscr{C}\backslash\mathscr{K}) \cup \{N\}$. *Then every point of* $\Pi\backslash\mathscr{K}'$ *lies on a bisecant of* $\mathscr{K}$.

Proof. Let $Q \in \Pi\backslash(\mathscr{K} \cup \mathscr{K}')$. The number of lines through Q to points of $\mathscr{K}'$ is at most $q + 2 - k < q/2 + 1 < k$ and each line contains at most one point of $\mathscr{K}$. So there is a point P of $\mathscr{K}$ on none of these lines. Hence PQ does not pass through N and so meets $\mathscr{C}$ in a further point of $\mathscr{K}$.□

COROLLARY 1. *If* $\mathscr{K}$ *is a* k*-arc in* $PG(2,q)$*,* q *even, such that* $|\mathscr{K} \cap \mathscr{C}| > q/2 + 1$*, where* $\mathscr{C}$ *is a conic with nucleus* N*, then* $\mathscr{K} \subset \mathscr{C} \cup \{N\} = \mathscr{O}$.

Proof. Let $\mathscr{K}_1 = \mathscr{K} \cap \mathscr{C}$. Then, by the lemma, every point of $\Pi\backslash\mathscr{O}$ lies on a chord of $\mathscr{K}_1$; so, no point of $\mathscr{K}$ does. Hence $\mathscr{K} \subset \mathscr{O}$.□

COROLLARY 2. *In* $PG(2,8)$*, a* k*-arc* $\mathscr{K}$ *not lying entirely on the oval* $\mathscr{C} \cup \{N\}$ *has at most five points in common with the conic* $\mathscr{C}$.□

9.2.5. THEOREM. *In* $PG(2,8)$,

(i) *there are three complete* 6*-arcs through a* 5*-arc;*

(ii) *the only complete arcs are* 6*-arcs and regular ovals.*

Proof. Let $\mathcal{K}$ be a 5-arc in $PG(2,8)$ and $\mathcal{C}$ the conic containing $\mathcal{K}$. Let $\mathcal{K} = \{P_1, \ldots, P_5\}$. The involution on $\mathcal{C}$ in which $\{P_1, P_2\}$ and $\{P_3, P_4\}$ are pairs is given by lines through $P_1P_2 \cap P_3P_4$. Then, exactly as for $PG(1,8)$, four points on $\mathcal{C}$ have an associated point, namely the fixed point of the three involutions, lemma 6.3.3. Also, given the five points P_i on $\mathcal{C}$, exactly one, say P_5, is the associated point of the other four, §6.4(vi). So P_5 lies on the line l of diagonal points of the tetrastigm $\mathcal{D}$ with vertices P_1, P_2, P_3, P_4. By theorem 9.2.3, there are two ovals through $\mathcal{K}$. The number of points on no bisecant of $\mathcal{K}$ is $l^*(5,8) = 13$, §7.1. Each of the two ovals containing $\mathcal{K}$ contains five of these points, leaving three, A_1, A_2, A_3. However, l contains P_5, one other point on each the two ovals and the three diagonal points of $\mathcal{D}$, leaving three, which are necessarily A_1, A_2, A_3. So there are three 6-arcs containing $\mathcal{K}$ which do not lie on an oval, namely $\mathcal{K}_i = \mathcal{K} \cup \{A_i\}$, $i = 1,2,3$. Now, by lemma 9.2.4, corollary 2, a 7-arc not on an oval can only have five points in common with it. So a 7-arc containing $\mathcal{K}$ but not an oval contains $\mathcal{K}_1$, say, and some other point. But A_2 and A_3 are the only possible candidates and they both lie on $l = A_1P_5$; so there is no such 7-arc. Thus each $\mathcal{K}_i$ is complete and there are no complete 7-arcs and no complete 8-arcs.□

COROLLARY. *The number of complete* 6-*arcs in* $PG(2,8)$ *is* $2^6 \cdot 3^2 \cdot 7^2 \cdot 73$.□

Proof. The number of 5-arcs in $PG(2,8)$ is $L(5,8) = 2^7 \cdot 3^2 \cdot 7^2 \cdot 73$, §7.1. So, as there are three complete 6-arcs containing a 5-arc, the required number is $3L(5,8)/6$.□

Theorem 9.2.2 gave an example of where there was no arc to match the parameters c_i, l_i. We can give a similar one also for 7-arcs. From theorem 9.1.12, there is no complete 7-arc in $PG(2,q)$ for $q \geq 14$. In fact, this can be improved.

9.2.6. THEOREM. *There is no complete* 7-*arc in* $PG(2,13)$.

Proof. For a complete 7-arc $\mathscr{K}$ in $PG(2,13)$, $c_0 = 0$, $c_1 = 129$, $c_2 = 18$, $c_3 = 29$. The equations (9.2_1), (9.2_2) are $l_1 + l_2 + l_3 = 12$, $l_2 + 2l_3 = 10$, whose solution in terms of l_1 is $l_3 = l_1 - 2$, $l_2 = 2(7 - l_1)$. Equation (9.3_1) is $129 = \sum_{j=1}^{21} l(j)_1$. So, if $l(j)_1 = 6$, for all j, we should have $129 = 126$. Hence, for some bisecant l, $l_1 = 7$ and so $l_3 = 5$. Then, the pentastigm formed by the five points of $\mathscr{K}$ other than those on l has five diagonal points collinear on l, which occurs only only if $x^2 = x + 1$ has roots in $GF(13)$, lemma 7.1.2. As 5 is a non-square in $GF(13)$, the equation has no roots. So the 7-arc $\mathscr{K}$ is not complete.□

9.3. Uniform arcs

A k-arc $\mathscr{K}$ is *uniform* if every external point has index $[(k-2)/2]$ or a constant m. Thus, through Q in $\Pi\backslash\mathscr{K}$, there pass two or $k-2m$ unisecants when k is even, and three or $k-2m$ unisecants when k is odd.

9.3.1. THEOREM. *If $\mathscr{K}$ is a uniform k-arc in $PG(2,q)$ with k even and $4 < k < q$, then $\mathscr{K}$ is complete.*

Proof. If $\mathscr{K}$ is not complete, then $c_0 = t_k \neq 0$. Hence, solving (9.5_1), (9.5_2), (9.5_3), we obtain

$$t_k = t(t-1),\quad t_2 = tk(k-1)/2,\quad t_0 = (k-1)(2k+4t-6-tk)$$

with all other $t_i = 0$. However, as $k < q$, so $t \geqslant 3$ and as $k > 4$, so $k \geqslant 6$. As $t_0 \geqslant 0$, $k \leqslant (4t-6)/(t-2) = 4 + 2/(t-2) \leqslant 6$. Hence $k = 6$, $t = 3$, $q = 7$, $t_0 = 0$, $t_2 = 45$, $t_6 = 6$. But, by the corollary to theorem 9.2.2, an incomplete 6-arc in $PG(2,7)$ lies on a conic. So $\mathscr{K}$ lies on a conic $\mathscr{C}$, but the 18 unisecants to $\mathscr{K}$ consist of the six unisecants common to $\mathscr{C}$ and $\mathscr{K}$ plus the joins of $\mathscr{K}$ to the two points of $\mathscr{C}\backslash\mathscr{K}$. Hence $t_6 = 2$, giving a contradiction.□

9.3.2. THEOREM. *If $\mathscr{K}$ is a uniform k-arc with k odd and $5 < k < q-1$, then $\mathscr{K}$ is complete.*

Proof. If $\mathscr{K}$ is not complete, then $t_k \neq 0$. Hence, solving (9.6_1), (9.6_2), (9.6_3), we obtain

$t_k = t^2 - 3t + 3$, $t_3 = k(k - 1)(t - 1)/2$, $t_1 = k\{k(3 - t) + 5t - 9\}/2$.

Since $t > 3$, so $t_1 \geqslant 0 \Rightarrow k \leqslant (5t - 9)/(t - 3) = 5 + 6/(t - 3)$. Since $k \geqslant 7$, so $t > 6 \Rightarrow k < 7$, which is impossible. So $4 \leqslant t \leqslant 6$ and there remain the following cases:

	t	k	q	t_1
(a)	4	11	13	0
(b)	4	9	11	9
(c)	4	7	9	14
(d)	6	7	11	0.

Equations (9.7_1), (9.7_2), (9.8_1) now read

$$m_1 + m_3 + m_k = q\ , \tag{9.7_1}$$

$$2m_3 + (k - 1)m_k = t(k - 1). \tag{9.7_2}$$

$$t_1 = \sum_j m(j)_1. \tag{9.8_1}$$

In cases (a), (b), (c), we have

$$m_1 + m_3 + m_k = k + 2\ ;$$

$$m_3 + (k - 1)m_k/2 = 2(k - 1).$$

There are three possible solutions:

	m_1	m_2	m_k
(i)	$k - 2$	0	4
(ii)	$(k - 1)/2$	$(k - 1)/2$	3
(iii)	1	$k-1$	2

Suppose there are N_1, N_2, N_3 solutions of the respective kinds; then, from (9.8_1),

$$N_1 + N_2 + N_3 = 4k,$$

$$(k - 2)N_1 + (k - 1)N_2/2 + N_3 = t_1.$$

So,

(a) $N_1 + N_2 + N_3 = 44$, $9N_1 + 5N_2 + N_3 = 0$;

(b) $N_1 + N_2 + N_3 = 36$, $7N_1 + 4N_2 + N_3 = 9$;

(c) $N_1 + N_2 + N_3 = 28$, $5N_2 + 3N_2 + N_3 = 14$.

In each case, there can be no solution in non-negative integers for N_1, N_2, N_3.

(d) Since $t = 0$, for any particular unisecant, $m_1 = 0$. So (9.7_1) and (9.7_2) become

$$m_3 + m_7 = 11, \quad m_3 + 5m_7 = 30;$$

these equations do not have a solution in non-negative integers.□

9.4. Examples of complete arcs

It is natural to ask if there exist complete arcs in $PG(2,q)$ other than ovals.

9.4.1. LEMMA. *In $PG(2,q)$ for $q = 2,3,4$, and 5, every complete arc is an oval.*

Proof. For $q = 2$ and 3, an oval is a 4-arc. For $q = 4$ and 5, an oval is a 6-arc. From §9.2, no 2-arc or 3-arc is complete and from lemma 9.2.1, a 4-arc is complete only for $q = 2$ and 3, whereas a 5-arc is never complete.□

9.4.2. LEMMA. *In $PG(2,q)$ with q odd, let the k-arc $\mathscr{K}$, where $k = (q + 5)/2$, have its points arbitrarily chosen from a conic $\mathscr{C}$. Then the bisecants of $\mathscr{K}$ contain all the points in the plane other than those of $\mathscr{C}\backslash\mathscr{K}$.*

Proof. Suppose Q, not on $\mathscr{C}$, does not lie on any bisecant of $\mathscr{K}$. There are at least $(q - 1)/2$ bisecants of $\mathscr{C}$ through Q and each of these bisecants can meet $\mathscr{K}$ in at most one point. So there are at least $(q - 1)/2$ points in $\mathscr{C}\backslash\mathscr{K}$. Hence $\mathscr{K}$ contains at most $(q + 1) - (q - 1)/2 = (q + 3)/2$ points, giving a contradiction.□

COROLLARY 1. *In $PG(2,q)$ with q odd, if $\mathscr{K}$ is a k-arc not contained in a conic, then $\mathscr{K}$ has at most $(q + 3)/2$ points in common with a conic.*

Proof. If there is a conic $\mathscr{C}$ containing $(q + 5)/2$ points of $\mathscr{K}$ and P is in $\mathscr{K}\backslash\mathscr{C}$, by the lemma there is a bisecant of $\mathscr{K}$ through P, giving a contradiction.□

9.4.3. LEMMA. *In $PG(2,q)$, q odd and $q \geq 9$, there exists a complete k-arc not contained in a conic.*

Proof. Let $\mathscr{C}$ be a conic in $PG(2,q)$ and Q an internal point of $\mathscr{C}$. Let the k-arc $\mathscr{K}$ consist of Q and one of the two points of $\mathscr{C}$ on each of the $(q + 1)/2$ bisecants of $\mathscr{C}$ through Q: hence $k = (q + 3)/2$. $\mathscr{K}$ can then be completed to $\bar{\mathscr{K}}$. Since two conics can meet in at most four points, if $k > 5$ and so $q \geq 9$, $\bar{\mathscr{K}}$ cannot be a conic.□

9.4.4. THEOREM. *In $PG(2,q)$, $q \equiv -1 \pmod 4$, there exists a complete k-arc $\mathscr{K}$ with $k = (q + 5)/2$.*

Proof. Let Q be an external point of the conic $\mathscr{C}$. The k-arc $\mathscr{K}$ will consist of Q, the two points of contact of the tangents to $\mathscr{C}$ through Q and one point of $\mathscr{C}$ on each of the $(q - 1)/2$ bisecants of $\mathscr{C}$ through Q. $\mathscr{K}$ can then be completed to $\bar{\mathscr{K}}$ for any odd q. If $k > 5$ and so $q \geq 7$, $\bar{\mathscr{K}}$ cannot be a conic. Now it will be shown that if $q \equiv -1 \pmod 4$, $\mathscr{K}$ can be chosen to coincide with $\bar{\mathscr{K}}$.

Let $\mathscr{C} = V(x_0^2 - x_1x_2)$, and let Q be U_0 so that the points of contact of the tangents to $\mathscr{C}$ through U_0 are U_1 and U_2. Let α be a primitive element of γ. Then the $q - 1$ points of

$\mathscr{C}\backslash\{U_1, U_2\}$ fall into two branches $\mathscr{S}$ and $\mathscr{N}$, where

$\mathscr{S} = \{P(1, \alpha^{2i}, \alpha^{-2i}) \mid i \in N_{(q-1)/2}\}$, the branch of the 'squares', and

$\mathscr{N} = \{P(1, \alpha^{2i-1}, \alpha^{-2i+1}) \mid i \in N_{(q-1)/2}\}$, the branch of the 'non-squares'.

The two points $P(1, \alpha^t, \alpha^{-t})$ and $P(1, -\alpha^t, -\alpha^{-t})$ belong to different branches if and only if -1 is a non-square, that is, $q \equiv -1 \pmod 4$. So, for these q, let $\mathscr{K} = \mathscr{S} \cup \{U_0, U_1, U_2\}$.

It remains to show that the bisecants of $\mathscr{K}$ fill the plane. The lines u_0, u_1, u_2 are bisecants of $\mathscr{K}$. The bisecants other than u_1 and u_2 through U_0 are

$$\{V(x_1 - \alpha^{4i}x_2) \mid i \in N_{(q-1)/2}\} = \{V(x_1 - \alpha^{2j}x_2) \mid j \in N_{(q-1)/2}\},$$

§1.5(x). Let $P = P(1, a_1, a_2)$ be a point on none of these bisecants of $\mathscr{K}$. The line $PU_0 = V(x_1 - \alpha^{2j-1}x_2)$, some j in $N_{(q-1)/2}$. Hence $a_1 = \alpha^{2j-1}a_2$, and so only one of a_1 and a_2 is a square. If a_1 is a square, P lies on the bisecant $V(x_1 - a_1x_0)$ through U_2 and $P(1, a_1, a_1^{-1})$: if a_2 is a square, P lies on the bisecant $V(x_2 - a_2x_0)$ through U_1 and $P(1, a_2^{-1}, a_2)$. So $\mathscr{K}$ is complete. In a similar way, we may take $\mathscr{N}$ instead of $\mathscr{S}$ and obtain the complete arc $\mathscr{K}' = \mathscr{S} \cup \{U_0, U_1, U_2\}$.

It may also be observed that any two points P_0, P_1 of $\mathscr{C}\backslash\{U_0, U_1\}$ belong to the same arc ($\mathscr{K}$ or $\mathscr{K}'$) or not as the cross-ratio $\{U_0, U_1; P_0, P_1\}$ is a square or not.□

COROLLARY. *In $PG(2,q)$, q odd, a complete arc other than an oval exists if and only if $q \geq 7$.*

Proof. This follows from lemmas 9.4.1 and 9.4.3 for $q \geq 9$. Then the theorem reduces this to $q \geq 7$.□

Now a comparable result will be shown for q even.

9.4.5. THEOREM. *In $PG(2,q)$, q even and $q \geq 16$, there exists a complete k-arc with $k = (q + 4)/2$.*

Proof. Let $\mathcal{O}$ be an oval and Q a point not on $\mathcal{O}$. As $\mathcal{O}$ has no tangents, there are $(q + 2)/2$ bisecants of $\mathcal{O}$ through Q. Let the k-arc $\mathcal{K}$ consist of Q and one of the points of $\mathcal{O}$ on each of these bisecants: then $k = (q + 4)/3$. It will be shown that $\mathcal{K}$ can be chosen to be complete.

Let $\mathcal{P}$ be the conic $\mathbf{V}(x_0{}^2 + x_1x_2)$; so $\mathcal{P} = \{P(t) = \mathbf{P}(t, t^2, 1) | t \in \gamma\} \cup \{\mathbf{U}_1\}$. Since t and t^2 belong both to $\mathcal{C}_1$ or both to $\mathcal{C}_0$, §1.4(iii), the point $P(t)$ is said to be of category zero or one as t is in $\mathcal{C}_0$ or $\mathcal{C}_1$. Hence there are $q/2$ points $P_0(t)$ of category zero and $q/2$ points $P_1(t)$ of category one on $\mathcal{P}$. The tangent at $\mathbf{U}_1$ to $\mathcal{P}$ is $\mathbf{u}_2$, which contains the nucleus $\mathbf{U}_0$ and $q - 1$ other points $Q(s) = \mathbf{P}(1,s,0)$ for s in γ_0. The latter comprise $(q - 2)/2$ points $Q_0(s)$ and $q/2$ points $Q_1(s)$, called respectively of category zero or one as s is in $\mathcal{C}_0$ or $\mathcal{C}_1$. The line $P(t_1)P(t_2)$ meets $\mathbf{u}_2$ in $Q(s)$, where $s = t_1 + t_2$. Hence $Q(s)$ is of category zero or one as $P(t_1)$ and $P(t_2)$ are of the same or different categories, §1.4(iii)(e).

Let $\mathcal{K}$ consist of $\mathbf{U}_0$, the $q/2$ points $P_1(t)$, and a point $Q' = Q_1(s')$. The line $\mathbf{u}_2$ is the bisecant $\mathbf{U}_0Q'$. For any point $P_0 = P_0(t_0)$, the line P_0Q' meets $\mathcal{P}$ in a point $P_1 = P_1(t_1)$. Thus P_0 lies on the bisecant P_1Q'.

It remains to consider a point $A = \mathbf{P}(a_0, a_1, 1)$ where $a_1 \neq a_0{}^2$. If the line $A\mathbf{U}_0$ meets $\mathcal{P}$ in a point $P_1(t)$, it is a bisecant of $\mathcal{P}$. Suppose that $A\mathbf{U}_0 = \mathbf{V}(x_1 - a_1x_2)$ meets $\mathcal{P}$ in a point $P_0(t)$; then $a_1 \in \mathcal{C}_0$. Now join A to the $q/2$ points $P_1(t)$. Either there are $q/2$ such lines or less. If less, then there is at least one line through A containing two points $P_1(t)$ and so A lies on a bisecant of $\mathcal{K}$. Thus we finally have the case that there are $q/2$ distinct lines $A\mathbf{P}_1(t)$, none of which is the tangent $A\mathbf{U}_0$ to $\mathcal{P}$. If it is shown that $AP_1(t) \neq A\mathbf{U}_1$ for all points $P_1(t)$, then each line $AP_1(t)$ meets $\mathcal{P}$ in a point $P_0(t)$ and therefore $\mathbf{u}_2$ in a point $Q_1(t)$. Thus each line $AP_1(t)$ will contain one of the $q/2$ points $Q_1(t)$ and, in particular, one line $AP_1(t)$ contains Q'; hence A will lie on a bisecant of $\mathcal{K}$.

It must therefore be shown that $AP_1(t) \neq A\mathbf{U}_1$ for all $P_1(t)$. So, suppose that $AP_1(t') = A\mathbf{U}_1$ and that $AP_1(t_1) \neq AP_1(t_2)$ for all t_1, t_2 in $\mathcal{C}_1\backslash\{t'\}$. Then $t' = a_0$ and

$$(t_1^2 + a_1)/(t_1 + a_0) \neq (t_2^2 + a_1)/(t_2 + a_0).$$

Further, $AP_1(t)$, with $t \neq t'$, meets $\mathscr{P}$ in $P_0(s)$, and hence

$$(t^2 + a_1)/(t + a_0) = t + s \in \mathscr{C}_1.$$

Thus, as t varies in $\mathscr{C}_1\backslash\{t'\}$, so $m(t) = (t^2 + a_1)/(t + a_0)$ takes all values but one in $\mathscr{C}_1$. However,

$$m(t) = (t + a_0) + (a_0^2 + a_1)/(t + a_0).$$

As $m(t)$, t and a_0 are all in $\mathscr{C}_1$, so $t + a_0$ is in $\mathscr{C}_0$ and $r = m(t) + t + a_0 = (a_0^2 + a_1)/(t + a_0)$ is in $\mathscr{C}_1$, §1.4(iii). As $t + a_0$ varies in $\mathscr{C}_0\backslash\{0\}$, r varies over all elements of $\mathscr{C}_1$ but one, say r_0. Put

$$v = t + a_0, \qquad c = a_0^2 + a_1.$$

Then $c\, v \neq 0$ and $c = r\, v$. As v is in $\mathscr{C}_0$, so $D_2(v) = 0$, §1.4. As $v \neq 0$, so $D_2(v)/v = 0$. Thus, with $q = 2^h$,

$$v^{2^{h-1}-1} + v^{2^{h-2}-1} + \ldots + v + 1 = 0. \tag{9.13}$$

Substituting $v = c/r$ in (9.13) and multiplying by $r^{2^{h-1}-1}$ gives

$$c^{2^{h-1}-1} + c^{2^{h-1}-1}\, r^{2^{h-2}} + \ldots + c\, r^{2^{h-2}-2} + r^{2^{h-1}-1} = 0. \tag{9.14}$$

Multiplying (9.14) by $r + r_0$ gives the following equation of degree $2^{h-1} = q/2$ in r:

$$r_0 c^{2^{h-1}-1} + c^{2^{h-1}-1} r + r_0 c^{2^{h-2}-1} r^{2^{h-2}} + \ldots + r^{2^{h-1}} = 0. \tag{9.15}$$

The solutions in r of (9.15) are all the elements of $\mathscr{C}_1$ and therefore satisfy $D_2(r) = 1$, whence

$$1 + r + r^2 + \ldots + r^{2^{h-1}} = 0. \tag{9.16}$$

The equations (9.15) and (9.16) must be identical apart from a factor of proportion. In (9.16) there is a term in r^2 with non-zero coefficient: in (9.15), the terms of lowest order in r are, omitting the coefficients, r^0, r^1, and r^{2h-2}. For $h > 3$, we have $2^{h-2} > 2$ and hence a contradiction. So $\mathscr{K}$ is complete.□

COROLLARY. *In $PG(2,q)$, q even, a complete arc other than an oval exists if and only if $q \geqslant 8$.*

Proof. From lemma 9.4.1, every complete arc is an oval for $q = 2$ or 4. From the theorem, a complete k-arc with $k = (q + 4)/2$ exists for $q \geqslant 16$. In theorem 9.2.5, it is shown that complete 6-arcs exist in $PG(2,8)$.□

We conclude this section with a list of all values of k for which a complete k-arc exists in $PG(2,q)$, $q \leqslant 11$ (Table 9.2).

TABLE 9.2

q	k
2	4
3	4
4	6
5	6
7	6,8
8	6,10
9	6,7,8,10
11	7,8,9,10,12

For $q \leqslant 9$, a projective classification of these arcs is given in Chapter 14.

9.5. Notes and references

§9.1. This is taken from Segre [1959a], except for lemma 9.1.4 and its corollaries, taken from Martin [1967].

§9.2. This proof of the regularity of ovals in $PG(2,8)$ is due to Sce [1960], p.253. This paper contains a wealth of results on small k-arcs. For another proof, see Segre [1957a]. Theorem 9.2.9 is taken from Segre [1959a].

§9.3. This is taken from Martin [1967].

§9.4. Theorem 9.4.4 is due to Lombardo-Radice [1956]. Theorem 9.4.5 comes from Segre [1967]. An alternative construction for complete $(q + 4)/2$-arcs when $q = 2^{2m+1}$ and $m \geqslant 2$ was given by Tallini Scafati [1964].

For examples of k-arcs, see also Di Comite [1962], [1963], [1964], [1969b], Di Comite and Peluso [1972], Lunelli and Sce [1958b], Sce [1958a], [1958b], Zirilli [1973].

10
ARCS IN OVALS

10.1. Plane curves

We recall some definitions from §2.6. As usual, let $\gamma = GF(q)$ and let $\bar{\gamma}$ be the algebraic closure of γ. A *plane curve of order n* is a variety $\mathscr{C}_n = V(F)$ in $PG(2,q)$, where F is a form of degree n in $\gamma[x_0, x_1, x_2]$. If F is irreducible over γ, then $\mathscr{C}_n$ is *irreducible*: if F is irreducible over $\bar{\gamma}$, then $\mathscr{C}_n$ is *absolutely irreducible*. For example, the conic $\mathscr{P}_2 = V(x_0{}^2 + x_1x_2)$ is absolutely irreducible for any γ. However, the cubic

$$\mathscr{C}_3 = V(x_0{}^3 + x_1{}^3 + x_2{}^3 + x_0{}^2x_1 + x_1{}^2x_2 + x_2{}^2x_0 + x_0x_1x_2)$$

is irreducible in $PG(2,2)$, but not absolutely irreducible, since in $PG(2,8)$,

$$\mathscr{C}_3 = V((x_0 + \varepsilon x_1 + \varepsilon^5 x_2)(x_0 + \varepsilon^2 x_1 + \varepsilon^3 x_2)(x_0 + \varepsilon^4 x_1 + \varepsilon^6 x_2)),$$

where, as in §1.7, $\varepsilon^3 + \varepsilon^2 + 1 = 0$.

If, for the plane curve $\mathscr{C}_n = V(F)$ in $PG(2,q)$, $F = F_1F_2 \ldots F_m$, where each F_i is irreducible in $\bar{\gamma}[x_0, x_1, x_2]$, then the $\mathscr{C}^i = V_{2,\bar{\gamma}}(F_i) \cap PG(2,q)$ are the *components* of $\mathscr{C}_n$. We call $\mathscr{C}^i$ *regular* or *irregular* as all the ratios of the coefficients of F_i lie in γ or not:

$$\mathscr{C}_n = \mathscr{C}^1 \cup \mathscr{C}^2 \cup \ldots \cup \mathscr{C}^m.$$

A component $\mathscr{C}^i$ is *linear* if $\deg F_i = 1$. The *order* or *degree* of $\mathscr{C}^i$ is the degree of F_i.

Firstly we require an estimate for the number of points on an irregular component of a curve.

10.1.1. LEMMA. *If $\mathscr{C}_r$ is an irreducible curve in $PG(2, \bar{\gamma})$ not defined over γ and $|\mathscr{C}_r \cap PG(2,q)| = E_r$, then $E_r \leqslant r^2$.*

Proof. Although a plane curve of order r is determined

by $r(r+3)/2$ points, any two curves of order r with no common component may intersect in up to r^2 distinct points. So there can be no two such curves through r^2+1 points. Thus a curve defined over $\bar{\gamma}$ and irreducible over $\bar{\gamma}$ and containing r^2+1 points of $PG(2,q)$ has the form $\mathbf{V}(F)$ with F in $\gamma[x_0, x_1, x_2]$. Hence $E_r \leqslant r^2$.□

Now we consider singular points of an absolutely irreducible curve $\mathscr{C}_n = \mathbf{V}(F)$ of order n in $PG(2,q)$. The approach here is slightly different from §2.6, but the definitions are equivalent. Write

$$F = f_0 x_0^{\,n} + f_1 x_0^{\,n-1} + \ldots + f_r x_0^{\,n-r} + \ldots + f_{n-1} x_0 + f_n, \qquad (10.1)$$

where f_i is a form in x_1 and x_2 of degree i.

(i) If $f_0 = f_1 = \ldots = f_{r-1} = 0$, $f_r \neq 0$, then $_n$ has a *point of multiplicity* r *at* $\mathbf{U}_0$; as in §2.6(xiii), write $m_{\mathbf{U}_0}(\mathscr{C}_n) = r$. Note than $\mathbf{U}_0$ is on $\mathscr{C}_n$ if and only if $r \geqslant 1$.

(a) For some $k \leqslant r$, $\mathbf{V}_{2,q^k}(f_r)$ consists of r lines in $PG(2,q^k)$, each of which is a *tangent* to $\mathscr{C}_n$ at $\mathbf{U}_0$.

(b) Each regular component of $\mathbf{V}_{2,q}(f_r)$ is a *real tangent* of $\mathscr{C}_n$; that is, if $f_r = L_1 L_2 \ldots L_r$ with deg $L_i = 1$ and L_i in $\bar{\gamma}[x_0, x_1, x_2]$, then $\mathbf{V}_{2,q}(L_i)$ is a real tangent of $\mathscr{C}_n$ if $\lambda L_i \in \gamma[x_0, x_1, x_2]$ for some λ in $\bar{\gamma}_0$.

(c) Suppose there are r' real tangents at $\mathbf{U}_0$, of which r'' are distinct; then r'' is the *real index* of $\mathscr{C}_n$ at $\mathbf{U}_0$. From the definitions, $r'' \leqslant r' \leqslant r$.

(ii) If $r = m_{\mathbf{U}_0}(\mathscr{C}_n) = 1$, then $\mathbf{U}_0$ is a *simple* point of $\mathscr{C}_n$: if $r > 1$, then $\mathbf{U}_0$ is a *singular* point of $\mathscr{C}_n$.

(iii) If $r = 1$, and $f_1 | f_2$, then $\mathbf{U}_0$ is a *point of inflexion* of $\mathscr{C}_n$. With $l = \mathbf{V}(f_1)$, the condition that $f_1 | f_2$ is equivalent to $m_{\mathbf{U}_0}(l, \mathscr{C}_n) \geqslant 3$, §2.6(xiii).

(iv) If $r > 1$ and all r tangents are distinct, then $\mathbf{U}_0$ is an *ordinary singularity*.

(v) If $r = 2$, then $\mathbf{U}_0$ is a double point and one of the following occurs:

(a) $r' = r'' = 2$, and $\mathbf{U}_0$ is a *node;*

(b) $r' = r'' = 0$, and $\mathbf{U}_0$ is an *isolated double point;*

(c) $r' = 2$, $r'' = 1$, and $\mathbf{U}_0$ is a *cusp*.

In other nomenclature, *crunode*, *acnode*, and *spinode* are used for double points of type (a), (b), and (c) respectively.

(vi) If P is a point of $\mathscr{C}_n$ and $\mathfrak{T}$ a projectivity of $PG(2,q)$ such that $P\mathfrak{T} = \mathbf{U}_0$ and $\mathscr{C}_n\mathfrak{T} = \mathscr{C}_n'$, then P has constants r, r', r'' with respect to $\mathscr{C}_n$ if and only if $\mathbf{U}_0$ has these constants with respect to $\mathscr{C}_n'$.

We have defined, both in §2.6 and above, the multiplicity of a point on a curve. Now we need to define the multiplicity of intersection of two curves at a point. We will not actually prove that such a number exists or that it is invariant, but we do give effective rules for calculating it.

Let $\mathscr{F} = V_{2,\gamma}(F)$, $\mathscr{G} = V_{2,\gamma}(G)$ with, as before, $\overline{\mathscr{F}} = V_{2,\bar{\gamma}}(F)$, $\overline{\mathscr{G}} = V_{2,\bar{\gamma}}(G)$. Then the *intersection multiplicity of* $\mathscr{F}$ *and* $\mathscr{G}$ *at* P, denoted $m_P(\mathscr{F},\mathscr{G})$, has the following properties.

I. $m_P(\mathscr{F},\mathscr{G}) = m_P(\overline{\mathscr{F}},\overline{\mathscr{G}}) = m_P(\overline{\mathscr{G}},\overline{\mathscr{F}})$.

II. (a) $m_P(\overline{\mathscr{F}}\overline{\mathscr{G}}) = 0$ if $P \notin \overline{\mathscr{F}} \cap \overline{\mathscr{G}}$;

(b) $m_P(\overline{\mathscr{F}},\overline{\mathscr{G}}) = \infty$ if $\overline{\mathscr{F}}$ and $\overline{\mathscr{G}}$ have a common component through P;

(c) $m_P(\overline{\mathscr{F}},\overline{\mathscr{G}}) \in \mathbf{N}$ otherwise.

III. For $\mathfrak{T}$ in $PGL(3,q)$ with $\overline{\mathscr{F}}\mathfrak{T} = \overline{\mathscr{F}}'$, $\overline{\mathscr{G}}\mathfrak{T} = \overline{\mathscr{G}}'$, $P\mathfrak{T} = P'$, $m_P(\overline{\mathscr{F}},\overline{\mathscr{G}}) = m_{P'}(\overline{\mathscr{F}}',\overline{\mathscr{G}}')$.

IV. $m_P(\overline{\mathscr{F}},\overline{\mathscr{G}}) \geq m_P(\overline{\mathscr{F}})\, m_P(\overline{\mathscr{G}})$ with equality if and only if $\overline{\mathscr{F}}$ and $\overline{\mathscr{G}}$ have no common tangent at P.

V. If $F = \prod_i F_i^{r_i}$, $G = \prod_j G_j^{s_j}$ with F_i, G_j forms in $\bar{\gamma}[x_0, x_1, x_2]$, then

$$m_P(\overline{\mathscr{F}},\overline{\mathscr{G}}) = \sum_{i,j} r_i s_j\, m_P(\overline{\mathscr{F}}_i,\overline{\mathscr{G}}_j),$$

where $\overline{\mathscr{F}}_i = V_{2,\bar{\gamma}}(F_i)$, $\overline{\mathscr{G}}_j = V_{2,\bar{\gamma}}(G_j)$.

VI. $m_P(\overline{\mathscr{F}},\overline{\mathscr{G}}) = m_P(\overline{\mathscr{F}},\overline{\mathscr{H}})$ where $\overline{\mathscr{H}} = V_{2,\bar{\gamma}}(H)$ with $H = G + EF$, where E is a form in $\bar{\gamma}[x_0, x_1, x_2]$ and $\deg E = \deg G - \deg F \geq 0$.

VII. (Bézout's theorem). If $\mathscr{F}$ of degree n_1 and $\mathscr{G}$ of degree n_2 have no common component, then

$$\sum_{P\in\mathscr{F}\cap\mathscr{G}} m_P(\mathscr{F},\mathscr{G}) = n_1 n_2.$$

These properties are consistent with the definition of $m_P(l,\mathscr{F})$ in §2.6. In practice, one uses III with $P' = \mathbf{U}_0$ and then V and VI till IV can give a final answer.

Finally, to contrast with §10.2, we give estimates for the number of points on $\mathscr{C}_n$ when it only has linear components.

10.1.2. LEMMA. (i) *If* $\mathscr{C}_n = \mathbf{V}_{2,q}(F)$ *consists of n lines in* $PG(2,q)$ *and N is the number of points on* $\mathscr{C}_n$, *then*

$$nq - n(n-3)/2 \leqslant N \leqslant nq + 1.$$

The lower limit is achieved if and only if the n lines form an n-side; the upper limit is achieved if and only if the n lines are concurrent.

(ii) *If* $\mathscr{C}_n = \mathbf{V}_{2,q}(F)$ *has n linear components and N is the number of points on* $\mathscr{C}_n$ *in* $PG(2,q)$, *then*

$$N \leqslant nq + 1.$$

Proof. (i) Let $P_1, P_2, \ldots, P_k$ be the points of $PG(2,q)$ on at least two lines of $\mathscr{C}_n$, and let r_i be the number of lines of $\mathscr{C}_i$ through P_i, $i \in \mathbf{N}_k$. Then

$$2 \leqslant r_i \leqslant n. \tag{10.2}$$

So $\mathscr{C}_n$ is an n-side if and only if $r_i = 2$, all i, and $\mathscr{C}_n$ consists of n lines of a pencil if and only if $r_i = n$; in the latter case, i is necessarily 1.

If we count in two ways the number of ordered pairs (l, l') of components of $\mathscr{C}_n$, we obtain

$$n(n-1) = \sum_{i=1}^{k} r_i(r_i - 1). \tag{10.3}$$

From (10.2) and (10.3),

$$2\,\Sigma\,(r_i - 1) \leqslant n(n-1) \leqslant n\,\Sigma\,(r_i - 1), \tag{10.4}$$

with the lower and upper equalities corresponding to the

extreme cases of an n-side and n lines of a pencil.

As each line contains $q+1$ points, $\mathscr{C}_n$ contains $n(q+1)$ points, in which each P_i has been counted r_i times. So

$$N = n(q+1) - \Sigma\,(r_i - 1). \tag{10.5}$$

The combination of (10.4) and (10.5) gives the result.

(ii) This now follows from the upper limit in (i) and lemma 10.1.1.□

10.2. The Hasse-Weil theorem

A deep theorem on the number of points lying on an algebraic curve of genus g in $PG(r,q)$ was proved originally by Hasse for $g = 1$ and then by Weil for arbitrary g. Here we will restrict our attention to plane curves. The proof of the theorem is well beyond the scope of this book. So the theorem will be stated, some particular cases examined and then some corollaries derived. The theorem concerns curves absolutely irreducible over γ, and was proved for non-singular models of such curves.

Let $\mathscr{C}_n = \mathbf{V}(F)$ be an absolutely irreducible curve in $PG(2,q)$; also let $\overline{\mathscr{C}}_n = \mathbf{V}_{2,\bar{\gamma}}(F)$. Suppose $\mathscr{C}_n$ has the singular points $P_1, \ldots, P_s$ of respective multiplicity $r_1, \ldots, r_s$ and real index $r''_1, \ldots, r''_s$. For $\mathscr{C}_n$ there are three point counts of interest: N is the number of simple points; $R = N + s$ is the number of *real points*, being the simple points plus the singular points each counted once; $M = N + r''_1 + \ldots r''_s$ is the number of *model points*, being the simple points plus the singular points counted according to their real index. Then

$$N \leqslant R \quad \text{and } N \leqslant M\ . \tag{10.6}$$

To show how the values of R, N, and M may vary, consider in $PG(2,q)$ the absolutely irreducible curve

$$\mathscr{C}_3 = \mathbf{V}(x_0(x_1{}^2 + x_2{}^2) + x_1{}^3).$$

There are three cases.

(i) $q \equiv 1 \pmod 4$. By §1.5(ix), there exists α in γ such that $\alpha^2 = -1$. The points of $\mathscr{C}_3$ are

$$P(t) = \mathbf{P}(1,\ 1 + t^2,\ t + t^3)$$

for $t \in \gamma^+$. So $P(\alpha) = P(-\alpha) = \mathbf{U}_0$, which is a node with tangents $\mathbf{V}(x_1 \pm \alpha x_2)$. Also $P(\infty) = \mathbf{U}_2$. Hence

$$N = q - 1,\ M = q + 1,\ R = q.$$

(ii) $q \equiv -1 \pmod 4$. By §1.5(x), -1 is a non-square in γ. The points of $\mathscr{C}_3$ are

$$\mathbf{U}_0,\ P(t) = \mathbf{P}(1,\ 1 + t^2,\ t + t^3)$$

for $t \in \gamma^+$. In this case $\mathbf{U}_0$ is an isolated double point. Hence

$$N = q + 1,\ M = q + 1,\ R = q + 2.$$

(iii) $q \equiv 0 \pmod 2$. The points of $\mathscr{C}_3$ are

$$P(t) = \mathbf{P}(1,\ 1 + t^2,\ t + t^3)$$

for $t \in \gamma^+$. Here $P(1) = \mathbf{U}_0$, which is a cusp with tangent $\mathbf{V}(x_1 + x_2)$. Hence

$$N = q,\ M = q + 1,\ R = q + 1.$$

To calculate the *genus* g of $\mathscr{C}_n$, we calculate the *genus* $\bar{g}$ of $\bar{\mathscr{C}}_n$ and define $g = \bar{g}$. If $\bar{\mathscr{C}}_n$ has $\bar{s}$ singularities, and they are all ordinary or cuspidal, then

$$\bar{g} = (n - 1)(n - 2)/2 - \sum_{i=1}^{\bar{s}} r_i(r_i - 1)/2\,, \tag{10.7}$$

where $r_1, r_2, \ldots, r_{\bar{s}}$ are the multiplicities of the singularities. So we may make the following statements about the genus g of $\mathscr{C}_n$.

(i) $s = 0$. $\mathscr{C}_n$ is non-singular and

$$g \leqslant (n - 1)(n - 2)/2 \tag{10.8}$$

with equality if and only if $\bar{s} = 0$.

(ii) $s > 0$. $\mathscr{C}_n$ is singular and

$$g + s \leqslant (n - 1)(n - 2)/2 \tag{10.9}$$

(a) if all the singular points are ordinary ones or cusps, then

$$g \leqslant (n - 1)(n - 2)/2 - \sum_1^s r_i(r_i - 1)/2 \tag{10.10}$$

with equality if and only if $s = \bar{s}$;

(b) if all the singular points are double points, we can do no better than (10.9), namely

$$g \leqslant (n - 1)(n - 2) - s \tag{10.9}$$

with equality if and only if $s = \bar{s}$.

As an illustration, we consider a case in which $\bar{s} > s$. Let $q \equiv -1 \pmod 4$ and let

$$F = (x_0{}^2 + x_1{}^2)^2 + (x_0{}^2 - x_1{}^2)x_2{}^2 + x_2{}^4 .$$

$\mathbf{V}_{2,\mathbf{R}}(F)$ is familiar in classical geometry as a *bicircular quartic*. Here consider

$$\mathscr{C} = \mathbf{V}_{2,q}(F), \quad \mathscr{C}' = \mathbf{V}_{2,q^2}(F).$$

Then $\mathscr{C}$ has no singular points, but $\mathscr{C}'$ has nodes at $\mathbf{P}(\alpha, 1, 0)$ and $\mathbf{P}(-\alpha, 1, 0)$, where $\alpha^2 = -1$. Hence $g = 1$.

It can be shown that $g \geqslant 0$. So, from (10.9), a cubic curve has at most one double point. It cannot therefore happen that an absolutely irreducible curve $\mathscr{C}_3$ in $PG(2,q)$ is non-singular, while $\overline{\mathscr{C}}_3$ is singular. For, if $P = \mathbf{P}(a_0, a_1, a_2)$ were a singular point of $\overline{\mathscr{C}}_3$, then so would $P^q = \mathbf{P}(a_0^q, a_1^q, a_2^q)$.

As $\overline{\mathscr{C}}_3$ can have at most one singular point, $P = P^q$ and hence is a point of $\mathscr{C}_3$.

The general statement of the Hasse-Weil theorem is the following.

10.2.1. THEOREM. *For an absolutely irreducible curve $\mathscr{C}_n$ of genus g in $PG(2,q)$,*

$$|M - (q + 1)| \leqslant 2g\sqrt{q} . \tag{10.11}$$

Proof. See §10.5.□

COROLLARY 1. *If $\overline{\mathscr{C}}_n$ is non-singular,*

$$q + 1 - (n - 1)(n - 2)\sqrt{q} \leqslant R = N = M \leqslant q + 1 + (n - 1)(n - 2)\sqrt{q}.\square \tag{10.12}$$

COROLLARY 2. *If $\bar{s} > 0$,*

$$N \leqslant M \leqslant q + 1 + 2g\sqrt{q} < q + 1 + (n - 1)(n - 2)\sqrt{q}, \tag{10.13}$$

$$R = N + s \leqslant q + 1 + 2g\sqrt{q} + s < q + 1 + (n - 1)(n - 2)\sqrt{q}. \tag{10.14}$$

Proof. (10.13) follows from (10.6), (10.8), (10.9), and (10.11). If $s = 0$, (10.14) is just (10.13). If $s > 0$, then, from (10.9),

$$s \leqslant (n - 1)(n - 2)/2 - g < 2\sqrt{q}[(n - 1)(n - 2)/2 - g].$$

Hence

$$\begin{aligned} q + 1 + 2g\sqrt{q} + s &< q + 1 + 2g\sqrt{q} + 2\sqrt{q}[(n - 1)(n - 2)/2 - g] \\ &= q + 1 + (n - 1)(n - 2)\sqrt{q}. \ \square \end{aligned}$$

To show how good the theorem is, consider the Hermitian curve $\mathscr{U}_2 = \mathrm{V}_{2,p^{2m}}(F)$, where

$$F = x_0^{p^m+1} + x_1^{p^m+1} + x_2^{p^m+1} ,$$

as in §7.3. Here $q = p^{2m}$, $n = p^m + 1$ and $N = p^{3m} + 1$. Also $\mathscr{U}_2$ is non-singular. Now,

$$\begin{aligned} q + 1 + (n - 1)(n - 2)\sqrt{q} &= p^{2m} + 1 + p^m(p^m - 1)p^m \\ &= p^{3m} + 1 \\ &= N. \end{aligned}$$

On the other hand, for the sub-Hermitian curve $\mathscr{U}'_2 = V_{2,p^{4m}}(F)$, we have $q = p^{4m}$, $n = p^m + 1$ and $N = p^{3m} + 1$. It is non-singular and

$$\begin{aligned} q + 1 - (n - 1)(n - 2)\sqrt{q} &= p^{4m} + 1 - p^m(p^m - 1)p^{2m} \\ &= p^{3m} + 1 \\ &= N. \end{aligned}$$

So $\mathscr{U}_2$ achieves the upper limit and $\mathscr{U}'_2$ the lower limit in (10.12).

Theorem 10.2.1 and its corollaries can be applied to give estimates for the number of points on a plane curve $\mathscr{C}_n$ or, dually, the number of lines belonging to an algebraic envelope Γ_n, when they are not necessarily absolutely irreducible.

10.2.2. THEOREM. *Let N be the number of simple points on a plane algebraic curve $\mathscr{C}_n$ in $PG(2,q)$ with no regular linear components.*
If

$$\sqrt{q} > n - 1, \tag{10.15}$$

then

$$N < n(q + 2 - n). \tag{10.16}$$

Proof. (i) *$\mathscr{C}_n$ absolutely irreducible.*
Suppose (10.16) does not hold. Then, by (10.12) and (10.13),

$$n(q + 2 - n) \leqslant N \leqslant q + 1 + (n - 1)(n - 2)\sqrt{q}.$$

Hence $$(n - 1)(\sqrt{q} + 1)(\sqrt{q} + 1 - n) \leqslant 0.$$

So $$\sqrt{q} \leqslant n - 1,$$

contradicting (10.15).

(ii) $\mathscr{C}_n$ *not absolutely irreducible.*

Suppose $\mathscr{C}_n = \mathscr{C}^1 \cup \ldots \cup \mathscr{C}^m$, where the $\mathscr{C}^i$ are the components of $\mathscr{C}_n$. Suppose $\mathscr{C}^i$ has order n_i and contains N_i simple points. Since $N \leqslant \Sigma N_i$, it suffices to show that

$$N_i < n_i(q + 2 - n) \text{ for } i \text{ in } \mathbf{N}_m. \tag{10.17}$$

Then $N \leqslant \Sigma N_i < \Sigma n_i(q + 2 - n) = n(q + 2 - n)$.

Two cases must be distinguished.

(a) $\mathscr{C}^i$ *regular.*

By hypothesis, $n_i \leqslant 2$ and, since $\mathscr{C}_n$ could not have just one irregular linear component, $n \geqslant n_i + 2 \geqslant 4$.

Suppose, (10.17) false. Then, by (10.12) and (10.13),

$$n_i(q + 2 - n) \leqslant N \leqslant (q + 1) + (n_i - 1)(n_i - 2)\sqrt{q}$$
$$\Longleftrightarrow \quad q(n_i - 1) - \sqrt{q}(n_i - 1)(n_i - 2) \leqslant n_i(n - 2) + 1$$
$$\Longleftrightarrow \quad \sqrt{q}(n_i - 1)(\sqrt{q} - n_i + 2) \leqslant n_i(n - 2) + 1$$
$$\Rightarrow \quad (n - 1)(n_i - 1)(n - n_i + 1) < n_i(n - 2) + 1 \text{ (using (10.15))}$$
$$\Rightarrow \quad (n - 1)(n_i - 1)(n - n_i + 1) < n_i(n - 1)$$
$$\Rightarrow \quad (n_i - 1)(n - n_i + 1) < n_i$$
$$\Rightarrow \quad 3(n_i - 1) < n_i$$
$$\Rightarrow \quad 2n_i < 3\text{: contradiction.}$$

So $N_i < n_i(q + 2 - n)$.

(b) $\mathscr{C}^i$ *irregular.*

By lemma 10.1.1, $N_i \leqslant n_i^2$.

Now,

$$
\begin{aligned}
& (\sqrt{q} - 1)^2 \geq 0 \\
\Longleftrightarrow\quad & 2\sqrt{q} + 1 \leq q + 2 \\
\Rightarrow\quad & 2n - 1 < q + 2 \text{ (using (10.15))} \\
\Rightarrow\quad & n + n_i < q + 2 \\
\Longleftrightarrow\quad & n_i < q + 2 - n \\
\Longleftrightarrow\quad & n_i^2 < n_i(q + 2 - n) \\
\Rightarrow\quad & N_i < n_i(q + 2 - n). \square
\end{aligned}
$$

Note: It is not true that $n^2 \leq q + 1 + (n - 1)(n - 2)\sqrt{q}$ for all values of n and q.

10.2.3. THEOREM. *Let $\mathscr{C}_n$ be a component of a plane curve $\mathscr{D}_{2t}$ in $PG(2,q)$ and let $n \geq 3$ if $\mathscr{C}_n$ is regular. Further, over γ, $\mathscr{C}_n$ has N simple points and s singular points, of which d are double points. If*

$$\sqrt{q} > 2t + n - 1, \tag{10.18}$$

then

$$L = N + d < n(q - t + 2)/2. \tag{10.19}$$

Proof. (i) *$\mathscr{C}_n$ regular and $n \geq 3$.*

Suppose (10.19) false. Then, by (10.12) and (10.13),

$$n(q - t + 2)/2 \leq L \leq N + s \leq (q + 1) + (n - 1)(n - 2)\sqrt{q}$$

$$
\begin{aligned}
\Rightarrow\quad & n(q - t + 2) \leq 2(q + 1) + 2(n - 1)(n - 2)\sqrt{q} \\
\Longleftrightarrow\quad & (n - 2)q - 2(n - 1)(n - 2)\sqrt{q} \leq n(t - 2) + 2 \\
\Longleftrightarrow\quad & (n - 2)\sqrt{q}[\sqrt{q} - 2(n - 1)] \leq n(t - 2) + 2 \\
\Rightarrow\quad & (n - 2)(2t + n - 1)(2t - n + 1) < n(t - 2) + 2 \text{ (using(10.18))} \\
\Longleftrightarrow\quad & (n - 2)[4t^2 - (n - 1)^2] < n(t - 2) + 2 \\
\Rightarrow\quad & (n - 2)[4t^2 - (2t - 1)^2] < n(t - 2) + 2 \text{ (since } 2t \leq n) \\
\Longleftrightarrow\quad & (n - 2)(4t - 1) < n(t - 2) + 2 \\
\Longleftrightarrow\quad & (3n - 8)t + n < 0,
\end{aligned}
$$

which is impossible as $n \geq 3$.

(ii) $\mathscr{C}_n$ *irregular.*

Suppose the result false. Then, by lemma 10.1.1,

$$n(q - t + 2)/2 \leq L \leq n^2$$

$\Rightarrow \quad q - t + 2 \leq 2n$

$\Rightarrow \quad (2t + n - 1)^2 - (t - 2) < 2n \quad$ (using (10.18))

$\Rightarrow \quad (2t + n - 1)^2 - (2t - 1)^2 < 2n$

$\iff \quad n(4t + n - 2) < 2n$

$\iff \quad 4t + n - 2 < 2$

$\iff \quad 4(t - 1) + n < 0,$

which is impossible as $t \geq 1$ and $n \geq 1$.□

10.3. Complete arcs for q even

Our aim is to consider conditions on the size of a k-arc for any complete arc containing it to be an oval or not. Firstly, theorem $2.8.4_k$ is used to associate an algebraic envelope to a k-arc $\mathscr{K}$. Then, the Hasse-Weil theorem in the guise of theorems 10.2.2 and 10.2.3 is used to obtain the desired conditions. The cases of even and odd characteristic of γ must be distinguished and we begin with the even case.

10.3.1. THEOREM. *Let $\mathscr{K}$ be a k-arc in $PG(2,q)$ with q even. Then the tk unisecants of $\mathscr{K}$ where $t = q + 2 - k$ belong to an algebraic envelope Γ_t of class t with the properties:*

(i) *Γ_t is unique if $k > t$, that is, $k > q/2 + 1$;*

(ii) *Γ_t contains no bisecant of $\mathscr{K}$ and so no pencil with vertex P in $\mathscr{K}$;*

(iii) *each unisecant of $\mathscr{K}$ is counted exactly once in Γ_t.*

Proof. By lemma 8.2.2, valid for q even or odd, $\Pi d = 1$ for each triangle of $\mathscr{K}$ where the product ranges over all the unisecants to $\mathscr{K}$ at the vertices of the triangle and d is the coordinate of such a unisecant. By theorem $2.8.4_k$, there exists an envelope Γ_t, which is unique if $k > t$. Also by definition of the family $\mathscr{N}_H$ in theorem $2.8.4_k$, no member of

$\mathcal{N}_H$ contains a bisecant of $\mathcal{K}$. So Γ_t contains no bisecant of $\mathcal{K}$ and so no pencil with vertex P in $\mathcal{K}$. Hence through a point P of $\mathcal{K}$, there pass at most t lines of Γ_t. As each of the t unisecants to $\mathcal{K}$ through P is in Γ_t, none is counted twice.□

COROLLARY. *In $PG(2,q)$ with q even, a $(q + 1)$-arc $\mathcal{K}$ is incomplete. The $q + 1$ unisecants to $\mathcal{K}$ are concurrent and $\mathcal{K}$ lies in a unique $(q + 2)$-arc.*

Proof. Since $k + t = q + 2$, so $t = 1$. Therefore Γ_t is a pencil, whose vertex Q added to $\mathcal{K}$ gives a $(q + 2)$-arc.□

This elementary result was already obtained in lemma 8.1.4. At least the above machinery is shown to be working.

10.3.2. LEMMA. *If $\mathcal{K}$ is an incomplete k-arc with $3 \leq k \leq q + 1$ in $PG(2,q)$, q even, and if $k > (q + 2)/2$, then*

(i) *there is a point Q on no bisecant of $\mathcal{K}$;*

(ii) *the pencil with vertex Q is a component of any envelope Γ_t containing all the unisecants to $\mathcal{K}$;*

(iii) *Γ_t is unique.*

Proof. (i) Since $\mathcal{K}$ is incomplete, there is such a Q.

(ii) By theorem 10.3.1, there exists an envelope Γ_t such that the k distinct lines QP, for P in $\mathcal{K}$, belong to Γ_t. Since Γ_t has class t, the pencil with vertex Q is a component if $k > t$. But $k > t$ if and only if $k > (q + 2)/2$.

(iii) Again by theorem 10.3.1, since $k > t$, Γ_t is unique. □

10.3.3. THEOREM. *If $\mathcal{K}$ is a complete k-arc with $3 \leq k \leq q + 1$ in $PG(2,q)$, q even, then $k \leq q - \sqrt{q} + 1$.*

Proof. Since $k + t = q + 2$, so $k \leq q - \sqrt{q} + 1$ if and only if $\sqrt{q} \leq t - 1$. By theorem 10.3.1, the unisecants to belong to an envelope Γ_t. Let N be the number of simple lines of Γ_t; then, by theorem 10.3.1(iii), all the unisecants to are simple lines of Γ_t. So $N \geq kt$.

Suppose Γ_t contains a regular linear component. Then there is a point Q in $PG(2,q)$ such that the pencil of lines with vertex Q lies in Γ_t. Since, by theorem 10.3.1(ii), no bisecant of $\mathscr{K}$ of lies in Γ_t, all the lines QP with P in $\mathscr{K}$ are unisecants to $\mathscr{K}$. Hence $\mathscr{K} \cup \{Q\}$ is a $(k + 1)$-arc, which is not possible as $\mathscr{K}$ is complete. So Γ_t contains no regular linear component.

Let us then suppose that $\sqrt{q} > t - 1$. By the dual of theorem 10.2.2, $N < t(q + 2 + t) = kt$, contradicting the previous estimate of N. So $\sqrt{q} \leqslant t - 1$.□

COROLLARY 1. *Let $\mathscr{K}$ be a k-arc with $3 \leqslant k \leqslant q + 1$ in $PG(2,q)$, q even. If $k > q - \sqrt{q} + 1$, then*

(i) *there is a point Q on no bisecant of $\mathscr{K}$;*

(ii) *the pencil with vertex Q is a component of any envelope Γ_t containing all the unisecants to $\mathscr{K}$;*

(iii) *Γ_t is unique.*

Proof. By the theorem, $\mathscr{K}$ is incomplete. Also

$$q - \sqrt{q} + 1 \geqslant (q + 2)/2 \Longleftrightarrow q - 2\sqrt{q} \geqslant 0 \Longleftrightarrow q \geqslant 4.$$

For $q = 2$, the smallest integer greater than $q - \sqrt{q} + 1$ is 2, which also equals $(q + 2)/2$. So, for all even q, the inequalities $k \geqslant 3$ and $k > q - \sqrt{q} + 1$ imply $k > (q + 2)/2$. Hence (i), (ii), and (iii) follow from lemma 10.3.2.□

COROLLARY 2. *Let $\mathscr{K}$ be a k-arc in $PG(2,q)$, q even. If $k > q - \sqrt{q} - 1$, then*

(i) *any complete arc containing $\mathscr{K}$ is an oval $\mathscr{O}$;*

(ii) *$\mathscr{O}$ is unique except when $k = q = 2$, in which case there are two ovals containing $\mathscr{K}$;*

(iii) *for $q > 2$, the only points Q for which $\mathscr{K} \cup \{Q\}$ is a $(k + 1)$-arc are the $q + 2 - k$ points of $\mathscr{O} \backslash \mathscr{K}$.*

Proof. (i) This follows from the theorem.

(ii) By lemma 8.1.5, no two ovals can have more than half their points in common. As in corollary 1, the inequalities $k \geqslant 3$ and $k > q - \sqrt{q} + 1$ together imply that $k > (q + 2)/2$. So, in this case, $\mathscr{O}$ is unique.

If $k > q + \sqrt{q} - 1$ and $k < 3$, then $k = q = 2$. In $PG(2,2)$, every 3-arc is in a unique 4-arc and every 2-arc is in four 3-arcs. So every 2-arc is contained in two 4-arcs.

(iii) If Q is any point on no bisecant of $\mathscr{K}$, then $\mathscr{K}_1 = \mathscr{K} \cup \{Q\}$ is a $(k + 1)$-arc. $\mathscr{K}_1$ now lies in some complete arc, which by (i) and (ii) is the oval $\mathscr{O}$. Hence $Q \in \mathscr{O}\backslash K$.□

COROLLARY 3. *In $PG(2,q)$, q even, a q-arc is contained in an oval.*

Proof. $q > q - \sqrt{q} + 1$ when $q > 1$.□

This was also shown in theorem 8.7.2.

COROLLARY 4. *In $PG(2,8)$, a complete k-arc has $k = 6$ or $k = 10$.*

Proof. By the theorem, a complete k-arc which is not an oval has $k \leqslant 9 - \sqrt{8} < 7$; so $k \leqslant 6$. From §9.2, no k-arc with $k = 3$, 4, or 5 is complete in $PG(2,8)$. So complete arcs in this plane are ovals or 6-arcs. This was also shown in theorem 9.2.5. In fact complete 6-arcs are classified in §14.6. It was shown in theorem 9.2.3 that each oval is regular.□

10.4. Complete arcs for q odd

For q even, $-1 = 1$: for q odd, $(-1)^2 = 1$. This is the simple reason that the results are weaker in the odd case: lemma 8.2.2 can no longer be directly combined with theorem $2.8.4_k$.

10.4.1. THEOREM. *Let $\mathscr{K}$ be a k-arc in $PG(2,q)$, q odd. Then the tk unisecants of $\mathscr{K}$ where $t = q + 2 - k$ belong to an envelope Γ_{2t} of class $2t$ with the properties:*

(i) *Γ_{2t} is unique if $k > 2t$, that is, $k > (2q + 4)/3$;*

(ii) *Γ_{2t} contains no bisecant of $\mathscr{K}$ and so no pencil with vertex P in $\mathscr{K}$;*

(iii) *each unisecant to $\mathscr{K}$ is counted exactly twice in Γ_{2t};*

(iv) *Γ_{2t} may contain components of multiplicity at most two, but does not consist entirely of double components.*

Proof. By lemma 8.2.2, $\Pi d = -1$ for each triangle of $\mathscr{K}$, where the product ranges over all the unisecants to $\mathscr{K}$ at the vertices of the triangle and d is the coordinate of such a unisecant. So $\Pi d^2 = 1$. By theorem $2.8.4_k$, the envelope exists and is unique if $k > 2t$. Also, by the definition of $\mathscr{N}_H$ in theorem $2.8.4_k$, no bisecant of $\mathscr{K}$ is in Γ_{2t} and so no pencil with vertex P in $\mathscr{K}$ is a component of Γ_{2t}. Each unisecant to $\mathscr{K}$ has been counted twice in Γ_{2t}: this accounts for the $2t$ lines of Γ_{2t} through any point of $\mathscr{K}$ and, by the above, there cannot be more. As each component of Γ_{2t} contains some unisecant to $\mathscr{K}$, there are no components of multiplicity three. Finally, if Γ_{2t} consisted entirely of double components, the unisecants to $\mathscr{K}$ would belong to an envelope of class t and the condition $\Pi d = 1$ would be satisfied for any triangle of $\mathscr{K}$.□

COROLLARY. *In $PG(2,q)$, q odd, every $(q + 1)$-arc $\mathscr{K}$ is a conic.*

Proof. Since $k = q + 1$, so $t = 1$. Hence, by the theorem, the unisecants to $\mathscr{K}$ belong to a unique envelope Γ_2, which does not consist of two pencils. So Γ_2 is the envelope of a conic $\mathscr{C}_2$. But, from part (iii) of the theorem, if P is in $\mathscr{K}$, the two unisecants from P to $\mathscr{C}_2$ coincide with the unisecant to $\mathscr{K}$ at P. So $P \in \mathscr{C}_2$ and $\mathscr{K} \subset \mathscr{C}_2$. As $\mathscr{K}$ and $\mathscr{C}_2$ have the same number of points, they coincide.□

10.4.2. LEMMA. *Let $\mathscr{K}$ be an incomplete k-arc with $3 \leqslant k < q + 1$ in $PG(2,q)$, q odd. If $k > 2(q + 2)/3$, that is $k > 2t$, then*

(i) *there is a point Q on no bisecant of $\mathscr{K}$;*

(ii) *the pencil with vertex Q is a component of any envelope Δ_{2t} containing all the unisecants to $\mathscr{K}$;*

(iii) *Δ_{2t} is unique.*

Proof. (i) Since $\mathscr{K}$ is incomplete, there is such a Q.

(ii) By theorem 10.4.1, there exists an envelope Δ_{2t} in which the k distinct lines QP, for P in $\mathscr{K}$, are double lines. Since Δ_{2t} has class $2t$, the pencil with vertex Q is a component if $k > 2t$. Since $k + t = q + 2$, so $k < 2(q + 2)/3$ if and only if $k > 2t$.

(iii) By theorem 10.4.1, since $k > 2t$, Δ_{2t} is unique.□

10.4.3. THEOREM. *Let $\mathscr{K}$ be a k-arc with $3 \leqslant k \leqslant q + 1$ in $PG(2,q)$, q odd, and let Δ_{2t} be an envelope containing the unisecants to $\mathscr{K}$. If*

(i) Δ_{2t} has the envelope of a conic $\mathscr{C}$ as component,

(ii) $k > (3q + 5)/4$,

then $\mathscr{K}$ is contained in $\mathscr{C}$.

Proof. $\mathscr{K} \cap \mathscr{C}$ is a k_1-arc, $\mathscr{K}\backslash\mathscr{C}$ is a k_2-arc, and $\mathscr{C}\backslash\mathscr{K}$ is a k_3-arc. Since, from each point of $\mathscr{K}\backslash\mathscr{C}$, there are at most two tangents to $\mathscr{C}$, we have $k_1 + 2k_2 \leqslant q + 1$, $k_1 + k_2 = k = q + 2 - t$, $k_1 + k_3 = q + 1$. Hence $k_2 \leqslant t - 1$, $k_1 \geqslant q + 3 - 2t$; $k_3 \leqslant 2t - 2$. For Q not on $\mathscr{C}$, there are at least $(q - 1)/2$ bisecants of $\mathscr{C}$ through Q. Also $k > (3q + 5)/4$ implies $(q - 1)/2 > 2t - 2$. As $k_3 \leqslant 2t - 2$, there is a bisecant m of $\mathscr{C}$ through Q containing no point of $\mathscr{C}\backslash\mathscr{K}$. So m is a bisecant of $\mathscr{C} \cap \mathscr{K}$. As Q is not in $\mathscr{C}$, so Q is not in $\mathscr{K}$. Therefore $\mathscr{K}$ is contained in $\mathscr{C}$.□

10.4.4. THEOREM. *Let $\mathscr{K}$ be a k-arc in $PG(2,q)$, q odd. If $k > q - \sqrt{q}/4 + 7/4$, that is $\sqrt{q} > 4t - 1$, then $\mathscr{K}$ is contained in a unique conic.*

Proof. $k > q - \sqrt{q}/4 + 7/4$ implies $k \geqslant 5$. So, if $\mathscr{K}$ is contained in a conic, the conic is unique.

By theorem 10.4.1, the unisecants to $\mathscr{K}$ are contained in an envelope Δ_{2t}, which has a simple component Γ_n, where $n \leqslant 2t$. For $t = 1$, $\mathscr{K}$ is a $(q + 1)$-arc and so a conic. Let $t \geqslant 2$ and distinguish the following three cases:

(i) Γ_n is a regular linear component;

(ii) Γ_n is a regular component of class two;

(iii) Γ_n is regular of class at least three or Γ_n is irregular.

Case (i)

Γ_n is a pencil with vertex Q which is not in $\mathscr{K}$. So $\mathscr{K}' = \mathscr{K} \cup \{Q\}$ is a $(k + 1)$-arc. If the envelope Γ'_n associated with $\mathscr{K}'$ is again of type (i) the process is continued until the arc becomes a conic or the respective envelope becomes one of the other two types.

Case (ii)

Γ_n is the envelope of a conic $\mathscr{C}$. Also $k > q - \sqrt{q}/4 + 7/4 \Leftrightarrow q > 4t - 1 \Rightarrow q > 4t - 3 \Leftrightarrow k > (3q + 5)/4$. So, by theorem 10.4.3, $\mathscr{K}$ is contained in $\mathscr{C}$.

Case (iii)

Γ_n is regular with $3 \leq n \leq 2t$ or Γ_n is irregular. Suppose Γ_n has N simple lines and d double lines: $L = N + d$. Then, by the dual of theorem 10.2.3,

$$L < n(q - t + 2)/2 = nk/2$$

By theorem 10.4.1, each of the t unisecants of $\mathscr{K}$ through a point P of $\mathscr{K}$ is a double line of Δ_{2t}. So, through P, there are n lines of Γ_n, where each line is counted once or twice. Hence there are at least $n/2$ distinct lines of Γ_n through P. Therefore,

$$L \geqslant kn/2,$$

contradicting the previous estimate.

Thus $\mathscr{K}$ is always contained in a unique conic.□

COROLLARY 1. *If $\mathscr{K}$ is a complete k-arc in $PG(2,q)$, q odd, other than a conic, then $k \leqslant q - \sqrt{q}/4 + 7/4$.*□

COROLLARY 2. *If $\mathscr{K}$ is a k-arc in $PG(2,q)$, q odd, with $k > q - \sqrt{q}/4 + 7/4$, then the only points for which $\mathscr{K} \cup \{Q\}$ is a $(k + 1)$-arc are the $q + 1 - k$ points of $\mathscr{C} \backslash \mathscr{K}$, where $\mathscr{C}$ is the unique conic containing $\mathscr{K}$.*

Proof. If $\mathcal{K} \cup \{Q\}$ is a $(k + 1)$-arc, then it lies in some complete arc, which is necessarily $\mathcal{C}$, by the theorem. Hence $Q \in \mathcal{C}\backslash\mathcal{K}$.□

As a comparison with theorem 8.6.10, we have the following.

COROLLARY 3. *In* $PG(2,q)$, *q odd and* $q > 49$, *a q-arc is contained in a conic.*

Proof. $q > q - \sqrt{q}/4 + 7/4$ when $q > 49$.□

10.5. Notes and references

§10.1. The axioms for $m_P(\mathcal{F},\mathcal{G})$ essentially follow W. Fulton [1969]. Lemma 10.1.2 is due to Segre [1959a]. Plücker's equations in the finite case were considered by Campbell [1928c], who glossed over the exceptions.

As in §2.9, it would be more accurate to define the curve determined by the form F in $\gamma[x_0, x_1, x_2]$ as a pair $(\mathcal{P}, \mathcal{I})$, where $\mathcal{P} = \{\mathbf{P}(X) \mid F(X) = 0\}$ and $\mathcal{I} = \{\lambda F \mid \lambda \in \gamma_0\}$.

§10.2. Theorem 10.2.1 was proved by Hasse [1934], [1943] for $g = 1$ and in general by Weil [1948a], [1948b]. For a detailed exposition of these results, see Eichler [1966], Joly [1973]. An elementary proof in certain cases was given by Stepanov [1970], [1972], whose methods were generalized to the whole theorem by Schmidt [1973], [1974], [1976]. See also Zimmer [1970], [1971] for the case $g = 1$. For other elementary results, see Hua and Vandiver [1949], Ireland and Rosen [1972], Joly [1971], Segre [1960c], [1962a], [1963], [1964a]. Stark [1973] improved the theorem for $g = 2$.

Weil [1949] conjectured a generalization of theorem 10.2.1, which implied that if $\mathcal{V}$ is a non-singular variety of dimension d that is a complete intersection in $PG(r,q)$, then

$$\big|\,|\mathcal{V}| - \theta(d)\,\big| \leqslant b\; q^{d/2}\,,$$

where b is a certain well-defined constant. Some progress was made on different aspects of the conjectures by Dwork [1960], [1965], Lang and Weil [1954]. The conjectures were proved by Deligne [1974]. For discussions of the proof, see Mazur [1975], Serre [1975].

On related problems, see Ax [1964], Berthelot [1971], Swinnerton-Dyer [1967], Waterhouse and Milne [1971].

Theorems 10.2.2, 10.2.3 are taken from Segre [1967].

As mentioned in the Preface, it is incongruous that there should be such a wide gap in knowledge between the particular results described in this work and the deep results of the Hasse-Weil-Deligne theorem, proved with the whole apparatus of modern algebraic geometry.

§§10.3, 10.4. These are adapted from Segre [1967]. For stronger results when q is small, see Segre [1959a]. See also Korchmáros [1974a]. For another application of theorem 10.2.1, see Hayes [1967].

11
CUBIC CURVES

11.1. Configurations

Over $\gamma = GF(q)$, $q = p^h$, plane cubic curves have properties familiar from the classical theory over the real and complex numbers. When $q \equiv 1 \pmod 3$, their properties resemble those of cubic curves over the complex numbers. When $q \equiv -1 \pmod 3$, the real numbers are the better analogy. When $q \equiv 0 \pmod 3$, there is no suitable classical model for the following reason. If $\mathscr{F} = \mathbf{V}(F)$ is a cubic curve, then, since

$$3F = x_0 \frac{\partial F}{\partial x_0} + x_1 \frac{\partial F}{\partial x_1} + x_2 \frac{\partial F}{\partial x_2},$$

the three partial derivatives can be zero at a point without F being zero there as well. For any γ, the point $Q = \mathbf{P}(y_0, y_1, y_2)$ is a *nucleus* of $\mathscr{F}$ if the three partial derivatives are zero at Q. For $q \equiv 0 \pmod 3$, there are cubics with nuclei not on $\mathscr{F}$: for other q, a nucleus is just a singular point of $\mathscr{F}$.

Although a superficial examination indicates that characteristic two is a special case, the results do not bear this out; slightly different methods are required however to obtain the results.

Apart from the value of q modulo 3, the value modulo 4 turns out to be also relevant. In fact, the value of q modulo 12 provides the separation of prime powers needed and it is given in Table 11.1.

TABLE 11.1

$q = p^h$	p	>3				3			2	
$q \equiv c \pmod 3$	c	1	1	-1	-1	0	0	1	-1	-1
$q \equiv d \pmod 4$	d	1	-1	1	-1	1	-1	0	0	2
$q \equiv m \pmod{12}$	m	1	7	5	11	9	3	4	8	2

It will be shown below that the inflexions of an absolutely irreducible cubic curve $\mathscr{F}$ over γ satisfy the following two conditons:

(i) two inflexions of $\mathscr{F}$ are collinear with a third;

(ii) either (a) $\mathscr{F}$ has at most nine inflexions;
or (b) all points of $\mathscr{F}$ but one are inflexions.

11.1.1. THEOREM. *The configurations in* $PG(2,q)$ *of at most nine points in which every two are collinear with a third are as follows:* (i) *zero points;* (ii) *one point;* (iii) *three collinear points;* (iv) *the tactical configuration* $(7_3, 7_3)$, *whose incidences are those of* $PG(2,2)$ *and which therefore occurs only for* q *even;* (v) *the tactical configuration* $(9_4, 12_3)$, *whose incidences are those of the affine geometry* $AG(2,3)$ *and which occurs when* $(q+1, 3) = 1$; *that is, when* $q \equiv 0$ *or* $1 \pmod 3$.

Proof. Given three collinear points P_1, P_2, P_3 and a fourth point Q, there is a point Q_i on P_iQ, $i = 1,2,3$. If the configuration is closed, it is a $(7_3, 7_3)$, which only exists for q even.

If the configuration does not close with seven points, it closes with nine. Let the points be labelled 1 to 9; then the lines are the rows, columns, and determinantal products in the matrix

$$\begin{bmatrix} 1 & 2 & 3 \\ 4 & 5 & 6 \\ 7 & 8 & 9 \end{bmatrix}$$

To attach coordinates, choose 1,2,4, and 5 as $\mathbf{U}_0$, $\mathbf{U}_1$, $\mathbf{U}_2$, and $\mathbf{U}$ respectively; then the coordinate vectors of the nine points are as follows:

1	2	3	4	5	6	7	8	9
$(1,0,0)$	$(0,1,0)$	$(t,1,0)$	$(0,0,1)$	$(1,1,1)$	$(1-t,1-t,1)$	$(1-t,0,1)$	$(1,1-t,1)$	$(0,1,1)$

where 3 was chosen as $\mathbf{P}(t,1,0)$, then 7 as the intersection of 147 and 357, then 6 as $456 \cap 267$, then 9 as

159 ∩ 249, then 8 as 168 ∩ 258. This accounts for all collinearities bar three: 348, 369, and 789, all of which occur if $t^2 - t + 1 = 0$. The quadratic $x^2 - x + 1$ has two, one, or zero roots in γ as $q \equiv 1$, 0, or -1 (mod 3), §1.5(xi)-(xiii).□

COROLLARY. *In* $PG(2,4)$, *the points of a* $(9_4, 12_3)$ *configuration form a Hermitian curve* $\mathcal{U}_{2,4}$.

Proof. Below in (11.1), the nine points are those of $V(x_0^3 + x_1^3 + x_2^3, x_0x_1x_2)$, which for $GF(4)$ is just $V(x_0^3 + x_1^3 + x_2^3) = \mathcal{U}_{2,4}$, §7.3.□

Let us examine the $(9_4, 12_3)$ configuration separately in the cases that $q \equiv 0$ and 1 (mod 3).

(i) *$q \equiv 0$ (mod 3).* The configuration is exactly the affine plane $AG(2,3)$ obtained as a subgeometry of $PG(2,3)$ with the four points of $l \cap PG(2,3)$ removed, where l is any line of $PG(2,q)$ meeting $PG(2,3)$ in four points. So each of the four sets of three lines containing the nine points is concurrent at a point on $l \cap PG(2,3)$. The figure is determined by four of its points.

(ii) *$q \equiv 1$ (mod 3).* If ω is a root of $x^2 + x + 1$, then we firstly observe that the configuration is projectively unique, although it is determined by five rather than four of its points. For, if the configurations are named $C(-\omega)$ and $C(-\omega^2)$ corresponding to $t = -\omega$ and $t = -\omega^2$ respectively, then the projectivity given by $x_0' = x_1$, $x_1' = x_0$, $x_2' = -\omega^2 x_2$ transforms $C(-\omega)$ to $C(-\omega^2)$. More symmetrically, the nine points may be given as $V(x_0^3 + x_1^3 + x_2^3, x_0x_1x_2)$ and have coordinate vectors

$$\begin{array}{lll} (0,1,-1) & (-1,0,1) & (1,-1,0) \\ (0,1,-\omega) & (-\omega,0,1) & (1,-\omega,0) \\ (0,1,-\omega^2) & (-\omega^2,0,1) & (1,-\omega^2,0) \end{array} \qquad (11.1)$$

There are four triangles each of which contains the nine points on its sides. The cubic curves formed by the triangles are

$$\begin{aligned}
&V(x_0x_1x_2),\\
&V(x_0 + x_1 + x_2)(x_0 + \omega^2x_1 + \omega x_2)(x_0 + \omega x_1 + \omega^2x_2)),\\
&V((\omega x_0 + x_1 + x_2)(x_0 + \omega x_1 + x_2)(x_0 + x_1 + \omega x_2)), \qquad (11.2)\\
&V((\omega^2x_0 + x_1 + x_2)(x_0 + \omega^2x_1 + x_2)(x_0 + x_1 + \omega^2x_2)).
\end{aligned}$$

As regards this and any other configuration, there are three groups of interest: (i) the *permutation group* consisting of all the permutations of the points which preserve the collinearities; (ii) the *projective group* consisting of all the elements of the permutation group which can be effected by projectivities; (iii) the *collineation group* consisting of all the elements of the permutation group which can be effected by collineations.

There is a risk of confusion in both (ii) and (iii) as there may be more than one projectivity or collineation effecting the same permutation on a certain configuration.

11.1.2. THEOREM. *The orders g_1, g_2, and g_3 of the permutation group, projective group, and collineation group of the $(9_4, 12_3)$ configuration are as follows.*

	g_1	g_2	g_3
(i) $q \equiv 0$ (*mod* 3)	432	432	432
(ii) q *square*, $\sqrt{q} \equiv -1$ (*mod* 3)	432	216	432
(iii) $q \equiv 1$ (*mod* 3), q *non-square* or $\sqrt{q} \equiv 1$ (*mod* 3)	432	216	216

Proof. To calculate g_1, we first consider the operations fixing each of three collinear points, say 1, 2, and 3, as in theorem 11.1.1. Apart from the identity, they are the permutations (47)(58)(69), (48)(59)(67), (49)(57)(68), (456)(798), (465)(789), which with the identity form a subgroup S_3. So there are six operations fixing each of 1, 2, and 3. Therefore, as the group is transitive, there are 36 operations fixing the set of three points on a line, 108 operations fixing a set of three lines containing the nine points and as there are a four sets of three lines, 432 operations fixing the set of 12 lines.

In case (i), the stabilizer in $PGL(3,3)$ of a line in

$PG(2,3)$ has order $p(3,3)/13 = 3^3 \cdot 26 \cdot 8/13 = 432$. So all the permutations can be effected by projectivities.

In cases (ii) and (iii), there are only three projectivities fixing each of three collinear points; namely, the operations (456)(798), (465)(789), and the identity. So $g_2 = g_1/2 = 216$. In case (ii), the involutory automorphism ϕ given by $x\phi = x^{\sqrt{q}}$ transposes $-\omega$ and $-\omega^2$. So, in this case, the other three permutations fixing each of three collinear points are effected by a collineation. Hence $g_3 = 432$. However, in case (iii), $g_3 = g_2 = 216$.□

11.2. Double points, points of inflexion, and nuclei

The curve to be considered throughout this section is $\mathscr{F} = \mathbf{V}(F)$, where F in $\gamma[x_0, x_1, x_2]$ is an absolutely irreducible cubic form. Let $Q = \mathbf{P}(Y)$, with $Y = (y_0, y_1, y_2)$. Then, following (2.12),

$$F(Y + tX) = F^{(0)}(X) + tF^{(1)}(X) + t^2F^{(2)}(X) + t^3F^{(3)}(X), \quad (11.3)$$

where $F^{(i)}$ is a form of degree i in x_0, x_1, x_2 and degree $3 - i$ in y_0, y_1, y_2. For $i = 1,2$, the curve $\mathscr{F}_{3-i} = \mathbf{V}(F^{(3-i)})$ is the *i-th polar of Q with respect to $\mathscr{F}$*. The first polar $\mathscr{F}_2$ is also called the *polar quadric of Q with respect to $\mathscr{F}$*, or, when appropriate, the *polar conic of Q with respect to $\mathscr{F}$*. The second polar $\mathscr{F}_1$ is also called the *polar line of Q with respect to $\mathscr{F}$*. When $F^{(2)} = 0$, $\mathscr{F}_2$ is called *indeterminate*, since $PG(2,q)$ is not included as a plane quadric in theorem 7.2.1. If $p > 2$, then, for $i = 1, 2$,

$$\begin{aligned} F^{(i)} &= (y_0\partial/\partial x_0 + y_1\partial/\partial x_1 + y_2\partial/\partial x_2)^{3-i} F(x_0, x_1, x_2) \\ &= (x_0\partial/\partial y_0 + x_1\partial/\partial y_1 + x_2\partial/\partial y_2)^{i} F(y_0, y_1, y_2). \end{aligned}$$

This formula generalizes to forms of degree n when $p > n$. For $p \leqslant n$, there exist differential operators for which the generalized formula remains valid. In fact, Taylor's theorem can be generalized to a commutative ring of arbitrary characteristic. Since we are dealing with curves of order three, the paraphernalia of partial derivatives is somewhat superfluous. Nevertheless, for arbitrary characteristic, the following hold and will sometimes be employed:

$$\begin{aligned} F^{(0)}(X) &= F(Y), \\ F^{(1)}(X) &= (x_0\partial/\partial y_0 + x_1\partial/\partial y_1 + x_2\partial/\partial y_2)F(Y), \\ F^{(2)}(X) &= (y_0\partial/\partial x_0 + y_1\partial/\partial x_1 + y_2\partial/\partial x_2)F(X); \\ F^{(3)}(X) &= F(X). \end{aligned} \tag{11.4}$$

If Q is a point of $\mathscr{F}$ and $R = \mathbf{P}(z_0, z_1, z_2)$, then $m_Q(RQ,\mathscr{F}) \geq 2$ if $F^{(1)}(Z) = 0$, where $Z = (z_0, z_1, z_2)$; then RQ is a tangent to $\mathscr{F}$ at Q. If the tangent is unique, then Q is a simple point and $RQ = \mathscr{F}_1$; otherwise Q is a double point, §10.1. If Q is simple and $F^{(1)}(Z) = F^{(2)}(Z) = 0$, then Q is a (point of) inflexion, §10.1, and RQ is the *inflexional tangent* with $m_Q(RQ, \mathscr{F}) = 3$.

11.2.1. LEMMA. (i) *Q is a double point of $\mathscr{F}$ if*

$$\frac{\partial F}{\partial x_2} = \frac{\partial F}{\partial x_2} = \frac{\partial F}{\partial x_2} = 0 \text{ at } Q .$$

(ii) *If Q is a simple point of $\mathscr{F}$, then the tangent to $\mathscr{F}$ at Q is the second polar $\mathscr{F}_1$ of Q with respect to $\mathscr{F}$.* □

11.2.2. LEMMA. *If Q is a double point of $\mathscr{F}$, then the polar quadric $\mathscr{F}_2$ is a pair of lines, a repeated line, or the single point Q as Q is a node, a cusp, or an isolated double point.*

Proof. Let $Q = \mathbf{U}_0$. Then

$$F = x_0 f_2(x_1, x_2) + f_3(x_1, x_2)$$

where f_i is a form of degree i. Now the result follows from the definitions in §10.1, since $\mathscr{F}_2 = \mathbf{V}(f_2)$. □

11.2.3. LEMMA. *Let Q be a simple point of the cubic curve $\mathscr{F}$ in $PG(2,q)$, q odd. Then the following are equivalent:*

(i) *Q is a point of inflexion;*

(ii) *the polar quadric $\mathscr{F}_2$ of Q consists of*

the tangent at Q and a line not through Q;

(iii) *the polar quadric $\mathscr{F}_2$ is a line pair.*

Proof. Let $Q = \mathbf{U}_0$ and let the tangent $\mathscr{F}_1$ at Q be $\mathbf{u}_1$. Then

$$F = x_0{}^2x_1 + x_0(a_2x_1{}^2 + b_2x_1x_2 + c_2x_2{}^2) + f_3(x_1,x_2).$$

So

$$F^{(2)} = 2x_0x_1 + a_2x_1{}^2 + b_2x_1x_2 + c_2x_2{}^2.$$

If $\mathbf{u}_1$ is an inflexional tangent, then $c_2 = 0$ and $F^{(2)} = x_1(2x_0 + a_2x_1 + b_2x_2)$. So (i) implies (ii). *A fortiori*, (ii) implies (iii). Finally, if $\mathscr{F}_2$ is reducible, then, from theorem 7.2.4, we have $4c_2 = 0$; so (iii) implies (i).□

11.2.4. LEMMA. *Let Q be a simple point of the cubic $\mathscr{F}$ in $PG(2,q)$, q even. Then*

(i) *the polar quadric $\mathscr{F}_2$ of Q is singular or indeterminate;*

(ii) *the following are equivalent:*

(a) *Q is a point of inflexion,*

(b) *$\mathscr{F}_2$ consists of the tangent at Q and another line or is indeterminate.*

Proof. As in the previous lemma, let $Q = \mathbf{U}_0$ and let $\mathscr{F}_1 = \mathbf{u}_1$. Then

$$F = x_0{}^2x_1 + x_0(a_2x_1{}^2 + b_2x_1x_2 + c_2x_2{}^2) + f_3(x_1,x_2),$$

and

$$F^{(2)} = a_2x_1{}^2 + b_2x_1x_2 + c_2x_2{}^2.$$

So $\mathscr{F}_2$ is singular or $a_2 = b_2 = c_2 = 0$. Now, Q is an inflexion if and only if $c_2 = 0$. Then, either $\mathbf{u}_1$ is a component of $\mathscr{F}_2$ or $a_2 = b_2 = 0$ and $\mathscr{F}_2$ is indeterminate.□

For both odd and even q, the component of the polar quadric at an inflexion Q other than the tangent is the *harmonic polar* of Q.

For $p > 2$, the *Hessian* of the curve $\mathscr{F} = \mathbf{V}(F)$ is the curve $\mathscr{H} = \mathbf{V}(H)$, where, if we write $F_{(i)} = \partial F/\partial x_i$ and $F_{(ij)} = \partial^2 F/\partial x_i \partial x_j$,

$$H = \begin{vmatrix} F_{(00)} & F_{(01)} & F_{(02)} \\ F_{(10)} & F_{(11)} & F_{(12)} \\ F_{(20)} & F_{(21)} & F_{(22)} \end{vmatrix}.$$

If $H = 0$, the Hessian is called *indeterminate*.

11.2.5. LEMMA. *If $\mathscr{F} = \mathbf{V}(F)$ is a cubic curve in $PG(2,q)$, q odd, and $\mathscr{H} = \mathbf{V}(H)$ is its Hessian, then*

(i) $\mathbf{V}(F, H)$ *is the set of double points and inflexions;*

(ii) *the simple points of $\mathscr{F}$ on $\mathbf{V}(F, H^{(0)}, H^{(1)}, H^{(2)})$ are the inflexions, where*

$$H^{(i)} = F_{(j)}{}^2 F_{(kk)} + F_{(k)}{}^2 F_{(jj)} - 2F_{(j)}F_{(k)}F_{(jk)}$$

and $\{i, j, k\} = \{0, 1, 2\}$.

Proof. By lemmas 11.2.2 and 11.2.3, the points of $\mathscr{F}$ for which the polar quadric is singular are the double points and the inflexions. Since q is odd, the condition that the polar quadric at a point P is singular is exactly that P is on $\mathscr{H}$.

The determinant H can be manipulated to give $H^{(0)}$ as follows:

$$H = \begin{vmatrix} F_{(00)} & F_{(01)} & F_{(02)} \\ F_{(10)} & F_{(11)} & F_{(12)} \\ F_{(20)} & F_{(21)} & F_{(22)} \end{vmatrix} = \frac{2}{x_0}\begin{vmatrix} F_{(0)} & F_{(01)} & F_{(02)} \\ F_{(1)} & F_{(11)} & F_{(12)} \\ F_{(2)} & F_{(21)} & F_{(22)} \end{vmatrix} = \frac{2}{x_0^2}\begin{vmatrix} F & F_{(1)} & F_{(2)} \\ F_{(1)} & F_{(11)} & F_{(12)} \\ F_{(2)} & F_{(21)} & F_{(22)} \end{vmatrix}.$$

If we put $F = 0$, then $H = (-2/x_0^2)H^{(0)}$. So, providing none of the inflexions lie on $\mathbf{u}_0$, $\mathbf{u}_1$, or $\mathbf{u}_2$, the result is proved. In fact, from (11.4), with all the derivatives evaluated at $Q = \mathbf{P}(Y)$,

$$F^{(1)} = F_{(0)}x_0 + F_{(1)}x_1 + F_{(2)}x_2\,,$$

$$F^{(2)} = F_{(00)}x_0^2 + F_{(11)}x_1^2 + F_{(22)}x_2^2 + 2F_{(01)}x_0x_1 + $$
$$+ 2F_{(02)}x_0x_2 + 2F_{(12)}x_1x_2.$$

If, instead of applying the condition that $\mathscr{F}_2$ is singular, we apply the condition that $\mathscr{F}_1$ is a component of $\mathscr{F}_2$, then, from theorem 7.2.5, we obtain that $H^{(0)} = H^{(1)} = H^{(2)} = 0$ at Q.□

In general, to find the inflexions of $\mathscr{F}$, it suffices to use just one of the conditions $H^{(i)} = 0$ with $F = 0$. This also gives a suitable characteristic-free method of finding the inflexions of $\mathscr{F}$. As in (11.4), find $\mathscr{F}_1$ and $\mathscr{F}_2$, and then use the conditions of theorem 7.2.5 to express that $F^{(1)}$ divides $F^{(2)}$.

In subsequent sections, it is convenient for the purpose of classification to consider *complex inflexions* of $\mathscr{F}$, namely inflexions of $\bar{\mathscr{F}} = \mathbf{V}_{2,\bar{\gamma}}(F)$.

11.2.6. THEOREM. *Any two inflexions of an absolutely irreducible cubic* $\mathscr{F} = \mathbf{V}(F)$ *are collinear with a third inflexion.*

Proof. Let $\mathbf{U}_0$ and $\mathbf{U}_1$ be inflexions whose tangents meet at $\mathbf{U}_2$. Since $\mathbf{U}_0$ is an inflexion with tangent $\mathbf{u}_1$ and since $\mathbf{U}_1 \in \mathscr{F}$,

$$F = x_0^2x_1 + x_0(a_2x_1^2 + b_2x_1x_2) + (a_3x_1^2x_2 + b_3x_1x_2^2 + d_3x_2^3).$$

Since $\mathbf{u}_0$ is the inflexional tangent at $\mathbf{U}_1$, so $a_3 = b_3 = 0$. Let $\mathbf{u}_2$ meet $\mathscr{F}$ again at $P = \mathbf{P}(1,-1,0)$. Then $a_2 = 1$ and

$$F = x_0^2x_1 + x_0x_1^2 + b_2x_0x_1x_2 + d_3x_2^3.$$

The tangent at P is $\mathbf{V}(x_0 + x_1 + b_2x_2)$ which meets $\mathscr{F}$ three times there. So P is an inflexion.□

COROLLARY. *An absolutely irreducible cubic* $\mathscr{F} = \mathbf{V}(F)$ *with three collinear inflexions has the canonical form*

(i) $F = x_0 x_1(x_0 + x_1) + ex_2^3$

or (ii) $F = x_0 x_1(x_0 + x_1 + x_2) + ex_2^3$

as the three inflexional tangents are concurrent or not.

Proof. From the form for $\mathscr{F}$ in the theorem, if $b_2 = 0$ then the tangents are concurrent. If $b_2 \neq 0$, the replacement of $b_2 x_2$ by x_2 and d_3 by eb_2^3 gives the result.□

For $p = 3$, we list some of the properties of the nuclei of a curve $\mathscr{F} = \mathbf{V}(F)$, that is, points P for which $\partial F/\partial x_0 = \partial F/\partial x_1 = \partial F/\partial x_2 = 0$. If a nucleus P lies on $\mathscr{F}$, then it is a double point.

11.2.7. THEOREM. *In* $PG(2,3^h)$, *an absolutely irreducible cubic* $\mathscr{F}$ *with Hessian* $\mathscr{H}$ *has the following properties:*

(i) *if* $\mathscr{F}$ *is non-singular with two nuclei, it has a third;*

(ii) *if* $\mathscr{F}$ *has three nuclei on a line* l, *then every point of* l *is a nucleus;*

(iii) *the polar quadric of a nucleus is singular or indeterminate;*

(iv) *a point* P *has indeterminate polar quadric if and only if* $\mathscr{F}$ *has a line of nuclei containing* P;

(v) *if* $\mathscr{F}$ *has a line* l *of nuclei, then every point of* $\mathscr{F}\backslash l$ *is an inflexion and* $\mathscr{H}$ *is indeterminate;*

(vi) *every nucleus lies on* $\mathscr{H}$;

(vii) *every component of the polar quadric of a nucleus is a component of* $\mathscr{H}$;

(viii) *the tangent at an inflexion is a component of* $\mathscr{H}$;

(ix) *if* $\mathscr{F}$ *is non-singular and has an inflexion, it has a nucleus.*

Proof.

Let $F = a_0 x_0^3 + x_0^2(a_1 x_1 + b_1 x_2) + x_0(a_2 x_1^2 + b_2 x_1 x_2 + c_2 x_2^2)$

$$+ a_3 x_1^3 + b_3 x_1^2 x_2 + c_3 x_1 x_2^2 + d_3 x_2^3 .$$

(i) If $\mathbf{U}_0$ and $\mathbf{U}_1$ are nuclei of $\mathscr{F}$, then $a_1 = b_1 = a_2 = b_3 = 0$. So $\partial F/\partial x_0 = b_2 x_1 x_2 + c_2 x_2^2$, $\partial F/\partial x_1 = b_2 x_0 x_2 + c_3 x_2^2$, $\partial F/\partial x_2 = b_2 x_0 x_1 - c_2 x_0 x_2 - c_3 x_1 x_2$. Hence $Q = \mathbf{P}(c_3, c_2, -b_2)$ is also a nucleus of $\mathscr{F}$. If $Q = \mathbf{U}_0$ or $\mathbf{U}_1$ then $\mathscr{F}$ is singular.

(ii) If, in (i), Q lies on $\mathbf{u}_2$, then $b_2 = 0$. So $\partial F/\partial x_0 = c_2 x_2^2$, $\partial F/\partial x_1 = c_3 x_2^2$, $\partial F/\partial x_2 = -x_2(c_2 x_0 + c_3 x_1)$ and every point of $\mathbf{u}_2$ is a nucleus.

(iii) If $\mathbf{U}_0$ is a nucleus, then $a_1 = b_1 = 0$; so, with $Q = \mathbf{U}_0$ in (11.3), we obtain $F^{(2)} = a_2 x_1^2 + b_2 x_1 x_2 + c_2 x_2^2$. So, either $\mathscr{F}_2$ is singular or, if $a_2 = b_2 = c_2 = 0$, then $\mathscr{F}_2$ is indeterminate.

(iv) If $P = \mathbf{U}_0$ has indeterminate polar quadric then $a_1 = b_1 = a_2 = b_2 = c_2 = 0$ and

$$F = a_0 x_0^3 + a_3 x_1^3 + b_3 x_1^2 x_2 + c_3 x_1 x_2^2 + d_3 x_2^3.$$

So $\partial F/\partial x_0 = 0$, $\partial F/\partial x_1 = 2b_3 x_1 x_2 + c_3 x_2^2$, $\partial F/\partial x_2 = b_3 x_1^2 + 2c_3 x_1 x_2$, whence $\mathbf{V}(b_3 x_1 - c_3 x_2)$ is a line of nuclei containing P.

If $\mathscr{F}$ has a line of nuclei, then taking $\mathscr{F}$ as in (ii), we have $\mathbf{u}_2$ as the line and

$$F = a_0 x_0^3 + c_2 x_0 x_2^2 + a_3 x_1^3 + c_3 x_1 x_2^2 + d_3 x_2^3.$$

With $Q = \mathbf{P}(y_0, y_1, 0)$, we obtain from (11.3) that $F^{(2)} = $ $= c_2 x_2^2 y_0 + c_3 x_2^2 y_1$. So $\mathbf{P}(c_3, -c_2, 0)$ has indeterminate polar quadric.

(v) Let $\mathbf{u}_2$ be the line of nuclei and from (iv) let $\mathbf{U}_0$ be the point with indeterminate polar quadric. Then

$$F = a_0 x_0^3 + a_3 x_1^3 + c_3 x_1 x_2^2 + d_3 x_2^3.$$

By a projectivity, we can change $\mathscr{F}$ to $\mathscr{G} = \mathbf{V}(G)$ where $G = x_0^3 + x_1 x_2^2$. $\mathscr{G}$ also has $\mathbf{u}_2$ as line of nuclei and $\mathbf{U}_0$ as the point with indeterminate polar quadric. Then $\mathbf{u}_2 \cap \mathscr{G} = \mathbf{U}_1$ is a cusp with tangent $\mathbf{u}_2$ and every other tangent $\mathbf{V}(x_1 - tx_2)$, t in γ, is inflexional.

(vi) This follows from (iii).

(vii) From (iii) with $\mathbf{U}_0$ as nucleus, a little manipulation gives that, with $F^{(2)} = a_2 x_1^2 + b_2 x_1 x_2 + c_2 x_2^2$ as in (iii),

$$H = F^{(2)}\{(a_2 c_2 - b_2^2)x_0 + (a_2 c_3 - b_2 b_3)x_1 + (c_2 b_3 - b_2 c_3)x_2\}.$$

(viii) If $\mathbf{U}_0$ is an inflexion with tangent $\mathbf{u}_1$, then

$F = x_0^2x_1 + x_0x_1(a_2x_1 + b_2x_2) + a_3x_1^3 + b_3x_1^2x_2 + c_3x_1x_2^2 +$

$+ d_3x_2^3$, $F_{(2)} = b_2x_0x_1 + b_3x_1^2 - c_3x_1x_2$, $F_{(22)} = -c_3x_1$.

Then, from the proof of lemma 11.2.5, since $3F = 0$, $H = -(2/x_0^2)H^{(0)}$. As x_1 is a factor of $F_{(2)}$ and $F_{(22)}$, it is also a factor of $H^{(0)}$ and hence of H.

(ix) Let $\mathbf{U}_0$ be an inflexion with tangent $\mathbf{u}_0$ and harmonic polar $\mathbf{u}_0$. Then

$$F = x_0^2x_1 + a_3x_1^3 + b_3x_1^2x_2 + c_3x_1x_2^2 + d_3x_2^3.$$

So $\mathbf{P}(0, c_3, b_3)$ is a nucleus of $\mathscr{F}$.□

The *class* of a cubic $\mathscr{F} = \mathbf{V}_{2,q}(F)$ is the maximum number of distinct tangents in $PG(2, \bar{\gamma})$ to $\bar{\mathscr{F}} = \mathbf{V}_{2,\bar{\gamma}}(F)$ through an arbitrary point Q in $PG(2,q)$ not on $\mathscr{F}$.

11.2.8. LEMMA. *The class $\kappa = \kappa(\mathscr{F})$ of a cubic $\mathscr{F}$ in $PG(2,q)$ satisfies the following:*

$$\kappa \leqslant 6, \quad q \text{ odd},$$

$$\kappa \leqslant 3, \quad q \text{ even}.$$

Proof. Take $\mathbf{U}_0$ not on $\mathscr{F} = \mathbf{V}(F)$. Then

$$F = x_0^3 + x_0^2f_1(x_1, x_2) + x_0f_2(x_1, x_2) + f_3(x_1, x_2).$$

The lines through $\mathbf{U}_0$ are $\mathbf{V}(x_1 - tx_2)$, $t \in \gamma^+$. From §1.8, the discriminant of F as a polynomial in x_0 is

$$\Delta(F) = f_1^2f_2^2 - 4f_2^3 - 4f_1^3f_3 - 27f_3^2 + 18f_1f_2f_3.$$

Substituting $x_1 = tx_2$ makes $\Delta(F)$ a sextic polynomial $S(t)$. As $\mathbf{V}(x_1 - tx_2)$ is a tangent to $\mathscr{F}$ if $S(t) = 0$, so $\kappa \leqslant 6$. However, for q even, $\Delta(F) = (f_1f_2 + f_3)^2$, whence $S(t)$ is the square of a cubic in t; hence $\kappa \leqslant 3$.□

Now we state some properties of cubic curves similar to classical ones over $\mathbf{C}$. Let F, G, H be absolutely irreducible cubic forms in $\gamma[x_0, x_1, x_2]$; also let $\bar{\mathscr{F}} = V_{2,\bar{\gamma}}(F)$,

$\overline{\mathscr{G}} = V_{2,\bar{\gamma}}(G)$, $\overline{\mathscr{H}} = V_{2,\bar{\gamma}}(H)$. Suppose $\overline{\mathscr{F}}$ and $\overline{\mathscr{G}}$ have no common components. Then their *intersection cycle* is

$$\overline{\mathscr{F}} \cdot \overline{\mathscr{G}} = \sum_{i=1}^{9} P_i ,$$

where the points P_i are the common points of $\overline{\mathscr{F}}$ and $\overline{\mathscr{G}}$, and occur as often as their intersection multiplicity, §10.1. More formally,

$$\overline{\mathscr{F}} \cdot \overline{\mathscr{G}} = \sum_{P \in \mathscr{F} \cap \mathscr{G}} m_P(\mathscr{F}, \mathscr{G}) P .$$

11.2.9. THEOREM. (*Theorem of the nine associated points*) *If* $\overline{\mathscr{F}} \cdot \overline{\mathscr{G}} = \sum_{i=1}^{9} P_i$ *and* $\overline{\mathscr{F}} \cdot \overline{\mathscr{H}} = \sum_{i=1}^{8} P_i + Q$, *then* $Q = P_9$.

Proof. See §11.11.□

11.2.10. THEOREM. *If* $\overline{\mathscr{F}}$ *has class six, then there are four tangents to* $\overline{\mathscr{F}}$ *from a point* P *of* $\overline{\mathscr{F}}$ *other than the tangent at* P, *and the cross-ratio* λ *of the four tangents is constant for* P *on* $\overline{\mathscr{F}}$. *Also,* λ *is in* γ.

Proof. See §11.11.□

If, in this theorem, the four tangents through a point form a harmonic, equianharmonic, or superharmonic set, §6.1, then the cubic $\mathscr{F} = V_{2,q}(F)$ is similarly named. Although this concept is invalid for characteristic two, it is in fact appropriate to call $\mathscr{F}$ *equianharmonic* if the polar quadric of a general point is a repeated line. This is not as paradoxical as it may appear. For, if the polar quadric at a simple point P of $\mathscr{F}$ is a repeated line, then there is only one tangent through P to $\mathscr{F}$ other than the tangent at P. The quartic polynomial f which determines these tangents to $\mathscr{F}$ through P has the form

$$f(x) = (x + t)^4 .$$

which, from §1.11, is equianharmonic. If the polar quadric at P is not a repeated line,

$$f(x) = (x^2 + bx + x)^2, \; b \neq 0,$$

which is not equianharmonic.

A non-singular cubic which is not harmonic or equianharmonic is called *general*.

11.3. Classification of singular cubics

If $\mathscr{F}$ is an absolutely irreducible cubic in $PG(2,q)$, then any double point must lie in γ. For if $P = \mathbf{P}(y_0,y_1,y_2)$ is a double point, then so is $P^q = \mathbf{P}(y_0{}^q,y_1{}^q,y_2{}^q)$. As $\mathscr{F}$ can have only one double point $P = P^q$ and so lies in γ.

If $\mathscr{F} = \mathbf{V}(F)$ has a double point at $\mathbf{U}_0$, then

$$F = (b_0x_1{}^2 + b_1x_1x_2 + b_2x_2{}^2)x_0 + c_0x_1{}^3 + c_1x_1{}^2x_2 + c_2x_1x_2{}^2 + c_3x_2{}^3$$

and any inflexions of $\mathscr{F}$ lie on the line

$$\mathbf{V}((4b_0b_2 - b_1{}^2)x_0 + (b_0c_2 - b_1c_1 + 3b_2c_0)x_1 + (3b_0c_3 - b_1c_2 + b_2c_1)x_2). \quad (11.5)$$

$\mathbf{U}_0$ is either a node, a cusp, or an isolated double point and these three cases are separately considered. Throughout, γ' and γ'' are respectively quadratic and cubic extensions of γ.

I. Cubics with a node

Let the node be $\mathbf{U}_0$ and the tangents at $\mathbf{U}_0$ be $\mathbf{u}_1$ and $\mathbf{u}_2$.

11.3.1. LEMMA. *There are one or two projectively distinct cubics of type I according as* $(q-1,3) = 1$ *or* 3 *with the following canonical forms:*

(i) $(q-1,3) = 1$,

$$F = x_0x_1x_2 + x_1{}^3 + x_2{}^3;$$

(ii) $(q-1,3) = 3$,

$$F = x_0x_1x_2 + x_1{}^3 + x_2{}^3,$$

$$F' = x_0x_1x_2 + x_1{}^3 + \alpha x_2{}^3,$$

where α *is any non-cube in* γ.

Proof. Let $\mathscr{G} = V(G)$ be a cubic with node at U_0 and tangents $\mathbf{u}_1$, $\mathbf{u}_2$. Then

$$G = x_0x_1x_2 + c_0x_1^3 + c_1x_1^2x_2 + c_2x_1x_2^2 + c_3x_2^3.$$

Put $x_0 - c_1x_1 - c_2x_2$ for x_0; then

$$G = x_0x_1x_2 + c_0x_1^3 + c_3x_2^3.$$

(i) When $(q-1,3) = 1$, each element of γ has a unique cube root; so the form F is obtained.

(ii) When $(q-1,3) = 3$, either $c_3/c_0 = t^3$ and we obtain the form F, or $c_3/c_0 = \alpha t^3$ and we obtain the form F', or $c_3/c_0 = \alpha^2t^3$; so $c_0/c_3 = \alpha s^3$ and we obtain the form F' again.□

11.3.2. THEOREM. *Cubics with a node in* $PG(2,q)$, $q = p^h$, *have the following properties (Table* 11.2*).*

TABLE 11.2

$(q-1,3)$	p	Form	Complex inflexions in γ	$\gamma'\setminus\gamma$	$\gamma''\setminus\gamma$	Class	Number of points	Group
1	3	F	1	0	0	4	q	$\mathbf{Z}_2$
1	2	F	1	2	0	2	q	$\mathbf{Z}_2$
1	>3	F	1	2	0	4	q	$\mathbf{Z}_2$
3	2	F	3	0	0	2	q	$\mathbf{S}_3$
3	>3	F	3	0	0	4	q	$\mathbf{S}_3$
3	2	F'	0	0	3	2	q	$\mathbf{Z}_3$
3	>3	F'	0	0	3	4	q	$\mathbf{Z}_3$

Proof. The inflexions of the cubics $\mathscr{F}$ and $\mathscr{F}'$ (with forms F and F') all lie on $\mathbf{u}_0$. When $q \equiv 1 \pmod 3$, $x^2 + x + 1$ has roots ω and ω^2 in γ.

(i) $p = 3$: $\mathscr{F}$ has one inflexion at $\mathbf{P}(0, 1, -1)$.

(ii) $(q-1,3) = 1$, $p \neq 3$: $\mathscr{F}$ has an inflexion at $\mathbf{P}(0, 1, -1)$ and two in γ' at $\mathbf{P}(0, 1, -\omega)$, $\mathbf{P}(0, 1, -\omega^2)$.

(iii) $(q-1,3) = 3$: $\mathscr{F}$ has the three inflexions $P(0, 1, -1)$, $P(0, 1, -\omega)$, $P(0, 1, -\omega^2)$, all in γ.

(iv) $(q-1,3) = 3$: $\mathscr{F}'$ has no inflexions over γ, but three over γ''. Suppose $\theta \in \gamma''$ with $\theta^{-3} = \alpha$. Then the inflexions are $P(0, 1, -\theta)$, $P(0, \omega, -\theta)$, $P(0, \omega^2, -\theta)$.

(v) The q points of $\mathscr{F}$ are $\{P(-(1 + t^3), t, t^2) \mid t \in \gamma\}$ and the q points of $\mathscr{F}'$ are $\{P(-(1 + \alpha t^3), t, t^2) \mid t \in \gamma\}$.

If the projectivities $\mathfrak{A}$ preserving the curves are given by $X' = XA$, then the groups $G(\mathscr{F})$ in the respective cases are as follows. Let $A_1 = \text{diag}(1,1,1,)$ $A_2 = \text{diag}(1,\omega,\omega^2)$, $A_3 = \text{diag}(1,\omega^2,\omega)$, $A_4 = \begin{bmatrix} 1 & 0 & 0 \\ 0 & 0 & 1 \\ 0 & 1 & 0 \end{bmatrix}$, $A_5 = \begin{bmatrix} 1 & 0 & 0 \\ 0 & 0 & \omega^2 \\ 0 & \omega & 0 \end{bmatrix}$,

$A_6 = \begin{bmatrix} 1 & 0 & 0 \\ 0 & 0 & \omega \\ 0 & \omega^2 & 0 \end{bmatrix}$.

(a) When $(q-1,3) = 1$, $G(\mathscr{F}) = \{A_1, A_4\} \cong \mathbf{Z}_2$.
(b) When $(q-1,3) = 3$, $G(\mathscr{F}) = \{A_i \mid i \in \mathbf{N}_6\} \cong \mathbf{S}_3$.
(c) When $(q-1,3) = 3$, $G(\mathscr{F}') = \{A_1, A_2, A_3\} \cong \mathbf{Z}_3$.

COROLLARY. *For $p = 3$, the nuclei of $\mathscr{F}$ are the node $\mathbf{U}_0$ on $\mathscr{F}$ and the points $\mathbf{U}_1, \mathbf{U}_2$ off $\mathscr{F}$, where the inflexional tangent $\mathbf{u}_0$ meets the tangents $\mathbf{u}_1, \mathbf{u}_2$ at $\mathbf{U}_0$.* □

II. Cubics with a cusp

Let the cusp be at $\mathbf{U}_0$ with cuspidal tangent $\mathbf{u}_1$.

11.3.3. LEMMA. *There are one or two projectively distinct cubics of type II in $PG(2,q)$ according as $(q,3) = 1$ or 3, and they have the following canonical forms:*

(i) $(q,3) = 1$,

$$F = x_0 x_1^2 + x_2^3;$$

(ii) $(q,3) = 3$,

$$F' = x_0 x_1^2 + x_2^3,$$

$$F' = x_0x_1^2 + x_1x_2^2 + x_2^3 .$$

Proof. If $\mathscr{G}$ is the cubic with cusp at $\mathbf{U}_0$ and tangent $\mathbf{u}_1$ there, then

$$G = x_0x_1^2 + c_0x_1^3 + c_1x_1^2x_2 + c_2x_1x_2^2 + c_3x_2^3.$$

(i) If $p \neq 3$, putting $x_2 - c_2x_1/3$ for x_2 gives

$$G = x_0x_1^2 + d_0x_1^3 + d_1x_1^2x_2 + d_3x_1^3 .$$

Then, putting $d_3x_0 - d_0x_1 - d_1x_2$ for x_0 and $F = G/d_3$ gives

$$F = x_0x_1^2 + x_2^3.$$

(ii) If $p = 3$, putting $x_0 - c_0x_1 - c_1x_2$ for x_0 and $x_2/c_3^{1/3}$ for x_2 gives

$$G = x_0x_1^2 + cx_1x_2^2 + x_2^3.$$

Either $c = 0$ and writing $F = G$,

$$F = x_0x_1^2 + x_2^3;$$

or, if $c \neq 0$, putting c^3x_0 for x_0, cx_2 for x_2 and $F' = G/c^3$ gives

$$F' = x_0x_1^2 + x_1x_2^2 + x_2^3.\square$$

11.3.4. THEOREM. *Cubics with a cusp in* $PG(2,q)$, $q = p^h$, *have the following properties (Table* 11.3*).*

TABLE 11.3

$(q,3)$	p	Form	Complex inflexions in γ	$\gamma'\backslash\gamma$	$\gamma''\backslash\gamma$	Class	Number of points	Group
1	2	F	1	0	0	1	$q+1$	$\mathbf{Z}_{q-1}$
1	>3	F	1	0	0	3	$q+1$	$\mathbf{Z}_q$
3	3	F	q	q^2-q	q^3-q	1	$q+1$	$\mathbf{Z}_q\mathbf{Z}_{q-1}$
3	3	F'	0	0	0	3	$q+1$	$\mathbf{Z}_q$

Proof.

(i) $(q,3) = 1$: the inflexions lie on $\mathbf{u}_2$. So there is just the inflexion $\mathbf{U}_1$. When $p = 2$, the tangents at all the points other than $\mathbf{U}_0$ are concurrent at $\mathbf{U}_1$.

(ii) $(q,3) = 3$; all the points of $\mathscr{F}$ other than $\mathbf{U}_0$ are inflexions. The tangents at all the points are concurrent at $\mathbf{U}_2$.

(iii) $(q,3) = 3$: the inflexions of $\mathscr{F}'$ lie on $\mathbf{u}_1$ and so there are none.

(iv) The $q + 1$ points of $\mathscr{F}$ are $\{\mathbf{P}(-t^3, 1, t) \mid t \in \gamma^+\}$ and the $q + 1$ points of $\mathscr{F}'$ are $\{\mathbf{P}(-t^2 - t^3, 1, t) \mid t \in \gamma^+\}$.

Let the elements $\mathfrak{A}$ of the groups of projectivities be given by $X' = XA$. When $(q,3) = 1$, the group $G(\mathscr{F}) = \{\text{diag}\,(1, t^3, t^2) \mid t \in \gamma_0\} \cong \mathbf{Z}_{q-1}$, the multiplicative group of γ. When $(q,3) = 3$,

$$G(\mathscr{F}) = \left\{ \begin{bmatrix} t^3 & 0 & 0 \\ -s^3 & 1 & s \\ 0 & 0 & t \end{bmatrix} \mid s \in \gamma,\ t \in \gamma_0 \right\} \cong \mathbf{Z}_q \mathbf{Z}_{q-1},$$

since

$$\begin{bmatrix} t^3 & 0 & 0 \\ -s^3 & 1 & s \\ 0 & 0 & t \end{bmatrix} = \begin{bmatrix} t^3 & 0 & 0 \\ 0 & 1 & 0 \\ 0 & 0 & t \end{bmatrix} \begin{bmatrix} 1 & 0 & 0 \\ -s^3 & 1 & s \\ 0 & 0 & 1 \end{bmatrix}$$

and

$$\left\{ \begin{bmatrix} 1 & 0 & 0 \\ -s^3 & 1 & s \\ 0 & 0 & 1 \end{bmatrix} \right\}$$ is a normal subgroup.

So $G(\mathscr{F})$ is the semi-direct product of $\mathbf{Z}_q$ by $\mathbf{Z}_{q-1}$ acting as indicated. When $(q,3) = 3$,

$$G(\mathscr{F}') = \left\{ \begin{bmatrix} 1 & 0 & 0 \\ -t^2 - t^3 & 1 & t \\ t & 0 & 1 \end{bmatrix} \mid t \in \gamma \right\} \cong \mathbf{Z}_q,$$ the additive group of γ. □

COROLLARY. *When $p = 3$, the nuclei of $\mathcal{F}$ and $\mathcal{F}'$ are as follows:*

(i) *The nuclei of $\mathcal{F}$ are all points of the cuspidal tangent $\mathbf{u}_1$; $\mathbf{U}_2$ is the point of $\mathbf{u}_1$ with indeterminate polar quadric as in theorem* 11.2.7(v).

(ii) *The only nucleus of $\mathcal{F}'$ is the cusp $\mathbf{U}_0$.* □

III. Cubics with an isolated double point

Let $\mathcal{F}$ have an isolated double point at $\mathbf{U}_0$ and let $\mathbf{u}_0$ be its line of inflexions. Then

$$F = (b_0x_1^2 + b_1x_1x_2 + b_2x_2^2)x_0 + c_0x_1^3 + c_1x_1^2x_2 + c_2x_1x_2^2 + c_3x_2^3$$

and, from (11.5), we have

$$b_0c_2 - b_1c_1 + 3b_2c_0 = 3b_0c_3 - b_1c_2 + b_2c_1 = 0. \qquad (11.6)$$

11.3.5. LEMMA. *Let $\mathcal{F}$ be a cubic of type III in $PG(2,q)$. Then its inflexions are described as follows:*

	Complex inflexions in γ	$\gamma'\backslash\gamma$	$\gamma''\backslash\gamma$
$q \equiv 0 \pmod 3$	1	0	0
$q \equiv 1 \pmod 3$	1	2	0
$q \equiv -1 \pmod 3$	0	0	3
	3	0	0 .

Proof. (i) When $p = 3$, then eqns (11.6) become $b_0c_2 - b_1c_1 = -b_1c_2 + b_2c_1 = 0$. As $\mathbf{U}_0$ is an isolated double $b_1^2 - 4b_0b_2 = b_1^2 - b_0b_2 \neq 0$. So $c_1 = c_2 = 0$. Hence

$$F = (b_0x_1^2 + b_1x_1x_2 + b_2x_2^2)x_0 + c_0x_1^3 + c_3x_2^3,$$

which means that $\mathbf{u}_0$ meets $\mathcal{F}$ in just the one real point and no complex ones.

(ii) When $p \neq 3$, $\mathbf{u}_0$ meets $\mathcal{F}$ where

$$f = c_0x_1^3 + c_1x_1^2x_2 + c_2x_1x_2^2 + c_3x_2^3 = 0.$$

Since $\mathbf{U}_0$ is isolated, $b_0b_2 \neq 0$. Also, since $g = b_0x_1^2 + b_1x_1x_2 + b_2x_2^2$ is irreducible, both $\Delta(g) = b_1^2 - 4b_0b_2$ and $\partial(g) = b_0b_2/b_1^2$ are of category one, §1.8. To determine the roots of f, we find its Hessian h as in §1.10, where

$$h = A_0x_1^2 + A_1x_1x_2 + A_2x_2^2.$$

By using theorem 1.10.1, and (11.6) to eliminate c_0 and c_3, we obtain

$$A_0 = 3c_0c_2 - c_1^2 = (-b_2c_1^2 + b_1c_1c_2 - b_0c_2^2)/b_2\ ,$$

$$A_1 = 9c_0c_3 - c_1c_2 = (-b_1b_2c_1^2 + b_1^2c_1c_2 - b_0b_1c_2^2)/(b_0b_2),$$

$$A_2 = 3c_1c_3 - c_2^2 = (-b_2c_1^2 + b_1c_1c_2 - b_0c_2^2)/b_0.$$

So, writing $R = (-b_2c_1^2 + b_1c_1c_2 - b_0c_2^2)/(b_0b_2)$, we have

$$A_0 = b_0R\ ,\ A_1 = b_1R\ ,\ A_2 = b_2R\ .$$

Then $\Delta(h) = A_1^2 - 4A_0A_2 = R^2(b_1^2 - 4b_0b_2)$

and $\partial(h) = A_0A_2/A_1^2 = b_0b_2/b_1^2.$

So h has both roots in $\gamma'\backslash\gamma$. Then the result follows from theorem 1.10.4.□

11.3.6. LEMMA. *Cubics of type III with at least one real inflexion are projectively unique and have the following canonical forms:*

(i) $p = 3$,

$$F = (x_1^2 - \nu x_2^2)x_0 + x_1^3;$$

(ii) $p > 3$,

$$F = (x_1^2 - \nu x_2^2)x_0 + x_1(x_1^2 + 3\nu x_2^2);$$

(iii) $p = 2$,

$$F = (x_1^2 + x_1x_2 + \nu x_2^2)x_0 + x_1((1+\nu)x_1^2 + \nu x_1x_2 + \nu^2x_2^2),$$

where ν *is of category one. When* $q \equiv -1 \pmod 4$, ν *may be taken as* -1 *in* (i) *and* (ii). *When* $q \equiv -1 \pmod 3$, ν *may be taken as* -3 *in* (ii) *and as* 1 *in* (iii).

Proof. In each case, we take U_0 as the isolated double point, u_0 as the line of inflexions, and U_2 as one of the inflexions.

(i) For $p = 3$, let the tangents at U_0 be $V(x_1^2 - \nu x_2^2)$, where ν is a particular non-square. Then from the proof of lemma 11.3.5,

$$F = (x_1^2 - \nu x_2^2)x_0 + c_0x_1^3 + c_3x_2^3 .$$

Since U_2 lies on the curve, the required form is obtained.

(ii) For $p > 3$, again let the tangents at U_0 be $V(x_1^2 - \nu x_2^2)$ so that $b_0 = 1$, $b_1 = 0$, $b_2 = -\nu$ in F, ν still being a particular non-square. As U_2 is an inflexion, $c_3 = 0$; so, from (11.4), $c_1 = 0$ and $c_2 - 3\nu c_0 = 0$. So we may take $c_0 = 1$, $c_2 = 3\nu$ giving the required form.

(iii) For $p = 2$, let the tangents be $V(x_1^2 + x_1x_2 + \nu x_2^2)$, where ν is of category one. As U_2 is an inflexion, $c_3 = 0$; so, from (11.4), $c_2 = \nu c_1$ and $\nu c_0 = c_1 + c_2$. So, taking $c_0 = 1 + \nu$, $c_1 = \nu$, $c_2 = \nu^2$, we have the required form.

(iv) When $q \equiv -1 \pmod 4$, $x^2 + 1$ is irreducible and -1 is a non-square. When $q \equiv -1 \pmod 3$, $x^2 + x + 1$ is irreducible and so, for q odd, -3 is a non-square.□

11.3.7. LEMMA. *A cubic of type III with no real inflexions has the following canonical forms:*

(i) $p > 3$,

$$F' = (x_1^2 + x_2^2)x_0 + x_1(x_1^2 - 9x_2^2) + \alpha x_2(x_1^2 - x_2^2),$$

where $x^3 + \alpha x^2 - 9x - \alpha$ *is irreducible;*

(ii) $p = 2$,

$$F' = (x_1^2 + x_1x_2 + x_2^2)x_0 + (x_1^3 + x_1^2x_2 + x_2^3) + \alpha(x_1^3 + x_1x_2^2 + x_2^3),$$

where $(1 + \alpha)x^3 + x^2 + \alpha x + (1 + \alpha)$ *is irreducible.*

Proof. From lemma 11.3.5, $q \equiv -1 \pmod 3$ and so $x^2 + x + 1$ is irreducible over γ, §1.5(xii).

(i) $p > 3$: hence -3 is a non-square. Then, from (11.6), since $b_0 = 1$, $b_1 = 0$, and $b_2 = 3$, we have $c_2 + 9c_0 = c_3 + c_1 = 0$. So, take $c_0 = 1$, $c_2 = -9$, $c_1 = \alpha$, $c_3 = -\alpha$.

(ii) $b_0 = b_1 = b_2 = 1$. So, from (11.4), $c_2 + c_1 + c_0 = c_3 + c_2 + c_1 = 0$. Hence, $c_1 = 1$, $c_2 = \alpha$, $c_0 = c_3 = 1 + \alpha$.□

Note: It has not yet been shown that the above forms are unique, but this will be done below.

For $q \equiv -1 \pmod 3$, the number N of cubics with an isolated double point will be found, as well as the orders g_1 and g_2 of the groups of the cubics with three and no real inflexions respectively. If there is only one projectively distinct type of the latter as well as the former, then

$$N = p(3,q)\left(\frac{1}{g_1} + \frac{1}{g_2}\right)$$

A projectivity $\mathfrak{T}$ which leaves both $\mathbf{U}_0$ and $\mathbf{u}_0$ fixed has the form

$$X' = XT, \text{ where } T = \begin{bmatrix} 1 & 0 & 0 \\ 0 & t_{11} & t_{12} \\ 0 & t_{21} & t_{22} \end{bmatrix}$$

From §6.2, the order of $PGO_-(2,q)$ is $2(q + 1)$; that is, there are $2(q + 1)$ projectivities leaving an elliptic quadric $\mathscr{E}_1$ on $PG(1,q)$ fixed. When $q \equiv -1 \pmod 3$, the elements of $PGO_-(2,q)$ for $\mathscr{E}_1$ are given by $(x_1,x_2) = (x_1',x_2')T$ where, when $p > 3$ and $\mathscr{E}_1 = \mathbf{V}(x_1^2 + 3x_2^2)$,

T is $S_1(t) = \begin{bmatrix} t & -1 \\ 3 & t \end{bmatrix}$ or $S_2(t) = \begin{bmatrix} t & 1 \\ 3 & -t \end{bmatrix}$,

$t \in \gamma^+$; whereas, when $p = 2$ and $\mathscr{E}_1 = \mathbf{V}(x_1{}^2 + x_1x_2 + x_2{}^2)$,

T is $R_1(t) = \begin{bmatrix} 1 & t \\ t & 1+t \end{bmatrix}$ or $R_2(t) = \begin{bmatrix} 1 & t \\ 1+t & 1 \end{bmatrix}$,

$t \in \gamma^+$. In particular, $S_1(\infty) = \begin{bmatrix} 1 & 0 \\ 0 & 1 \end{bmatrix}$, $S_2(\infty) = \begin{bmatrix} 1 & 0 \\ 0 & -1 \end{bmatrix}$,

$R_1(\infty) = \begin{bmatrix} 0 & 1 \\ 1 & 1 \end{bmatrix}$, $R_2(\infty) = \begin{bmatrix} 0 & 1 \\ 1 & 0 \end{bmatrix}$.

11.3.8. LEMMA. *For $q \equiv -1 \pmod 3$, the projective group of a cubic $\mathscr{F}$ in $PG(2,q)$ with an isolated double point is $\mathbf{S}_3$ or $\mathbf{Z}_3$ according as $\mathscr{F}$ has three or no real inflexions.*

Proof. (i) When $p > 3$ and $\mathscr{F}$ has three inflexions,

$$F = (x_1{}^2 + 3x_2{}^2)x_0 + x_1(x_1{}^2 - 9x_2{}^2).$$

Substituting $(x_1,x_2) = (x_1',x_2')T$ where $T = S_1(t)$ or $S_2(t)$ and omitting the dashes, we obtain the form

$$F' = (x_1{}^2 + 3x_2{}^2)x_0 + t(t^2 - 9)\, x_1(x_1{}^2 - 9x_2{}^2) + 27(t^2 - 1)x_2(x_1{}^2 - x_2{}^2).$$

So $\mathscr{F} = \mathscr{F}'$ for $t = 1, -1, \infty$. Hence the group $G(\mathscr{F})$ of the cubic is $\{S_1(1), S_1(-1), S_1(\infty), S_2(1), S_2(-1), S_2(\infty)\}$. $S_1(\infty)$ is the identity; $S_1(1)$ and $S_1(-1)$ have order three; $S_2(1)$, $S_2(-1)$, and $S_2(\infty)$ have order two. So $G(\mathscr{F}) \cong \mathbf{S}_3$.

(ii) When $p > 3$ and $\mathscr{F}$ has no real inflexions,

$$F = (x_1{}^2 + 3x_2{}^2)x_0 + x_1(x_1{}^2 - 9x_2{}^2) + \alpha x_2(x_1{}^2 - x_2{}^2),$$

where $f(x) = x^3 + \alpha x^2 - 9x - \alpha$ is irreducible over γ. Substituting $(x_1,x_2) = (x_1{}',x_2{}')S_1(t)$ and omitting the dashes, we obtain the form

$$F' = (x_1{}^2 + 3x_2{}^2)x_0 + x_1(x_1{}^2 - 9x_2{}^2) + \beta_1 x_2(x_1{}^2 - x_2{}^2),$$

where $\beta_1 = [27(t^2 - 1) + \alpha t(t^2 - 9)]/[t(t^2 - 9) - \alpha(t^2 - 1]$. So $\mathscr{F} = \mathscr{F}'$ when $\beta_1 = \alpha$, that is, $(\alpha^2 + 27)(t^2 - 1) = 0$. Hence $S_1(1)$, $S_1(-1)$, and $S_1(\infty)$ are all in the projective group of $\mathscr{F}$.

If, now, $(x_1,x_2) = (x_1{}',x_2{}')S_2(t)$ is substituted and the dashes omitted, the following form is obtained:

$$F' = (x_1{}^2 + 3x_2{}^2) + x_1(x_1{}^2 - 9x_2{}^2) + \beta_2 x_2(x_1{}^2 - x_2{}^2),$$

where $\beta_2 = [27(t^2 - 1) - \alpha t(t^2 - 9)]/[t(t^2 - 9) + \alpha(t^2 - 1)]$. So $\mathscr{F} = \mathscr{F}'$ when $\beta_2 = \alpha$, that is,

$$g(t) = t(t^2 - 9) + \frac{\alpha^2 - 27}{2\alpha}(t^2 - 1) = 0.$$

It will be shown that g is irreducible. From above,

$$f(t) = t(t^2 - 9) + \alpha(t^2 - 1)$$

is irreducible. The Hessian $h(t)$ of both f and g is, modulo a constant, the same; namely

$$h(t) = t^2 + 3.$$

So, by theorem 1.10.4, g has three roots or none in γ. Let $\eta = f(\sqrt{-3})/f(-\sqrt{-3})$ and $\rho = g(\sqrt{-3})/g(-\sqrt{-3})$; then

$$\eta = \frac{\alpha + 3\sqrt{-3}}{\alpha - 3\sqrt{-3}} \quad \text{and} \quad \rho = \left(\frac{\alpha + 3\sqrt{-3}}{\alpha - 3\sqrt{-3}}\right)^2 .$$

Again by theorem 1.10.4, since f is irreducible, η is a non-cube in γ'. Since γ' has order q^2 and $q^2 \equiv 1 \pmod 3$, the cubes in γ' are given by θ^{3i}, where θ is a primitive root of γ'. So $\eta = \theta^{3i+1}$ or θ^{3j-1}. Hence $\rho = \theta^{6i+2}$ or θ^{6j-2}; thus ρ is a non-cube. Therefore g is irreducible over γ and $G(\mathscr{F}) = \{S_1(1), S_1(-1), S_1(\infty)\} \cong \mathbf{Z}_3$.

(iii) A similar process must be carried out for $p = 2$. In this case

$$F = (x_1^2 + x_1x_2 + x_2^2)x_0 + (x_1^3 + x_1^2x_2 + x_2^3) + s(x_1^3 + x_1x_2^2 + x_2^3),$$

where $\mathscr{F}$ has three inflexions when $s = 1$ and $\mathscr{F}$ has no real inflexions when $s = \alpha$ and $f(x) = (1 + \alpha)x^3 + x^2 + \alpha x + (1 + \alpha)$ is irreducible. Putting $(x_1,x_2) = (x_1',x_2')R_1(t)$ and omitting the dashes, the form F' is obtained with s_1 for s, where

$$s_1 = [t(1 + t) + s(1 + t + t^3)]/[1 + t^2 + t^3 + st(1 + t)].$$

$\mathscr{F} = \mathscr{F}'$ when $s = s_1$, that is, $(s^2 + s + 1)t(t + 1) = 0$. So $t = 0, 1, \infty$. Hence $R_1(0)$, $R_1(1)$, and $R_1(\infty)$ all belong to $G(\mathscr{F})$ both for three and no real inflexions.

(iv) Now substituting $(x_1,x_2) = (x_1',x_2')R_2(t)$ in F and omitting the dashes, the form F' is obtained with s_2 for s, where

$$s_2 = [1 + t^2 + t^3 + st(1 + t)]/[t(1 + t) + s(1 + t + t^3)].$$

$\mathscr{F} = \mathscr{F}'$ when $s = s_2$, that is,

$$g(t) = (1 + s)t^3 + t^2 + st + (1 + s) = 0.$$

When $s = 1$, $t(t + 1) = 0$ and $t = 0, 1$, or ∞. So the group $G(\mathscr{F})$ when the curve has three real inflexions is $\{R_1(0), R_1(1), R_1(\infty), R_2(0), R_2(1), R_2(\infty)\}$ and, as for $p > 3$, $G \cong \mathbf{S}_3$. When $s = \alpha$, $g(t) = f(t)$ and so g is irreducible. Thus $G(\mathscr{F})$ when the curve has no real inflexions is $\{R_1(0), R_1(1), R_1(\infty)\}$ and, as for $p > 3$, $G(\ ) \cong \mathbf{Z}_3$.□

COROLLARY. *In $PG(2,q)$, a cubic with an isolated double point and no real inflexions is projectively unique and has the canonical form given in lemma* 11.3.7.

Proof. This case only occurs for $q \equiv -1 \pmod 3$. By theorem 11.3.11, the number of cubics with an isolated double point is $N_7 = q^3(q^3 - 1)(q^2 - 1)/2$. The order of $PGL(3,q)$ is $p(3,q) = q^3(q^3 - 1)(q^2 - 1)$. The cubics with three inflexions are projectively unique and have groups of order six. If there are n projectively distinct cubics with no real inflexions, each has a group of order three by the previous lemma.

Hence $N_7 = p(3,q)\left(\frac{1}{6} + \frac{n}{3}\right)$. But $N_7 = p(3,q)/2$. So $n = 1$.□

11.3.9. THEOREM. *Cubics of type III in* $PG(2,q)$, $q = p^h$, *have the following properties, where the forms are those of lemmas* 11.3.6 *and* 11.3.7 *(Table* 11.4*).*

TABLE 11.4

$(q-1,3)$	p	Form	Inflexions in γ	$\gamma'\backslash\gamma$	$\gamma''\backslash\gamma$	Class	Number of points	Group
1	3	F	1	0	0	4	$q + 2$	$\mathbf{Z}_2$
1	2	F	3	0	0	2	$q + 2$	$\mathbf{S}_3$
1	> 3	F	3	0	0	4	$q + 2$	$\mathbf{S}_3$
1	2	F'	0	0	3	2	$q + 2$	$\mathbf{Z}_3$
1	> 3	F'	0	0	3	4	$q + 2$	$\mathbf{Z}_3$
3	2	F	1	2	0	2	$q + 2$	$\mathbf{Z}_2$
3	> 3	F	1	2	0	4	$q + 2$	$\mathbf{Z}_2$

Proof. For $q \equiv -1 \pmod 3$, the groups have been determined. For $q \equiv 0 \pmod 3$, the group $G(\mathscr{F}) = \{\mathrm{diag}\ (1,1,\pm1)\} \cong \mathbf{Z}_2$. For $q \equiv 1 \pmod 3$ and $p = 2$,

$$G(\mathscr{F}) = \left\{\begin{bmatrix}1 & 0 & 0\\0 & 1 & 0\\0 & 0 & 1\end{bmatrix} \quad \begin{bmatrix}1 & 0 & 0\\0 & 1 & \nu^{-1}\\0 & 0 & 1\end{bmatrix}\right\} \cong \mathbf{Z}_2 .$$

For $q \equiv 1 \pmod 3$ and $p \neq 2$,

$$G(\mathscr{F}) = \{\mathrm{diag}\ (1,\ 1,\ \pm1)\} \cong \mathbf{Z}_2.\square$$

COROLLARY. *When* $p = 3$, $\mathscr{F}$ *has the nucleus* $\mathbf{U}_0$ *plus two complex nuclei where the inflexional tangent* $\mathbf{u}_0$ *meets the complex tangents* $\mathbf{V}(x_1{}^2 - \nu x_2{}^2)$ *at* $\mathbf{U}_0$.□

To summarize singular cubics, let $\mathscr{N}_i^j$ indicate an absolutely irreducible singular cubic $\mathscr{F}$ with i real inflexions and j real tangents at the double point. The notation $\mathcal{N}^j$ for a cubic with j real tangents at the double point will also be used.

11.3.10. THEOREM. *For absolutely irreducible singular cubics over $\gamma = GF(q)$, the complex inflexions are all on a line, except when there are more than three. The complex inflexions which are not real lie over a quadratic extension γ' or a cubic extension γ''. The type of singularity and the number of real inflexions projectively determine the cubic as in Tables 11.5 and 11.6, where $|\mathcal{F}|$ is the number of points on $\mathcal{F}$ and $|G(\mathcal{F})|$ is the order of the projective group of $\mathcal{F}$.*

TABLE 11.5

Singularity		$q \equiv c$ (mod 3)	Complex inflexions over			$\lvert\mathcal{F}\rvert$	$\lvert G(\mathcal{F})\rvert$
		c	γ	$\gamma'\backslash\gamma$	$\gamma''\backslash\gamma$		
Node	$\mathcal{N}_1^2$	0	1	0	0	q	2
	$\mathcal{N}_1^2$	-1	1	2	0	q	2
	$\mathcal{N}_3^2$	1	3	0	0	q	6
	$\mathcal{N}_0^2$	1	0	0	3	q	3
Cusp	$\mathcal{N}_0^1$	0	0	0	0	$q+1$	q
	$\mathcal{N}_q^1$	0	q	q^2-q	q^3-q	$q+1$	$q(q-1)$
	$\mathcal{N}_1^1$	-1	1	0	0	$q+1$	$q-1$
	$\mathcal{N}_1^1$	1	1	0	0	$q+1$	$q-1$
Isolated double point	$\mathcal{N}_1^0$	0	1	0	0	$q+2$	2
	$\mathcal{N}_3^0$	-1	3	0	0	$q+2$	6
	$\mathcal{N}_0^0$	-1	0	0	3	$q+2$	3
	$\mathcal{N}_1^0$	1	1	2	0	$q+2$	2

TABLE 11.6

	$q \equiv m \pmod{12}$ m	Canonical form
$\mathcal{N}_1^2$	3,9,2,8,5,11	$x_0x_1x_2+x_1^3+x_2^3$
$\mathcal{N}_3^2$	4,1,7	$x_0x_1x_2+x_1^3+x_2^3$
$\mathcal{N}_0^2$	4,1,7	$x_0x_1x_2+x_1^3+\alpha x_2^3$, α non-cube
$\mathcal{N}_0^1$	3,9	$x_0x_1^2+x_1x_2^2+x_2^3$
$\mathcal{N}_q^1$	3,9	$x_0x_1^2+x_2^3$
$\mathcal{N}_1^1$	2,8,4,1,7,5,11	$x_0x_1^2+x_2^3$
$\mathcal{N}_1^0$	3,9	$(x_1^2-\nu x_2^2)x_0+x_1^3$, ν category one
$\mathcal{N}_3^0$	5,11	$(x_1^2+3x_2^2)x_0+x_1(x_1^2-9x_2^2)$
$\mathcal{N}_3^0$	2,8	$(x_1^2+x_1x_2+x_2^2)x_0+x_1x_2(x_1+x_2)$
$\mathcal{N}_0^0$	5,11	$(x_1^2+3x_2^2)x_0+x_1(x_1^2-9x_2^2)+\alpha x_2(x_1^2-x_2^2)$, $x^3+\alpha x^2-9x-\alpha$ irreducible
$\mathcal{N}_0^0$	2,8	$(x_1^2+x_1x_2+x_2^2)x_0+(x_1^3+x_1^2x_2+x_2^3)+\alpha(x_1^3+x_1x_2^2+x_2^3)$, $(1+\alpha)x^3+x^2+\alpha x+(1+\alpha)$ irreducible
$\mathcal{N}_1^0$	1,7	$(x_1^2-\nu x_2^2)x_0+x_1(x_1^2+3\nu x_2^2)$, ν category one
$\mathcal{N}_1^0$	4	$(x_1^2+x_1x_2+\nu x_2^2)x_0+x_1((1+\nu)x_1^2+\nu x_1x_2+\nu^2x_2^2)$, ν category one

There is now enough information to count the different types of cubic curves in $PG(2,q)$, which may be divided into the following types:

(1) three real lines;

(2) one real line and two conjugate complex lines;

(3) three conjugate complex lines;

(4) a line and a conic;

(5) an absolutely irreducible cubic with a node;

(6) an absolutely irreducible cubic with a cusp;

(7) an absolutely irreducible cubic with an isolated double point;

(8) a non-singular cubic.

Let N_i denote the number of cubics of type i and write $N = \Sigma N_i$.

11.3.11. THEOREM

$$N = (q^{10} - 1)/(q - 1);$$
$$N_1 = (q^2 + q + 1)(q^2 + q + 2)(q^2 + q + 3)/6;$$
$$N_2 = q(q^3 - 1)(q^2 + q + 1)/2;$$
$$N_3 = q(q^2 - 1)(q^3 + q + 1)/3;$$
$$N_4 = q^2(q^3 - 1)(q^2 + q + 1);$$
$$N_5 = q^3(q^3 - 1)(q^2 - 1)/2;$$
$$N_6 = q^3(q^3 - 1)(q + 1);$$
$$N_7 = q^3(q^3 - 1)(q^2 - 1)/2;$$
$$N_8 = q^4(q^3 - 1)(q^2 - 1).$$

Proof.

$$N_1 = \mathbf{h}(q^2 + q + 1, 3) = (q^2 + q + 1)(q^2 + q + 2)(q^2 + q + 3)/6.$$

Through each point of the plane, there are $(q^2 - q)/2$ pairs of conjugate complex lines. Hence

$$N_2 = (q^2 + q + 1)^2(q^2 - q)/2 = q(q^3 - 1)(q^2 + q + 1)/2.$$

The lines of $PG(2,q^3)\backslash PG(2,q)$ form conjugate triples over $GF(q)$. Hence

$$\begin{aligned} N_3 &= [(q^6 + q^3 + 1) - (q^2 + q + 1)]/3 \\ &= (q^6 + q^3 - q^2 - q)/3 \\ &= q(q^2 - 1)(q^3 + q + 1)/3. \end{aligned}$$

From §7.2, the number of conics is $q^5 - q^2$. Hence

$$N_4 = (q^2 + q + 1)(q^5 - q^2) = q^2(q^3 - 1)(q^2 + q + 1).$$

The number of quadrics on a line are $q(q + 1)/2$ hyperbolic, $q + 1$ parabolic, and $q(q - 1)/2$ elliptic, §6.2. If the double point is $\mathbf{U}_0$ and the tangents there are given by $f_2(x_1,x_2) = 0$, the cubic $\mathscr{F}$ is given by

$$F = f_2(x_1,x_2)x_0 + f_3(x_1,x_2).$$

$\mathscr{F}$ is absolutely irreducible if f_2 and f_3 have no common factor. When $f_2(x_1,x_2) = x_1x_2$ and $f_3(x_1,x_2) = c_0x_1^3 + c_1x_1^2x_2 + c_2x_1x_2^2 + c_3x_2^3$, we require $c_0c_3 \neq 0$: this gives

$$(q - 1)[(q^3 + q^2 + q + 1) - 2(q^2 + q + 1) + (q + 1)] = q^2(q - 1)^2$$

choices for f_3. When $f_2(x_1, x_2) = x_1^2$, we require $c_3 \neq 0$: this gives $q^3(q - 1)$ choices for f_3. When f_2 is irreducible, we require that f_2 should not divide f_3: this gives $q^4 - q^2$ choices for f_3. Hence

$$\begin{aligned} N_5 &= (q^2 + q + 1)[q(q + 1)/2]q^2(q - 1)^2 \\ &= q^3(q^3 - 1)(q^2 - 1)/2, \\ N_6 &= (q^2 + q + 1)(q + 1)q^3(q - 1) \\ &= q^3(q^3 - 1)(q + 1), \\ N_7 &= (q^2 + q + 1)[q(q - 1)/2](q^4 - q^2) \\ &= q^3(q^3 - 1)(q^2 - 1)/2. \end{aligned}$$

Therefore

$$\begin{aligned} N_8 &= N - \sum_{i=1}^{7} N_i \\ &= q^4(q^3 - 1)(q^2 - 1). \\ &= qp(3,q). \square \end{aligned}$$

11.4. Properties of non-singular cubics

In this section we begin the classification of non-singular cubics.

11.4.1. LEMMA. *There exists a non-singular cubic curve $\mathscr{F}$ with nine inflexions in $PG(2,q)$ if and only if $q \equiv 1 \pmod 3$. Then $\mathscr{F}$ has canonical form*

$$\mathscr{F}(c) = \mathbf{V}(x_0^3 + x_1^3 + x_2^3 - 3cx_0x_1x_2).$$

Proof. From theorem 11.1.1, the $(9_4, 12_3)$ configuration exists if and only if $q \equiv 0$ or $1 \pmod 3$. From the corollary to lemma 11.2.6, if $\mathscr{F}$ has three collinear inflexions,

$\mathscr{F} = V(x_0x_1(x_0 + x_1) + ex_2{}^3)$ or $\mathscr{F} = V(x_0x_1(x_0 + x_1 + x_2) + ex_2{}^3)$ in a suitable coordinate system. When $q \equiv 0 \pmod 3$, then in the first case $\mathscr{F}$ has a cusp at $P(1,1,e^{-1/3})$ and in the second case, by a projectivity, $\mathscr{F}$ becomes $\mathscr{F}' = V(x_0x_1x_2 + e(x_0 + x_1 + x_2)^3)$ which has exactly the three inflexions $P(1,-1,0)$, $P(0,1,-1)$, $P(-1,0,1)$.

When $q \equiv 1 \pmod 3$ and the $(9_4, 12_3)$ configuration is taken in canonical form as in (11.1), then the pencil of curves containing the nine points is the required one, and the nine points are inflexions of each non-singular curve of the pencil.□

COROLLARY 1. *When $q \equiv 1 \pmod 3$, $\mathscr{F}$ is singular if and only if $c = \infty$, 1, ω, or ω^2; in each of these cases, $\mathscr{F}(c)$ is a set of three lines through the nine inflexions.*

Proof. $F_{(0)} = 3(x_0{}^2 - cx_1x_2)$, $F_{(1)} = 3(x_1{}^2 - cx_0x_2)$, $F_{(2)} = 3(x_2{}^2 - cx_0x_1)$. The only points of $\mathscr{F}(c)$ on $\mathbf{u}_0$, $\mathbf{u}_1$, or $\mathbf{u}_2$ are the inflexions. So $F_{(0)} = F_{(1)} = F_{(2)} = 0$ at a point of $\mathscr{F}(c)$ if $c^3 = 1$. Hence $c = \infty$, 1, ω, or ω^2. But $\mathscr{F}(\infty)$, $\mathscr{F}(1)$, $\mathscr{F}(\omega)$, $\mathscr{F}(\omega^2)$ are just those cubics given by (11.2) in §11.1.□

If a cubic $\mathscr{F}$ has nine inflexions, the pencil of cubics through the inflexions is the *syzygetic pencil* determined by $\mathscr{F}$. Since a singular curve in the pencil is a set of three lines through the nine inflexions, it will be called an *inflexional triangle*.

COROLLARY 2. *The singular cubics in the syzygetic pencil determined by a non-singular cubic $\mathscr{F}$ with nine inflexions form an equianharmonic set.*□

In fact, this corollary holds for any non-singular cubic $\mathscr{F}$ providing γ has characteristic other than three, where "inflexions" is replaced by "complex inflexions". The quartic equation determining the singular cubics will have coefficients in γ. Before we state this as generally as possible, we still require a result on the maximum number of possible inflexions of a non-singular cubic.

When $p > 3$, we should expect this to be nine, in a manner

analogous to the classical case, since a cubic and its Hessian can meet in at most nine points. However, deg $H^{(i)} = 5$, lemma 11.2.5, and, when $p = 2$, a similar variety to $\mathbf{V}(H^{(i)})$ of order five is obtained when the condition that the second polar of a point is a component of the first polar is applied, theorem 7.2.5.

To verify this we begin from the canonical forms for cubics with three collinear inflexions given by the corollary to theorem 11.2.6.

11.4.2. LEMMA. *In $PG(2,q)$ there are $(q-1,3)$ projectively distinct cubic curves with three collinear inflexions such that the inflexional tangents are concurrent. They have the following canonical forms:*

(i) $(q-1,3) = 1$,

$$F = x_0x_1(x_0 + x_1) + x_2^3;$$

(ii) $(q-1,3) = 3$,

$$F = x_0x_1(x_0 + x_1) + x_2^3,$$
$$F' = x_0x_1(x_0 + x_1) + \alpha x_2^3,$$
$$F'' = x_0x_1(x_0 + x_1) + \alpha^2 x_2^3,$$

where α is a primitive element of γ.

Proof. This is merely a repetition of one part of the corollary to theorem 11.2.6.□

11.4.3. THEOREM. *Non-singular cubics in $PG(2,q)$ with three collinear inflexions and concurrent inflexional tangents have three or nine inflexions as in Table* 11.7.

Proof. The cubics can be written as $\mathscr{G} = \mathbf{V}(G) = \mathbf{V}(x_0x_1(x_0 + x_1) + ex_2^3)$. When $q \equiv 0 \pmod 3$, $\mathscr{G}$ is singular. So let $q \not\equiv 0 \pmod 3$. Then, as in (11.3),

$$G^{(2)} = x_0^2y_1 + 2x_0x_1(y_0 + y_1) + x_1^2y_0 + 3ex_2^2y_2,$$

$$G^{(1)} = x_0(y_1^2 + 2y_0y_1) + x_1(y_0^2 + 2y_0y_1) + 3ex_2y_2^2.$$

The condition that $G^{(1)}$ divides $G^{(2)}$ is, by theorem 7.2.5,

$$y_0(3ey_2^2)^2 + 3ey_2(y_0^2 + 2y_0y_1)^2 = 0\ ;$$

hence

$$y_0y_2(3ey_2^3 + y_0(y_0 + 2y_1)^2) = 0\ .$$

For $P(y_0,y_1,y_2)$ on $\mathscr{G}$,

$$y_0y_2(-3y_0y_1(y_0 + y_1) + y_0(y_0 + 2y_1)^2) = 0\ ,$$

whence

$$y_0^2y_2(y_0^2 + y_0y_1 + y_1^2) = 0\ .$$

$y_0^2y_2 = 0$ gives the three inflexions $U_0, U_1, P(1,-1,0)$.

When $q \equiv -1 \pmod 3$, $x^2 + x + 1$ is irreducible and its roots ω, ω^2 lie in $\gamma'\backslash\gamma$. So the inflexions of $\mathscr{F}$ over $\gamma'\backslash\gamma$ have coordinate vectors

$$(1,\ \omega,\ 1),\ (1,\ \omega,\ \omega),\ (1,\ \omega,\ \omega^2),$$
$$(1,\ \omega^2,\ 1), (1,\ \omega^2,\ \omega),\ (1,\ \omega^2,\ \omega^2).$$

When $q \equiv 1 \pmod 3$, $\mathscr{F}$ has nine inflexions, all over γ. When $\mathscr{G} = \mathscr{F}'$ or $\mathscr{F}''$, the inflexions other than the original ones on $\mathbf{u}_2$ lie on $V(\omega x_0 - x_1)$ or $V(\omega^2x_0 - x_1)$, both of which meet $\mathscr{G}$ in $V(x_0^3 - \alpha^i x_2^3)$, where $i = 1$ or 2 as $\mathscr{G} = \mathscr{F}'$ or $\mathscr{F}''$. If θ^{-1} is a (primitive) element of γ'' such that $\theta^{-3} = \alpha$, then $x_0^3 - \alpha^i x_2^3 = (x_0 - \theta^{-i}x_2)(x_0 - \omega\theta^{-i}x_2)(x_0 - \omega^2\theta^{-i}x_2)$. So the

TABLE 11.7

$q \equiv c \pmod 3$ c	Form	Complex inflexions in γ	$\gamma'\backslash\gamma$	$\gamma''\backslash\gamma$
-1	F	3	6	0
1	F	9	0	0
1	F', F''	3	0	6

inflexions of $\mathscr{G}$ over $\gamma''\backslash\gamma$ have coordinate vectors

$$(1,\ \omega,\ \theta^i),\quad (1,\ \omega,\ \omega\theta^i),\quad (1,\ \omega,\ \omega^2\theta^i),$$
$$(1,\ \omega^2,\ \theta^i),\ (1,\ \omega^2,\ \omega\theta^i),\ (1,\ \omega^2,\ \omega^2\theta^i)\ ,$$

where $i = 1,2$ as $\mathscr{G} = \mathscr{F}'$ or $\mathscr{F}''$.□

11.4.4. THEOREM. *Non-singular cubics in* $PG(2,q)$ *with three collinear inflexions and non-concurrent inflexional tangents have three or nine inflexions and canonical form* $\mathscr{F} = \mathrm{V}(x_0x_1x_2 + e(x_0 + x_1 + x_2)^3)$, $e \neq 0$ *or* $-1/27$ (*Table* 11.8).

TABLE 11.8

$q \equiv c \pmod 3$	e	Complex inflexions in		
c		γ	$\gamma'\backslash\gamma$	$\gamma''\backslash\gamma$
0		3	0	0
-1		3	6	0
1	$(t^3 - 1)/27$	9	0	0
1	$\neq (t^3 - 1)/27$	3	0	6

Proof. A transformation of the cubic in the corollary to theorem 11.2.6 gives $\mathscr{F}$. When $e = -1/27$, $\mathscr{F} = \mathrm{V}(F)$ is irreducible but singular: in this case $\mathscr{F}$ has a node or an isolated double point as $q \equiv 1$ or $-1 \pmod 3$. When $q \equiv 0 \pmod 3$, $\mathscr{F}$ is non-singular for all e in γ_0.

$\mathscr{F}$ has inflexions at $\mathrm{P}(0,\ 1,\ -1)$, $\mathrm{P}(-1,\ 0,\ 1)$ and $\mathrm{P}(1,\ -1,\ 0)$. As in (11.3),

$$F^{(2)} = x_0x_1y_2 + x_0x_2y_1 + x_1x_2y_0 + 3e(x_0 + x_1 + x_2)^2(y_0 + y_1 + y_2).$$

$$F^{(1)} = x_0y_1y_2 + x_1y_0y_2 + x_2y_0y_1 + 3e(x_0 + x_1 + x_2)(y_0 + y_1 + y_2)^2\ .$$

(i) When $p = 3$, $F^{(2)} = x_0x_1y_2 + x_0x_2y_1 + x_1x_2y_0$, which is reducible only when $y_0y_1y_2 = 0$. So the only inflexions are just the three on $\mathbf{u}$. The vertices of the triangle of reference, which are the meets of pairs of inflexional tangents, are the nuclei of $\mathscr{F}$.

(ii) When $p \neq 3$, let ω and ω^2 be the roots of $x^2 + x + 1$. Put

$$x_0 = x_0' + x_1' + x_2' ,$$
$$x_1 = x_0' + \omega x_1' + \omega^2 x_2' ,$$
$$x_2 = x_0' + \omega^2 x_1' + \omega x_2' ,$$

and omit the dashes; then F becomes G, where

$$G = (27e + 1)x_0^3 + x_1^3 + x_2^3 - 3x_0x_1x_2 .$$

Also

$$G^{(2)} = 3((27e + 1)x_0^2y_0 + x_1^2y_1 + x_2^2y_2 - x_0x_1y_2 - x_0x_2y_1 - x_1x_2y_0).$$

$$G^{(1)} = 3(x_0((27e + 1)y_0^2 - y_1y_2) + x_1(y_1^2 - y_0y_2) + x_2(y_2^2 - y_0y_1)).$$

By theorem 7.2.5, if $G^{(2)} = \Sigma' a_{ij}x_ix_j$, $G^{(1)} = \Sigma b_ix_i$ and $G^{(1)}|G^{(2)}$, then $a_{11}b_2^2 + a_{22}b_1^2 = a_{12}b_1b_2$. So

$$y_1(y_2^2 - y_0y_1)^2 + y_2(y_1^2 - y_0y_2)^2 = -y_0(y_1^2 - y_0y_2)(y_2^2 - y_0y_1).$$

Simplifying, we obtain

$$y_1y_2(y_0^3 + y_1^3 + y_2^3 - 3y_0y_1y_2) = 0 .$$

So all inflexions lie on $\mathbf{u}_0$, $\mathbf{u}_1$, and $\mathbf{u}_2$.

(iii) When $p \neq 3$ and $(q-1,3) = 1$, then ω lies in a quadratic extension γ' of γ. $\mathscr{G}$ is singular if $27e + 1 = 0$. Otherwise, $\mathscr{G}$ is non-singular (since $e \neq 0$) and has three inflexions over γ and six over $\gamma'\backslash\gamma$. The same is true for $\mathscr{F}$, despite the apparent illegitimacy of the transformation used.

(iv) When $p \neq 3$ and $(q-1,3) = 3$, then ω lies in γ. $\mathscr{G}$ is singular if $27e + 1 = 0$. Otherwise $\mathscr{G}$ has nine or three inflexions in γ as $27e + 1$ is a non-zero cube or not (always with $e \neq 0$). In the latter case, the inflexions of $\mathscr{G}$ on $\mathbf{u}_1$ and $\mathbf{u}_2$

lie in a cubic extension γ'' of γ.□

11.4.5. THEOREM. *Non-singular cubics in $PG(2,q)$ have zero, one, three, or nine inflexions. The possibilities are*

$$q \equiv 0 \pmod 3 \quad : \quad 0,\ 1,\ 3\ ,$$
$$q \equiv -1 \pmod 3 \quad : \quad 0,\ 1,\ 3\ ,$$
$$q \equiv 1 \pmod 3 \quad : \quad 0,\ 1,\ 3,\ 9\ .$$

Proof. This follows from the previous two theorems and theorem 11.2.6.□

Note: The $(7_3,7_3)$ configuration allowed by theorem 11.1.1 does not in fact occur.

11.4.6. THEOREM. *Every non-singular cubic in $PG(2,q)$, $q \not\equiv 0$* (mod 3), *has nine complex inflexions.*

Proof. This again follows from theorems 11.4.3 and 11.4.4.□

Let $\bar{\gamma}$ be the algebraic closure of γ. Then, for a non-singular cubic $\mathscr{F}$ in $PG(2,q)$, $q \not\equiv 0 \pmod 3$, the *full syzygetic pencil* determined by $\mathscr{F}$ is the pencil over $\bar{\gamma}$ of cubics through the nine complex inflexions of $\mathscr{F}$. A set of three lines over $\bar{\gamma}$ through the nine complex inflexions of $\mathscr{F}$ is called an *inflexional triangle* as in the case that all nine inflexions are real.

COROLLARY. *In $PG(2,q)$, $q \not\equiv 0 \pmod 3$, the singular cubics in the full syzygetic pencil determined by a non-singular cubic $\mathscr{F}$ are inflexional triangles and form an equi-anharmonic set. The possible numbers of these singular cubics lying in γ are as follows:*

$$q \equiv -1 \pmod 3 : 0,\ 2;$$
$$q \equiv 1 \pmod 3 : 0,\ 1,\ 4.$$

Proof. From corollary 2 to theorem 11.4.1, the singular cubics form an equianharmonic set determined by a quartic equation over γ. The possible numbers of roots of the quartic are then given by theorem 1.11.7.□

A non-singular cubic $\mathscr{F}$ will be denoted as $\mathscr{F}_n^r$ where n is the number of real inflexions and r the number of its real inflexional triangles. $\mathscr{F}_n^r = \mathscr{G}_n^r, \mathscr{E}_n^r, \mathscr{H}_n^r$ when $\mathscr{F}$ is respectively general, equianharmonic, harmonic. The cubics will be classified according to the values of n and r. An *inflexional line* of $\mathscr{F}$ will be one of the 12 lines, defined over γ or some extension, containing three complex inflexions.

11.4.7. LEMMA. *If $\mathscr{F}$ is a non-singular cubic in $PG(2,q)$, $q \not\equiv 0$ (mod 3), and P is a real inflexion of $\mathscr{F}$, then an inflexional line l through P belongs to the same field as the inflexional triangle containing l.*

Proof. Let l_1, l_2, l_3, l_4 be the four inflexional lines through P and let T_1, T_2, T_3, T_4 be the four inflexional triangles such that $l_i \in T_i$, $i \in \mathbf{N}_4$. If both the full syzygetic pencil Φ and the pencil of lines Φ' over $\bar{\gamma}$ through P are considered as representations of $PG(1, \bar{\gamma})$, then the mapping $\mathfrak{T} : \Phi \to \Phi'$ which associates to every cubic of Φ the tangent at P of the cubic is a projectivity in which $T_i\mathfrak{T} = l_i$, $i \in \mathbf{N}_4$. As $P \in PG(2,\gamma)$, the projectivity $\mathfrak{T}$ is real; that is, the coefficients in its equations lie in γ.□

11.4.8. THEOREM. *In $PG(2,q)$, $q \not\equiv 0$ (mod 3), the possible types of non-singular cubics are as follows:*

$$q \equiv -1 \pmod 3) \;:\; \mathscr{F}_3^{\,2},\ \mathscr{F}_1^{\,0},\ \mathscr{F}_0^{\,2};$$
$$q \equiv 1 \pmod 3) \;:\; \mathscr{F}_9^{\,4},\ \mathscr{F}_3^{\,1},\ \mathscr{F}_1^{\,4},\ \mathscr{F}_1^{\,1},\ \mathscr{F}_1^{\,0},\ \mathscr{F}_0^{\,4},\ \mathscr{F}_0^{\,1}\ .$$

Proof. From theorem 11.4.5 and its corollary, the possibilities for $\mathscr{F}_n^r$ when $q \equiv -1$ (mod 3) are $n = 0, 1, 3$ and $r = 0, 2$. When $q \equiv 1$ (mod 3) the possibilities are $n = 0, 1, 3, 9$ and $r = 0, 1, 4$. The previous lemma excludes those not on the list.□

11.5. Non-singular cubics with nine inflexions

From §11.4, a non-singular cubic with nine inflexions exists in $PG(2,q)$ if and only if $q \equiv 1 \pmod 3$ and then has canonical form

$$\mathscr{F}(c) = V(x_0^3 + x_1^3 + x_2^3 - 3cx_0x_1x_2).$$

11.5.1. LEMMA. *In $PG(2,q)$, $q \equiv 1 \pmod 3$, with ω a root of $x^2 + x + 1$,*

(i) *$\mathscr{F}(c)$ is an inflexional triangle for $c = \infty, 1, \omega, \omega^2$;*

(ii) *$\mathscr{F}(c)$ is equianharmonic for $c = 0, -2, -2\omega, -2\omega^2$;*

(iii) *$\mathscr{F}(c)$ is harmonic for $c = 1 \pm \sqrt{3}, (1 \pm \sqrt{3})\omega, (1 \pm \sqrt{3})\omega^2$.* □

We note that when q is even $\mathscr{F}(c)$ is equianharmonic if and only if $c = 0$. Also, up to a constant, the *absolute invariant* of $\mathscr{F}(c)$ is $c^3(c^3 + 8)^3/(c^6 - 20c^3 - 8)^2$, which is 0 or ∞ as $\mathscr{F}(c)$ is equianharmonic or harmonic.

11.5.2. LEMMA. (i) *Under transformation of the triangle of reference to another inflexional triangle or itself, $\mathscr{F}(c)$ is projectively equivalent to $\mathscr{F}(c')$, where*

$$c' = \lambda,\ \lambda\left(\frac{c+2}{c-1}\right),\quad \lambda\left(\frac{\omega c+2}{\omega c-1}\right),\quad \lambda\left(\frac{\omega^2 c+2}{\omega^2 c-1}\right) \text{ with } \lambda^3 = 1.$$

(ii) *$c' = c$ for $c = 0, \lambda, -2\lambda, (1 + \sqrt{3})\lambda, (1 - \sqrt{3})\lambda$ with again $\lambda^3 = 1$.* □

COROLLARY. *The values of c for which $c' = c$ are as follows, always with $\lambda^3 = 1$:*

(i) *$q \equiv 1 \pmod{12}$, $c = 0, \lambda, -2\lambda, (1 \pm \sqrt{3})\lambda$, (13 values);*

(ii) *$q \equiv 7 \pmod{12}$, $c = 0, \lambda, -2\lambda$, (7 values);*

(iii) *$q \equiv 4 \pmod{12}$, $c = 0, \lambda$, (4 values).*

Proof. When $q \equiv 1 \pmod 3$ and odd, -3 is a square. As -1 is a square or not as $q \equiv 1$ or $-1 \pmod 4$, so 3 is a square or not as $q \equiv 1$ or $7 \pmod{12}$. □

Let us recall from §11.1 what the inflexional triangles are among the pencil of curves $\mathscr{F}(c)$, $c \in \gamma^+$.

$$\mathscr{F}(\infty) = V(x_0x_1x_2),$$
$$\mathscr{F}(1) = V((x_0 + x_1 + x_2)(x_0 + \omega x_1 + \omega^2 x_2)(x_0 + \omega^2 x_1 + \omega x_2)),$$
$$\mathscr{F}(\omega) = V((\omega x_0 + x_1 + x_2)(x_0 + \omega x_1 + x_2)(x_0 + x_1 + \omega x_2)),$$
$$\mathscr{F}(\omega^2) = V((\omega^2 x_0 + x_1 + x_2)(x_0 + \omega^2 x_1 + x_2)(x_0 + x_1 + \omega^2 x_2)).$$

11.5.3. LEMMA. *The transformations of the inflexional triangle* $\mathscr{F}(\infty)$ *leaving* $\mathscr{F}(c)$ *fixed are given in Table* 11.9.

TABLE 11.9

$q \equiv m \pmod{12}$ m	Type of $\mathscr{F}(c)$	Transformations
1, 7	*general*	$\mathscr{F}(\infty) \to \mathscr{F}(\infty)$
1, 7	*equianharmonic*	$\mathscr{F}(\infty) \to \mathscr{F}(\infty)$, $\mathscr{F}(\omega)$, $\mathscr{F}(\omega^2)$
1	*harmonic*	$\mathscr{F}(\infty) \to \mathscr{F}(\infty)$, $\mathscr{F}(1)$
4	*general*	$\mathscr{F}(\infty) \to \mathscr{F}(\infty)$
4	*equianharmonic*	$\mathscr{F}(\infty) \to \mathscr{F}(\infty)$, $\mathscr{F}(1)$, $\mathscr{F}(\omega)$, $\mathscr{F}(\omega^2)$.□

11.5.4. THEOREM. *In* $PG(2,q)$, $q \equiv 1 \pmod 3$, *the number* n *of projectively distinct non-singular cubics of each type with nine inflexions, the order* o *of the corresponding group of projectivities of each curve and the total number* n_9 *of projectively distinct types are as follows, where* $\mathscr{G}_9^4$, $\mathscr{E}_9^4$ *and* $\mathscr{H}_3^4$ *are respectively the general, equianharmonic, and harmonic types.*

TABLE 11.10

$q \equiv m \pmod{12}$	$\mathscr{G}_9^4$		$\mathscr{E}_9^4$		$\mathscr{H}_9^4$		
m	n	o	n	o	n	o	n_9
1	$(q-13)/12$	18	1	54	1	36	$(q+11)/12$
7	$(q-7)/12$	18	1	54	0	—	$(q+5)/12$
4	$(q-4+/12$	18	1	216	0	—	$(q+8)/12$

Proof. From lemma 11.5.2(i), a general $\mathscr{F}(c)$ can be transformed to 12 other $\mathscr{F}(c')$ including $\mathscr{F}(c)$ itself. By the corollary, the numbers of $\mathscr{F}(c)$ to be excluded from the general case are 13, 7, and 4 according as m is 1, 7, and 4. This gives the first column of Table 11.10 for n.

If the projectivities preserving $\mathscr{F}(c)$ are given by $X' = XA$, then the group of those fixing each of $\mathbf{u}_0$, $\mathbf{u}_1$, $\mathbf{u}_2$ is

$$\{\text{diag}\ (1,1,1),\ \text{diag}\ (1,\omega,\omega^2),\ \text{diag}\ (1,\omega^2,\omega)\} \cong \mathbf{Z}_3.$$

Permutations of $\mathbf{u}_0$, $\mathbf{u}_1$, $\mathbf{u}_2$ give the whole group $G(\mathscr{G}_9{}^4) \cong \mathbf{S}_3\mathbf{Z}_3$, where $\mathbf{Z}_3 \lhd \mathbf{S}_3\mathbf{Z}_3$.

The rest of Table 11.10 is a consequence of theorem 11.1.2. The order of the projective group of the configuration of the nine inflexions is 216. When q is even, there is only one c for which $\mathscr{F}(c)$ is equianharmonic: when q is odd, there are four values of c. So o is correspondingly 216 and $216/4 = 54$. There are six values of c for which $\mathscr{F}(c)$ is harmonic when $q \equiv 1 \pmod{12}$ and so $o = 216/6 = 36$.□

11.6. Non-singular cubics with three inflexions

The classification of non-singular cubics with exactly three inflexions can be easily completed on the basis of the previous sections. From theorems 11.4.3 and 11.4.4, such a cubic $\mathscr{F}$ has canonical form

$$\mathbf{V}(x_0x_1(x_0 + x_1) + ex_2{}^3) \text{ or } \mathbf{V}(x_0x_1x_2 + e(x_0 + x_1 + x_2)^3)$$

as the three inflexional tangents are concurrent or not. The former is always equianharmonic.

11.6.1. LEMMA. $\mathbf{V}(x_0x_1x_2 + e(x_0 + x_1 + x_2)^3)$ *is*

(i) *singular and irreducible if* $e = -1/27$;

(ii) *equianharmonic if* $e = -1/24$, *which is* $(t^3 - 1)/27$ *for* $t = -1/6$;

(iii) *harmonic if* $216e^2 + 36e + 1 = 0$, *which has two roots when* 3 *is a square*.□

11.6.2. THEOREM. *In* $PG(2,q)$, *the number* n *of projectively distinct non-singular cubics of each type with exactly three inflexions, the order* o *of the corresponding group of projectivities and the total number* n_3 *of projectively distinct types are as follows, where* $\mathscr{G}$, $\mathscr{E}$, *and* $\mathscr{H}$ *are respectively the general,* *equianharmonic, and harmonic types when the inflexional tangents are not concurrent, and $\mathscr{E}$ the type when they are concurrent. For $q \equiv 0$ (mod 3), each curve has three nuclei at the meets of pairs of the inflexional tangents (Table 11.11).*

TABLE 11.11

$q \equiv m$ (mod 12)	$\mathscr{G}$		$\mathscr{E}$		$\mathscr{H}$		$\bar{\mathscr{E}}$		
m	n	o	n	o	n	o	n	o	n_3
3, 9	$q-1$	6	0	—	0	—	0	—	$q-1$
2, 8	$q-2$	6	0	—	0	—	1	6	$q-1$
4	$2(q-1)/3$	6	0	—	0	—	2	18	$2(q+2)/3$
1	$2(q-1)/3$	6	0	—	0	—	2	18	$2(q+2)/3$
7	$2(q-1)/3$	6	0	—	0	—	2	18	$2(q+2)/3$
5	$q-3$	6	1	6	0	—	1	6	$q-1$
11	$q-5$	6	1	6	2	6	1	6	$q-1$

Proof. Consider the cubics $\bar{\mathscr{E}}$ first. From lemma 11.4.2, there are $(q-1, 3)$ projectively distinct curves of the form $\bar{\mathscr{E}} = \mathbf{V}(x_0x_1(x_0 + x_1) + ex_2^{\,3})$. From theorem 11.4.3, when $q \equiv 0$ (mod 3), $\bar{\mathscr{E}}$ is singular. When $q \equiv 1$ (mod 3), $\bar{\mathscr{E}}$ has nine inflexions when $e = 1$ and there are two types with $e = \alpha, \alpha^2$, where α is a primitive element of γ, with three inflexions. When $q \equiv -1$ (mod 3), $\bar{\mathscr{E}}$ has three inflexions. In this case, the group $G(\bar{\mathscr{E}}) \cong \mathbf{S}_3$, which effects all permutations on $\mathbf{u}_0$, $\mathbf{u}_1$, and $\mathbf{V}(x_0 + x_1)$. When $q \equiv 1$ (mod 3), $G(\bar{\mathscr{E}}) \cong \mathbf{S}_3 \times \mathbf{Z}_3$, a direct product where $\mathbf{Z}_3 \cong \{\text{diag }(1,1,1), \text{diag }(1,1,\omega), \text{diag }(1,1,\omega^2)\}$.

For $\mathscr{F} = \mathbf{V}(x_0x_1x_2 + e(x_0 + x_1 + x_2)^3)$, the group is always $\mathbf{S}_3$, which permutes the inflexional tangents and leaves the line of inflexions fixed.

When $q \equiv 0$ (mod 3), the previous lemma means that none of the curves for $e \neq 0$ are other than general. So there are $q-1$ types of $\mathscr{G}$, each of which has three nuclei at the meets $\mathbf{U}_0$, $\mathbf{U}_1$, $\mathbf{U}_2$ of pairs of the inflexional tangents.

When $q \equiv 2$ or 8 (mod 12), $\mathscr{F}$ is reducible for $e = 0$ and singular for $e = 1$, but otherwise always has three inflexions and is never equianharmonic; so $n = q-2$.

When $q \equiv 4$ (mod 12), for $(q-1)/3$ values of e in γ_0, $\mathscr{F}$ has nine inflexions. For the odd values of q with $q \equiv 1$ (mod

3), the equianharmonic $\mathscr{F}$ with $e = -1/24$ also has nine inflexions. For $q \equiv 1$ or $11 \pmod{12}$, 3 is a square and hence there are two harmonic curves. For $q \equiv 1 \pmod{12}$, both of these harmonic curves have nine inflexions and are equivalent. So $n = 2(q-1)/3$ when $q \equiv 1 \pmod 3$.

For $q \equiv 5 \pmod{12}$, $\mathscr{F}$ is reducible for $e = 0$, singular for $e = -1/27$, and equianharmonic for $e = -1/24$. So when $\mathscr{F} = \mathscr{G}$, $n = q-3$. For $q \equiv 11 \pmod{12}$, the two harmonic curves make $n = q-5$ for $\mathscr{F} = \mathscr{G}$.□

11.7. Non-singular cubics with one inflexion

11.7.1. THEOREM. *A non-singular cubic curve* $\mathscr{F} = \mathbf{V}(F)$ *in* $PG(2,q)$, $q = p^h$, *with at least one inflexion has one of the following canonical forms.*

(i) $p \neq 2,3$,

$$F = x_2{}^2x_1 + x_0{}^3 + cx_0x_1{}^2 + dx_1{}^3,$$

where $4c^3 + 27d^2 \neq 0$.

(ii) $p = 3$,

$$F = x_2{}^2x_1 + x_0{}^3 + bx_0{}^2x_1 + dx_1{}^3,$$

where $bd \neq 0$;

$$F' = x_2{}^2x_1 + x_0{}^3 + cx_0x_1{}^2 + dx_1{}^3,$$

where $c \neq 0$.

(iii) $p = 2$,

$$F = x_2{}^2x_1 + x_0x_1x_2 + x_0{}^3 + bx_0{}^2x_1 + cx_0x_1{}^2,$$

where $b = 0$ *or* b *is a particular element of* $\mathscr{C}_1$, *and* $c \neq 0$;

$$F' = x_2{}^2x_1 + x_2x_1{}^2 + ex_0{}^3 + cx_0x_1{}^2 + dx_1{}^3,$$

where $e = 1$ *when* $(q-1,3) = 1$ *and* $e = 1$, α, *or* α^2 *when*

$(q-1, 3) = 3$ *with* α *a primitive element of* γ; *also* $d = 0$ *or a particular element of* $\mathscr{C}_1$.

Proof. Let the cubic $\mathscr{F}$ have inflexion $\mathbf{U}_2$ and tangent $\mathbf{u}_1$ there. Then

$$F = x_2^2 x_1 + x_2 x_1 (a_1 x_0 + b_1 x_1) + f_3(x_0, x_1).$$

Hence the polar quadric $\mathscr{F}_2 = \mathbf{V}(F^{(2)})$ at $\mathbf{U}_2$ is given by

$$F^{(2)} = 2x_1 x_2 + x_1(a_1 x_0 + b_1 x_1).$$

When $p > 2$, let $\mathbf{u}_2$ be the harmonic polar of $\mathbf{U}_2$, so that $a_1 = b_1 = 0$ and

$$F^{(2)} = 2x_1 x_2.$$

When $p = 2$, there are three cases:

(a) if $a_1 = b_1 = 0$, then $F^{(2)} = 0$ and $\mathscr{F}_2$ is indeterminate;

(b) if $a_1 = 0$ and $b_1 \neq 0$, then $\mathscr{F}_2$ is a repeated line and the substitution of x_1/b_1 for x_1 makes $F^{(2)} = x_1^2$;

(c) if $a_1 \neq 0$, then $\mathscr{F}_2$ is a line pair and the substitution of $a_1 x_0 + b_1 x_1$ for x_0 makes $F^{(2)} = x_0 x_1$.

(i) $p > 3$. From (11.7) with $a_1 = b_1 = 0$,

$$F = x_2^2 x_1 + a_3 x_0^3 + b_3 x_0^2 x_1 + c_3 x_0 x_1^2 + d_3 x_1^3. \tag{11.8}$$

As $a_3 \neq 0$, put $x_0 - b_3 x_1/3$ for x_0 and $a_3 x_1$ for x_1. Then

$$F = x_2^2 x_1 + x_0^3 + c x_0 x_1^2 + d x_1^3.$$

$\mathscr{F}$ is non-singular if $4c^3 + 27d^2 \neq 0$.

(ii) $p = 3$. As in (11.8),

$$F = x_2^2 x_1 + a_3 x_0^3 + b_3 x_0^2 x_1 + c_3 x_0 x_1^2 + d_3 x_1^3.$$

If $b_3 \neq 0$, put $a_3^{-1/3}(x_0 + c_3 x_1/b_3)$ for x_0. Then

$$F = x_2^2 x_1 + x_0^3 + b x_0^2 x_1 + d x_1^3.$$

$\mathscr{F}$ is non-singular when $bd \neq 0$. If $b_3 = 0$, put $a_3^{-1/3}x_0$ for x_0. Then F becomes

$$F' = x_2^2x_1 + x_0^3 + cx_0x_1^2 + dx_1^3.$$

$\mathscr{F}' = \mathbf{V}(F')$ is non-singular when $c \neq 0$.

(iii) $p = 2$. Consider the above three cases.

(a) With the same substitution as in (i), we obtain

$$\mathscr{F} = \mathbf{V}(x_2^2x_1 + x_0^3 + cx_0x_1^2 + dx_1^3),$$

which has a singularity at $\mathbf{P}(\sqrt{c}, 1, \sqrt{d})$.

(b) $F^{(2)} = x_1^2$ and

$$F = x_2^2x_1 + x_2x_1^2 + a_3x_0^3 + b_3x_0^2x_1 + c_3x_0x_1^2 + d_3x_1^3.$$

The substitution of $x_0 + b_3x_1/a_3$ for x_0 gives

$$F = x_2^2x_1 + x_2x_1^2 + A_3x_0^3 + C_3x_0x_1^2 + D_3x_1^3.$$

When $(q-1, 3) = 1$, each element in γ has a unique cube root, §1.5(iv), and so we can make $A_3 = 1$. When $(q-1, 3) = 3$, we can make $A_3 = 1$, α, or α^2, where α is a primitive element of γ. Also the substitution of $x_2 + Ax_1$ for x_2 changes the coefficient of x_1^3 to $A^2 + A + D_3$. So, if $D_3 \in \mathscr{C}_0$, we select A so that $A^2 + A + D_3 = 0$: if $D_3 \in \mathscr{C}_1$, we select A so that $A^2 + A + D_3$ is some fixed element of $\mathscr{C}_1$. Hence F becomes

$$F' = x_2^2x_1 + x_2x_1^2 + ex_0^3 + cx_0x_1^2 + dx_1^3$$

with e and d as stated. $\mathscr{F}' = \mathbf{V}(F')$ is non-singular.

(c) $F^{(2)} = x_0x_1$ and

$$F = x_2^2x_1 + x_0x_1x_2 + a_3x_0^3 + b_3x_0^2x_1 + c_3x_0x_1^2 + d_3x_1^3.$$

The substitution of a_3x_1 for x_1 and $x_2 + a_3\sqrt{d_3}x_1 + Ax_0$ for x_2 gives the form

$$F = x_2{}^2x_1 + x_0x_1x_2 + x_0{}^3 + bx_0{}^2x_1 + cx_0x_1{}^2,$$

where a suitable selection of A makes b equal to 0 or a particular element of $\mathscr{C}_1$. Also, if $c \neq 0$, then $\mathscr{F}$ is non-singular.□

To continue the classification of cubics with at least one inflexion, we separate the process into the three cases dictated by the theorem:

$$\text{I. } p > 3; \quad \text{II. } p = 3; \quad \text{III. } p = 2.$$

I. $p > 3$

$$\mathscr{F} = \mathbf{V}(F), \; F = x_2{}^2x_1 + x_0{}^3 + cx_0x_1{}^2 + dx_1{}^3 \tag{11.9}$$

$\mathscr{F}$ is general when $cd \neq 0$, harmonic when $c \neq 0$ and $d = 0$, equianharmonic when $c = 0$ and $d \neq 0$, and singular when $4c^3 + 27d^2 = 0$. Write

$$\tilde{\mathscr{F}} = \mathbf{V}(\tilde{F}), \; \tilde{F} = x_2{}^2x_1 + x_0{}^3 + \tilde{c}x_0x_1{}^2 + \tilde{d}x_1{}^3. \tag{11.10}$$

11.7.2. LEMMA. *If $\mathscr{F}$ and $\tilde{\mathscr{F}}$ are general, then they are projectively equivalent if and only if $c^3/d^2 = \tilde{c}^3/\tilde{d}^2$ and $d/\tilde{d}$ is a square. In other words, the projective class of $\mathscr{F}$ is determined by the value of c^3/d^2 and the category of d.*

Proof. If the projectivity $\mathfrak{T}$ for which $\mathbf{P}(X') = \mathbf{P}(X)\mathfrak{T}$ fixes $\mathbf{U}_2$, $\mathbf{u}_1$, and $\mathbf{u}_2$, then

$$x_0 = a_0x_0 + a_{01}x_1, \; x_1 = a_1x_1', \; x_2 = x_2'.$$

So that no term in $x_0{}^2x_1$ appears under this transformation, we have $a_{01} = 0$. Hence

$$\mathscr{F}\mathfrak{T} = \mathbf{V}(a_1x_2{}^2x_1 + a_0{}^3x_0{}^3 + ca_0a_1{}^2x_0x_1{}^2 + da_1{}^3x_1{}^3).$$

If $\mathscr{F}\mathfrak{T} = \tilde{\mathscr{F}}$, then

$$a_0{}^3 = a_1,\ ca_0a_1 = \tilde{c},\quad da_1 = \tilde{d}.$$

So $ca_0{}^4 = \tilde{c}$, $da_0{}^6 = \tilde{d}$, whence $c^3/d^2 = \tilde{c}^3/\tilde{d}^2$. Conversely, if $d/\tilde{d} = t^2$ and $c^3/d^2 = \tilde{c}^3/\tilde{d}^2$, then $c^3/\tilde{c}^3 = t^4$, whence $t = s^3$ for some s in γ; now, choose $a_0 = s^{-1}$, $a_1 = s^{-3}$.□

COROLLARY. *For each value of c^3/d^2 other than 0 and -27/4, there are two distinct types of general non-singular cubic.*□

Two general non-singular cubics are *complementary* if they have the same invariant c^3/d^2 with d of different cateogires.

11.7.3. THEOREM. *If $\mathscr{G}_n^r$ and $\mathscr{G}_{n'}^{r'}$ are complementary with n, $n' \geq 1$, then*

(i) $r = r'$;

(ii) $n + n' = 2(r + 1)$.

Proof. Let $\mathscr{G}_n^r = \mathscr{F}$ and $\mathscr{G}_{n'}^{r'} = \tilde{\mathscr{F}}$ as in (11.9) and (11.10). A short calculation shows that the four inflexional lines of $\mathscr{F}$ through U_2 are

$$V(3x_0{}^4 + 6cx_0{}^2x_1{}^2 + 12dx_0x_1{}^3 - c^2x_1{}^4).$$

Since $c^3/d^2 = \tilde{c}^3/\tilde{d}^2$, there exists α in γ such that $\tilde{c}/c = \alpha^2$, $\tilde{d}/d = \alpha^3$. So the inflexional lines of $\tilde{\mathscr{F}}$ through U_2 are

$$V(3x_0{}^4 + 6c\alpha^2x_0{}^2x_1{}^2 + 12d\alpha^3x_0x_1{}^3 - c^2\alpha^4x_1{}^4).$$

Hence the number of real inflexional lines through U_2 is the same for $\mathscr{F}$ and $\tilde{\mathscr{F}}$. Therefore $r = r'$ by lemma 11.4.7.

If $V(x_0 - tx_1)$ is an inflexional line of $\mathscr{F}$, then $V(x_0 - t\alpha x_1)$ is an inflexional line of $\tilde{\mathscr{F}}$. Now, $V(x_0 - tx_1)$ meets $\mathscr{F}$ in two or no points apart from U_2 as $-(t^3 + ct + d)$ is a square or not; also, $V(x_0 - t\alpha x_1)$ meets $\tilde{\mathscr{F}}$ in two or no points apart from U_2 as $-[(\alpha t)^3 + \tilde{c}(\alpha t) + \tilde{d}] = -\alpha^3(t^3 + ct + d)$ is a square or not. As $\mathscr{F}$ and $\tilde{\mathscr{F}}$ are complementary, so α is a non-square. Hence, if $V(x_0 - tx_1)$ meets $\mathscr{F}$ in three real inflexions, then $V(x_0 - t\alpha x_1)$ meets $\tilde{\mathscr{F}}$ in one real and two 2-complex conjugate inflexions. So, let m and m' be the number of lines through U_2 meeting $\mathscr{F}$ and $\tilde{\mathscr{F}}$ respectively in three real inflexions. Then

$$m + m' = r, \quad 2m + 1 = n, \quad 2m' + 1 = n',$$

whence $n + n' = 2(r + 1)$.□

COROLLARY. *There is a one-to-one correspondence between the following types of pairs of cubics:*

(i) $\mathscr{G}_0^4$ *and* $\mathscr{G}_1^4$;
(ii) $\mathscr{G}_3^1$ *and* $\mathscr{G}_1^1$.□

Let $N(\mathscr{F}_n^r)$ be the number of projectively distinct cubics of type $\mathscr{F}_n^r$.

11.7.4. THEOREM. *In* $PG(2,q)$, $q = p^h$ *and* $p > 3$, *the numbers* $N(\mathscr{G}_1^r)$ *of projectively distinct, general non-singular cubics with exactly one inflexion are given by Table* 11.12. *Each curve has a projective group isomorphic to* $\mathbf{Z}_2$.

TABLE 11.12

$q \equiv m \pmod{12}$ m	$\mathscr{G}_1^4$	$\mathscr{G}_1^1$	$\mathscr{G}_1^0$	Total
1	$(q-13)/12$	$2(q-1)/3$	$(q-1)/2$	$(5q-9)/4$
7	$(q-7)/12$	$2(q-1)/3$	$(q-3)/2$	$(5q-11)/4$
5	0	0	$q-1$	$q-1$
11	0	0	$q+1$	$q+1$

Proof. By lemma 11.7.2, there are $2(q - 2)$ projectively distinct curves of the form (11.9). For $m = 1$, $N(\mathscr{G}_1^4) = N(\mathscr{G}_9^4) = (q - 13)/12$ by theorem 11.5.4, and $N(\mathscr{G}_1^1) = N(\mathscr{G}_3^1) = 2(q - 1)/3$ by theorem 11.6.2. So,

$$N(\mathscr{G}_1^0) = 2(q - 2) - 2(q - 13)/12 - 2 \cdot 2(q - 1)/3 = (q - 1)/2.$$

Similarly, for $m = 7$, $N(\mathscr{G}_1^4) = N(\mathscr{G}_9^4) = (q - 7)/12$, $N(\mathscr{G}_1^1) = N(\mathscr{G}_3^1) = 2(q - 1)/3$, and

$$N(\mathscr{G}_1^{\,0}) = 2(q - 2) - 2(q - 7)/12 - 2\cdot 2(q - 1)/3 = (q - 3)/2.$$

For $m = 5$, $N(\mathscr{G}_3^{\,2}) = q - 3$; so $N(\mathscr{G}_1^{\,0}) = q - 1$.
For $m = 11$, $N(\mathscr{G}_3^{\,2}) = q - 5$; so $N(\mathscr{G}_1^{\,0}) = q + 1$.

The projectivities fixing $\mathscr{F}$ are given by $x_0 \to x_0$, $x_1 \to x_1$, $x_2 \to \pm x_2$.□

The facts about harmonic and equianharmonic cubics with one inflexion can be deduced in the same way as the general case from theorem 11.7.3. However, it is not difficult to obtain the results directly from the canonical form and this we prefer to do.

(a) $\mathscr{F}$ *equianharmonic*

$\mathscr{F} = \mathscr{E} = \mathbf{V}(E)$, $E = x_2^{\,2}x_1 + x_0^{\,3} - dx_1^{\,3}$.

11.7.5. LEMMA. *Over $\gamma = GF(q)$, $q = p^h$ and $p > 3$, with α a primitive element of γ, the number of projectively distinct curves $\mathscr{E}$ is as follows:*

(i) $q \equiv 1 \pmod 3$: *six types with* $d = 1, \alpha, \alpha^2, \alpha^3, \alpha^4, \alpha^5$;

(ii) $q \equiv -1 \pmod 3$: *two types with* $d = 1, \alpha$.

Proof. As in lemma 11.6.2, we have $\tilde{d} = da_0^{\,6}$.□

11.7.6. LEMMA. (i) *The inflexional lines of $\mathscr{E}$ through $\mathbf{U}_2$ are* $\mathbf{V}(x_0(x_0^{\,3} - 4dx_1^{\,3}))$.

(ii) *The pairs of complex points other than $\mathbf{U}_2$ in which the inflexional lines through $\mathbf{U}_2$ meet $\mathscr{E}$ are as follows:*

(a) $\mathbf{V}(x_0,\ x_2^{\,2} - dx_1^{\,2})$;

(b) $\mathbf{V}(x_0 - tx_1,\ x_2^{\,2} + 3dx_1^{\,2})$ *for each component* $\mathbf{V}(x_0 - tx_1)$ *of* $\mathbf{V}(x_0^{\,3} - 4dx_1^{\,3})$.□

11.7.7. THEOREM. *In $PG(2,q)$, $q = p^h$ and $p > 3$, the number of projectively distinct equianharmonic curves $\mathscr{E}$ with at least one inflexion is as follows:*

(i) $q \equiv 1 \pmod 3$:

$d = \alpha, \alpha^3, \alpha^5$: $\quad 1\,\mathscr{E}^{4}_{1},\quad 2\,\mathscr{E}^{1}_{1}$;

$d = 1, \alpha^2, \alpha^4$: $\quad 1\,\mathscr{E}^{4}_{9},\quad 2\,\mathscr{E}^{1}_{3}$;

(ii) $q \equiv -1 \pmod 3$: $\quad 2\,\mathscr{E}^{2}_{3}$.

Proof. (i) To obtain only one inflexion, d is a non-square; so $d = \alpha, \alpha^3, \alpha^5$. Then there are four real inflexional triangles or one according as $4d$ is a cube in γ or not. But exactly one of 4α, $4\alpha^3$, $4\alpha^5$ is a cube. So there is one $\mathscr{E}_1^4$ and two $\mathscr{E}_1^1$.

When d is a square, both $\mathbf{V}(x_2^2 - dx_1^2)$ and $\mathbf{V}(x_2^2 + 3dx_1^2)$ are a real pair of lines. Here exactly one of 4, $4\alpha^2$, $4\alpha^4$ is a cube. So there is one $\mathscr{E}_9^4$ and two $\mathscr{E}_3^1$.

(ii) This follows from the previous two lemmas.□

(b) $\mathscr{F}$ *harmonic*

$$\mathscr{F} = \mathscr{H} = \mathbf{V}(H),\ H = x_2{}^2x_1 + x_0{}^3 + ex_0x_1{}^2.$$

11.7.8. LEMMA. *Over* $\gamma = GF(q)$, $q = p^h$ *and* $p > 3$, *with* α *a primitive root of* γ, *the number of projectively distinct curves* $\mathscr{H}$ *is as follows:*

(i) $q \equiv 1 \pmod 4$: *four types with* $d = 1, \alpha, \alpha^2, \alpha^3$;

(ii) $q \equiv -1 \pmod 4$: *two types with* $d = 1, \alpha$.

Proof: As in lemma 11.7.2, $\tilde{c} = ca_0^4$.□

11.7.9. LEMMA. (i) *The inflexional lines through* $\mathbf{U}_2$ *of* $\mathscr{H}$ *are*

$$\mathbf{V}(3x_0{}^4 + 6cx_0{}^2x_1{}^2 - c^2x_1{}^4) = \mathbf{V}\left(\prod_{i=1}^{4} (x_0 - t_ix_1)\right),$$

$\{t_1, t_2, t_3, t_4\} = \{\pm\surd[c(-1 \pm 2\surd 3/3]\}$.

(ii) *The pairs of complex points other than* $\mathbf{U}_2$ *in which the inflexional line* $\mathbf{V}(x_0 - t_ix_1)$ *meets* $\mathscr{H}$ *lie on* $\mathbf{V}(x_2{}^2 - s_ix_1{}^2)$, *where*

$$s_i = \pm\surd[c(-1 \pm 2\surd 3/3)] \cdot 2c\surd 3/3.□$$

11.7.10. THEOREM. *In* $PG(2,q)$, $q = p^h$ *and* $p > 3$, *the number of projectively distinct harmonic cubics* $\mathscr{H}$ *with at least one inflexion is as follows:*

(i) $q \equiv 1 \pmod{12}$: $1\mathscr{H}_9^4$, $1\mathscr{H}_1^4$, $2\mathscr{H}_1^0$;

(ii) $q \equiv 7 \pmod{12}$: $2\mathscr{H}_1^0$;

(iii) $q \equiv 5 \pmod{12}$: $4\mathscr{H}_1^0$;

(iv) $q \equiv 11 \pmod{12}$: $2\mathscr{H}_3^2$.

Proof: $(-1 + 2\sqrt{3}/3)/(-1 - 2\sqrt{3}/3) = -3(-1 + 2\sqrt{3}/3)^2$.

(i) $q \equiv 1 \pmod 3$, $q \equiv 1 \pmod 4$. So -3, -1, and 3 are all squares in γ. Then $c(-1 + 2\sqrt{3}/3)$ and $c(-1 - 2\sqrt{3}/3)$ are both squares or both non-squares. So, for two values of c there are no real inflexional triangles and hence two $\mathscr{H}_1^0$, and for two values of c there are four real inflexional triangles. When there are four real triangles, the four values of s_i are all squares or all non-squares. However, the two values of c involved are 1 and α^2 or α and α^3. As $c\sqrt{c}$ appears in the formula for the s_i, for one of the c all the s_i are squares and for the other all non-squares. Hence there is one $\mathscr{H}_9^4$ and one $\mathscr{H}_1^4$.

(ii) $q \equiv 1 \pmod 3$, $q \equiv -1 \pmod 4$. So -3 is a square, but both 3 and -1 are non-squares. Hence none of the t_i lie in γ and there are just two $\mathscr{H}_1^0$, by lemma 11.7.8.

(iii) $q \equiv -1 \pmod 3$, $q \equiv 1 \pmod 4$. So -1 is a square, but both 3 and -3 are non-squares. So again none of the t_i lie in γ and there are four $\mathscr{H}_1^0$.

(iv) $q \equiv -1 \pmod 3$, $q \equiv -1 \pmod 4$. So 3 is a square, but both -1 and -3 are non-squares. So for each of the two values of c, there are two t_i in γ. The corresponding s_i have sum zero. As -1 is a non-square, only one of the two s_i is a square and so only one of the real inflexional lines meets $\mathscr{H}$ in three real inflexions. So there are two $\mathscr{H}_3^2$.□

11.7.11. THEOREM. *In $PG(2,q)$, $q = p^h$ and $p > 3$, the classification of non-singular cubics with exactly one inflexion is as follows, where n is the number of projectively distinct cubics of each type, o is the order of the corresponding projective group and n_1 is the total number of projectively distinct types (Table* 11.13*).*

Proof. Table 11.13 collects the results of theorems 11.7.4, 11.7.7, and 11.7.10. The only figures that remain to be established are the orders of the projective groups in the equianharmonic and harmonic cases.

TABLE 11.13

$q \equiv m \pmod{12}$	m	1	7	5	11
$\mathcal{G}^4_1$	n	$(q-13)/12$	$(q-7)/12$	0	0
	o	2	2	-	-
$\mathcal{E}^4_1$	n	1	1	0	0
	o	6	6	-	-
$\mathcal{H}^4_1$	n	1	0	0	0
	o	4	-	-	-
$\mathcal{G}^1_1$	n	$2(q-1)/3$	$2(q-1)/3$	0	0
	o	2	2	-	-
$\mathcal{E}^1_1$	n	2	2	0	0
	o	6	6	-	-
$\mathcal{H}^1_1$	n	0	0	0	0
	o	-	-	-	-
$\mathcal{G}^0_1$	n	$(q-1)/2$	$(q-3)/2$	$q-1$	$q+1$
	o	2	2	2	2
$\mathcal{E}^0_1$	n	0	0	0	0
	o	-	-	-	-
$\mathcal{H}^0_1$	n	2	2	4	0
	o	4	2	4	-
	n_1	$5(q+3)/4$	$(5q+9)/4$	$q+3$	$q+1$

(a) *Equianharmonic*

$$E = x_2{}^2 x_1 + x_0{}^3 - dx_1{}^3.$$

Projectivities fixing $\mathcal{E}$ are given by $X = X'A$ where $A = \operatorname{diag}(a_0, a_1, 1)$ with $a_0{}^3 = a_1$ and $a_1{}^2 = 1$. As $\mathcal{E}$ only exists for $q \equiv 1 \pmod 3$, if we take ω as a root of $x^2 + x + 1$, then $G(\mathcal{E}) = \{\operatorname{diag}(\lambda, 1, 1) \mid \lambda = 1, \omega, \omega^2\} \cup \{\operatorname{diag}(-\lambda, -1, 1) \mid \lambda = 1, \omega, \omega^2\} \cong \mathbf{Z}_6$.

(b) *Harmonic*

$$H = x_2{}^2x_1 + x_0{}^3 + cx_0x_1{}^2$$

$A = \mathrm{diag}(a_0, a_1, 1)$ with $a_0{}^3 = a_1$ and $a_0a_1 = 1$. So $a_0{}^4 = 1$, whence, if $q \equiv 1 \pmod 4$, $a_0 = 1, -1, \theta, -\theta$ with $\theta^2 = -1$, and, if $q \equiv -1 \pmod 4$, $a_0 = 1, -1$. Therefore,

(i) when $q \equiv 1 \pmod 4$,

$$G(\mathscr{H}) = \{\mathrm{diag}(1,1,1), \mathrm{diag}(-1,-1,1), \mathrm{diag}(\theta,-\theta,1), \mathrm{diag}(-\theta,\theta,1)\} \cong \mathbf{Z}_4;$$

(ii) when $q \equiv -1 \pmod 4$,

$$G(\mathscr{H}) = \{\mathrm{diag}\ (1,1,1), \mathrm{diag}\ (-1,-1,1)\} \cong \mathbf{Z}_2. \square$$

II. $p = 3$

(a) $\mathscr{F} = \mathbf{V}(F)$, $F = x_2{}^2x_1 + x_0{}^3 + bx_0{}^2x_1 + dx_1{}^3$.

$\mathscr{F}$ is non-singular if $bd \neq 0$, and is never harmonic. Write

$$\tilde{\mathscr{F}} = \mathbf{V}(\tilde{F}),\ \tilde{F} = x_2{}^2x_1 + x_0{}^3 + \tilde{b}x_0{}^2x_1 + \tilde{d}x_1{}^3.$$

11.7.12. LEMMA. (i) *$\mathscr{F}$ has three inflexions or one as $-b$ is a square or not;*

(ii) *$\mathscr{F}$ and $\tilde{\mathscr{F}}$ are projectively equivalent if and only if $b^3/d = \tilde{b}^3/\tilde{d}$ and $b/\tilde{b}$ is a square.*

Proof. (i) $F^{(2)} = bx_0{}^2y_1 + x_2{}^2y_1 - x_1x_2y_2 - bx_0x_1y_0$,

$$F^{(1)} = 2by_0y_1x_0 + (by_0{}^2 + y_2{}^2)x_1 - y_1y_2x_2.$$

So, by theorem 7.2.5, the inflexions of $\mathscr{F}$ are on $\mathbf{V}(x_1(x_0{}^3 + dx_1{}^3))$. So the triple line $\mathbf{V}(x_0{}^3 + dx_1{}^3)$ meets $\mathscr{F}$ in $\mathbf{U}_2$ and in $\mathbf{V}(x_2{}^2 + bx_0{}^2)$, which is a pair of real or complex lines as $-b$ is a square or not. $\mathscr{F}$ has three inflexions or one accordingly.

(ii) If the projectivity $\mathfrak{T}$ is given by $X = X'A$ and $\mathscr{F}\mathfrak{T} = \tilde{\mathscr{F}}$ with $\mathbf{U}_2$ fixed, then A is diagonal. So, with $A = \mathrm{diag}\ (a_0, a_1, 1)$,

$$\mathscr{F}\mathfrak{T} = \mathbf{V}(a_1 x_2^2 x_1 + a_0^3 x_0^3 + b a_0^2 a_1 x_0^2 x_1 + d a_1^3 x_1^3),$$

So, if $\mathscr{F}\mathfrak{T} = \tilde{\mathscr{F}}$, then $a_0^3 = a_1$, $ba_0^2 = \tilde{b}$, and $da_1^2 = \tilde{d}$. Hence, $b^3/d = \tilde{b}^3/\tilde{d}$ and $b/\tilde{b}$ is a square. Conversely, if $\tilde{b}^3/\tilde{d} = b^3/d$ and $b/\tilde{b} = s^2$ with $s \in \gamma$, then take $a_0 = 1/s$ and $a_1 = 1/s^3$ to obtain $\mathfrak{T}$.□

COROLLARY. *$\mathscr{F}$ has the same number of nuclei as inflexions.*

Proof. From the form $F^{(1)}$ in the proof of the theorem, all nuclei lie on $\mathbf{V}(x_0 x_1, x_1 x_2, bx_0^2 + x_2^2) = \{\mathbf{U}_1\} \cup \mathbf{V}(x_1, bx_0^2 + x_2^2)$. Hence $\mathscr{F}$ has three nuclei if $-b$ is a square and only one if $-b$ is a non-square.□

We will describe $\mathscr{F}$ as $\mathscr{G}_1$, a general cubic with one inflexion, or $\mathscr{G}_3$, a general cubic with three inflexions.

11.7.13. THEOREM. *In $PG(2,q)$, $q = 3^h$, the classification of general non-singular cubics with exactly one inflexion is as follows, where n is the number of projectively distinct cubics and o the order of the projective group:*

$$n = q-1, \quad o = 2.$$

Proof. From the lemma, there is one cubic for each value of b^3/d in γ_0. As for the group, in the notation of the lemma, $a_0^3 = a_1$, $a_0^2 = 1$, $a_1^2 = 1$. So

$$G(\mathscr{G}_1) = \{\mathrm{diag}(1,1,1), \mathrm{diag}(-1,-1,1)\} \cong \mathbf{Z}_2.□$$

(b) $\mathscr{F}' = \mathbf{V}(F')$, $F' = x_2^2 x_1 + x_0^3 + c x_0 x_1^2 + d x_1^3$.

$\mathscr{F}'$ is non-singular if $c \neq 0$ and is then always harmonic and so superharmonic. All the inflexions lie on $\mathbf{u}_1$ and so $\mathbf{U}_2$ is the only inflexion. Also $\mathbf{U}_0$ is the only nucleus. The harmonic polar $\mathbf{u}_2$ of $\mathbf{U}_2$ meets $\mathscr{F}'$ in zero, one, or three points as $x^3 + cx + d$ has zero, one, or three roots in γ. $\mathscr{F}'$ is accordingly described as $\mathscr{S}_1^0$, $\mathscr{S}_1^1$, $\mathscr{S}_1^3$.

11.7.14. LEMMA. *Let $\mathscr{F}'$ be non-singular over $\gamma = GF(q)$, $q = 3^h$. Then, with $D_3(x) = x + x^3 + \ldots + x^{3^{h-1}}$,*

(i) *if $-c$ is a non-square, $\mathscr{F}'$ is $\mathscr{S}_1^{\,1}$;*

(ii) *if $-c = s^2$, s in γ_0, and $D_3(d/s^3) \neq 0$, $\mathscr{F}'$ is $\mathscr{S}_1^{\,0}$;*

(iii) *if $-c = s^2$, s in γ_0, and $D_3(d/s^3) = 0$, $\mathscr{F}'$ is $\mathscr{S}_1^{\,3}$.*

Proof. This is just a restatement of theorem 1.9.3, corollary 1, for the polynomial $x^3 + cx + d$.□

11.7.15. LEMMA. *Over $\gamma = GF(q)$, $q = 3^h$, with α a primitive root of γ, the number of projectively distinct curves $\mathscr{F}'$ is as follows:*

		$\mathscr{S}_1^{\,0}$	$\mathscr{S}_1^{\,1}$	$\mathscr{S}_1^{\,3}$
(i) $q \equiv 1 \pmod 4$	n	2	2	2
	c	$-1, -\alpha^2$	$-\alpha, -\alpha^3$	$-1, -\alpha^2$
(ii) $q \equiv -1 \pmod 4$	n	2	1	1
	c	-1	$-\alpha$	-1

Proof. Let $\mathscr{F}' = V(F')$, $F' = x_2^{\,2}x_1 + x_0^{\,3} + \tilde{c}x_0x_1^{\,2} + \tilde{d}x_1^{\,3}$. If $\mathscr{F}'\mathfrak{T} = \tilde{\mathscr{F}}'$, then $\mathfrak{T}$ fixes $\mathbf{U}_2$, $\mathbf{u}_1$, and $\mathbf{u}_2$; so it is given by $x_0 = a_0x_0' + a_{01}x_1'$, $x_1' = a_1x_1$, $x_2' = x_2$. So $\mathscr{F}'\mathfrak{T} = \tilde{\mathscr{F}}'$ if and only if there exist a_0, a_1, a_{01} with $a_1 = a_0^{\,3}$, $ca_1a_0 = \tilde{c}$ and $a_{01}^3 + ca_1^{\,2}a_{01} + da_1^{\,3} = \tilde{d}a_1$; that is, $ca_0^{\,4} = \tilde{c}$ and $(a_{01}/a_1)^3 + c(a_{01}/a_1) + d = \tilde{d}/a_1^{\,2}$. So, when $q \equiv 1 \pmod 4$, there are four possible values for c, which may be taken as -1, $-\alpha$, $-\alpha^2$, $-\alpha^3$. Then $-c$ is a square for the two values $c = -1, -\alpha^2$, and $-c$ is a non-square for the two values $c = -\alpha, -\alpha^3$. When $q \equiv -1 \pmod 4$, there are two possible values for c which may be taken as -1, $-\alpha$. Then $-c$ is a square for $c = -1$ and a non-square for $c = -\alpha$. When $-c$ is a non-square, there is always a solution for a_{01} by theorem 1.9.3, corollary 1. So there are two $\mathscr{S}_1^{\,1}$ when $q \equiv 1 \pmod 4$ and one $\mathscr{S}_1^{\,1}$ when $q \equiv -1 \pmod 4$.

When $-c = s^2$, there is a solution for a_{01} if and only if $D_3((d - \tilde{d}/a_1^{\,2})/s^3) = 0$. Take $\tilde{c} = c$; then $a_{00}^4 = 1$. When $q \equiv -1 \pmod 4$, $a_1 = a_{00} = \pm 1$. So there is a solution for a_{01} if $D_3((d - \tilde{d})/s^3) = 0$. If $\mathscr{F}'$ and $\tilde{\mathscr{F}}'$ are $\mathscr{S}_1^3$, then

$D_3(d/s^3) = D_3(\tilde{d}/s^3) = D_3((d - \tilde{d})/s^3) = 0$. If $\mathscr{F}'$ and $\tilde{\mathscr{F}}'$ are $\mathscr{S}_1^0$, there is a solution for a_{01} only when $D_3(d/s^3)$ and $D_3(\tilde{d}/s^3)$ have the same sign. So, for $q \equiv -1 \pmod 4$, there are two $\mathscr{S}_1^0$ and one $\mathscr{S}_1^3$.

When $-c = s^2$ and $q \equiv 1 \pmod 4$, there is a solution for a_{01} if and only if $D_3((d \pm \tilde{d})/s^3) = 0$; that is, $D_3(d/s^3) = \pm D_3(d/s^3)$.

As $D_3(d/s^3) = 0$, 1, or -1, there is always a solution for a_{01}. Hence there are two $\mathscr{S}_1^0$ and two $\mathscr{S}_1^3$. □

11.7.16. LEMMA. *The orders of the projective groups of the harmonic cubics over* $\gamma = GF(q)$, $q = 3^h$, *as follows:*

	$\mathscr{S}_1^0$	$\mathscr{S}_1^1$	$\mathscr{S}_1^3$
$q \equiv -1 \pmod 4$	6	2	6
$q \equiv 1 \pmod 4$	6	4	12

Proof. If $\mathscr{F}'\mathfrak{T} = \mathscr{F}'$, then $\mathfrak{T}$ is given by $x_0 = a_0x_0' + a_{01}x_1'$, $x_1 = a_1x_1'$, $x_2 = x_2'$, and

$$a_1 = a_0^3 = a_1^2a_0,\ a_{01}^3 + ca_1^2a_{01} + da_1^3 = da_1.$$

So $a_1^4 = a_0^4 = 1$. Therefore,

(i) if $q \equiv -1 \pmod 4$, $a_1 = \pm 1$;

(ii) if $q \equiv 1 \pmod 4$, $a_1 = \pm 1, \pm\theta$ where $\theta^2 = -1$.

Write $f(x) = x^3 + ca_1^2x + d(a_1^3 - a_1)$. When $a_1 = \pm 1$, $f(x) = x^3 + cx$, which has three roots or one as $-c$ is square or not. When $a_1 = \pm\theta$, $f(x) = x^3 - cx + a_1d$. So when $-c$ is a non-square, f has one root t in γ. When $-c = s^2$, $c = (a_1s)^2$ and $D_3(a_1d/(a_1s)^3) = -D_3(d/s^3)$; so f has three roots or none as $\mathscr{F}'$ is $\mathscr{S}_1^3$ or $\mathscr{S}_1^0$. Write $\tau, \tau + 1, \tau - 1$ for the three roots of $x^3 - x + d/s^3$. Then the roots of f are $-a_1s\tau$, $-a_1s(\tau + 1)$, $-a_1s(\tau - 1)$.

(i) $q \equiv -1 \pmod 4$.

(a) $-c$ non-square.

$$G(\mathscr{S}_1^1) = \{\text{diag}(1,1,1),\ \text{diag}(-1,-1,1)\} \cong \mathbf{Z}_2.$$

(b) $-c = s^2$.

$$G(\mathscr{S}_1^0) \cong G(\mathscr{S}_1^3) = G_6 \cong \mathbf{Z}_6, \text{ where } G_6 = \left\{\text{diag}(B,I_1) \mid B = \begin{bmatrix}1&0\\0&1\end{bmatrix}, \begin{bmatrix}-1&0\\0&-1\end{bmatrix}, \begin{bmatrix}1&0\\s&1\end{bmatrix}, \begin{bmatrix}-1&0\\-s&1\end{bmatrix}, \begin{bmatrix}-1&0\\s&-1\end{bmatrix}, \begin{bmatrix}-1&0\\-s&-1\end{bmatrix}\right\}.$$

(ii) $q \equiv 1 \pmod 4$.

(a) $-c$ non-square.

$$G(\mathscr{S}_1^1) = \left\{\text{diag}(B,I_1) \mid B = \begin{bmatrix}1&0\\0&1\end{bmatrix}, \begin{bmatrix}-1&0\\0&-1\end{bmatrix}, \begin{bmatrix}\theta&0\\t&-\theta\end{bmatrix}, \begin{bmatrix}-\theta&0\\t&\theta\end{bmatrix}\right\} \cong \mathbf{Z}_4.$$

(b) $-c = s^2$, $D_3(d/s^3) \neq 0$.

$$G(\mathscr{S}_1^0) = G_6 \cong \mathbf{Z}_6.$$

(c) $-c = s^2$, $D_3(d/s^3) = 0$.

$$G_6' = \left\{\text{diag}(B,I_1) \mid B = \begin{bmatrix}\theta&0\\\theta st&-\theta\end{bmatrix}, \begin{bmatrix}\theta&0\\-\theta st&\theta\end{bmatrix}, \begin{bmatrix}\theta&0\\\theta s(\tau+1)&-\theta\end{bmatrix}, \begin{bmatrix}-\theta&0\\-\theta s(\tau+1)&\theta\end{bmatrix}, \begin{bmatrix}\theta&0\\\theta s(\tau-1)&-\theta\end{bmatrix}, \begin{bmatrix}-\theta&0\\-\theta s(\tau-1)&\theta\end{bmatrix}\right\}.$$

$G(\mathscr{S}_1^3) = G_6 \cup G_6'$, the unique group of order 12 with one subgroup $\mathbf{Z}_6$, three $\mathbf{Z}_4$, one $\mathbf{Z}_3$, one $\mathbf{Z}_2$. It is also the fifth group of order 12, apart from the two abelian groups, $\mathbf{A}_4$ and the dihedral group $\mathbf{D}_6$.□

11.7.17. THEOREM. *In $PG(2,q)$, $q = 3^h$, the classification of non-singular cubics with exactly one inflexion is as follows, where n is the number of projectively distinct curves of each type, o the order of the corresponding projective group, and n_1 the total number of projectively distinct curves (Table 11.14).*□

III. $p = 2$

(a) $\mathscr{F} = \mathbf{V}(F)$, $F = x_2{}^2x_1 + x_0x_1x_2 + x_0{}^3 + bx_0{}^2x_1 + cx_0x_1{}^2$.

TABLE 11.14

	$\mathscr{G}_1$ n	$\mathscr{G}_1$ o	$\mathscr{S}_1^0$ n	$\mathscr{S}_1^0$ o	$\mathscr{S}_1^1$ n	$\mathscr{S}_1^1$ o	$\mathscr{S}_1^3$ n	$\mathscr{S}_1^3$ o	n_1
$q \equiv -1 \pmod 4$	$q-1$	2	2	6	1	2	1	6	$q+3$
$q \equiv 1 \pmod 4$	$q-1$	2	2	6	2	4	2	12	$q+5$

$\mathscr{F}$ is non-singular if $c \neq 0$, and is never equianharmonic.

$$\tilde{\mathscr{F}} = \mathbf{V}(\tilde{F}),\ \tilde{F} = x_2{}^2 x_1 + x_0 x_1 x_2 + x_0{}^3 + \tilde{b} x_0{}^2 x_1 + \tilde{c} x_0 x_1{}^2.$$

11.7.18. LEMMA. *$\mathscr{F}$ is projectively equivalent to $\tilde{\mathscr{F}}$ if and only if $c = \tilde{c}$, and b and $\tilde{b}$ are of the same category.*

Proof. From theorem 11.7.1, $\mathbf{u}_1$ and $\mathbf{u}_0$ are respectively the tangent and harmonic polar at $\mathbf{U}_2$. So a projectivity $\mathfrak{T}$ fixing these has the form

$$x_0 = a_0 x_0',\ x_1 = a_1 x_1',\ x_2 = a_{02} x_0' + a_{12} x_1' + x_2'.$$

So $\mathscr{F}\mathfrak{T} = \mathbf{V}(a_1(a_{02}x_0 + a_{12}x_1 + x_2)^2 x_1 + a_0 a_1 x_0 x_1 (a_{02}x_0 + a_{12}x_1 + x_2) + a_0{}^3 x_0{}^3 + b a_0{}^2 a_1 x_0{}^2 x_1 + c a_0 a_1{}^2 x_0 x_1{}^2)$.

If $\mathscr{F}\mathfrak{T} = \tilde{\mathscr{F}}$, then $a_{12} = 0$ and $a_1 = a_0 a_1 = a_0{}^3$, so that $a_0 = a_1 = 1$. Hence $\mathfrak{T}$ is given by

$$x_0 = x_0',\ x_1 = x_1',\ x_2 = a_{02} x_0' + x_2',$$

and $\mathscr{F}\mathfrak{T} = \mathbf{V}(x_2{}^2 x_1 + x_0 x_1 x_2 + x_0{}^3 + (a_{02}^2 + a_{02} + b) x_0{}^2 x_1 + c x_0 x_1{}^2)$. Hence $\mathfrak{T}\mathscr{F} = \tilde{\mathscr{F}}$ if and only if $c = \tilde{c}$ and $a_{02}^2 + a_{02} + b = \tilde{b}$. □

COROLLARY 1. *If $\mathscr{F}$ has only one inflexion, its projective group is $\mathbf{Z}_2$.*

Proof. From the lemma, if $b = \tilde{b}$, then $a_{02} = 0$ or 1. So, for $\mathscr{F} = \mathscr{G}_1$,

$$G(\mathscr{G}_1) = \left\{ \begin{bmatrix} 1 & 0 & 0 \\ 0 & 1 & 0 \\ 0 & 0 & 1 \end{bmatrix}, \begin{bmatrix} 1 & 0 & 1 \\ 0 & 1 & 0 \\ 0 & 0 & 1 \end{bmatrix} \right\} \cong \mathbf{Z}_2 . \square$$

COROLLARY 2. *There are* $2(q - 1)$ *projectively distinct non-singular cubics* $\mathscr{F}$. □

11.17.19. LEMMA. *The inflexional lines through* $\mathbf{U}_2$ *of* $\mathscr{F}$ *are* $\mathbf{V}(x_0{}^4 + x_0{}^3 x_1 + c^2 x_1{}^4)$.

Proof.

$$F^{(2)} = x_0{}^2 y_0 + by_1 + x_1{}^2 cy_0 + x_2{}^2 y_1 + x_0 x_1 y_2 + x_0 x_2 y_1 + x_1 x_2 y_0,$$

$$F^{(1)} = x_0(y_0{}^2 + cy_1{}^2 + y_1 y_2) + x_1(by_0{}^2 + y_2{}^2 + y_0 y_2) + x_2 y_0 y_1 .$$

Again, theorem 7.2.5 gives the result.□

Let $f(x) = c^2x^4 + x + 1$. If $c^2 = s^3$, then $f(x) = [(sx)^4 + sx + s]/s$. So $sf(x/s) = x^4 + x + s = g(x)$. Let τ be a root in γ of g if it exists, and similarly, let t be a root in γ of f, if it exists. Then we use theorem 1.9.4 and its corollaries to elicit the properties of $\mathscr{F}$ as in the following.

11.7.20. THEOREM. *Over* $GF(q)$, $q = 2^h$, *the curves* $\mathscr{F}$ *have the following properties where* $c^2 = s^3$ *and* $s \in K$, *where* $K = \gamma$ *or* $\gamma''\backslash\gamma$ *with* γ'' *a cubic extension of* γ; n *is the number of projectively distinct curves of each type (Table* 11.15*)*.

TABLE 11.15

$(q-1,3)$	K	$D_2(s)$	$D_4(s)^3$	$D_2(1/t+b)$	$\mathscr{F}$	n
1	γ	0			$\mathscr{G}_3^2$	$q-2$
1	γ	1			$\mathscr{G}_1^0$	q
3	γ		0	0	$\mathscr{G}_9^4$	$(q-4)/12$
3	γ		0	1	$\mathscr{G}_1^4$	$(q-4)/12$
3	γ		1		$\mathscr{G}_1^0$	$q/2$
3	$\gamma''\backslash\gamma$			0	$\mathscr{G}_3^1$	$2(q-1)/3$
3	$\gamma''\backslash\gamma$			1	$\mathscr{G}_1^1$	$2(q-1)/3$.

Proof. (i) If $(q-1,3) = 1$, then f has two roots or none in γ as $D_2(s) = 0$ or 1; the two roots are τ/s and $(\tau + 1)/s$.

(ii) If $(q-1,3) = 3$ and $c^2 = s^3$, s in γ, then f has four roots or none in γ as $D_4(s) = 0$ or 1; the four are τ/s, $(\tau + 1)/s$, $(\tau + \omega)/s$, $(\tau + \omega^2)/s$ where $\omega^2 + \omega + 1 = 0$.

(iii) If $(q-1,3) = 3$ and c is not a cube in γ, then f has exactly one root in γ by theorem 1.9.3(i).

$\mathbf{V}(x_1 - tx_0)$ meets $\mathscr{F}$ in $\mathbf{U}_2$ and $\mathbf{V}(tx_2{}^2 + tx_0x_2 + (1 + bt + ct^2)x_0{}^2)$. By §1.4, $\mathbf{V}(x_1 - tx_0)$ meets $\mathscr{F}$ in three points or one as $D_2(1/t + b + ct) = 0$ or 1. If t is a root of f, then $c^2t^4 = t + 1$, whence $c^2t^2 = 1/t + 1/t^2$; so $D_2(ct) = D_2(c^2t^2) = 0$ and $D_2(1/t + b + ct) = D_2(1/t) + D_2(b)$.

(i) $(q-1,3) = 1$.

(a) $D_2(s) = 1$. Then f has no roots and $\mathbf{U}_2$ is the only inflexion. Hence $\mathscr{F}$ is a $\mathscr{G}_1^0$. For each s in $\mathscr{C}_1$ and each category of b, there is a distinct $\mathscr{G}_1^0$; so $n = 2(q/2) = q$.

(b) $D_2(s) = 0$. The roots t of f in γ are τ/s, $(\tau + 1)/s$. When $t = \tau/s$, we have

$$D_2(1/t) = D_2(s/\tau) = D_2(\tau^3 + 1) = D_2(\tau^3) + 1.$$

When $t = (\tau + 1)/s$, we have

$$D_2(1/t) = D_2(s/(\tau + 1)) = D_2(\tau^3 + \tau^2 + \tau) = D_2(\tau^3).$$

So, for one value of t, the line $\mathbf{V}(x_1 - tx_0)$ meets $\mathscr{F}$ in three points, and, for the other value, in one point. So $\mathscr{F}$ is a $\mathscr{G}_3{}^2$. For each s in $\mathscr{C}_0\backslash\{0\}$ and each category of b, there is a distinct $\mathscr{G}_3{}^2$; so $n = 2(q/2 - 1) = q - 2$.

(ii) $(q-1,3) = 3$, $c^2 = s^3$ for s in γ.

(a) If $D_4(s) = 0$, the four roots t of f in γ are τ/s, $(\tau + 1)/s$, $(\tau + \omega)/s$, $(\tau + \omega^2)/s$, where, as above, $\tau^4 + \tau = s$. For $t = (\tau + \lambda)/s$ with $\lambda^4 = \lambda$,

$$\begin{aligned} D_2(1/t) = D_2(s/(\tau + \lambda)) &= D_2((\tau^4 + \tau)/(\tau + \lambda)) \\ &= D_2(\tau^3 + \lambda\tau^2 + \lambda^2\tau + \lambda^3 + 1) \\ &= D_2(\tau^3 + \lambda^3 + 1), \end{aligned}$$

since $D_2(\lambda^2\tau) = D_2((\lambda^2\tau)^2) = D_2(\lambda\tau^2)$. However, $\lambda^3 = 0$ or 1, and $D_2(1) = D_2(0) = 0$. So $D_2(\tau^3 + \lambda^3 + 1) = D_2(\tau^3)$. Therefore all four lines $\mathbf{V}(x_1 - tx_0)$ meet $\mathscr{F}$ in $\mathbf{U}_2$ and two other points, or all four lines meet $\mathscr{F}$ in $\mathbf{U}_2$ only. Hence, for one value of b, the cubic $\mathscr{F}$ is a $\mathscr{G}_9{}^4$, and, for the other value, a $\mathscr{G}_1{}^4$.

(b) If $D_4(s) \neq 0$ and so $D_4(s)^3 = 1$, the polynomial f has no roots in γ, and $\mathbf{U}_2$ is the only inflexion of $\mathscr{F}$. So $\mathscr{F}$ is a $\mathscr{G}_1{}^0$.

$D_4(s) = 0$ for $q/4$ elements of γ and so for $q/4 - 1$ elements of γ_0. As $\omega^4 = \omega$, so $D_4(s) = 0$ implies $D_4(\omega s) = D_4(\omega^2 s) = 0$. So there are $(q/4 - 1)/3 = (q - 4)/12$ non-zero cubes c for which $D_4(s) = 0$. For such a cube c, one category of b gives a $\mathscr{G}_9{}^4$ and the other a $\mathscr{G}_1{}^4$. Hence, for both $\mathscr{G}_9{}^4$ and $\mathscr{G}_1{}^4$, we have $n = (q - 4)/12$. For each of the remaining non-zero cubes c, of which there are $(q - 1)/3 - (q - 4)/12 = q/4$, there are two types of b corresponding to its category. Hence, for $\mathscr{F} = \mathscr{G}_1{}^0$, we have $n = 2(q/4) = q/2$.

(iii) $(q-1,3) = 3$, c a non-cube in γ.

The polynomial f has one root t in γ. So, for one value of b, we have $D_2(1/t + b) = 0$, and, for the other value of b, we have $D_2(1/t + b) = 1$; then $\mathscr{F}$ is accordingly a $\mathscr{G}_3{}^1$ or a $\mathscr{G}_1{}^1$. As there are $2(q - 1)/3$ non-cubes c, for both $\mathscr{G}_3{}^1$ and $\mathscr{G}_1{}^1$ we have $n = 2(q - 1)/3$.□

Note: The values for n accord with those for cubics with three and nine inflexions obtained in theorems 11.5.4 and 11.6.2. In fact, theorem 11.7.3 could have been established for q even and the earlier results used to obtain all n. The more direct method was favoured here at the expense of some brevity.

(b) It remains to consider, for characteristic two, equianharmonic cubics $\mathscr{F}' = \mathbf{V}(F')$ where

$$F' = x_2{}^2x_1 + x_2x_1{}^2 + ex_0{}^3 + cx_0x_1{}^2 + dx_1{}^3;$$

here $e = 1$ when $(q-1,3) = 1$, and $e = 1, \alpha$, or α^2 when $(q-1,3) = 3$

and d is defined up to its category; that is, d is 0 or a particular element of $\mathscr{C}_1$.

$$F'^{(2)} = ey_0x_0{}^2 + (cy_0 + dy_1 + y_2)x_1{}^2 + y_1x_2{}^2,$$

$$F'^{(1)} = (ey_0{}^2 + cy_1{}^2)x_0 + (dy_1{}^2 + y_2{}^2)x_1 + y_1{}^2x_2.$$

11.7.21. LEMMA. *The inflexional lines through* $\mathbf{U}_2$ *of* $\mathscr{F}'$ *are* $\mathbf{V}(e^2x_0{}^4 + ex_0x_1{}^3 + c^2x_1{}^4)$.□

11.7.22. THEOREM. *Over* $\gamma = GF(q)$, $q = 2^h$, *the equianharmonic cubics are given in Table* 11.16, *where* t *is a root of* $e^2x^4 + ex + c^2$ *and* α *is a primitive root of* γ.

TABLE 11.16

$(q-1,3)$	e	$D_2(c)$	$D_4(c)^3$	$D_2(et^3+d)$	$\mathscr{F}'$
1	1	0			$\mathscr{E}_3^2$
1	1	1			$\mathscr{E}_1^0$
3	1		0	0	$\mathscr{E}_9^4$
3	1		0	1	$\mathscr{E}_1^4$
3	1		1		$\mathscr{E}_1^0$
3	α,α^2			0	$\mathscr{E}_3^1$
3	α,α^2			1	$\mathscr{E}_1^1$

Proof. (i) When $(q-1,3) = 1$, $x^4 + x + c^2$ has two or zero roots in γ as $D_2(c) = 0$ or 1.

(ii) When $(q-1,3) = 3$, $x^4 + x + c^2$ has four or zero roots in γ as $D_4(c) = 0$ or 1.

(iii) When $(q-1,3) = 3$ and $e = \alpha$ or α^2, then $e^2x^4 + ex + c^2$ has exactly one root in γ.

$\mathbf{V}(x_0 - tx_1)$ meets $\mathscr{F}'$ in $\mathbf{V}(x_2{}^2 + x_1x_2 + (et^3 + ct + d)x_1{}^2)$. If t is a root of $e^2x^4 + ex + c^2$, then $c = et^2 + \sqrt{(et)}$ and $D_2(ct) = D_2(et^3 + t\sqrt{(et)}) = D_2((t\sqrt{(et)})^2 + t\sqrt{(et)}) = 0$. So

$D_2(et^3 + ct + d) = D_2(et^3 + d)$. Therefore an inflexional line $\mathbf{V}(x_0 - tx_1)$ meets $\mathscr{F}'$ in three inflexions or one as $D_2(et^3 + d) = 0$ or 1.

(i) $(q-1,3) = 1$.

(a) $D_2(c) = 1$. So $x^4 + x + c^2$ has no roots in γ. Hence $\mathscr{F}'$ is an $\mathscr{E}_1{}^0$.

(b) $D_2(c) = 0$. Let $t = \tau, \tau + 1$ be the roots in γ of $x^4 + x + c^2$. Now, $D_2((\tau + 1)^3 + d) = D_2(\tau^3 + \tau^2 + \tau + 1 + d) = D_2(\tau^3 + d) + D_2(1) = D_2(\tau^3 + d) + 1$. So for the two values of t, one of the lines $\mathbf{V}(x_0 - tx_1)$ meets ' in three inflexions. Hence $\mathscr{F}'$ is an $\mathscr{E}_3{}^2$.

(ii) $(q-1,3) = 3$.

(a) $e = 1$. If $D_4(c) \neq 0$, $x^4 + x + c^2$ has no roots in γ and ' is $\mathscr{E}_1{}^0$. If $D_4(c) = 0$, then $t = \tau, \tau + 1, \tau + \omega, \tau + \omega^2$ are the roots of $x^4 + x + c^2$ in γ. For $\lambda^4 = \lambda$,

$$\begin{aligned} D_2((\tau + \lambda)^3 + d) &= D_2(\tau^3 + \lambda\tau^2 + \lambda^2\tau + \lambda^3 + d) \\ &= D_2(\tau^3 + d) + D_2(\lambda^4\tau^2 + \lambda^2\tau) + D_2(\lambda^3) \\ &= D_2(\tau^3 + d) + D_2(\lambda^3) = D_2(\tau^3 + d). \end{aligned}$$

So all four inflexional lines $\mathbf{V}(x_0 - tx_1)$ meet $\mathscr{F}'$ in the same number of points. So $\mathscr{F}'$ is an $\mathscr{E}_9{}^4$ or $\mathscr{E}_1{}^4$ as $D_2(\tau^3 + d) = 0$ or 1.

(b) $e = \alpha, \alpha^2$. As $e^2x^4 + ex + c^2$ has one root τ in γ for each c, so $\mathscr{F}'$ is an $\mathscr{E}_3{}^1$ or $\mathscr{E}_1{}^1$ as $D_2(e\tau^3 + d) = 0$ or 1.

11.7.23. THEOREM. *The number n of projectively distinct cubics $\mathscr{F}'$ is as follows, where $q = 2^h$:*

h	odd	odd	even	even	even	even	even
$\mathscr{F}'$	$\mathscr{E}_3{}^2$	$\mathscr{E}_1{}^0$	$\mathscr{E}_9{}^4$	$\mathscr{E}_1{}^4$	$\mathscr{E}_1{}^0$	$\mathscr{E}_3{}^1$	$\mathscr{E}_1{}^1$
n	1	2	1	1	1	2	2

Proof. Let $\tilde{\mathscr{F}}' = \mathbf{V}(\tilde{F}')$ where $\tilde{F}' = x_2{}^2x_1 + x_1x_2{}^2 + ex_0{}^3 +$

$+ \tilde{c}x_0x_1^2 + \tilde{d}x_1^3$. If $\mathscr{F}'\mathfrak{T} = \tilde{\mathscr{F}}'$. then since $\mathbf{U}_2$ and $\mathbf{u}_1$ are fixed, $\mathfrak{T}$ is given by

$$x_0 = a_0x_0' + a_{10}x_1', \; x_1 = a_1x_1', \; x_2 = a_{02}x_0' + a_{12}x_1' + x_2' .$$

It follows immediately that $a_0^3 = a_1 = 1$. Then

$$\mathscr{F}'\mathfrak{T} = \mathbf{V}(x_2^2x_1 + x_2x_1^2 + ex_0^3 + (a_{02}^2 + ea_0^2a_{10})x_0^2x_1$$
$$+ (a_{02} + ea_0a_{10}^2 + ca_0)x_0x_1^2 + (a_{12}^2 + a_{12} +$$
$$+ ea_{10}^3 + ca_{10} + d)x_1^3.$$

Therefore, $\mathscr{F}\mathfrak{T} = \tilde{\mathscr{F}}'$ if and only if there exist $a_0, a_{10}, a_{02}, a_{12}$ with

$$a_0^3 = 1,$$
$$a_{02}^2 + ea_0^2a_{10} = 0,$$
$$a_{02} + a_0(ea_{10}^2 + c) = \tilde{c},$$
$$a_{12}^2 + a_{12} + ea_{10}^3 + ca_{10} + d = \tilde{d}. \tag{11.11}$$

These equations imply

$$e^2a_{10}^4 + ea_{10} + c^2 = \tilde{c}^2a_0. \tag{11.12}$$

If, for $a_0^3 = 1$, eqns (11.11) and (11.12) have solutions for a_{10} and a_{12}, then $\mathscr{F}'$ and $\tilde{\mathscr{F}}'$ are projectively equivalent.

(i) $(q-1,3) = 1$, $e = 1$, $a_0 = 1$.

By theorem 1.9.4, eqn (11.12) has two solutions τ, $\tau + 1$ for a_{10}. However,

$$D_2((\tau + 1)^3 + c(\tau + 1)) = D_2(\tau^3 + \tau^2 + \tau + 1 + c\tau + c)$$
$$= D_2(\tau^3 + c\tau) + D_2(\tau^2 + \tau) + D_2(1) + D_2(c)$$
$$= D_2(\tau^3 + c\tau) + D_2(c) + 1.$$

So, when $D_2(c) = 1$, there is a solution for a_{12} only if $D_2(\tau^3 + c\tau + d) = D_2(\tilde{d})$; but, when $D_2(c) = 0$, there is always a solution for a_{12}. Hence there are two $\mathscr{E}_1^{\,0}$ and one $\mathscr{E}_3^{\,2}$.

(ii) $(q-1,3) = 3$, $e = 1$.

There is a solution for a_{10} if $D_4(c) = a_0{}^2 D_4(\tilde{c})$. If $D_4(c) = D_4(\tilde{c}) = 0$, then, with $a_{01} = \tau$, there is a solution for a_{12} if $D_2(\tau^3 + c\tau + d) = D_2(\tilde{d})$. So there is one $\mathscr{E}_1^{\,4}$ and one $\mathscr{E}_9^{\,4}$. If $D_4(c) \neq 0$, then, for some a_0, we have $D_4(c) = a_0{}^2 D_4(\tilde{c})$. So there are four solutions $t = \tau, \tau + 1, \tau + \omega, \tau + \omega^2$ for a_{10}. With $\lambda^4 = \lambda$,

$$D_2((\tau + \lambda)^3 + c(\tau + \lambda)) = D_2(\tau^3 + c\tau) + D_2(c\lambda)$$
$$= D_2(\tau^3 + c\tau) + \lambda D_4(c) + \lambda^2 D_4(c)^2.$$

So, for the four values of t, $D_2(t^3 + ct)$ is twice zero and twice one. So there is always a solution for a_{12} and so one $\mathscr{E}_1^{\,0}$

(iii) $(q-1,3) = 3$, $e = \alpha, \alpha^2$.

There is always a solution τ for a_{10}. Hence there are two $\mathscr{E}_1^{\,1}$ and two $\mathscr{E}_3^{\,1}$.

It finally remains to find the projective groups of the curves $\mathscr{F}'$ with just one inflexion.

11.7.24. LEMMA. *The orders o of the projective groups of the equianharmonic cubics $\mathscr{F}'$ over $GF(q)$, $q = 2^h$, are as follows:*

h	*odd*	*even*	*even*	*even*
$\mathscr{F}$	$\mathscr{E}_1^{\,0}$	$\mathscr{E}_1^{\,0}$	$\mathscr{E}_1^{\,1}$	$\mathscr{E}_1^{\,4}$
o	4	4	6	24

Proof. If $\mathscr{F}'\mathfrak{T} = \mathscr{F}'$, then $\mathfrak{T}$ is given by $X = X'A$ with

$$A = \begin{bmatrix} a_0 & 0 & a_{02} \\ a_{10} & 1 & a_{12} \\ 0 & 0 & 1 \end{bmatrix},$$

where, from the previous theorem a_0, a_{10}, a_{02}, a_{12} are

$$a_0{}^3 = 1,$$

$$a_{02}{}^2 = ea_0{}^2 a_{10},$$

$$e^2 a_{10}{}^4 + ea_{10} = c^2(1 + a_0),$$

$$a_{12}{}^2 + a_{12} = ea_{10}{}^3 + ca_{10}.$$

(i) $(q-1,3) = 1$. $e = 1$, $a_0 = 1$, $a_{10}{}^4 = a_{10}$; so $a_{10} = 0$ or 1 and $a_{02} = a_{10}$. When $a_{10} = 0$, $a_{12} = 0$ or 1. From theorem 11.7.22, $\mathscr{F}'$ has one inflexion only when $D_2(c) = 1$. So when $a_{10} = 1$, $a_{12}{}^2 + a_{12} = 1 + c$; as $D_2(1 + c) = D_2(1) + D_2(c) = 1 + 1 = 0$, so a_{12} has two values λ, $\lambda + 1$. Hence

$$G(\mathscr{E}_1{}^0) = \left\{ \begin{bmatrix} 1 & 0 & 0 \\ 0 & 1 & 0 \\ 0 & 0 & 1 \end{bmatrix}, \begin{bmatrix} 1 & 0 & 0 \\ 0 & 1 & 1 \\ 0 & 0 & 1 \end{bmatrix}, \begin{bmatrix} 1 & 0 & 1 \\ 1 & 1 & \lambda \\ 0 & 0 & 1 \end{bmatrix}, \begin{bmatrix} 1 & 0 & 1 \\ 1 & 1 & \lambda+1 \\ 0 & 0 & 1 \end{bmatrix} \middle| \ \lambda(\lambda+1) = c+1 \right\} \cong \mathbf{Z}_4 .$$

(ii) $(q-1,3) = 3$.

(a) $e = 1$.

$a_0{}^3 = 1$; so $a_0 = 1, \omega, \omega^2$.

(1) $a_0 = 1$. So $a_{10}{}^4 = a_{10}$. When $a_{10} = 0$, $a_{02} = 0$ and $a_{12} = 0$ or 1.

When $a_{10}{}^3 = 1$, $a_{12}{}^2 + a_{12} = 1 + ca_{10}$. For $\lambda^4 = \lambda$, $D_2(\lambda c) = \lambda D_4(c) + \lambda^2 D_4(c)^2$. So, if $D_4(c) = 0$, then $D_2(c) = D_2(\omega c) = D_2(\omega^2 c) = 0$. So for each of the three values of a_{10}, there are two values of a_{12}. If $D_4(c) \neq 0$, then exactly one of $D_2(c)$, $D_2(\omega c)$, $D_2(\omega^2 c)$ is zero; see theorem 1.9.4, corollary 2. Also $D_2(1) = 0$. So for one value of a_{10}, there are two values of a_{12}. Hence

$$G(\mathscr{E}_1{}^4) \supset G_8 = \left\{ \begin{bmatrix} 1 & 0 & 0 \\ 0 & 1 & 0 \\ 0 & 0 & 1 \end{bmatrix}, \begin{bmatrix} 1 & 0 & 0 \\ 0 & 1 & 1 \\ 0 & 0 & 1 \end{bmatrix}, \begin{bmatrix} 1 & 0 & 1 \\ 1 & 1 & \lambda_1 \\ 0 & 0 & 1 \end{bmatrix}, \begin{bmatrix} 1 & 0 & 1 \\ 1 & 1 & \lambda_1+1 \\ 0 & 0 & 1 \end{bmatrix}, \right.$$

$$\left. \begin{bmatrix} 1 & 0 & 1 \\ 1 & 1 & \lambda_2 \\ 0 & 0 & 1 \end{bmatrix}, \begin{bmatrix} 1 & 0 & \omega^2 \\ \omega & 1 & \lambda_2 \\ 0 & 0 & 1 \end{bmatrix}, \begin{bmatrix} 1 & 0 & \omega \\ \omega^2 & 1 & \lambda_3 \\ 0 & 0 & 1 \end{bmatrix}, \begin{bmatrix} 1 & 0 & \omega \\ \omega^2 & 1 & \lambda_3+1 \\ 0 & 0 & 1 \end{bmatrix} \middle| \lambda_i(\lambda_i+1) = 1 + c\omega^{i-1} \right\},$$

which is isomorphic to the quarternion group;

$$G(\mathscr{E}_1{}^0) \supset G_4 = \left\{ \begin{bmatrix} 1 & 0 & 0 \\ 0 & 1 & 0 \\ 0 & 0 & 1 \end{bmatrix}, \begin{bmatrix} 1 & 0 & 0 \\ 0 & 1 & 1 \\ 0 & 0 & 1 \end{bmatrix}, \begin{bmatrix} 1 & 0 & \lambda^2 \\ \lambda & 1 & \mu \\ 0 & 0 & 1 \end{bmatrix}, \begin{bmatrix} 1 & 0 & \lambda^2 \\ \lambda & 1 & \mu+1 \\ 0 & 0 & 1 \end{bmatrix} \middle| \lambda^3 = 1,\right.$$

$$\left.\mu(\mu + 1) = 1 + c\lambda \right\} \cong \mathbf{Z}_4.$$

(2) $a_0 = \omega$. So $a_{10}{}^4 + a_{10} + c^2\omega^2 = 0$ $a_{12}{}^2 + a_{12} + a_{10}{}^3 + ca_{10} = 0$. When $D_4(c) \neq 0$, there are no solutions for a_{10}. When $D_4(c) = 0$, there are four solutions t for a_{10}; namely, $t = \omega^2\tau$, $\omega^2\tau + 1$, $\omega^2\tau + \omega$, $\omega^2\tau + \omega^2$, where τ is a root of $x^4 + x + c^2$. However, $D_2(t^3 + ct) = D_2(t^3 + t(\omega^2t^2 + \omega^2\surd t)) = D_2(\omega t^3 + \omega^2 t\surd t) = 0$. So, for each value of t, there are two solutions for a_{12}.

(3) $a_0 = \omega^2$. So $a_{10}{}^4 + a_{10} + c^2\omega = 0$, $a_{12}{}^2 + a_{12} + a_{10}{}^3 + ca_{10} = 0$. This gives the same result as (2). For $D_4(c) \neq 0$, there are no solutions for a_{10}. For $D_2(c) = 0$, there are four solutions for a_{10}; namely, $t = \omega\tau$, $\omega\tau + 1$, $\omega\tau + \omega$, $\omega\tau + \omega^2$, and for each of these there are two values for a_{12}. Hence

$$G(\mathscr{E}_1{}^0) = G_4 \cong \mathbf{Z}_4,$$

and $G(\mathscr{E}_1{}^4)$ is a group of order 24, consisting of the identity, one element of order two, six of order four, eight of order three, and eight of order six.

(b) $e = \alpha, \alpha^2$.

$a_0{}^3 = 1$; so $a_0 = 1, \omega, \omega^2$.

(1) $a_0 = 1$. So $ea_{10}{}^4 + a_{10} = 0$. Then $a_{10} = 0$ and $a_{12} = 0$ or 1.

(2) $a_0 = \omega$. So $ea_{10}{}^4 + ea_{10} + c^2\omega^2 = 0$, $a_{12}{}^2 + a_{12} + ea_{10}{}^3 + ca_{10} = 0$. Then $a_{10} = \omega^2\tau$, where τ is the root in γ of $e^2x^4 + ex + c^2$. So $ex_{10}{}^3 + ca_{10} = e\tau^3 + c\omega^2\tau$ and

$$\begin{aligned} D_2(e\tau^3 + c\omega^2\tau) &= D_2(e\tau^3 + \omega^2\tau(e\tau^2 + \surd(e\tau)) \\ &= D_2(\omega e\tau^3 + \omega^2\tau\surd(e\tau)) \\ &= 0. \end{aligned}$$

So there are two solutions for a_{12}, namely $\omega^2\tau\surd(e\tau)$ and $\omega^2\tau\surd(e\tau) + 1$.

(3) $a_0 = \omega^2$. As in (2), $a_{10} = \omega\tau$ and $a_{12} = \omega\tau\surd(e\tau)$ and $\omega\tau\surd(e\tau) + 1$.

$$G(\mathscr{E}_1^{\ 1}) = \left\{ \begin{bmatrix} 1&0&0\\0&1&0\\0&0&1 \end{bmatrix}, \begin{bmatrix} 1&0&0\\0&1&1\\0&0&1 \end{bmatrix}, \begin{bmatrix} \omega&0&\mu_1\\ \omega^2\tau&1&\tau\mu_1\\0&0&1 \end{bmatrix}, \begin{bmatrix} \omega&0&\mu_1\\ \omega^2\tau&1&\tau\mu_1+1\\0&0&1 \end{bmatrix}, \right.$$

$$\left. \begin{bmatrix} \omega^2&0&\mu_2\\ \omega\tau&1&\tau\mu_2\\0&0&1 \end{bmatrix} \begin{bmatrix} \omega^2&0&\mu_2\\ \omega\tau&1&\tau\mu_2+1\\0&0&1 \end{bmatrix} \middle| \ \mu_1 = \omega^2\surd(e\tau),\ \mu_2 = \omega\surd(e\tau) \right\} \cong \mathbf{Z}_6. \square$$

11.7.25. THEOREM. *In* $PG(2,q)$, $q = 2^h$, *the classification of cubics with exactly one inflexion is given in Table* 11.17, *where* n *is the number of projectively distinct curves of each type,* o *the order of the corresponding projective group, and* n_1 *the total number of projectively distinct curves:*

TABLE 11.17

	$\mathscr{G}_1^{\ 4}$		$\mathscr{E}_1^{\ 4}$		$\mathscr{G}_1^{\ 1}$		$\mathscr{E}_1^{\ 1}$		$\mathscr{G}_1^{\ 0}$		$\mathscr{E}_1^{\ 0}$		
	n	o	n	o	n	o	n	o	n	o	n	o	n_1
$q\equiv -1 \pmod 3$	0	–	0	–	0	–	0	–	q	2	2	4	$q+2$
$q\equiv 1 \pmod 3$	$(q-4)/12$	2	1	24	$2(q-1)/3$	2	2	6	$q/2$	2	1	4	$(5q+12)/4$

Proof. This theorem collects the results of the previous theorems and lemmas in case III.□

11.8. Non-singular cubics with no inflexions

In this section a summary of the results for cubics with no inflexions is given.

I. $q \equiv -1 \pmod 3$

11.8.1. LEMMA. *Over* $GF(q)$, $q \equiv -1 \pmod 3$, *a non-singular cubic* $\mathscr{F} = \mathbf{V}(F)$ *with no inflexions has canonical form*

$$F = x_2^3 - 3c(x_0^2 - dx_0x_1 + x_1^2)x_2 - (x_0^3 - x_0x_1^2 + dx_1^3),$$

where $x^3 - 3x + d$ *is irreducible.*

Proof. $\mathscr{F}$ may be given the form $\mathbf{V}(F)$ with

$$F = x_2^3 - f(x_0,x_1) + \mu x_2 h(x_0,x_1)$$

where f is irreducible over γ, and h is the Hessian of f. Then, for a suitable fixed d, take

$$f(x_0,x_1) = x_0^3 - 3x_0x_1^2 + dx_1^3;$$

so $$h(x_0,x_1) = -9(x_0^2 - dx_0x_1 + x_1^2).\square$$

11.8.2. LEMMA. (i) *The absolute invariant of* $\mathscr{F}$ *is*

$$e^3c^3(e^3c^3 - 8)^3/(e^6c^6 + 20e^3c^3 - 8)^2 \text{ where } e^3 = d^2 - 4.$$

(ii) $\mathscr{F}$ *is equianharmonic for* $c = 0, 2/e$; *harmonic for* $c = (-1 \pm \sqrt{3})/e$; *and an inflexional triangle for* $c = \infty, -1/e$.$\square$

11.8.3. THEOREM. *The classification of non-singular cubics with no inflexions in* $PG(2,q)$, $q \equiv -1 \pmod 3$, *is given in Table* 11.18, *where* n *is the number of projectively distinct curves of each type,* o *is the order of the corresponding group, and* n_0 *the total number of projectively distinct curves.*$\square$

TABLE 11.18

$q \equiv m \pmod{12}$	$\mathscr{G}_0^2$		$\mathscr{E}_0^2$		$\mathscr{H}_0^2$		
m	n	o	n	o	n	o	n_0
5	$q-3$	3	2	3	0	-	$q-1$
11	$q-5$	3	2	3	2	3	$q-1$
2,8	$q-2$	3	1	3	0	-	$q-1$

II. $q \equiv 1 \pmod 3$

From theorem 11.4.8, $\mathscr{F}$ is an $\mathscr{F}_0^4$ or an $\mathscr{F}_0^1$.

11.8.4. LEMMA. *Every* $\mathscr{F}_0^4$ *has canonical form* $\mathscr{F} = \mathbf{V}(F)$,

$$F = x_0^3 + \alpha x_1^3 + \alpha^2 x_2^3 - 3c x_0 x_1 x_2,$$

where α *is a primitive root of* $\gamma = GF(q)$.□

11.8.5. LEMMA. (i) *The absolute invariant of* $\mathscr{F}$ *is*

$$c^3(c^3 + 8\alpha^3)^3/(c^6 - 20\alpha^3 c^3 - 8\alpha^6)^2.$$

(ii) *With* $\lambda^3 = 1$, $\mathscr{F}$ *is*

(a) *equianharmonic for* $c = 0, -2\alpha\lambda$;

(b) *harmonic for* $c = (1 \pm \surd 3)\alpha\lambda$;

(c) *an inflexional triangle for* $c = \alpha\lambda$.□

When $c \neq 0$ and $\mathscr{F}$ is equianharmonic, write $\mathscr{F} = \mathscr{E}_0^4$; when $c = 0$ and $\mathscr{F}$ is equianharmonic, write $\mathscr{F} = \overline{\mathscr{E}}_0^4$. For q odd, the Hessian of $\mathscr{E}_0^4$ is a set of 3-complex conjugate lines, whereas the Hessian of $\overline{\mathscr{E}}_0^4$ consists of the triangle of reference.

11.8.6. LEMMA. *Every* $\mathscr{F}_0^1$ *has canonical form* $\mathscr{F} = \mathbf{V}(F)$ *where, with* α *a primitive root of* γ,

$$F = x_0 x_1^2 + x_0^2 x_2 + e x_1 x_2^2 - c(x_0^3 + e x_1^3 + e^2 x_2^3 - 3e x_0 x_1 x_2),$$

$e = \alpha$ or α^2.□

11.8.7. LEMMA. *$\mathscr{F}$ is equianharmonic for $c = 0$.*□

11.8.8. THEOREM. *The classification of non-singular cubics in $PG(2,q)$, $q \equiv 1 \pmod 3$, is given in Table 11.19, where n is the number of projectively distinct curves of each type, o is the order of the corresponding projective group and n_0 is the total number of projectively distinct curves.*□

TABLE 11.19

$q \equiv m \pmod{12}$	m	1	7	4
$\mathscr{G}_0^4$	n	$(q-13)/3$	$(q-7)/3$	$(q-4)/3$
	o	9	9	9
$\mathscr{E}_0^4$	n	1	1	0
	o	9	9	-
$\bar{\mathscr{E}}_0^4$	n	1	1	1
	o	27	27	27
$\mathscr{H}_0^4$	n	2	0	0
	o	9	-	-
$\mathscr{G}_0^1$	n	$2(q-1)/3$	$2(q-1)/3$	$2(q-1)/3$
	o	3	3	3
$\mathscr{E}_0^1$	n	2	2	2
	o	9	9	9
	n_0	$q+1$	$q+1$	$q+1$

III. $q \equiv 0 \pmod 3$

11.8.9. LEMMA. *Over $GF(q)$, $q \equiv 0 \pmod 3$, a non-singular cubic with no inflexions has no nuclei and has canonical form $\mathscr{F} = \mathbf{V}(F)$,*

$$F = x_0^3 + x_1^3 + cx_2^3 + dx_0^2x_2 + dx_0x_1^2 + d^2x_0x_2^2 + dx_1x_2^2,$$

where $c \neq 1$ and $x^3 + dx - 1$ is a fixed irreducible polynomial.□

11.8.10. THEOREM. *In $PG(2,q)$, $q = 3^h$, the classification of non-singular cubics $\mathcal{F}$ with no inflexions is as follows, where n is the number of distinct types and o the order of the projective group*

$\mathcal{F}$	n	o
$\mathcal{G}_0$	$q-1$	3 □

11.9. Numbers of points on cubic curves

If $\mathcal{F}$ is an absolutely irreducible cubic curve in $PG(2,q)$, let $|\mathcal{F}|$ denote the number of distinct points on $\mathcal{F}$. For $\mathcal{F}$ singular, it was shown that $|\mathcal{F}| = q$, $q + 1$ or $q + 2$. For $\mathcal{F}$ non-singular, from theorem 10.2.1, corollary 1,

$$(\sqrt{q} - 1)^2 \leqslant |\mathcal{F}| \leqslant (\sqrt{q} + 1)^2.$$

It is still a difficult question to discover what values $|\mathcal{F}|$ can actually take. Let $M_{\min}$ denote the least integer greater than or equal to $(\sqrt{q} - 1)^2$ and $M_{\max}$ the integral part of $(\sqrt{q} + 1)^2$ (Table 11.20).

TABLE 11.20

q	$M_{\min}$	$M_{\max}$
2	1	5
3	1	7
4	1	9
5	2	10
7	3	13
8	4	14
9	4	16
11	6	18
13	7	21
16	9	25

11.9.1. THEOREM. *For $2 \leqslant q \leqslant 7$, the number of points on an absolutely irreducible cubic $\mathcal{F}$ in $PG(2,q)$ can take all values* *between* M_{min} *and* M_{max} *inclusive; for* $q = 8$, *only the values* 7 *and* 11 *are excluded.*

Proof. The list of possible values of $|\mathcal{F}|$ is given in Table 11.21 for $2 \leq q \leq 7$. For $q = 8$, the details are omitted. □

TABLE 11.21

	$q=2$ $\lvert\mathcal{F}\rvert$	n	$q=3$ $\lvert\mathcal{F}\rvert$	n	$q=4$ $\lvert\mathcal{F}\rvert$	n	$q=5$ $\lvert\mathcal{F}\rvert$	n	$q=7$ $\lvert\mathcal{F}\rvert$	n
$\mathcal{N}0$	4	2	5	1	6	1	7	2	9	1
$\mathcal{N}1$	3	1	4	2	5	1	6	1	8	1
$\mathcal{N}2$	2	1	3	1	4	2	5	1	7	2
$\mathcal{G}_9$										
$\mathcal{E}_9$					9	1			9	1
$\mathcal{H}_9$										
$\mathcal{G}_3$			3,6	2	6	2	3,9	2	6,9,12	4
$\mathcal{E}_3$							6	1		
$\bar{\mathcal{E}}_3$	3	1			3	2	6	1	3,12	2
$\mathcal{H}_3$										
$\mathcal{G}1$			2,5	2						
$\mathcal{S}_1^0$			1,7	2						
$\mathcal{S}_1^1$			4	1						
$\mathcal{S}_1^3$			4	1						
$\mathcal{G}_1^4$										
$\mathcal{E}_1^4$					1	1			7	1
$\mathcal{H}_1^4$										
$\mathcal{G}_1^1$					4	2			4,7,10	4
$\mathcal{E}_1^1$					7	2			4,13	2
$\mathcal{G}_1^0$	2,4	2			2,8	2	4,5,7,8	4	5,11	2
$\mathcal{E}_1^0$	1,5	2			5	1				

TABLE 11.21 (contd)

	$q=2$		$q=3$		$q=4$		$q=5$		$q=7$	
	$\|\mathscr{F}\|$	n	$\|\mathscr{F}\|$	n	$\|\mathscr{F}\|$	n	$\|\mathscr{F}\|$	n	$\|\mathscr{F}\|$	n
$\mathscr{H}_1^{\;0}$							2,4,8,10	4	8	2
$\mathscr{G}_0$			3,6	2						
$\mathscr{G}_0^{\;2}$							3,9	2		
$\mathscr{E}_0^{\;2}$	3	1					6	2		
$\mathscr{H}_0^{\;2}$										
$\mathscr{G}_0^{\;4}$										
$\mathscr{E}_0^{\;4}$									9	1
$\bar{\mathscr{E}}_0^{\;4}$					9	1			9	1
$\mathscr{H}_0^{\;4}$										
$\mathscr{G}_0^{\;1}$					6	2			6,9,12	4
$\mathscr{E}_0^{\;1}$					3	2			3,12	2
Σn		10		14		22		20		30

11.10. Classification of non-singular cubics

The material of §§11.5 - 11.9 is summarized in Tables 11.22 and 11.23. Table 11.22 lists the order of the projective group of each type of non-singular cubic curve and the number of projectively distinct cubics of that type. A check is provided by the final column T: this number was also calculated in theorem 11.3.11. Table 11.23 lists the canonical forms for each each type of cubic.

11.11. Notes and references

The results for $p > 3$ are based on Cicchese [1965], [1971], summaries of which are given in Cicchese [1962], [1970]. For $p = 3$, see De Groote [1973], [1974]. The case $p = 2$ is discussed by Boughon, Nathan, and Samuel [1955a], Campbell [1926],

TABLE 11.22

n = number of projectively distinct cubics of each type: $n_k = \Sigma n$, with the sum taken for those cubics with exactly k inflexions; $t_k = \Sigma(n/o)$ = total number of cubics with k inflexions$/p(3,q)$, where $p(3,q) = |PGL(3,q)| = q^3(q^2-1)(q^3-1)$;
$o = |G(\mathcal{F})|$, where $G(\mathcal{F}) < PGL(3,q)$ is the group of the cubic $\mathcal{F}$.
$N = \Sigma n_k$; $T = \Sigma t_k$; $\mathcal{G}$ = general; $\mathcal{E}$ = equianharmonic; $\mathcal{H}$ = harmonic; $\mathcal{S}$ = superharmonic.
$q \equiv m \pmod{12}$.

	Nine inflexions								Three inflexions									
	$\mathcal{G}_9$		$\mathcal{E}_9$		$\mathcal{H}_9$				$\mathcal{G}_3$		$\mathcal{E}_3$		$\bar{\mathcal{E}}_3$		$\mathcal{H}_3$			
m	n	o	n	o	n	o	n_9	t_9	n	o	n	o	n	o	n	o	n_3	t_3
3							0	0	$q-1$	6							$q-1$	$\frac{q-1}{6}$
9							0	0	$q-1$	6							$q-1$	$\frac{q-1}{6}$
2,8							0	0	$q-2$	6			1	6			$q-1$	$\frac{q-1}{6}$
4	$\frac{q-4}{12}$	18	1	216			$\frac{q+8}{12}$	$\frac{q-3}{216}$	$\frac{2q-2}{3}$	6			2	18			$\frac{2q+4}{3}$	$\frac{q}{9}$
1	$\frac{q-13}{12}$	18	1	54	1	36	$\frac{q+11}{12}$	$\frac{q-3}{216}$	$\frac{2q-2}{3}$	6			2	18			$\frac{2q+4}{3}$	$\frac{q}{9}$
7	$\frac{q-7}{12}$	18	1	54			$\frac{q+5}{12}$	$\frac{q-3}{216}$	$\frac{2q-2}{3}$	6			2	18			$\frac{2q+4}{3}$	$\frac{q}{9}$
5							0	0	$q-3$	6	1	6	1	6			$q-1$	$\frac{q-1}{6}$
11							0	0	$q-5$	6	1	6	1	6	2	6	$q-1$	$\frac{q-1}{6}$

	One inflexion									
	$\mathcal{G}_1$		$\mathcal{S}_1^0$		$\mathcal{S}_1^1$		$\mathcal{S}_1^3$			
m	n	o	n	o	n	o	n	o		
3	$q-1$	2	2	6	1	2	1	6		
9	$q-1$	2	2	6	2	4	2	12		
	$\mathcal{G}_1^4$		$\mathcal{E}_1^4$		$\mathcal{H}_1^4$		$\mathcal{G}_1^1$		$\mathcal{E}_1^1$	
2,8	n	o	n	o	n	o	n	o	n	o
4	$\frac{q-4}{12}$	2	1	24			$\frac{2q-2}{3}$	2	2	6
1	$\frac{q-13}{12}$	2	1	6	1	4	$\frac{2q-2}{3}$	2	2	6
7	$\frac{q-7}{12}$	2	1	6			$\frac{2q-2}{3}$	2	2	6
5										
11										

Lower suffix on $\mathscr{G}$, $\mathscr{E}$, $\mathscr{H}$, $\mathscr{S}$ is the number of inflexions; the inflexional tangents of $\bar{\mathscr{E}}_3$ are concurrent; on $\mathscr{S}_1^r$, r is the number of points in which the harmonic polar meets the curve; $\mathscr{F}_0^4$ occurs only for $m = 1,4,7$; $\mathscr{F}_0^2$ occurs only for $m = 2,5,8,11$; $\bar{\mathscr{E}}_0^4$ is given by a form which is a linear combination of cubes of linear forms.

								Zero inflexions																			
								$\mathscr{G}_0$																			
						n_1	t_1	n	o															n_0	t_0	N	T
						$q+3$	$\frac{q+1}{2}$	$q-1$	3															$q-1$	$\frac{q-1}{3}$	$3q+1$	q
						$q+5$	$\frac{q+1}{2}$	$q-1$	3															$q-1$	$\frac{q-1}{3}$	$3q+3$	q
$\mathscr{G}_1^0$		$\mathscr{E}_1^0$		$\mathscr{H}_1^0$				$\mathscr{G}_0^4, \mathscr{G}_0^2$		$\mathscr{E}_0^4, \mathscr{E}_0^2$		$\bar{\mathscr{E}}_0^4$		$\mathscr{H}_0^4, \mathscr{H}_0^2$		$\mathscr{G}_0^1$		$\mathscr{E}_0^1$									
n	o	n	o	n	o			n	o	n	o	n	o	n	o	n	o	n	o								
q	2	2	4			$q+2$	$\frac{q+1}{2}$	$q-2$	3	1	3													$q-1$	$\frac{q-1}{3}$	$3q$	
$\frac{q}{2}$	2	1	4			$\frac{5q+12}{4}$	$\frac{5q+1}{8}$	$\frac{q-4}{3}$	9			1	27			$\frac{2q-2}{3}$	3	2	9					$q+1$	$\frac{7q-3}{27}$	$3q+6$	q
$\frac{q-1}{2}$	2			2	4	$\frac{5q+15}{4}$	$\frac{5q+1}{8}$	$\frac{q-13}{3}$	9	1	9	1	27	2	9	$\frac{2q-2}{3}$	3	2	9					$q+1$	$\frac{7q-3}{27}$	$3q+7$	q
$\frac{q-3}{2}$	2			2	2	$\frac{5q+9}{4}$	$\frac{5q+1}{8}$	$\frac{q-7}{3}$	9	1	9	1	27			$\frac{2q-2}{3}$	3	2	9					$q+1$	$\frac{7q-3}{27}$	$3q+5$	q
$q-1$	2			4	4	$q+3$	$\frac{q+1}{2}$	$q-3$	3	2	3													$q-1$	$\frac{q-1}{3}$	$3q+1$	q
$q+1$	2					$q+1$	$\frac{q+1}{2}$	$q-5$	3	2	3			2	3									$q-1$	$\frac{q-1}{3}$	$3q-1$	q

TABLE 11.23

	$q \equiv m \pmod{12}$ m	Canonical form	Conditions
$\mathscr{G}_9^{\,4}$	4, 1, 7	$x_0^3 + x_1^3 + x_2^3 - 3cx_0x_1x_2$	$c^3 \neq 0, 1, -8, (1\pm\sqrt{3})^3$
$\mathscr{E}_9^{\,4}$	4, 1, 7	$x_0^3 + x_1^3 + x_2^3$	
$\mathscr{H}_9^{\,4}$	1	$x_0^3 + x_1^3 + x_2^3 - 3(1 + \sqrt{3})x_0x_1x_2$	
$\mathscr{G}_3$	3, 9	$x_0x_1x_2 + e(x_0 + x_1 + x_2)^3$	$e \neq 0$
$\mathscr{G}_3^{\,2}$	8, 5, 11	$x_0x_1x_2 + e(x_0 + x_1 + x_2)^3$	$e \neq 0, -1/27, -1/24, (-3\pm\sqrt{3})/36$
$\mathscr{G}_3^{\,1}$	4, 1, 7	$x_0x_1x_2 + e(x_0 + x_1 + x_2)^3$	$e \neq (t^3 - 1)/27$
$\mathscr{E}_3^{\,2}$	5, 11	$x_0x_1x_2 - (x_0 + x_1 + x_2)^3/24$	
$\overline{\mathscr{E}}_3^{\,2}$	2, 8, 5, 11	$x_0x_1(x_0 + x_1) + x_2^3$	
$\overline{\mathscr{E}}_3^{\,1}$	4, 1, 7	$x_0x_1(x_0 + x_1) + ex_2^3$	$e = \alpha, \alpha^2$; α *primitive*
$\mathscr{H}_3^{\,2}$	11	$x_0x_1x_2 + e(x_0 + x_1 + x_2)^3$	$e = (-3\pm\sqrt{3})/36$
$\mathscr{G}_1$	3, 9	$x_2^2x_1 + x_0^3 + bx_0x_1^2 + dx_1^3$	$d \neq 0$; $-b \in \mathscr{C}_1$
$\mathscr{I}_1^{\,0}$	3, 9	$x_2^2x_1 + x_0^3 + cx_0x_1^2 + dx_1^3$	$c \neq 0$; $x^3 + cx + d$ *irreducible*
$\mathscr{I}_1^{\,1}$	3, 9	$x_2^2x_1 + x_0^3 + cx_0x_1^2 + dx_1^3$	$c \neq 0$; $x^3 + cx + d$ *has* 1 *root in* γ
$\mathscr{I}_1^{\,3}$	3, 9	$x_2^2x_1 + x_0^3 + cx_0x_1^2 + dx_1^3$	$c \neq 0$; $x^3 + cx + d$ *has* 3 *roots in* γ
$\mathscr{G}_1^{\,4}$	1, 7	$x_2^2x_1 + x_0^3 + cx_0x_1^2 + dx_1^3$	$cd \neq 0$; $3x^4 + 6cx^2 + 12dx - c^2$ *has four roots in* γ; *for each* t, $-(t^3 + ct + d) \in \mathscr{C}_1$
$\mathscr{G}_1^{\,4}$	4	$x_2^2x_1 + x_0x_1x_2 + x_0^3 + bx_0^2x_1 + cx_0x_1^2$	$c \neq 0$; $c^2x^4 + x + 1$ *has four roots* t *in* γ; *for each* t, $1/t + b \in \mathscr{C}_1$
$\mathfrak{E}_1^{\,4}$	1, 7	$x_2^2x_1 + x_0^3 - dx_1^3$	$d \neq 0$; $d \in \mathscr{C}_1$; $4d$ *is a cube*
$\mathfrak{E}_1^{\,4}$	4	$x_2^2x_1 + x_2x_1^2 + x_0^3 + cx_0x_1^2 + dx_1^3$	$x^4 + x + c^2$ *has four roots* t *in* γ; *for each* t, $t^3 + d \in \mathscr{C}_1$

TABLE 11.23 (continued)

	$q \equiv m \pmod{12}$ m	Canonical form	Conditions
$\mathscr{H}_1^{\ 4}$	1	$x_2^2x_1 + x_0^3 + cx_0x_1^2$	$c \neq 0$; $3x^4 + 6cx^2 - c^2$ *has four roots* t *in* γ; *for each* t, $t^3 + ct \in \mathscr{C}_1$
$\mathscr{G}_1^{\ 1}$	1, 7	$x_2^2x_1 + x_0^3 + cx_0x_1^2 + dx_1^3$	$cd \neq 0$; $3x^4 + 6cx^2 + 12dx - c^2$ *has one root* t *in* γ; $-(t^3 + ct + d) \in \mathscr{C}_1$
$\mathscr{G}_1^{\ 1}$	4	$x_2^2x_1 + x_0x_1x_2 + x_0^3 + bx_0^2x_1 + cx_0x_1^2$	$c \neq 0$, $c^2x^4 + x + 1$ *has one root* t *in* γ; $1/t + b \in \mathscr{C}_1$
$\mathscr{E}_1^{\ 1}$	1, 7	$x_2^2x_1 + x_0^3 - dx_1^3$	$d \neq 0$; $d \in \mathscr{C}_1$; $4d$ *is a non-cube*
$\mathscr{E}_1^{\ 1}$	4	$x_2^2x_1 + x_2x_1^2 + ex_0^3 + cx_0x_1^2 + dx_1^3$	$e = \alpha, \alpha^2$; α *primitive*; $e^2x^4 + ex + c^2$ *has one root* t *in* γ; $et^3 + d \in \mathscr{C}_1$
$\mathscr{G}_1^{\ 0}$	1, 7, 5, 11	$x_2^2x_1 + x_0^3 + cx_0x_1^2 + dx_1^3$	$cd \neq 0$; $3x^4 + 6cx^2 + 12dx - c^2$ *has no roots in* γ
$\mathscr{G}_1^{\ 0}$	2, 8, 4	$x_2^2x_1 + x_0x_1x_2 + x_0^3 + bx_0^2x_1 + cx_0x_1^2$	$c \neq 0$, $c^2x^4 + x + 1$ *has no roots in* γ
$\mathscr{E}_1^{\ 0}$	2, 8, 4	$x_2^2x_1 + x_2x_1^2 + x_0^3 + cx_0x_1^2 + dx_1^3$	$x^4 + x + c^2$ *has no roots in* γ
$\mathscr{H}_1^{\ 0}$	1, 7, 5	$x_2^2x_1 + x_0^3 + cx_0x_1^2$	$c \neq 0$, $3x^4 + 6cx^2 - c^2$ *has no roots in* γ
$\mathscr{G}_0$	3, 9	$x_0^3 + x_1^3 + cx_2^3 + dx_0^2x_2 + dx_0x_1^2 + dx_0x_2^2 + dx_1x_2^2$	$c \neq 1$; $x^3 + dx - 1$ *irreducible*

TABLE 11.23 (continued)

	$q \equiv m \pmod{12}$ m	Canonical form	Conditions
$\mathscr{G}_0^2$	8, 5, 11	$x_2^3 - 3c(x_0^2 - dx_0x_1 + x_1^2)x_2 - (x_0^3 - 3x_0x_1^2 + dx_1^3)$	$x^3 - 3x + d$ *irreducible* $c \neq 0,\ 2/e,\ -1/e,\ (-1\pm\sqrt{3})/e$ *with* $e^3 = d^2 - 4$
$\mathscr{F}_0^2$	2, 8, 5, 11	$x_2^3 - 3c(x_0^2 - dx_0x_1 + x_1^2)x_2 - (x_0^3 - 3x_0x_1^2 + dx_1^3)$	$c = 0,\ 2/e$; $x^3 - 3x + d$ *irreducible with* $e^3 = d^2 - 4$
$\mathscr{H}_0^2$	11	$x_2^3 - 3c(x_0^2 - dx_0x_1 + x_1^2)x_2 - (x_0^3 - 3x_0x_1^2 + dx_1^3)$	$c = (-1\pm\sqrt{3})/e$ *with* $e^3 = d^2 - 4$, $x^3 - 3x + d$ *irreducible*
$\mathscr{G}_0^4$	4, 1, 7	$x_0^3 + \alpha x_1^3 + \alpha^2 x_2^3 - 3cx_0x_1x_2$	α *primitive*; $c^3 \neq 0,\ -8\alpha^3,\ \alpha^3,\ (1\pm\sqrt{3})^3\alpha^3$
$\mathscr{E}_0^4$	1, 7	$x_0^3 + \alpha x_1^3 + \alpha^2 x_2^3 + 6\alpha x_0x_1x_2$	α *primitive*
$\overline{\mathscr{E}}_0^4$	4, 1, 7	$x_0^3 + \alpha x_1^3 + \alpha^2 x_2^3$	α *primitive*
$\mathscr{H}_0^4$	1	$x_0^3 + \alpha x_1^3 + \alpha^2 x_2^3 - 3cx_0x_1x_2$	α *primitive*; $c = (1\pm\sqrt{3})\alpha$
$\mathscr{G}_0^1$	4, 1, 7	$x_0x_1^2 + x_0^2x_2 + ex_1x_2^2 - c(x_0^3 + ex_1^3 + e^2x_2^3 - 3ex_0x_1x_2)$	$c \neq 0,\ e = \alpha,\ \alpha^2$; α *primitive*
$\mathscr{E}_0^1$	4, 1, 7	$x_0x_1^2 + x_0^2x_2 + ex_1x_2^2$	$e = \alpha,\ \alpha^2$; α *primitive*

Dickson [1914a], [1915b], Segre [1956].

For a general treatment of cubics over $\bar{\gamma}$, see Bretagnolle-Nathan [1958]. See also Châtelet [1947].

§11.2. Boughon [1955] gives Taylor's theorem over a ring of characteristic $p > 0$. For polar curves, see Campbell [1928a]. For a generalization of the nucleus to arbitrary characteristic, see D'Orgeval [1975b]. Theorems 11.2.9, 11.2.10 are proved by Bretagnolle-Nathan [1958].

For classical accounts of the cubic, see Enriques [1915], Hilton [1920], Salmon [1879], Seidenberg [1968], R.J. Walker [1950].

For quartic curves in the finite case, see Campbell [1933c], Casse [1976], De Resmini [1970], [1971], Dickson [1915c], [1915d], [1915e].

12
PLANE $(k;n)$-ARCS

12.1. Diophantine equations

A $(k;n)$-arc $\mathcal{K}$ in $PG(2,q)$ is a set of k points such that some line of the plane meets $\mathcal{K}$ in n points but such that no line meets $\mathcal{K}$ in more than n points, where $n \geq 2$.

Although irreducible algebraic curves of order n in $PG(2,q)$ give examples of $(k;n')$-arcs with $n' \leq n$, very little is known about $(k;n)$-arcs in general. In fact, the maximum value that k can take for given q and n is known for comparatively few cases, as shall be seen below. Also, most of the known results are true for any projective plane of order q and not merely the Desarguesian one.

Throughout, Π will denote $PG(2,q)$. A line l of Π is an *i-secant* of a $(k;n)$-arc $\mathcal{K}$ if $|l \cap \mathcal{K}| = i$. Let τ_i denote the total number of i-secants to $\mathcal{K}$ in Π, $\rho_i = \rho_i(P)$ the number of i-secants to $\mathcal{K}$ through a point P of $\mathcal{K}$ and $\sigma_i = \sigma_i(Q)$ the number of i-secants to $\mathcal{K}$ through a point Q of $\Pi\backslash\mathcal{K}$. A $(k;n)$-arc is *complete* if there is no $(k+1;\ n)$-arc containing it.

12.1.1. LEMMA. *For a $(k;n)$-arc $\mathcal{K}$, the following equations hold:*

$$\sum_{i=0}^{n} \tau_i = q^2 + q + 1, \tag{12.1$_1$}$$

$$\sum_{i=1}^{n} i\tau_i = k(q+1), \tag{12.1$_2$}$$

$$\sum_{i=2}^{n} i(i-1)\tau_i/2 = k(k-1)/2; \tag{12.1$_3$}$$

$$\sum_{i=1}^{n} \rho_i = q + 1, \tag{12.2$_1$}$$

$$\sum_{i=2}^{n} (i-1)\rho_i = k - 1; \tag{12.2$_2$}$$

$$\sum_{i=1}^{n} \sigma_i = q + 1, \tag{12.3$_1$}$$

$$\sum_{i=1}^{n} i\sigma_i = k; \tag{12.3$_2$}$$

$$i\tau_i = \sum_P \rho_i; \tag{12.4$_i$}$$

$$(q + 1 - i)\tau_i = \sum_Q \sigma_i. \tag{12.5$_i$}$$

Proof. As for the comparable results for k-arcs in §9.1, the cardinality of appropriate sets is calculated in two different ways.□

If the only i are for which $\tau_i \neq 0$ are $m_1, m_2, \ldots, m_r, n$, then $\mathscr{K}$ is of *type* $(m_1, \ldots, m_r, n)$. Without ambiguity we can also say that $\mathscr{K}$ is of *type* m where $m = m_1$; that is, m is the smallest integer i for which $\tau_i \neq 0$.

12.1.2. LEMMA. *If $\mathscr{K}$ is a $(k;n)$-arc of type $m \neq 0$, then $k \leqslant (n - 1)q + m$.*

Proof. If P in $\mathscr{K}$ lies on an m-secant, then every other line through P contains at most n-1 points of $\mathscr{K}$ distinct from P. Hence the $q + 1$ lines through P contain at most $(n - 1)q + m$ points of $\mathscr{K}$.

Let us repeat the argument using eqns (12.2). From (12.2$_1$),

$$\sum_{i=m+1}^{n} \rho_i = q + 1 - \rho_m.$$

From (12.2$_2$),

$$\begin{aligned} k &= 1 + (m - 1)\rho_m + \sum_{i=m+1}^{n} (i - 1)\rho_i \\ &\leqslant 1 + (m - 1)\rho_m + (n - 1) \sum_{i=m+1}^{n} \rho_i \\ &= 1 + (m - 1)\rho_m + (n - 1)(q + 1 - \rho_m) \\ &= (n - 1)q - \rho_m(n - m) + n. \end{aligned}$$

Since $\rho_m \geq 1$,

$$k \leq (n - 1)q - (n - m) + n = (n - 1)q + m. \square$$

COROLLARY *If* $\mathscr{K}$ *is a* $(k;n)$*-arc, then* $k \leq (n - 1)q + n$. □

12.1.3. LEMMA. *If* $\mathscr{K}$ *is a* $(k;n)$*-arc of type* m*, and* $n \leq q$*, then* $m \leq n - 2$.

Proof. Let Q be a point of $\Pi \backslash \mathscr{K}$. Then, if $m > n - 2$, eqns (12.3) become

$$\sigma_{n-1} + \sigma_n = q + 1,$$

$$(n - 1)\sigma_{n-1} + n\sigma_n = k.$$

Hence $\sigma_n = k - (n - 1)(q + 1)$. As $\sigma_n \geq 1$ by definition,

$$k \geq (n - 1)(q + 1) + 1 = (n - 1)q + n.$$

If $m = n$, then $\tau_n = q^2 + q + 1$ and $n = q + 1$, a contradiction. So $m < n$ and $k > (n - 1)q + m$, contradicting the previous lemma. □

Before proceeding to any substantive results, we will give the other diophantine equations that can be used. There are two distinct approaches: the first uses equations (12.6)-(12.13), the second (12.14)-(12.16).

For the first approach, suppose that there are M distinct solutions

$$R_j = (\rho_{oj}, \ldots, \rho_{nj}), \; j = 1, 2, \ldots, M,$$

for eqns (12.2), and L distinct solutions

$$S_j = (\sigma_{oj}, \ldots, \sigma_{nj}), \; j = 1, 2, \ldots, L,$$

for eqns (12.3). So there are M different types of points on $\mathscr{K}$ and L different types in $\Pi \backslash \mathscr{K}$. Suppose there are r_j points with solution R_j and s_j points with solution S_j.

12.1.4. LEMMA.

$$\sum_{j=1}^{M} r_j \rho_{ij} = i\tau_i, \qquad (12.6_i)$$

$$\sum_{j=1}^{M} r_j = k, \qquad (12.7)$$

$$\sum_{j=1}^{L} s_j \sigma_{ij} = (q+1-i)\tau_i, \qquad (12.8_i)$$

$$\sum_{j=1}^{L} s_j = q^2 + q + 1 - k. \qquad (12.9)$$

Proof. These equations are just slight variants of (12.4) and (12.5).□

To complement these equations, we require equations connecting parameters associated with a particular line in the plane. Suppose a d-secant l of $\mathscr{K}$ contains u_j points of $\mathscr{K}$ with solution R_j and v_j points of $\Pi\backslash\mathscr{K}$ with solution S_j.

12.1.5. LEMMA.

$$\sum_{j=1}^{M} u_j = d, \qquad (12.10)$$

$$\sum_{j=1}^{L} v_j = q + 1 - d, \qquad (12.11)$$

$$\sum_{j=1}^{M} (\rho_{dj} - 1)u_j + \sum_{j=1}^{L} (\sigma_{dj} - 1)v_j = \tau_d - 1, \qquad (12.12)$$

$$\sum_{j=1}^{M} \rho_{ij}u_j + \sum_{j=1}^{L} \sigma_{ij}v_j = \tau_i \text{ for } i \neq d. \qquad (12.13_i)$$

Proof. Equations (12.10), (12.11), (12.12), (12.13_i) respectively count the sets $l \cap \mathscr{K}$, $l \cap (\Pi\backslash\mathscr{K})$, {$d$-secants other than l}, {i-secants, $i \neq d$}.□

For the second approach let ν_{id} and ν'_{id} be respectively the numbers of points of Π and $\Pi\backslash\mathscr{K}$ through which the number of d-secants is i exactly. For a particular d-secant l, let μ_{ij} be the number of points of l through which the number of j-secants is i exactly. Also, let θ_j be the maximum number of j-secants through a point of Π; for convenience, write $\theta_0 = \theta$.

12.1.6. LEMMA.

$$\sum_{i=0}^{\theta_d} \nu_{id} = q^2 + q + 1 , \qquad (12.14_1)$$

$$\sum_i i\nu_{id} = (q + 1)\tau_d , \qquad (12.14_2)$$

$$\sum_i i(i - 1)\nu_{id} = \tau_d(\tau_d - 1).\square \qquad (12.14_3)$$

COROLLARY. *If $\mathcal{K}$ has 0-secants then $\nu_{00} \geqslant k$ and $\nu'_{00} = \nu_{00} - k$. Then (12.14_1) becomes, for $d = 0$,*

$$\nu'_{00} + \nu_{10} + \ldots + \nu_{\theta 0} = q^2 + q + 1 - k.\square \qquad (12.14_4)$$

12.1.7. LEMMA. *Let μ_{ij} be constants for a d-secant l of the $(k;n)$-arc $\mathcal{K}$.*

(i) *For $j \neq d$,*

$$\sum_{i=0}^{\theta_j} \mu_{ij} = q + 1 , \qquad (12.15_1)$$

$$\sum_{i=1}^{\theta_j} i\mu_{ij} = \tau_j , \qquad (12.15_2)$$

$$\sum_l \mu_{ij} = (q + 1 - i)\nu_{ij} ; \qquad (12.15_3)$$

(ii) *For $j = d$,*

$$\sum_{i=0}^{\theta_d} \mu_{id} = q + 1, \qquad (12.16_1)$$

$$\sum_{i=0}^{\theta_d} (i - 1)\mu_{id} = \tau_d - 1 , \qquad (12.16_2)$$

$$\sum_l \mu_{id} = i\nu_{id} ; \qquad (12.16_3)$$

the summation in (12.15_3) and (12.16_3) is over all d-secants l.$\square$

12.2. Maximal arcs

For a $(k;n)$-arc $\mathcal{K}$, the corollary to lemma 12.1.2 tells us that $k \leqslant (n - 1)q + n$; if equality is achieved, $\mathcal{K}$ is a *maximal arc*.

12.2.1. THEOREM. *If $\mathcal{K}$ is a maximal $(k;n)$-arc in $\Pi = PG(2,q)$, then*

(i) *if $n = q + 1$, $\mathcal{K} = \Pi$;*
(ii) *if $n = q$, $\mathcal{K} = \Pi \backslash l$ where l is a line;*
(iii) *if $2 \leq n < q$, then $n|q$ and the dual of the complement of $\mathcal{K}$ forms a $(q(q + 1 - n)/n; q/n)$-arc, also maximal.*

Proof. Equations (12.2) are

$$\rho_1 + \rho_2 + \ldots + \rho_n = q + 1,$$

$$\rho_2 + 2\rho_3 + \ldots + (n - 1)\rho_n = (n - 1)(q + 1),$$

whence $\rho_n = q + 1$, $\rho_{n-1} = \ldots = \rho_1 = 0$. So $\tau_{n-1} = \ldots = \tau_1 = 0$, and eqns (12.1) give

$$\tau_n = (q + 1)[(n - 1)q + n]/n,$$

$$\tau_0 = q(q + 1 - n)/n.$$

Also $\sigma_{n-1} = \ldots = \sigma_1 = 0$, and eqns (12.3) give

$$\sigma_n = q + 1 - q/n,$$

$$\sigma_0 = q/n.$$

(i) When $n = q + 1$, $k = q^2 + q + 1$. So $\mathcal{K} = \Pi$.
(ii) When $n = q$, $k = q^2$ and $\tau_0 = 1$. So $\mathcal{K}$ is the complement of a line l in Π.
(iii) When $2 \leq n < q$, there exists Q in $\Pi \backslash \mathcal{K}$. So, as $\sigma_0 = q/n$, it follows that $n|q$. The dual of the complement of $\mathcal{K}$ is a $(\tau_0;\sigma_0)$-arc.□

COROLLARY. *A $(k;n)$-arc $\mathcal{K}$ is maximal if and only if every line in Π is a 0-secant or an n-secant.*

Proof. If $\mathcal{K}$ is maximal then the result was proved in the theorem. If $\tau_1 = \tau_2 = \ldots = \tau_{n-1} = 0$, then eqns (12.1_2) and (12.1_3) become

$$n\tau_n = k(q+1),$$

$$n(n-1)\tau_n = k(k-1).$$

So $n\tau_n/k = q + 1 = (k-1)/(n-1)$, whence $k = (n-1)q + n$.□

For $q = 2^h$, there are examples of maximal $(k;n)$-arcs for every n dividing q. For odd q, no example is known for $n < q$. In fact, it will be shown that, for $n = 3$, no maximal arc exists. However, firstly the examples for $q = 2^h$ will be given.

A maximal arc for $n = 2$ has $k = q + 2$ and so is an oval. For $n = q/2$, a maximal arc has $k = q(q-1)/2$. Therefore, since an oval exists, part (iii) of theorem 12.2.1 shows that a maximal $(q(q-1)/2;\ q/2)$-arc exists.

Now we describe the construction of a maximal $(k;n)$-arc in $PG(2,q)$ for q even; that is, $k = (n-1)q + n$ and $n|q$.

Let $x^2 + bx + 1$ be an irreducible quadratic over $\gamma = GF(q)$, q even, and let $\mathscr{L}$ be the pencil of quadrics $\mathscr{F}_\lambda = \mathbf{V}(F_\lambda)$, λ in γ^+, given by

$$F_\lambda = x_0{}^2 + bx_0x_1 + x_1{}^2 + \lambda x_2{}^2,\ \lambda \in \gamma,$$

$$F_\infty = x_2{}^2.$$

Then $\mathscr{F}_0$ is just the point $\mathbf{U}_2$, $\mathscr{F}_\infty$ is the line $\mathbf{u}_2$ and, for λ in γ_0, $\mathscr{F}_\lambda$ is a conic with nucleus $\mathbf{U}_2$. Any two members of $\mathscr{L}$ are disjoint. Hence the members of the pencil contain $1 + (q+1) + (q-1)(q+1) = q^2 + q + 1$ points, that is, all points of Π.

Since γ under addition is a 2-group, it contains subgroups of every order dividing $q = 2^h$. Let H be one such subgroup of order n.

12.2.2. THEOREM. *In $PG(2,2^h)$, let $\mathscr{K}$ be the set of points on all $\mathscr{F}_\lambda$ for λ in H. Then $\mathscr{K}$ is a maximal $(k;n)$-arc.*

Proof. A line l through $\mathbf{U}_2$ is tangent to each $\mathscr{F}_\lambda$ for λ in γ_0, and so meets each $\mathscr{F}_\lambda$ in exactly one point. Hence

$|l \cap \mathscr{K}| = n$.

A line l not through $\mathbf{U}_2$ is not a tangent to any conic $\mathscr{F}_\lambda$. So the $q + 1$ points of l consist of one point on $\mathbf{u}_2$ and two each on $q/2$ conics of $\mathscr{L}$. If $l = \mathbf{V}(a_0x_0 + a_1x_1 + x_2)$, then it meets $\mathscr{F}_\lambda$ in no points if and only if $(1 + \lambda a_0{}^2)x^2 + bx + (1 + \lambda a_1{}^2)$ is irreducible; that is

$$D_2(A + B\lambda + C\lambda^2) = 1, \tag{12.17}$$

where $A = 1/b^2$, $B = (a_0{}^2 + a_1{}^2)/b^2$, $C = a_0{}^2a_1{}^2/b^2$, §1.4. As $x^2 + bx + 1$ is irreducible, $D_2(A) = 1$. So (12.17) becomes

$$D_2(B\lambda + C\lambda^2) = 0.$$

However, if $\lambda = \lambda_1 + \lambda_2$, then

$$D_2(B\lambda_1 + C\lambda_1{}^2) + D_2(B\lambda_2 + C\lambda_2{}^2) = D_2(B\lambda + C\lambda^2).$$

Thus the λ in γ for which l does not meet $\mathscr{F}_\lambda$ form a subgroup G of index two in γ, and the λ in γ for which l meets $\mathscr{F}_\lambda$ comprise $G' = \gamma\backslash G$. Now, if H is a subgroup of G, then $G' \cap H = \emptyset$ and $|l \cap \mathscr{K}| = 0$. If H is not a subgroup of G, then $HG = \gamma$. Since $HG/G \cong H/(G \cap H)$, so $G \cap H$ is of index two in H and $|G \cap H| = |G' \cap H| = n/2$. Thus l meets $\mathscr{K}$ in n points.

It has therefore been shown that every line of Π meets $\mathscr{K}$ in 0 or n points, whence the result follows by the corollary to theorem 12.2.1.□

By comparison with §7.4, $\mathscr{P}(4)$, a projectivity $\mathfrak{T} = \mathbf{M}(T)$, where

$$T = \begin{bmatrix} 0 & 1 & 0 \\ 1 & b & 0 \\ 0 & 0 & 1 \end{bmatrix},$$

leaves each $\mathscr{F}_\lambda$ and so $\mathscr{K}$ fixed. $\mathfrak{T}$ has order τ, where τ is the order of a root α of $x^2 + bx + 1$ and, in fact τ divides $q + 1$. Thus, for $q = 2, 4, 16, 256$, and 65536, τ is necessarily $q + 1$. In any case, if b is chosen so that α is primitive, then $\tau = q + 1$.

Before considering the existence and properties of other maximal $(k;n)$-arcs, we will prove the following result concerning $(k;n)$-arcs with one point less than a maximal arc.

12.2.3. THEOREM. *If $\mathcal{K}$ is an $((n-1)q+n-1;\ n)$-arc, it is incomplete and can be uniquely completed to a maximal $((n-1)q+n;\ n)$-arc by adding the meet of all its $(n-1)$-secants.*

Proof. Equations (12.2) are

$$\rho_1 + \rho_2 + \ldots + \rho_n = q + 1 ,$$

$$\rho_2 + 2\rho_3 + \ldots + (n-1)\rho_n = (n-1)(q+1) - 1 ;$$

these have the unique solution

$$\rho_n = q,\ \rho_{n-1} = 1,\ \rho_{n-2} = \ldots = \rho_1 = 0.$$

Equations (12.1) then give

$$\tau_n = (n-1)(q+1)q/n,\ \tau_{n-1} = q+1,\ \tau_0 = q(q+1-n)/n,\ \tau_{n-2} = \ldots = \tau_1 = 0.$$

Equations (12.3) are now

$$\sigma_0 + \sigma_{n-1} + \sigma_n = q + 1,$$

$$(n-1)\sigma_{n-1} + n\sigma_n = (n-1)(q+1).$$

These have L possible solutions S_j:

$$\sigma_{0j} = j - 1,\ \sigma_{n-1,j} = q + 1 - (j-1)n,\ \sigma_{nj} = (j-1)(n-1),$$

where $j = 0,1,\ldots,L-1$ and $L = [(q+1)/n] + 1$; here $[r]$ is the integral part of r.

We now apply (12.11) and (12.12) to an $(n-1)$-secant l of $\mathcal{K}$. There are v_j points of $\Pi\backslash\mathcal{K}$ with solution σ_{0j}, $\sigma_{n-1,j}$, σ_{nj} on l. All points of $\mathcal{K}$ are of the same type: so $M = 1$ and $u_1 = n - 1$. So (12.11) and (12.12) become

$$\sum_{j=1}^{L} v_j = q - n + 2 \,, \tag{12.11'}$$

$$\sum_{j=1}^{L} [q - n(j - 1)]v_j = q \,. \tag{12.12'}$$

Now, suppose that $\sigma_{n-1} = 0$ for some point of $\Pi\backslash\mathcal{K}$. Then $\sigma_{n-1} = \sigma_{n-1,L} = q + 1 - (L - 1)n = 0$ giving $L = (q + 1)/n + 1$. But, from (12.12'), the non-negative coefficient of v_L is $q - n(L - 1)$; so $L \leqslant q/n + 1$, a contradiction. Thus $\sigma_{n-1} \neq 0$ for all points of $\Pi\backslash\mathcal{K}$. But $\rho_{n-1} = 1$ for all points of $\mathcal{K}$. Therefore the $\tau_{n-1} = q + 1$ lines $(n - 1)$-secant to $\mathcal{K}$ contain all points of the plane. Thus, by the dual of theorem 3.2.1 in the plane, the $(n - 1)$-secants of $\mathcal{K}$ are concurrent at a point Q. Hence $\mathcal{K} \cup \{Q\}$ is maximal.

To complete the calculation for l, Q is a point with $\sigma_0 = 0$, $\sigma_{n-1} = q + 1$, $\sigma_n = 0$ and so corresponds to a solution S_1. So $v_1 > 0$, and (12.12') gives that $v_1 = 1$, $v_2 = \ldots = v_{L-1} = 0$ with $L = q/n + 1$; hence, from (12.11'), $v_L = q - n + 1$. Thus, apart from Q, the remaining $q - n + 1$ points of $\Pi\backslash\mathcal{K}$ on l have $\sigma_0 = q/n$, $\sigma_{n-1} = 1$, $\sigma_n = q(n - 1)/n$.□

COROLLARY 1. *In $PG(2,q)$, a $(k;n)$-arc $\mathcal{K}$ for which n does not divide q satisfies $k \leqslant (n - 1)q + n - 2$.*□

COROLLARY 2. *In $PG(2,q)$, a $(k;3)$-arc $\mathcal{K}$ for which $q \neq 3^h$ satisfies $k \leqslant 2q + 1$.*□

An example of a $(2q + 1; 3)$-arc is the set of points on the Hermitian curve $\mathcal{U}_{2,4}$, which is a $(9;3)$-arc, §7.3. In fact, a $(k;n)$-arc $\mathcal{K}$ in $PG(2,q)$ is a *Hermitian arc* if q is a square, $n = \sqrt{q} + 1$ and, for each point P of $\mathcal{K}$, we have $\rho_1 = 1$, $\rho_n = q$. A Hermitian curve is a complete Hermitian arc, §7.3.

It is reasonable to make the following

CONJECTURE. *In $PG(2,q)$, a $(k;n)$-arc for which $q \not\equiv 0$ (mod n) and $n < q$ satisfies $k \leqslant (n - 1)q + 1$.*

An alternative conjecture would be to make the inequality $k \leqslant (n - 1)q + (n,q)$.

The following are examples in which the upper limit in the conjecture is achieved.

(i) $\mathscr{K} = \mathscr{U}_{2,q}$, $k = q\sqrt{q} + 1$, $n = \sqrt{q} + 1$, $\tau_1 = k$, $\tau_n = q(q + 2 - n)$.

(ii) $\mathscr{K}$ is the complement of a triangle, $k = (q - 1)^2$, $n = q - 1$, $\tau_0 = 3$, $\tau_{q-2} = (q - 1)^2$, $\tau_{q-1} = 3(q - 1)$.

(iii) For q odd, let $\mathscr{I}$ be the set of internal points of a conic $\mathscr{C}$, §8.2, and let P be a point $\mathscr{C}$. Each bisecant of $\mathscr{C}$ contains $(q - 1)/2$ points of $\mathscr{I}$, and $|\mathscr{I}| = q(q - 1)/2$.

(a) $\mathscr{K} = \mathscr{I} \cup \{P\}$, $k = q(q - 1)/2 + 1$, $n = (q + 1)/2$, $\tau_0 = q$, $\tau_1 = 1$, $\tau_{n-1} = q(q - 1)/2$, $\tau_n = q(q + 1)/2$;

(b) $\mathscr{K} = \mathscr{I} \cup \mathscr{C}$, $k = q(q + 1)/2 + 1$, $n = (q + 3)/2$, $\tau_1 = q + 1$, $\tau_{n-1} = q(q - 1)/2$, $\tau_n = q(q + 1)/2$.

From lemma 12.1.2, a $(k;3)$-arc in $PG(2,q)$ satisfies $k \leq 2q + 3$. If $q \neq 3^h$, then by theorem 12.2.3, corollary 2, we have $k \leq 2q + 1$. If $\mathscr{K}$ is a $(2q + 2;3)$-arc, it is contained in a $(2q + 3;3)$-arc, and $q = 3^h$. So, for $(k;3)$-arcs, one asks if $(2q + 3;3)$-arcs exist in $PG(2,3^h)$. Firstly we prove a general lemma for maximal $(k;n)$-arcs, similar to lemma 8.2.2 for k-arcs.

12.2.4. LEMMA. *Let $P_0P_1P_2$ be a triangle whose vertices belong to a $(k;n)$-arc $\mathscr{K}$ in $PG(2,q)$ with $k = (n - 1)q + n$. Then*

(i) *the product of the coordinates of the $3(n - 2)$ points of $\mathscr{K}$ on the sides of the triangle but excluding the vertices is 1;*

(ii) *the product of the coordinates of the $3(q + 1 - n)$ points of $\Pi\backslash\mathscr{K}$ on the sides of the triangle is -1.*

Proof. Let $P_rP_s \cap \mathscr{K} = \{P_r, P_s, P_{t1},\ldots,P_{t,n-2}\}$ and let $P_rP_s \cap (\Pi\backslash\mathscr{K}) = \{P_{t,n-1},\ldots,P_{t,q-1}\}$, where $\{r, s, t\} = \{0,1,2\}$. Now choose $P_0 = \mathbf{U}_0$, $P_1 = \mathbf{U}_1$, $P_2 = \mathbf{U}_2$. Then $P_0P_{0i} = \mathbf{V}(x_1 - d_{0i}x_2)$, $P_1P_{1i} = \mathbf{V}(x_2 - d_{1i}x_0)$, $P_2P_{2i} = \mathbf{V}(x_0 - d_{2i}x_1)$, $i \in \mathbf{N}_{q-1}$. So d_{0i}, d_{1i}, d_{2i} are the respective coordinates of P_{0i}, P_{1i}, P_{2i}, as in §2.8.

Let $\mathscr{K}' = \mathscr{K}\backslash\{\mathbf{U}_0 \cup \mathbf{U}_1 \cup \mathbf{U}_2\}$. For a point P of $\mathscr{K}'$. the lines PP_0, PP_1, PP_2 are respectively $\mathbf{V}(x_1 - e_0x_2)$, $\mathbf{V}(x_2 - e_1x_0)$, $\mathbf{V}(x_0 - e_2x_1)$ with $e_0e_1e_2 = 1$. Now, PP_r meets $\mathscr{K}'$ in $n - 2$ points if $PP_r \cap P_sP_t \in \mathscr{K}$ and in $n - 1$ points if $PP_r \cap P_sP_t \notin \mathscr{K}$. Thus, as P varies in $\mathscr{K}'$, e_r takes each value in $\{d_{r1},\ldots, d_{r,n-2}\}$ exactly $n - 2$ times and each value in

$\{d_{r,n-1},\ldots,d_{r,q-1}\}$ exactly $n-1$ times. Hence,

$$\prod_{P\in\mathcal{K}} e_0e_1e_2 = \Big(\prod_{i=1}^{n-2} d_{0i}d_{1i}d_{2i}\Big)^{n-2}\Big(\prod_{i=n-1}^{q-1} d_{0i}d_{1i}d_{2i}\Big)^{n-1}$$

So

$$1 = (-1)^{3(n-2)}\prod_{n-1}^{q-1} d_{0i}d_{1i}d_{2i}$$

and

$$\prod_{n-1}^{q-1} d_{0i}d_{1i}d_{2i} = (-1)^n = -1 ,$$

since, as $n|q$, n and q have the same parity. This proves (ii).

However,

$$\prod_{1}^{n-2} d_{0i}d_{1i}d_{2i} = \Big(\prod_{1}^{q-1} d_{0i}d_{1i}d_{2i}\Big)\Big/\Big(\prod_{n-1}^{q-1} d_{0i}d_{1i}d_{2i}\Big)$$

$$= (-1)^3/(-1) = 1,$$

which proves (i).□

COROLLARY 1. *There exists a curve* $\mathscr{C}_{q-n+1}$ *meeting the sides of* $P_0P_1P_2$ *in the* $3(q+1-n)$ *points not on* $\mathcal{K}$.

Proof. This follows directly from theorem $2.8.3_3$ and part (ii) of the lemma.□

COROLLARY 2. *If* $\mathcal{K}$ *is a* $(2q+3;3)$*-arc in* $PG(2,q)$, $q=3^h$, *containing the vertices of* $P_0P_1P_2$ *and hence the points* Q_0,Q_1,Q_2 *on the respective sides* P_1P_2, P_2P_0, P_0P_1, *then the triangles* $P_0P_1P_2$ *and* $Q_0Q_1Q_2$ *are in perspective.*

Proof. Let $P_i = \mathbf{U}_i$ for $i=0,1,2$. Then P_0Q_0, P_1Q_1, P_2Q_2 are respectively $\mathbf{V}(x_1-d_0x_2)$, $\mathbf{V}(x_2-d_1x_0)$, $\mathbf{V}(x_0-d_2x_1)$. But, by part (i) of the lemma, $d_0d_1d_2=1$, which is exactly the condition that P_0Q_0, P_1Q_1, and P_2Q_2 are concurrent.□

12.2.5. THEOREM. *In* $PG(2,q)$, $q=3^h$ *and* $h>1$, *there are no*

$(2q + 3;3)$-*arcs*.

Proof. The method is to show that a $(2q + 3;3)$-arc forms a 2-design, §2.2, with a parallelism and is in fact an affine space $AG(r,3)$, from which a numerical contradiction is obtained.

Since $\mathscr{K}$ is maximal, the $q + 1$ lines through a point of are all trisecants. A 2-$(2q + 3,3,1)$ design $\mathscr{S}$ is defined as follows. The points of $\mathscr{S}$ are the points of $\mathscr{K}$ and to each trisecant l of $\mathscr{K}$ there corresponds a block $l \cap \mathscr{K}$ of $\mathscr{S}$, while the incidence of $\mathscr{S}$ is that of the plane.

Now, it will be shown that the 2–(9,3,1) design $\mathscr{S}'$ determined by three non-collinear points P_0,P_1,P_2 is an affine plane $AG(2,3)$. Let $P_i = \mathbf{U}_i$, $i = 0,1,2$, and let $\mathbf{u}_0,\mathbf{u}_1,\mathbf{u}_2$ meet $\mathscr{K}$ in Q_0,Q_1,Q_2 respectively, where no Q_i is a P_j. Then, by corollary 2 to the previous lemma, P_0Q_0, P_1Q_1, and P_2Q_2 are concurrent at a point which may be chosen as the unit point $\mathbf{U}$. If, instead of $P_0P_1P_2$, we take the triangle $P_0Q_1Q_2$ and apply the corollary, then the second triangle is $R_0P_1P_2$, where $R_0 \in \mathscr{K}$, and the point of perspectivity is still $\mathbf{U}$; so $R_0 = P_0Q_0 \cap Q_1Q_2$ and R_1,R_2 are similarly obtained. Hence the nine points are $P_0 = \mathbf{U}_0$, $P_1 = \mathbf{U}_1$, $P_2 = \mathbf{U}_2$, $Q_0 = \mathbf{P}(0,1,1)$, $Q_1 = \mathbf{P}(1,0,1)$, $Q_2 = \mathbf{P}(1,1,0)$, $R_0 = \mathbf{P}(-1,1,1)$, $R_1 = \mathbf{P}(1,-1,1)$, $R_2 = \mathbf{P}(1,1,-1)$. So $\mathscr{S}'$ is the affine plane $AG(2,3)$ obtained as the complement of the line $\mathbf{u}$ in $PG(2,3)$. In Fig. 13, only the collinearities $P_0R_1R_2, P_1R_2R_0, P_2R_0R_1$ of $AG(2,3)$ are not indicated.

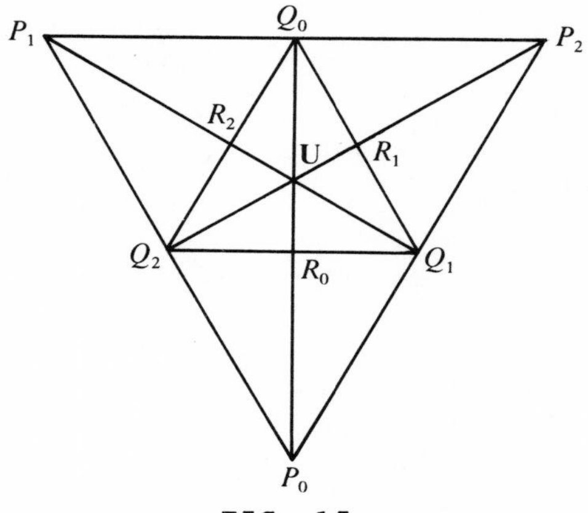

FIG. 13

Next a parallelism is defined: two blocks l, l' of $\mathscr{S}$ are *parallel* if they are non-intersecting blocks of a subsystem $\mathscr{S}'$ of $\mathscr{S}$ and we write $l \| l'$. This relation is reflexive and symmetric. To show that it is transitive, consider three blocks l_0, l_1, l_2 of $\mathscr{S}$ such that $l_0 \| l_1$ and $l_1 \| l_2$. For a point P of l_0, let $\mathscr{S}_1'$ be the subsystem of $\mathscr{S}$ defined by P and l_1, and $\mathscr{S}_2'$ the subsystem defined by P and l_2. We note that l_0 is the unique block of $\mathscr{S}_1'$ through P that is parallel to l_1. From the first part of the proof it can be assumed that $l_1 = \{P_0, P_1, Q_2\}$ and $l_2 = \{Q_0, Q_1, R_2\}$, where the points have coordinates as above. Let $P = \mathbf{P}(y_0, y_1, y_2)$. If $PP_0 \cap \mathscr{K} = \{P, P_0, S_0\}$ and $PP_1 \cap \mathscr{K} = \{P, P_1, S_1\}$, then the triangles PP_0P_1 and $Q_2S_1S_0$ are in perspective from $T = \mathbf{P}(t_0, t_1, t_2)$. As T is on PQ_2,

$$(y_0 - y_1)t_2 - y_2(t_0 - t_1) = 0.$$

Also $S_0 = \mathbf{P}(y_2t_0, y_1t_2, y_2t_2)$ and $S_1 = \mathbf{P}(y_0t_2, y_2t_1, y_2t_2)$. If $P_0S_1 \cap \mathscr{K} = \{P_0, S_1, R\}$, then the triangles PP_0S_1 and RP_1S_0 are in perspective. Hence $R = \mathbf{P}(-y_0t_2 - y_2t_0, y_2t_1, y_2t_2)$. At this point it may be noted that the block containing R and P contains none of P_0, P_1, or Q. So, by a previous remark, $R \in l_0 \backslash \{P\}$. If $PQ_1 \cap \mathscr{K} = \{P, Q_1, S\}$, then the triangles PP_1Q_1 and R_1SS_1 are in perspective. Hence S can be obtained and $S = \mathbf{P}(y_0t_2 + y_1t_2 - y_2t_1, y_1t_2, y_1t_2 - y_2t_1 + y_2t_2)$. However, the points Q_0, R, and S are collinear, since

$$\begin{vmatrix} 0 & 1 & 1 \\ -y_0t_2 - y_2t_0 & y_2t_1 & y_2t_2 \\ y_0t_2 + y_1t_2 - y_2t_1 & y_1t_2 & y_1t_2 - y_2t_1 + y_2t_2 \end{vmatrix}$$

$$= y_2(t_2 - t_1)[(y_0 - y_1)t_2 - y_2(t_0 - t_1)] = 0.$$

So $\{Q_0, R, S\}$ is a block of $\mathscr{S}'_2$; hence l_0 is the block of $\mathscr{S}_2'$ through P which is parallel to l_2, and so $l_0 \| l_2$. Hence parallelism is an equivalence relation on the blocks of $\mathscr{S}$ (Fig. 14).

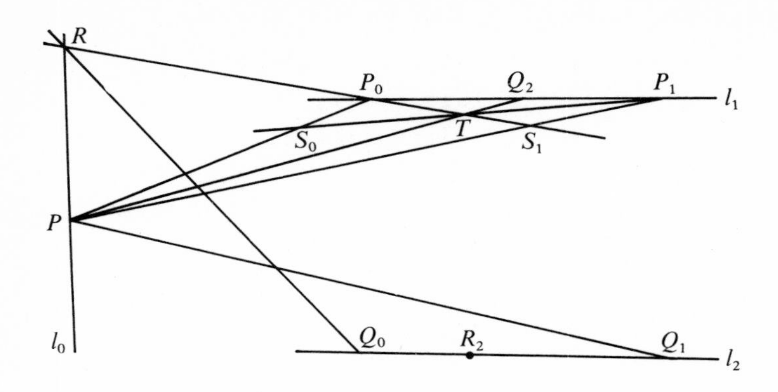

FIG. 14

It remains to show that $\mathscr{S}$ is an affine space.

(i) Any two points of $\mathscr{S}$ are contained in exactly one block.

(ii) For each point P and each block l, there is exactly one block l' such that $P \in l'$, and $l \| l'$. If $P \in l$, this is trivial; if $P \notin l$, then l' is the block of $\mathscr{S}'$ through P parallel to l, where $\mathscr{S}'$ is the subsystem of $\mathscr{S}$ defined by P and l.

(iii) Suppose that $l \| l'$, $P_1 \in l$, $P_2 \in l'$, $P_1P_2 \cap \mathscr{S} = \{P_1, P_2, P_3\}$ and $Q_2 \in l' \backslash \{P_2\}$; then l and Q_2P_3 have a point Q_1 of $\mathscr{S}$ in common. For, P_1, P_2, P_3, Q_2, l, and l' all belong to the subsystem $\mathscr{S}'$ defined by l and l' (Fig. 15).

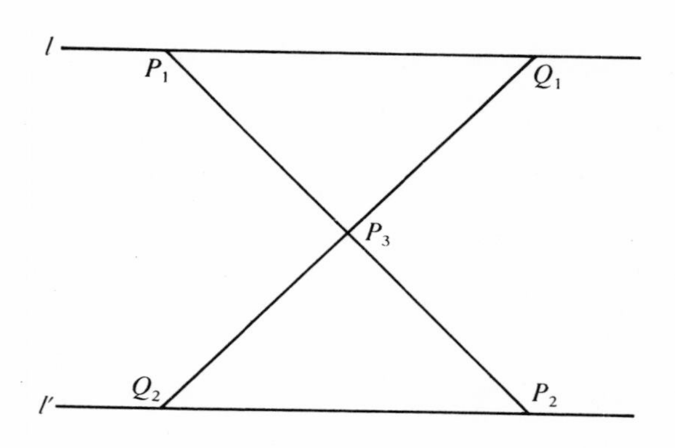

FIG. 15

(iv) Each block of $\mathscr{S}$ is incident with exactly three points.

(v) For a block l and a point P not on l, there are $q-3$ blocks l' for which $l \cap l' = \emptyset$ and $l \nparallel l'$.

Therefore, from §2.2, the points and blocks of $\mathscr{S}$ are the points and lines of an affine space $AG(r,3)$ with $r > 2$. Hence

$$2q + 3 = 2\cdot 3^h + 3 = 3^r,$$

which occurs only if $h = 1$ and $r = 2$, a contradiction.□

COROLLARY 1. *There do not exist any $(q(q-2)/3, q/3)$-arcs in $PG(2,q)$, $q = 3^h$ and $h > 1$.*□

COROLLARY 2. *Any $(k;3)$-arc in $PG(2,q)$, $q > 3$, satisfies $k \leqslant 2q + 1$.*□

It is feasible to make the following

CONJECTURE. *The only maximal $(k;n)$-arcs in $PG(2,q)$, q odd, are $PG(2,q)$ itself and the affine plane $AG(2,q) = PG(2,q)\backslash l$.*

12.3. Complete arcs of type (m,n)

In this section, a partial characterization of the $(k;n)$-arcs of the title will be given when $m \neq 0$. If $m = 0$, then a $(k;n)$-arc $\mathscr{K}$ of type $(0,n)$ has $k = (n-1)q + n$ and is maximal; these $(k;n)$-arcs were considered in §12.2.

12.3.1. LEMMA. *If $\mathscr{K}$ is a complete $(k;n)$-arc, then*

$$(q + 1 - n)\tau_n \geqslant q^2 + q + 1 - k$$

with equality if and only if $\sigma_n = 1$ for all Q in $\Pi\backslash\mathscr{K}$.

Proof. Since $\mathscr{K}$ is complete, $\sigma_n \geqslant 1$ for all Q in $\Pi\backslash\mathscr{K}$. But equation (12.5_n) states that

$$(q + 1 - n)\tau_n = \sum_Q \sigma_n \geqslant q^2 + q + 1 - k$$

with equality exactly when $\sigma_n = 1$, all Q.□

A $(k;n)$-arc $\mathscr{K}$ of type $m \neq 0$ is *maximal with respect to type* or *t-maximal* if $k = (n-1)q + m$.

12.3.2. LEMMA. *If $\mathscr{K}$ is a (k,n)-arc maximal with respect to type m, then it is complete if $\tau_m \geqslant 2$. If $\tau_m = 1$, $\mathscr{K}$ is either complete or can be completed by adding at most $n-m$ points on the unique m-secant.*

Proof. If $\mathscr{K}$ is not complete, it is contained in a $(k + 1;n)$-arc $\mathscr{K}'$ of type m', where $\mathscr{K}' = \mathscr{K} \cup \{Q\}$. So $k + 1 \leqslant \leqslant (n - 1)q + m'$ and, as $k = (n - 1)q + m$, it follows that $m' \geqslant m + 1$; hence $m' = m + 1$ since only one point has been added to $\mathscr{K}$ to form $\mathscr{K}'$. Therefore $\mathscr{K}'$ has no m-secant and every m-secant of $\mathscr{K}$ must contain Q. But each line through Q has at most n-1 points in common with $\mathscr{K}$. So, through Q, there is one m-secant and q lines $(n - 1)$-secant to $\mathscr{K}$.□

In fact, both complete and incomplete t-maximal $(k;n)$-arcs exist. Let P be a point of the conic $\mathscr{C}$ with tangent l at P. Then $(\Pi\backslash(\mathscr{C} \cup l)) \cup \{P\}$ is a $(q^2 - q + 1;q)$-arc maximal with respect to type one, which is complete if q is odd but incomplete if q is even.

12.3.3. THEOREM. *A complete $(k;n)$-arc $\mathscr{K}$ of type m satisfies*

$$k \geqslant (q^2 + q + 1)[m(q - n) + n]/[(q + 1)^2 - nq - m]$$

with equality if and only if $m \neq 0$, $\mathscr{K}$ is of type (m,n) and $\sigma_n = 1$ for each Q in $\Pi\backslash\mathscr{K}$.

Proof. Suppose first that $m \geqslant 1$. Since $i\tau_i \geqslant m\tau_i$ for $m \leqslant i \leqslant n - 1$,

$$\sum_{i=m}^{n-1} i\tau_i \geqslant m \sum_{i=m}^{n-1} \tau_i$$

with equality if and only if $\tau_j = 0$ for $j = m + 1,\ldots,n - 1$; that is, if and only if $\mathscr{K}$ is of type (m,n). From eqns (12.1),

$$\begin{aligned} k(q + 1) &= \sum_{m}^{n-1} i\tau_i + n\tau_n \geqslant m \sum_{m}^{n-1} \tau_i + n\tau_n \\ &= m(q^2 + q + 1 - \tau_n) + n\tau_n \\ &= \tau_n(n - m) + m(q^2 + q + 1); \end{aligned}$$

hence

$$\tau_n \leq [k(q+1) - m(q^2+q+1)]/(n-m).$$

So, from lemma 12.2.1,

$$(q^2+q+1-k)/(q+1-n) \leq [k(q+1) - m(q^2+q+1)]/(n-m),$$

whence the result follows.

Now, let $m = 0$. Then, from (12.1_2),

$$k(q+1) \geq n\tau_n$$

with equality if and only if $\tau_1 = \ldots = \tau_{n-1} = 0$; that is, $\mathcal{K}$ is of type $(0,n)$. Hence, from lemma 12.3.1,

$$(q^2+q+1-k)/(q+1-n) \leq k(q+1)/n,$$

whence

$$k \geq n(q^2+q+1)/[(q+1)^2 - nq] \tag{12.18}$$

with equality if and only if $\mathcal{K}$ is of type $(0,n)$ and $\sigma_n = 1$ for each Q in $\Pi\backslash\mathcal{K}$. But this would mean that $k = n$, which is contradicted by (12.18).□

In fact it will be shown below that equality holds in the theorem exactly when $\mathcal{K}$ is a subplane $PG(2,\sqrt{q})$ of Π or $\mathcal{K}$ is $\Pi\backslash\mathcal{H}$, where $\mathcal{H}$ is a Hermitian arc, namely a $(q\sqrt{q}+1;\sqrt{q}+1)$-arc of type $(1,\sqrt{q}+1)$.

Let $\mathcal{K}$ be a $(k;n)$-arc of type (m,n) with $m \neq 0$. Then eqns (12.1), (12.2), and (12.3) have the solution:

$$\begin{aligned}
\tau_m &= [n(q^2+q+1) - k(q+1)]/(n-m),\\
\tau_n &= [k(q+1) - m(q^2+q+1)]/(n-m),\\
\rho_m &= [n(q+1) - k - q]/(n-m),\\
\rho_n &= [k+q-m(q+1)]/(n-m),\\
\sigma_m &= [n(q+1) - k]/(n-m),\\
\sigma_n &= [k - m(q+1)]/(n-m);
\end{aligned} \tag{12.19}$$

$$k^2 - k[q(n + m - 1) + n + m] + mn(q^2 + q + 1) = 0. \quad (12.20)$$

12.3.4. LEMMA. *If $\mathcal{K}$ is a $(k;n)$-arc of type (m,n), $\mathcal{K}$ is complete and*

$$mq + n \leq k \leq (n - 1)q + m.$$

Proof. Since σ_n is constant, $\sigma_n \geq 1$ for all Q in $\Pi \backslash \mathcal{K}$ and so $\mathcal{K}$ is complete. Also, $\sigma_n \geq 1$ gives the lower estimate for $\mathcal{K}$, while the upper one was obtained in lemma 12.1.2.□

Any $(k;n)$-arc $\mathcal{K}$ of type m has been named t-maximal if $k = (n - 1)q + m$, and, naturally, this still applies when $\mathcal{K}$ is of type (m,n). Let us call $\mathcal{K}$ *t-minimal* when $k = mq + n$.

From any $(k;n)$-arc $\mathcal{K}$ of type (m,n), three new arcs can be formed.

(1) The complement of $\mathcal{K}$ in Π is a $(k';n)$-arc $\mathcal{K}'$ of type (m',n') where

$$k' = q^2 + q + 1, \; n' = q + 1 - m, \; m' = q + 1 - n;$$

write $\mathcal{K}' = \mathcal{K}\kappa$.

(2) The dual of the set of n-secants to $\mathcal{K}$ is a $(k_1;n_1)$-arc $\mathcal{K}_1$ of type (m_1,n_1), where

$$k_1 = \tau_n, \; n_1 = \rho_n, \; m_1 = \sigma_n \; ;$$

write $\mathcal{K}_1 = \mathcal{K}\partial_1$.

(3) The dual of the set of m-secants to $\mathcal{K}$ is a $(k_2;n_2)$-arc $\mathcal{K}_2$ of type (m_2,n_2), where

$$k_2 = \tau_m, \; n_2 = \sigma_m, \; m_2 = \rho_m \; ;$$

write $\mathcal{K}_2 = \mathcal{K}\partial_2$.

12.3.5. LEMMA. (i) *$\mathcal{K}'$ is t-maximal or t-minimal as $\mathcal{K}$ is t-minimal or t-maximal.*

(ii) *$\mathcal{K}_1$ and $\mathcal{K}_2$ are complements.*

(iii) *$\mathcal{K}_1$ is t-maximal if and only if $n = q$ and t-minimal if and only if $m = 1$.*

(iv) $\mathscr{K}_2$ *is t-minimal if and only if* $n = q$ *and t-maximal if and only if* $m = 1$.

(v) *The operators* κ, ∂_1, ∂_2, *and the identity* ι *satisfy*

$$\kappa^2 = \partial_1{}^2 = \partial_2{}^2 = \iota,\ \kappa\partial_1 = \partial_1\kappa = \partial_2,\ \kappa\partial_2 = \partial_2\kappa = \partial_1,$$

$$\partial_1\partial_2 = \partial_2\partial_1 = \kappa;$$

that is, they form a four-group isomorphic to $\mathbf{Z}_2 \times \mathbf{Z}_2$ (Fig. 16).□

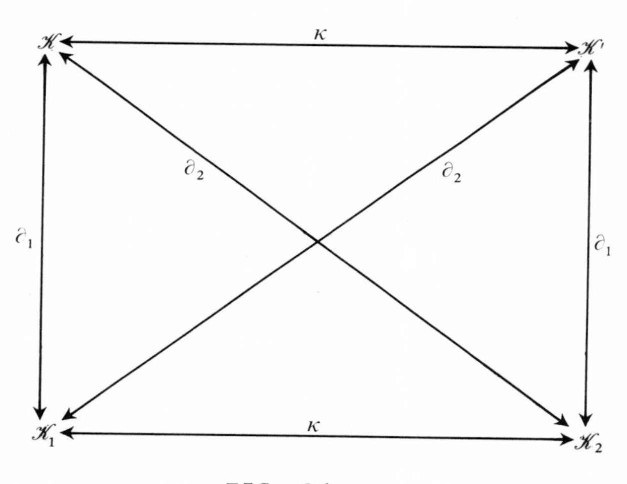

FIG. 16

Now, we give a classification of t-maximal and t-minimal $(k;n)$-arcs of (m,n).

12.3.6. THEOREM. *Let* $\mathscr{K}$ *be a* $(k;n)$*-arc of type* (m,n) *in* $PG(2,q)$ *with* $m \neq 0$.

(i) *If* $\mathscr{K}$ *is t-maximal, it is a Hermitian arc or the complement of a subplane* $PG(2,\sqrt{q})$.

(ii) *If* $\mathscr{K}$ *is t-minimal, it is a subplane* $PG(2,\sqrt{q})$ *or the complement of a Hermitian arc.*

Proof. It suffices to prove (i), as (ii) then follows from the previous lemma. So, let $\mathscr{K}$ be a $(k;n)$-arc of type

(m,n) with $k = (n - 1)q + m$. Then, from (12.19) and (12.20),

$$\rho_m = 1,\ \rho_n = q,\ \sigma_m = q/(n - m) + 1,\ \sigma_n = q - q/(n - m)\ ,$$

$$\tau_m = [q^2 - (m-1)q + n - m]/(n - m),\ \tau_n = q[(n - m - 1)q + n - 1]/(n - m),$$

$$m^2 - m(q + n) + n(n - 1) = 0.$$

Since $n-m$ divides q, write $q_1 = n-m$ and $q_2 = q/q_1$. Then

$$m^2 + m(q_1 - q - 1) + q_1(q_1 - 1) = 0. \tag{12.21}$$

So, if m_1 and m_2 are the solutions of (12.21),

$$m_1 + m_2 = q - q_1 + 1,\ m_1 m_2 = q_1(q_1 - 1).$$

By lemma 12.1.3, $q_1 \geqslant 2$. If $q_1 = 2$, then $m = 1$ or 2 and $q = 4$. If $q_1 > 2$, then m_1, say, is prime to p, and hence m_2 is a multiple of q_1. So put $m_2 = cq_1$, $cm_1 = q_1 - 1$. Thus, for $q_1 \geqslant 2$,

$$q_1 - 1 + c^2 q_1 = cm_1 + cm_2 = c(q - q_1 + 1)$$

and $c + 1 = q_1[c^2 - c(q_2 - 1) + 1] = q_1 d$ with $d \geqslant 1$. Therefore $c = q_1 d - 1$ and

$$m_1 = (q_1 - 1)/c = (q_1 - 1)/(q_1 d - 1).$$

Thus $d = 1$, $m_1 = 1$, $c = q_1 - 1$, $m_2 = q_1(q_1 - 1)$, and $q = q_1^2$.

There are now two possibilities. If $m = m_1 = 1$, then $n = \sqrt{q} + 1$ and $k = q\sqrt{q} + 1$. So $\mathcal{K}$ is a $(q\sqrt{q} + 1;\sqrt{q} + 1)$-arc of type $(1,\sqrt{q} + 1)$; that is, $\mathcal{K}$ is a Hermitian arc, an example of which is the Hermitian curve $\mathcal{U}_{2,q}$, §7.3.

If $m = m_2 = q - \sqrt{q}$, then $n = q$ and $k = q^2 - \sqrt{q}$. So $\mathcal{K}$ is a $(q^2 - \sqrt{q};q)$-arc of type $(q - \sqrt{q},q)$. Then its complement $\mathcal{K}'$ is a $(q + \sqrt{q} + 1;\sqrt{q} + 1)$-arc of type $(1,\sqrt{q} + 1)$. So the line joining any two points of $\mathcal{K}'$ is a $(\sqrt{q} + 1)$-secant of $\mathcal{K}'$. So any two $(\sqrt{q}+1)$-secants of $\mathscr{K}'$ meet at a point of $\mathscr{K}'$. So $\mathscr{K}'$ is a subplane $PG(2,\sqrt{q})$ whose lines are the $(\sqrt{q}+1)$-secants of $\mathscr{K}'$. Thus $\mathscr{K}$ is the complement of a subplane $PG(2,\sqrt{q})$.□

COROLLARY 1. *Let $\mathscr{K}$ be a $(k;n)$-arc of type (m,n) with $m \neq 0$ in $PG(2,q)$. Then,*

(i) *if $m = 1$, $\mathscr{K}$ is a Hermitian arc or a subplane $PG(2,\sqrt{q})$;*

(ii) *if $n = q$, $\mathscr{K}$ is the complement of one of these.*□

COROLLARY 2. *A $(k;n)$-arc $\mathscr{K}$ of type (m,n) in $\Pi = PG(2,q)$ such that $\sigma_n = 1$ for all Q in $\Pi\backslash\mathscr{K}$ is a subplane $PG(2,\sqrt{q})$ or the complement of a Hermitian arc.*□

COROLLARY 3. *A complete $(k;n)$-arc $\mathscr{K}$ of type m in $PG(2,q)$ satisfies*

$$k \geq (q^2+q+1)[m(q-n)+n]/[(q+1)^2-nq-m]$$

with equality if and only if $\mathscr{K}$ is a subplane $PG(2,\sqrt{q})$ or the complement of a Hermitian arc.

Proof. This is just a restatement of theorem 12.3.3 in the light of corollary 2.□

As regards the existence of $(k;n)$-arcs of type (m,n) with $m \geq 2$ and $n \leq q-1$, certain necessary conditions can be given.

12.3.7. LEMMA. *If $\mathscr{K}$ is a $(k;n)$-arc of type (m,n) then*

(i) $k^2 - k[q(n+m-1)+n+m] + mn(q^2+q+1) = 0$;

(ii) $\Delta = q^2[(n-m)^2 - 2(n+m)+1] + 2q[n(n-1) + m(m-1)] + (n-m)^2$ is a perfect square;

(iii) $n - m$ *divides* q;

(iv) $q \geq n(n-1)/m - (n-m)$.

Proof. Part (i) was previously obtained as (12.20). Then, Δ is the discriminant of the quadratic. If the roots of the quadratic are compared with the estimates for k of lemma 12.3.4, then (iv) is obtained.□

COROLLARY. *There is no $(k;4)$-arc of type $(2,4)$ and, for $q > 11$, no $(k;5)$-arc of type $(2,5)$.* □

In connection with these results, something can be said for a $(k;n)$-arc of type $(0,1,n)$.

12.3.8. LEMMA. *If $\mathcal{K}$ is a $(k;n)$-arc of type $(0,1,n)$ in $PG(2,q)$ with $n > 4$, either*

(i) $k < (3q - 1)/2$

or (ii) $k > [(2n - 3)q + 2n + 3]/2$.

Proof. From eqns (12.1),

$$\tau_0 = [k^2 - k(nq + n + 1) + n(q^2 + q + 1)]/n,$$
$$\tau_1 = k[q + 1 - (k - 1)/(n - 1)],$$
$$\tau_n = k(k - 1)/[n(n - 1)].$$

Since $\tau_0 \geq 0$,

$$k^2 - k(nq + n + 1) + n(q^2 + q + 1) \geq 0\,, \qquad (12.22)$$

and the discriminant Δ of this quadratic is

$$\Delta = n(n - 4)q^2 + 2n(n - 1)q + (n - 1)^2.$$

If $n > 4$, then $\Delta > [(n - 3)q + n + 2]^2$ and this inequality applied to the roots of the quadratic gives the result. □

12.3.9. THEOREM. *If $\mathcal{K}$ is a $(k;n)$-arc of type $(0,1,n)$ in $PG(2,q)$, then*

$$k < 1 + q\sqrt{q}.$$

Proof. Equations (12.2) give $\rho_n = (k - 1)/(n - 1)$. So, let us put $(k - 1)/(n - 1) = q - \alpha$ with $\alpha \geq 0$ and substitute for k in the inequality (12.22):

$$(n-1)^2(\alpha+1)(q-\alpha)-(n-1)\alpha q-q^2 \leqslant 0.$$

Let $\delta = \sqrt{[\alpha^2 + 4(\alpha+1)(q-\alpha)]}$. Then

$$\frac{q(\alpha-\delta)}{2(\alpha+1)(q-\alpha)} \leqslant n-1 \leqslant \frac{q(\alpha+\delta)}{2(\alpha+1)(q-\alpha)} .$$

The lower bound for $n-1$ is negative and so gives no information. From the upper bound,

$$k-1 \leqslant q(\alpha+\delta)/[2(\alpha+1)] .$$

If $\alpha = 0$, then $\delta = 2\sqrt{q}$ and $k \leqslant 1 + q\sqrt{q}$. However, if $k = 1 + q\sqrt{q}$, $n = 1 + \sqrt{q}$ and $\tau_0 = 0$; in fact, $\mathscr{K}$ is Hermitian and of type $(1,n)$.

Now let $\alpha > 0$. Then

$(\alpha+\delta)/[2(\alpha+1)] \geqslant \sqrt{q}$

$\Leftrightarrow \delta \geqslant 2(\alpha+1)\sqrt{q} - \alpha$

$\Rightarrow \alpha^2 + 4(\alpha+1)(q-\alpha) \geqslant 4(\alpha+1)^2 q - 4\alpha(\alpha+1)\sqrt{q} + \alpha^2$

$\Leftrightarrow q - \alpha \geqslant (\alpha+1)q - \alpha\sqrt{q}$

$\Leftrightarrow \alpha\sqrt{q} \geqslant \alpha(q+1)$

$\Rightarrow \sqrt{q} \geqslant q + 1$, a contradiction.

Hence $(\alpha+\delta)/[2(\alpha+1)] < \sqrt{q}$.□

It is opportune to mention here an example of an arc related to a subgeometry $\Pi' = PG(2,\sqrt{q})$ of $\Pi = PG(2,q)$, q square. Let l be a line of Π which is $(\sqrt{q}+1)$-secant to Π' and let $\mathscr{L}$ be a set of $\sqrt{q}$ points of $l\backslash\Pi'$. Then $\mathscr{K} = (\Pi'\backslash l) \cup \mathscr{L}$ is a complete $(q+\sqrt{q};\sqrt{q})$-arc which is of type $(0,1,2,\sqrt{q})$ when $q > 4$ and of type $(0,2)$ when $q = 4$; in the latter case $\mathscr{K}$ is just an oval.

To complete this section, something more will be proved about the connection between Hermitian arcs and Hermitian curves.

In $PG(2,q)$, q square, $\mathscr{K}$ is a Hermitian arc if it is a $(q\sqrt{q}+1;\sqrt{q}+1)$-arc of type $(1,\sqrt{q}+1)$. $\mathscr{K}$ is *reciprocal* if any four of its points P_1,P_2,P_3,P_4, no three of which are collinear, with respective 1-secants l_1, l_2, l_3, l_4, satisfy the condition that when $l_1 \cap l_2$ lies on P_3P_4 then $l_3 \cap l_4$ lies on P_1P_2.

12.3.9. LEMMA. *If $\mathcal{K}$ is a reciprocal Hermitian arc in $PG(2,q)$, then the $\sqrt{q}+1$ unisecants through a point Q of $\Pi\backslash\mathcal{K}$ meet $\mathcal{K}$ in $\sqrt{q}+1$ collinear points.*

Proof. For $q = 4$, a Hermitian arc is a Hermitian curve, theorem 11.1.1, corollary. Let $q \geq 9$ and suppose that through Q there are three unisecants l_1, l_2, l_2' meeting $\mathcal{K}$ at non-collinear points P_1, P_2, P_2'. Let l be one of the $q - \sqrt{q}$ lines through Q which are $(\sqrt{q}+1)$-secant to $\mathcal{K}$. Since $\sqrt{q}+1 \geq 4$, there are two points P_3, P_4 of l on $\mathcal{K}$ neither of which are on P_1P_2 or P_1P_2'. Then $P_1P_2P_3P_4$ and $P_1P_2'P_3P_4$ are both quadrangles. Let l_3 and l_4 be the unisecants to $\mathcal{K}$ at P_3 and P_4. Then, since $l_1 \cap l_2 = Q$ lies on $P_3P_4 = l$, we have that $l_3 \cap l_4$ lies on P_1P_2. Similarly $l_3 \cap l_4$ lies on P_1P_2'. So $l_3 \cap l_4 = P_1$, which is a contradiction as P_1 is on $\mathcal{K}$, and l_3 and l_4 are unisecants at P_3 and P_4. □

12.3.10. THEOREM. *In $PG(2,q)$, q square, a reciprocal Hermitian arc $\mathcal{K}$ is a Hermitian curve.*

Proof. It suffices to define a polarity for which $\mathcal{K}$ is the set of self-conjugate points. For P in $\mathcal{K}$, define the *polar of P* to be the unisecant to $\mathcal{K}$ at P; for Q in $\Pi\backslash\mathcal{K}$, define the *polar of Q* as the line of points of contact of the $\sqrt{q}+1$ unisecants to $\mathcal{K}$ through Q.

Conversely, if a line l of Π is unisecant to $\mathcal{K}$, its *pole* is the point of contact with $\mathcal{K}$. If l is a $(\sqrt{q}+1)$-secant, let P_1 and P_2 be two points of $l \cap \mathcal{K}$, let l_1 and l_2 be the unisecants at P_1 and P_2 and let $l_1 \cap l_2 = Q$. Then $Q \in \Pi\backslash\mathcal{K}$ and the polar of Q is l. Now, the $\sqrt{q}+1$ lines through Q which are unisecant to $\mathcal{K}$ meet $\mathcal{K}$ in the points of $l \cap \mathcal{K}$; that is, the $\sqrt{q}+1$ lines unisecant to $\mathcal{K}$ at the points of $l \cap \mathcal{K}$ meet at Q. So let Q be the *pole of l*. The correspondence between pole and polar is therefore a bijective function $\mathfrak{T}$ from the points to the lines of Π.

It remains to show that $\mathfrak{T}$ is a reciprocity: that is, as P varies on the line l, the polar of P describes a pencil through L, the pole of l: and vice versa. This is true by definition if l is a unisecant. So now let l be a $(\sqrt{q}+1)$-secant with P_3 and P_4 as two points of $l \cap \mathscr{K}$. The unisecants l_3 and l_4 at P_3 and P_4 pass through L as do all the other unisecants at points of $l \cap \mathscr{K}$. Let Q be a point of $l \backslash \mathscr{K}$ and let l' be the polar of Q. If $l \cap l' = Q'$ and Q' is a point of $\mathscr{K}$, then the polar of Q' contains Q and Q' and so would be l; hence Q' is not in $\mathscr{K}$. Therefore, take P_1 and P_2 as points of $l' \cap \mathscr{K}$ with respective unisecants l_1 and l_2. Then $l_1 \cap l_2 = Q$, which lies on P_3 and P_4; hence $l_3 \cap l_4$, that is L, lies on P_1P_2. So l' contains L. Therefore, as P varies among the $q+1$ points of l, its polar varies among the $q+1$ lines through L.

It has now been shown that $\mathfrak{T}$ is a polarity in the plane. As the number of self-conjugate points is $q\sqrt{q}+1$, the polarity is Hermitian, §§7.3, 8.3, and so $\mathscr{K}$ is a Hermitian curve.□

COROLLARY. *In $PG(2,q)$, a $(k;n)$-arc of type $(1,n)$ is a subgeometry $PG(2,\sqrt{q})$ or a Hermitian arc; if the latter is reciprocal, it is a Hermitian curve.*□

As was observed in the proof of lemma 12.3.9, a Hermitian arc in $PG(2,4)$ is a Hermitian curve $\mathscr{U}_{2,4}$. It had been conjectured that in $PG(2,q)$, q square, a Hermitian arc is always a Hermitian curve. This is not in fact true, and there is a counterexample for every square $q > 4$. This will be described in Volume II, as its construction depends on detailed properties of $PG(3,q)$ and $PG(4,q)$.

12.4. Arcs with small deficiency

In §12.2 it is conjectured that a $(k;n)$-arc $\mathscr{K}$ in $PG(2,q)$ with $q \not\equiv 0 \pmod{n}$ and $n < q$ satisfies $k \leqslant (n-1)q+1$. So we define the *deficiency* d of $\mathscr{K}$ as

$$d = (n-1)q + n - k\ .$$

and investigate those arcs for which

$$0 \leqslant d \leqslant n - 1,$$

that is, $(n - 1)q + 1 \leqslant k \leqslant (n - 1)q + n$. In particular, it will be shown that the conjecture is true for $n = 4$; that is, for q odd, a $(k;4)$-arc satisfies $k \leqslant 3q + 1$. Throughout this section, $n < q$.

12.4.1. LEMMA. *Let $\mathscr{K}$ be a $(k;n)$-arc with $k = (n - 1)q + n - d$ and $d < n - 1$. Then*

(i) $\tau_1 = \tau_2 = \ldots = \tau_{n-d-1} = 0$;

(ii) $\sigma_0 \leqslant (q + d)/n$.

Proof. (i) Equations (12.2) are

$$\rho_1 + \rho_2 + \ldots + \rho_n = q + 1,$$

$$\rho_2 + 2\rho_3 + \ldots + (n - 1)\rho_n = (n - 1)q + n - d - 1.$$

Hence

$$(n - 1)\rho_1 + (n - 2)\rho_2 + \ldots + \rho_{n-1} = d.$$

So, when $d < n - 1$,

$$\rho_1 = \rho_2 = \ldots = \rho_{n-d-1} = 0$$

and

$$\tau_1 = \tau_2 = \ldots = \tau_{n-d-1} = 0.$$

(ii) Equations (12.3) are

$$\sigma_0 + \sigma_1 + \ldots + \sigma_n = q + 1$$

$$\sigma_1 + 2\sigma_2 + \ldots + n\sigma_n = (n - 1)q + n - d.$$

Hence

$$n\sigma_0 + (n-1)\sigma_1 + \ldots + \sigma_{n-1} = q + d.$$

Thus

$$\sigma_0 \leqslant (q+d)/n. \square$$

Part (ii) of the lemma suggests that we put

$$q + d = \theta n + e, \; e < n \qquad (12.23)$$

where, as in §12.1, θ is the maximum number of 0-secants through any point of the plane. Then the dual of the set of 0-secants forms a $(\tau_0;\theta)$-arc, suggesting that, analogous to the expression for k above, we put

$$\tau_0 = (\theta - 1)q + \theta - d', \; 0 \leqslant d'. \qquad (12.24)$$

12.4.2. LEMMA. *If $\mathcal{K}$ is a $(k;n)$-arc of deficiency $d < n - 1$, then*

(i) $\tau_0 \geq [q^2 - q\{n - d - 1 + d(n-1)/(n-d)\}]/n = t_0$;
(ii) $q[e - d(n-1)/(n-d)] \leqslant d - e - nd'$.

Proof. Eliminating τ_n from equations (12.1) gives

$$\sum_{i=0}^{n-1} (n-i)\tau_i = q(q+1-n) + d(q+1), \qquad (12.25)$$

$$\sum_{i=0}^{n-1} i(n-i)\tau_i = kd. \qquad (12.26)$$

Subtracting (12.26) from N times (12.25) gives

$$\sum_{i=0}^{N-1} (N-i)(n-i)\tau_i + \sum_{i=N}^{n-1} (N-i)(n-i)\tau_i$$

$$= N[q(q+1-n) + d(q-1)] - kd.$$

Since none of the terms in the second sum are positive,

$$\sum_{i=0}^{N-1} (N-i)(n-i)\tau_i \geq N[q(q+1-n) + d(q-1)] - kd.$$

However, from lemma 12.4.1, $\tau_i = 0$ for $1 \leqslant i \leqslant n - d - 1$; so the choice $N = n - d$ gives

$$(n - d)n\tau_0 \geqslant (n - d)[q(q + 1 - n) + d(q - 1) - kd,$$

and this expression gives the result required. When the expression for τ_0 in (12.24) is used and then θ is eliminated using (12.23), the inequality for q given by (ii) is obtained.□

Now an upper bound for τ_0 must be obtained. This is done using eqns (12.14).

12.4.3. LEMMA.

(i) $\theta(\theta - 1)\nu'_{00} + \sum_{i=1}^{\theta} (\theta - i)(\theta - i - 1)\nu_{i0}$

$$= \theta(\theta - 1)[q^2 - (n - 2)q - (n - d - 1)] -$$

$$[2(\theta - 1)(q + 1) + 1]\tau_0 + \tau_0^2 = F(\tau_0);$$

(ii) $\tau_0 \leqslant (\theta - 1)q + \theta - (1 + \surd\Delta)/2 = t_2$, *when*

$$\Delta = 1 + 4(\theta - 1)[(q - 1)(d + e) - \theta d]$$

$$= 1 + 4(\theta - 1)[q\{(n - 1)d - ne\} + (n - d)(d - e)]/n$$

is positive and $(\tau_0, \theta) \neq (0,1)$.

Proof. If we take equations (12.14_2) and (12.14_3) with $d = 0$, and (12.14_4), then eliminate $\nu_{\theta 0}$ and $\nu_{\theta-1,0}$, we obtain (i). Now let Δ be the discriminant of the quadratic F in τ_0 in (i). Then $F(\tau_0) \geqslant 0$ from (i). If $\Delta \leqslant 0$, no further information is obtained. If Δ is positive, then either

$$\tau_0 \leqslant 1/2 + (q + 1)(\theta - 1) - \surd\Delta/2 = t_2$$

or

$$\tau_0 \geqslant 1/2 + (q + 1)(\theta - 1) + \surd\Delta/2 = t_3.$$

If $\tau_0 \geqslant t_3$, then from (12.24),

$$d' = (\theta - 1)q + \theta - \tau_0 \leqslant (1 - \sqrt{\Delta})/2,$$

whence $d' = 0$ and $\Delta = \theta = \tau_0 = 1$. So, unless $\theta = \Delta = 1$ and $\tau_0 = 0$,

$$\tau_0 \leqslant (\theta - 1)q + \theta - (1 + \sqrt{\Delta})/2 = t_2. \square$$

12.4.4. THEOREM. *If $\mathscr{K}$ is a $(k;n)$-arc of deficiency d with $(\tau_0, \theta) \neq (0,1)$ and $\Delta > 0$, then*

$$Aq^2 - Bq - C \leqslant 0 ,$$

where

$$A = n(d - e) - d - [d(n - 1)/(n - d) - e]^2,$$
$$B = (d - e)[n^2 - (n - 2)(d - e + 1) + 2n(d - 1)/(n - d)]$$
$$\qquad - n(d - e + 1) - n^2(d - 1)/(n - d),$$
$$C = (d - e)(n - d + e)(n - d - 1).$$

Proof. If $\Delta > 0$, then by lemma 12.4.3, $\tau_0 \leqslant t_2$. By lemma 12.4.2, $t_0 \leqslant \tau_0$. So $t_0 \leqslant t_2$ and manipulation gives the result. $\square$

This theorem will be used to lower the maximum estimate for k given by theorem 12.2.3, corollary 1, and to prove the conjecture that the maximum value for $q \not\equiv 0 \pmod{n}$ is $k = (n - 1)q + 1$ when $n = 4$.

12.4.5. THEOREM. *For $n \geqslant 5$ and $q \not\equiv 0 \pmod{n}$, there is no $(k;n)$-arc $\mathscr{K}$ of deficiency $d = 2$.*

Proof. With $d = 2$, lemma 12.4.2 (ii) becomes

$$q[e - 2(n - 1)/(n - 2)] \leqslant 2 - e - nd'.$$

For $e \geqslant 3$, the inequality becomes $Lq \leqslant R$ with $L > 0$ and $R < 0$, a contradiction. So $e = 0$, 1, or 2. When $e = 2$, then by (12.23), $q \equiv 0 \pmod{n}$. When $e = 0$, $q + 2 \equiv 0 \pmod{n}$, $\theta > 1$, and $\Delta > 0$. So theorem 12.4.4 gives

$$2(n - 1)[1 - 2(n - 1)/(n - 2)^2]q^2 - [n^2 - 5n + 7 + 2/(2 - n)]q$$
$$- (n - 2)(n - 3) \le 0.$$

For $n \geq 5$, this inequality cannot be satisfied.

When $e = 1$, $q + 1 \equiv 0 \pmod{n}$, $\theta > 1$, and $\Delta > 0$. So theorem 12.4.4 gives

$$[(n - 2) - n^2/(n - 2)^2]q^2 - (n - 1)(n - 4)q - (n - 1)(n - 3) \leqslant 0.$$

For $n \geqslant 6$, this is not satisfied for any q. For $n = 5$, it becomes $2q^2 - 36q - 72 \leqslant 0$, which is only satisfied for $q = 9$ or 19. However, with $n = 5$, $d = 2$, and $e = 1$, lemmas 12.4.2 and 12.4.3 give

$$\tau_0 \geqslant q(3q - 14)/15 ,$$

$$\tau_0 \leqslant \{2q^2 - 6q - 3 - \surd(12q^2 - 36q - 23)\}/10.$$

For $q = 19$, these imply that $\tau_0 \geqslant 55$ and $\tau_0 \leqslant 54$: a contradiction. For $q = 9$, they imply that $\tau_0 \geqslant 8$ and $\tau_0 \leqslant 8$: so $\tau_0 = 8$. Then, by lemma 12.4.1(i), $\tau_1 = \tau_2 = 0$. So eqns (12.1) become, with $q = 9$ and $k = 39$,

$$\tau_0 + \tau_3 + \tau_4 + \tau_5 = 91,\ 3\tau_3 + 4\tau_4 + 5\tau_5 = 390,\ 3\tau_3 + 6\tau_4 + 10\tau_5 = 741.$$

Hence, as $\tau_0 = 8$, we have $\tau_3 = 11$, $\tau_4 = 3$, $\tau_5 = 69$. By lemma 12.4.1(ii), $\sigma_0 \leqslant 2$ for any Q in $\Pi\backslash\mathscr{K}$. So no three of the 0-secants are concurrent. Thus there $8\cdot7/2 = 28$ points Q for which $\sigma_0 = 2$. For such a point, eqns (12.3) become

$$\sigma_3 + \sigma_4 + \sigma_5 = 8,\ 3\sigma_3 + 4\sigma_4 + 5\sigma_5 = 39.$$

If $\sigma_4 = 0$, σ_5 is not an integer: so $\sigma_4 \geqslant 1$. Hence eqn (12.5_4) becomes

$$18 = \Sigma\,\sigma_4 \geq 28\text{: a contradiction.}\square$$

This theorem shows that a $(k;n)$-arc with $5 \leqslant n < q$ and

$q \not\equiv 0 \pmod{n}$ satisfies $k \leqslant (n-1)q + n - 3$. Now, it will be shown that this holds true for $n = 4$ as well. Although the methods of the previous theorem do not work, the inequality obtained from theorem 12.4.4 with $n = 4$, $d = 2$, and $e = 1$ is satisfied for all q.

12.4.6. THEOREM. *For q odd, a $(k;4)$-arc satisfies $k \leq 3q + 1$.*

Proof. We suppose that there exists a $(k;4)$-arc of deficiency $d = 2$. Then, exactly as in previous proof, $e = 0$, 1, or 2. For $q = 0$ or 2, q is even. So there remains the case $e = 1$; that is, $q \equiv -1 \pmod 4$.

By lemma 12.4.1, $\tau_1 = 0$ and eqns (12.3) are

$$\sigma_0 + \sigma_2 + \sigma_3 + \sigma_4 = q + 1,$$

$$2\sigma_2 + 3\sigma_3 + 4\sigma_4 = 3q - 2.$$

Hence

$$4\sigma_0 + 2\sigma_2 + \sigma_3 = q + 2,$$

and so σ_3 is odd. For each fixed σ_0, the maximum value of σ_2 occurs when $\sigma_3 = 1$; then $\sigma_0 = (q + 1 - 4i)/4$, $\sigma_2 = 2i$, $i = 0,1,\ldots,(q + 1)/4$. With, as before, $\theta = (q + 1)/4$, this means that through each point which lies on exactly $\theta - i$ 0-secants, there pass at most $2i$ bisecants. But $\nu_{\theta-i,0}$ is the number of such points for $i < \theta$ and ν'_{00} the number for $i = \theta$. Hence

$$\sum_{Q} \sigma_2 \leq \Sigma\, 2i\nu_{\theta-i,0} + 2\theta\nu'_{00}.$$

Now (12.5_2) is $(q - 1)\tau_2 = \Sigma\sigma_2$, and eliminating $\nu_{\theta 0}$ from (12.14_2) with $d = 0$ and (12.14_4) gives

$$\Sigma\, i\nu_{\theta-i,0} + \theta\nu'_{00} = \theta(q^2 - 2q - 1) - (q + 1)\tau_0\,.$$

Hence

$$(q - 1)\tau_2 \leqslant 2\theta(q^2 - 2q - 1) - 2(q + 1)\tau_0.$$

As $\tau_1 = 0$, the elimination of τ_3 and τ_4 from eqns (12.1) gives

$$\tau_2 = (3q^2 - 9q + 2)/2 - 6\tau_0.$$

Eliminating τ_2 gives

$$\tau_0 \geqslant [q(q-2)(2q - 7) - 1]/[8(q - 2)].$$

Now $q \equiv c \pmod 8$ where $c = 3$ or 7. Hence

$$\tau_0 \geqslant (2q^2 - 7q + c)/8 = t_1.$$

For $q = 7$, $\theta = 2$, $d = 2$, $e = 1$. So lemma 12.4.3(ii) gives $\tau_0 \leqslant 9 - (1 + \sqrt{17})/2$ and hence $\tau_0 \leqslant 6$. However $t_1 = 7$ and so $\tau_0 \geqslant 7$. Thus there is no (23;4)-arc in $PG(2,7)$.

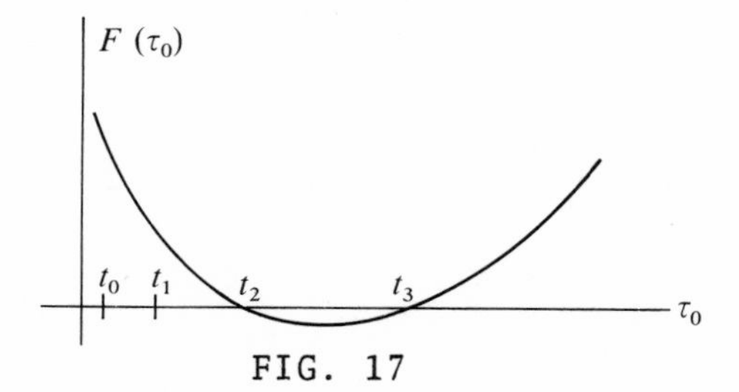

FIG. 17

In general, $t_1 \leqslant \tau_0 \leqslant t_2$ and so the maximum value of F is $F(t_1)$ (Fig. 17). If we write F_3 or F_7 for $F(t_1)$ as $c = 3$ or 7, then, from lemma 12.4.3(i),

$$F_3 = (q - 3)(q - 11)/64, \quad F_7 = (q - 7)(q - 31)/64 - 2.$$

This tells us that for $q = 23$ and 31, there is no $(3q + 2;4)$-arc.

From lemma 12.4.3,

$$\sum_{i=1}^{\theta} (\theta - i)(\theta - i - 1)\nu_{i0} + \theta(\theta - 1)\nu'_{00} \leq F_3 \text{ or } F_7;$$

hence

$$\nu'_{00} = \nu_{10} = \ldots = \nu_{(q-3)/8,0} = 0 \text{ for } c = 3$$

and

$$\nu'_{00} = \nu_{10} = \ldots = \nu_{(q+17)/8,0} = 0 \text{ for } c = 7.$$

With $r = (q + 5)/8$ when $c = 3$ and $r = (q + 25)/8$ when $c = 7$, eqns (12.14_4) and (12.14_2) with $d = 0$ become

$$\nu_{r0} + \ldots + \nu_{\theta 0} = q^2 - 2q - 1,$$

$$r\nu_{r0} + \ldots + \theta\nu_{\theta 0} = (q + 1)\tau_0 .$$

Eliminating $\nu_{\theta-1,0}$ gives

$$(\theta - r - 1)\nu_{r0} + \ldots + \nu_{\theta-2,0} - \nu_{\theta 0} = (q - 3)(q^2 - 2q - 1)/4 - (q + 1)\tau_0$$

and hence

$$(q + 1)\tau_0 - (q - 3)(q^2 - 2q - 1)/4 \leqslant \nu_{\theta 0}.$$

However, if for a point Q in $\Pi\backslash\mathscr{K}$, $\sigma_0 = \theta$, then $\sigma_2 = 0$. So, as ν'_{02} is the number of points of $\Pi\backslash\mathscr{K}$ through which there are no bisecants, $\nu_{\theta 0} \leqslant \nu'_{02}$. Thus

$$(q + 1)\tau_0 - (q - 3)(q^2 - 2q - 1)/4 \leqslant \nu'_{02}. \qquad (12.27)$$

To obtain an upper limit for ν'_{02}, we again use eqns (12.14) but adjust them slightly. Equations (12.2) are $\rho_2 + \rho_3 + \rho_4 = q + 1$, $\rho_2 + 2\rho_3 + 3\rho_4 = 3q + 1$. Hence $\rho_2 = 0$ or 1. Thus, through every point of $\mathscr{K}$, there is one or no 0-secant; so

$$\nu_{02} + \nu_{12} = k + \nu'_{02} + \nu'_{12}.$$

Also, the number of points of $\mathscr{K}$ through which there is exactly one bisecant is $2\tau_2$. Then (12.14) are

$$\nu'_{02} + \nu'_{12} + \nu_{22} + \ldots + \nu_{\theta 2} = q^2 - 2q - 1,$$

$$\nu'_{12} + 2\nu_{22} + \ldots + \theta_2\, \nu_{\theta_2 2} = (q - 1)\tau_2,$$

$$2\nu_{22} + \ldots + \theta_2(\theta_2 - 1)\nu_{\theta_2 2} = \tau_2(\tau_2 - 1).$$

The elimination of ν'_{12} and ν_{22} gives

$$2\nu'_{02} + 2\nu_{32} + \ldots + (\theta_2 - 1)(\theta_2 - 2)\nu_{\theta_2 2} = 2(q^2 - 2q - 1) - 2(q - 1)\tau_2 + \tau_2(\tau_2 - 1).$$

Hence

$$2\nu'_{02} \leqslant 2(q^2 - 2q - 1) - 2(q - 1)\tau_2 + \tau_2(\tau_2 - 1). \quad (12.28)$$

So, comparing (12.28) with the lower bound for ν'_{02} in (12.27) gives

$$2(q + 1)\tau_0 - (q - 3)(q^2 - 2q - 1)/4 \leqslant 2(q^2 - 2q - 1) - 2(q - 1)\tau_2 + \tau_2(\tau_2 - 1). \quad (12.29)$$

However, if τ_3 and τ_4 are eliminated from (12.1),

$$\tau_0 = (3q^2 - 9q + 2)/12 - \tau_2/6.$$

Substituting this in (12.29) gives

$$\tau_2^{\,2} - \tau_2(5q - 4)/3 + (q + 1)(3q - 5)/6 \geqslant 0.$$

Hence, either

$$6\tau_2 \leqslant 5q - 4 - \surd(7q^2 - 28q + 46)$$

or

$$6\tau_2 \geqslant 5q - 4 + \surd(7q^2 - 28q + 46),$$

which implies that

$$\tau_0 \geqslant [9q^2 - 32q + 10 + \sqrt{(7q^2 - 28q + 46)}]/36 \qquad (12.30)$$

or

$$\tau_0 \leqslant [9q^2 - 32q + 10 - \sqrt{(7q^2 - 28q + 46)}]/36. \qquad (12.31)$$

The comparison of (12.30) with $\tau_0 \leqslant t_2$ and of (12.31) with $\tau_0 \geqslant t_1$ both give a contradiction. Hence, there is no $(3q + 2;4)$-arc.□

COROLLARY. *A $(k;n)$-arc with $n \geqslant 4$ in $PG(2,q)$, $q \not\equiv 0 \pmod n$, satisfies*

$$k \leqslant (n - 1)q + n - 3.\square$$

An example of a $(3q + 1;4)$-arc is the Hermitian curve in $PG(2,9)$ which is a $(28;4)$-arc for which every point satisfies $\rho_1 = 1$, $\rho_2 = 0$, $\rho_3 = 0$, $\rho_4 = 9$.

TABLE 12.1

The full list of complete projectively distinct arcs in $PG(2,4)$

n	k	d	τ_0	τ_1	τ_2	τ_3	τ_4	τ_5	Structure
2	6	0	6	0	15	-	-	-	*oval* $\mathscr{O}$ = *conic* + *nucleus*
3	7	4	0	14	0	7	-	-	$PG(2,2)$
3	9	2	0	9	0	12	-	-	*Hermitian curve* $\mathscr{U}_2$
3	9	2	2	3	6	10	-	-	$\mathscr{K}_1 = \Pi\backslash(\mathscr{O} \cup l \cup l')$
3	9	2	3	0	9	9	-	-	$\mathscr{K}_2 = \Pi\backslash(l_1 \cup l_2 \cup l_3)$
4	12	4	0	0	12	0	9	-	$\Pi\backslash\mathscr{U}_2$
4	13	3	0	2	3	7	9	-	$\mathscr{K}_3 = \mathscr{K}_2 \cup (l_1\backslash P)$
4	14	2	0	0	7	0	14	-	$\Pi \backslash PG(2,2)$
4	16	0	1	0	0	0	20	-	$AG(2,4) = \Pi\backslash l$
5	21	0	0	0	0	0	0	21	$PG(2,4)$

Here, if $\mathscr{O}$ is an oval containing P,Q,Q' with $l = PQ$ and $l' = PQ'$, then $\mathscr{K}_1 = \Pi\backslash(\mathscr{O} \cup l \cup l')$. If l_1,l_2,l_3 form a triangle, $\mathscr{K}_2 = \Pi\backslash(l_1 \cup l_2 \cup l_3)$; if P is on l_1 but not a vertex, $\mathscr{K}_3 = \mathscr{K}_2 \cup (l_1\backslash P)$.

Some complete arcs in $PG(2,5)$ are given in Table 12.2.

TABLE 12.2

n	k	d	τ_0	τ_1	τ_2	τ_3	τ_4	τ_5	τ_6	Structure
2	6	1	10	6	15	-	-	-	-	conic $\mathscr{C}$
3	11	2	5	1	10	15	-	-	-	$\mathscr{K}_1 = \mathscr{I} \cup \{P\}$
3	11	2	4	4	7	16	-	-	-	$\mathscr{K}_2 = \Pi \backslash \mathscr{H}$
4	16	3	1	2	6	6	16	-	-	
4	16	3	1	3	3	9	15	-	-	
4	16	3	2	0	6	8	15	-	-	
4	16	3	2	1	3	11	14	-	-	
4	16	3	3	0	0	16	12	-	-	$\mathscr{K}_3 = \Pi \backslash (l_1 \cup l_2 \cup l_3)$
5	21	4	0	1	0	10	5	15	-	$\mathscr{K}_4 = (\Pi \backslash (\mathscr{C} \cup l)) \cup \{P\}$
5	25	1	1	0	0	0	0	30	-	$AG(2,5)$
6	31	0	0	0	0	0	0	0	31	$PG(2,5)$

Here, $\mathscr{I}$ is the set of internal points of $\mathscr{C}$; P is a point of $\mathscr{C}$ and l is the tangent to $\mathscr{C}$ at P. $\mathscr{H}$ is the set of points on the sides of a tetragram plus two of the vertices of the diagonal triangle. The lines l_1, l_2, l_3 form a triangle.

Some $(k;n)$-arcs for $q = 7$ with $k = (n - 1)q + 1$ are shown in Table 12.3.

TABLE 12.3

n	k	d	τ_0	τ_1	τ_2	τ_3	τ_4	τ_5	τ_6	τ_7	τ_8	Structure
2	8	1	21	8	28	-	-	-	-	-	-	conic
3	15	2	12	0	15	30	-	-	-	-	-	$\mathscr{K}_1$
4	22	3	6	2	3	16	30	-	-	-	-	
5	29	4	3	1	1	9	13	30	-	-	-	
5	29	4	3	0	4	6	14	30	-	-	-	
5	29	4	3	0	3	9	11	31	-	-	-	
6	36	5	3	0	0	0	0	36	18	-	-	$\mathscr{K}_2 = \Pi \backslash (l_1 \cup l_2 \cup l_3)$

If $\mathscr{D}$ is a general Desargues configuration, then each of the ten lines meets three others at no point of the configuration. The 15 points formed in this way are the points of $\mathcal{K}_1$. $\mathcal{K}_2$ is the complement of a triangle.

12.5. Arcs and algebraic curves

Firstly we begin with some results on $(k;3)$-arcs. From eqns (12.1),

$$\tau_1 = k(k - 4q - 5)/2 + 3(q^2 + q + 1 - \tau_0),$$

$$\tau_2 = k(3q - k + 4) - 3(q^2 + q + 1 - \tau_0),$$

$$\tau_3 = k(k - 2q - 3)/2 + (q^2 + q + 1 - \tau_0).$$

Now we construct some $(k;3)$-arcs with $\tau_0 = 0$.

In $PG(2,3)$, let $\mathcal{C}$ be a conic with tangent l at P; let l' be an external line to $\mathcal{C}$ and Q an external point on l'. Then $\mathcal{K}_1 = (\mathcal{C}\backslash P) \cup (l\backslash P)$ is a $(6;3)$-arc and $\mathcal{K}_2 = \mathcal{C} \cup (l'\backslash Q)$ is a $(7;3)$-arc; for both $\mathcal{K}_1$ and $\mathcal{K}_2$, $\tau_0 = 0$.

In $PG(2,4)$, let $\mathcal{K}_3$ be the points of a subgeometry $PG(2,2)$; then $\mathcal{K}_3$ is a $(7;3)$-arc. Let $\mathcal{K}_4$ be a Hermitian curve; then $\mathcal{K}_4$ is a $(9;3)$-arc. Both arcs have $\tau_0 = 0$. $\mathcal{K}_4$ may also be considered as the points on the sides of a triangle excluding the vertices.

For a $(k;3)$-arc $\mathcal{K}$ let ρ be the maximum number of unisecants through any point of $\mathcal{K}$.

12.5.1. THEOREM. *If $\mathcal{K}$ is a $(k;3)$-arc in $PG(2,q)$, then*

(i) $\tau_0 \geqslant \rho(\rho - 2)$;

(ii) *if* $q > 4$, $\tau_0 \geqslant 1$;

(iii) *if* $q \leqslant 4$ *and* $\tau_0 = 0$, *then* $\mathcal{K}$ *is one of* $\mathcal{K}_1, \mathcal{K}_2, \mathcal{K}_3, \mathcal{K}_4$.

Proof. See §12.6.□

12.5.2. THEOREM. *Let $\mathcal{C}$ be an absolutely irreducible cubic curve in $PG(2,q)$.*

(i) *If $\mathcal{C}$ is a k-arc, then*

(a) *for $\mathcal{C}$ non-singular, $k \leqslant 3$;*

(b) *for $\mathcal{C}$ singular, $k \leqslant 4$.*

(ii) *For $q \geqslant 8$, $\mathcal{C}$ is a $(k;3)$-arc.*

(iii) *For each* $q \leq 7$, *there exists a cubic curve which is a* k*-arc.*

Proof. If $\mathscr{C}$ is a $(k;2)$-arc and so has no trisecants, every bisecant must be a tangent at one of its two points of contact with $\mathscr{C}$. Suppose $k \geq 3$ and P_1, P_2 are simple points of $\mathscr{C}$ with P_1P_2 tangent at P_1. Let P_3 be a third simple point of $\mathscr{C}$. If P_2P_3 were tangent at P_3, then P_1P_3 would be a trisecant of $\mathscr{C}$. So P_2P_3 is a tangent at P_2 and P_3P_1 is tangent at P_3. Thus there are no other simple points on $\mathscr{C}$. As $\mathscr{C}$ can contain only one double point, (i) is established. If $\mathscr{C}$ is singular, it contains only q, $q + 1$, or $q + 2$ points as the singular point is a node, a cusp, or an isolated double point, theorem 11.3.10. So, when $q > 4$, every singular cubic is a $(k;3)$-arc. If $\mathscr{C}$ is non-singular, then by theorem 10.2.1, corollary 1, the number k of its points satisfies

$$(\sqrt{q} - 1)^2 \leq k \leq (\sqrt{q} + 1)^2.$$

If $q \geq 8$, then $k \geq 4$. Hence (ii) is established. For each $q \leq 7$, it is easy to construct a cubic whose points form a 3-arc: see theorem 11.9.1.□

12.5.3. THEOREM. *In* $PG(2,q)$, $q > 4$, *every absolutely irreducible cubic considered as a* $(k;n)$*-arc has* $\tau_0 \geq 1$. *Also, if* $q > 13$,

$$\tau_0 \geq (q - 5)(q - 4\sqrt{q} - 1)/4 + 4.$$

Proof. The first part is an immediate consequence of the previous two theorems. Now suppose the cubic $\mathscr{C}$ is a $(k;3)$-arc. Then, since through any point P of $\mathscr{C}$, there are at most four lines which are tangent to $\mathscr{C}$ at other points, $\rho_2 \leq 5$. Then, eqns (12.2) are

$$\rho_1 + \rho_2 + \rho_3 = q + 1, \quad \rho_2 + 2\rho_3 = k - 1.$$

Hence

$$\begin{aligned}\rho_1 &= (q+1) - (k+\rho_2-1)/2 \\ &\geq (q+1) - [(\sqrt{q}+1)^2 + 5 - 1]/2 \\ &= (\sqrt{q}+1)(\sqrt{q}-3)/2 .\end{aligned}$$

So

$$\rho \geq \rho_1 \geq (\sqrt{q}+1)(\sqrt{q}-3)/2.$$

Applying theorem 12.5.1, we have

$$\tau_0 \geq \rho(\rho-2) \geq (q-5)(q-4\sqrt{q}-1)/4 + 4.$$

This implies that $\tau_0 \geq 2$ for $q > 13$.□

Most of the results so far obtained on inequalities for the number of points of a $(k;n)$-arc in $PG(2,q)$ are true for any projective plane of order q. To obtain better results for $PG(2,q)$, it is necessary to use algebraic rather than combinatorial methods; in particular, it would seem sensible to relate $(k;n)$-arcs to algebraic curves. That this will not be as successful for general n as for $n = 2$ is more than likely, since a $(k;n)$-arc satisfies

$$k \leq (n-1)q + n,$$

whereas the number R of points of an absolutely irreducible $\mathscr{C}_n$ satisfies, by (10.14),

$$R \leq q + 1 + (n-1)(n-2)\sqrt{q}.$$

An *n-curve* is a pair $(\mathscr{K},\mu)$ where $\mathscr{K}$ is a $(k;n)$-arc with $k \geq (n+1)(n-2)/2$ and $\mu : \mathscr{K} \to \mathbf{N}$ is a map such that for every algebraic curve $\mathscr{F}_m$ of order $m < n$

$$\sum_{P\in\mathscr{K}\cap\mathscr{F}_m} \mu(P) \leq mn. \qquad (12.32)$$

$\mu(P)$ is called the *multiplicity* of P. The point P is *simple* or *multiple* as $\mu(P) = 1$ or $\mu(P) > 1$. Compare §10.1.

An n-curve $(\mathscr{K},\mu)$ is *complete* if there is no n-curve $(\mathscr{K}',\mu')$ such that $\mathscr{K}'$ properly contains $\mathscr{K}$ and $\mu(P) = \mu'(P)$ for all P in $\mathscr{K}$.

12.5.4. LEMMA. *For every n-curve* $(\mathscr{K},\mu)$ *in* $PG(2,q)$,

$$\mu(P) \leqslant n - 1, \text{ all } P \text{ in } \mathscr{K}.$$

Proof. If there is a point P_1 in $\mathscr{K}$ with $\mu(P_1) \geqslant n$, then, if $P_2 \in \mathscr{K}$ and $l = P_1P_2$, summation over $\mathscr{K} \cap l$ gives $\Sigma\, \mu(P) \geqslant \mu(P_1) + \mu(P_2) \geqslant n + 1$, contradicting that by definition $\Sigma\, \mu(P) \leqslant n$.□

For a 2-curve, it follows that every point is simple; thus μ is the constant map with value one. So the concept of 2-curve adds nothing to that of k-arc. Henceforth we will take $n \geqslant 3$.

12.5.5. LEMMA. *The maximum number of multiple points of an n-curve* $(\mathscr{K},\mu)$ *is* $(n - 1)(n - 2)/2$.

Proof. The freedom of a curve $\mathscr{F}_m$ is $m(m + 3)/2$; that is, through $m(m + 3)/2$ points of the plane, there is at least one algebraic curve of order m. Suppose that $(\mathscr{K},\mu)$ contains $(n - 1)(n - 2)/2 + 1$. multiple points. Then, since $\mathscr{K}$ is a $(k;n)$-arc and $k \geqslant (n + 1)(n - 2)/2$, there are n-3 points on $\mathscr{K}$ apart from the multiple ones. As $(n + 1)(n - 2)/2$ is the freedom of a curve of order $n - 2$, there is a curve $\mathscr{F}_{n-2}$ through the given multiple points and the n-3 other points of $\mathscr{K}$. So, summing over $\mathscr{K} \cap \mathscr{F}_{n-2}$ gives

$$\begin{aligned}\Sigma\, \mu(P) &\geqslant 2[(n - 1)(n - 2)/2 + 1] + n - 3\\ &= (n - 1)^2.\end{aligned}$$

On the other hand, from (12.32),

$$\Sigma\, \mu(P) \leqslant n(n - 2):$$

a contradiction.□

12.5.6. LEMMA. *If an n-curve* $(\mathscr{K},\mu)$ *contains a point of multiplicity* $n-1$, *then all its other points are simple and* $k = |\mathscr{K}| \leq q + 2$.

Proof. Let P in $\mathscr{K}$ have multiplicity $n - 1$. Then, by (12.32), each line through P can only meet $\mathscr{K}$ in one further point, which is necessarily simple.□

12.5.7. LEMMA. *If* $\mathscr{K}$ *is a* $(k;n)$*-arc and the* n*-curve* $(\mathscr{K},\mu)$ *has* $(n - 1)(n - 2)/2$ *double points, then* $k \leq q + (n + 1)(n - 2)/2$.

Proof. As in lemma 12.5.4, there are $n - 3$ simple points on $(\mathscr{K},\mu)$. Let $\mathscr{H}$ be the subset of $\mathscr{K}$ consisting of these simple points and the double points: so $|\mathscr{H}| = (n + 1)(n - 2)/2 - 1$. Thus there is at least a pencil $\mathscr{L}$ of algebraic curves of order $n - 2$ through the points of $\mathscr{H}$. Again, by (12.32), it follows that any $\mathscr{F}_{n-2}$ in $\mathscr{L}$ contains at most one point of $\mathscr{K}\backslash\mathscr{H}$.

If two curves of the pencil $\mathscr{L}$ meet in a point P of $\mathscr{K}\backslash\mathscr{H}$, then P will be a base point of $\mathscr{L}$. But, as every curve of $\mathscr{L}$ meets $(\mathscr{K},\mu)$ in at most one point of $\mathscr{K}\backslash\mathscr{H}$, it follows that $\mathscr{K} = \mathscr{H} \cup \{P\}$ and so $k = (n + 1)(n - 2)/2$.

If, on the other hand, there is only one curve of $\mathscr{L}$ through every point of $\mathscr{K}\backslash\mathscr{H}$, we can define a map $\psi:\mathscr{K}\backslash\mathscr{H} \to \mathscr{L}$ which associates to a point P the curve $\mathscr{F}(P)$ of $\mathscr{L}$ through it. By definition ψ is injective. So, as $|\mathscr{L}| = q + 1$, $|\mathscr{K}\backslash\mathscr{H}| \leq q + 1$ and the result follows.□

If q is sufficiently large with respect to n, then an absolutely irreducible curve $\mathscr{F}_n$ in $PG(2,q)$ contains at least $(n + 1)(n - 2)/2$ points and determines an n-curve $(\mathscr{K},\mu)$ where $\mathscr{K}$ is the $(k;n)$-arc consisting of all points of $\mathscr{F}_n$, and μ is the map that associates to each point P of $\mathscr{K}$ its multiplicity on $\mathscr{F}_n$. It is of interest to consider, conversely, whether a given n-curve determines an algebraic curve of order n.

12.5.8. THEOREM. *In* $PG(2,q)$ *with* q *odd and* $q > 11$, *let* $\mathscr{K}$ *be a* $(k;3)$*-arc containing four distinct points* D, P_1, P_2, P_3, such that

(i) *there are no trisecants of* $\mathscr{K}$ *through* D;

(ii) *every conic through* D *and one* P_i *contains at most three other points of* $\mathcal{K}$;

(iii) $k > q - \sqrt{q}/4 + 19/4$.

Then $\mathcal{K}$ *is contained in an absolutely irreducible cubic curve with double point* D.

Proof. Of the sides of the triangle DP_1P_2, only P_1P_2 can meet $\mathcal{K}$ in a point other than a vertex. So, if $\mathcal{H}$ is the set of points of $\mathcal{K}$ on no side of the triangle, $|\mathcal{H}| = k - d$, where $d = 3$ or 4 according as P_1P_2 is a bisecant or trisecant of $\mathcal{K}$. Let ϕ be the standard quadratic transformation of $PG(2,q)$ with DP_1P_2 as triangle of reference; see §12.6. By (ii), every conic through D, P_1, and P_2 meets $\mathcal{H}$ in at most two points; every line through P_1 or P_2 meets $\mathcal{H}$ in at most two points, while every line through $\mathcal{D}$ meets $\mathcal{H}$ in at most one other point. As ϕ transforms conics through D, P_1, and P_2 into lines not containing these points and transforms each of the pencils with centres D, P_1, and P_2 into itself, it follows that $\mathcal{H}\phi$ meets every line of the plane in at most two points and every line through D in at most one point. As the restriction of ϕ to the plane with the sides of the triangle DP_1P_2 removed is bijective, $\mathcal{H}\phi$ is an incomplete $(k - d)$-arc and $\bar{\mathcal{H}} = \mathcal{H}\phi \cup \{D\}$ is a $(k - d + 1)$-arc. By (iii),

$$k - d + 1 > q - \sqrt{q}/4 + 7/4.$$

Hence by theorem 10.4.4, $\bar{\mathcal{H}}$ is contained in a conic $\mathcal{F}_2$ not containing both P_1 and P_2; otherwise, $\mathcal{F}_2\phi^{-1}$ would be a line containing $\mathcal{H}$ and, by (iii), $k - d > 3$. Moreover, if $\mathcal{F}_2$ contains P_1 only, then $\mathcal{F}_2\phi^{-1}$ would be a conic containing $\mathcal{H}$, P_1, and D and so would meet $\mathcal{H}$ in at most three points, again contradicting that $k - d > 3$; so $\mathcal{F}_2$ contains neither P_1 nor P_2. If D is U_0, then $\mathcal{F}_2 = \mathrm{V}(F_2)$, with

$$F_2 = a_1x_1^2 + a_2x_2^2 + b_0x_1x_2 + b_1x_0x_2 + b_2x_0x_1.$$

So $\mathcal{F}_2\phi^{-1}$ is an absolutely irreducible cubic $\mathcal{F}_3 = \mathrm{V}(F_3)$, with

$$F_3 = (a_2x_1^2 + b_0x_1x_2 + a_1x_2^2)x_0 + x_1x_2(b_1x_1 + b_2x_2).$$

So $\mathscr{F}_3$ contains $\mathscr{H}$, P_1, P_2, and D as a double point.

To show that $\mathscr{F}_3$ contains $\mathscr{K}$, it remains to show that P_3 and the possible further intersection of P_1P_2 with $\mathscr{K}$ lie on $\mathscr{F}_3$. If the above argument is repeated for the triangle DP_1P_3, a cubic $\mathscr{F}'_3$ is obtained which contains the set $\mathscr{H}'$ of points of $\mathscr{K}$ on no side of the triangle as well as P_1, P_3, and D as a double point. So both cubics contain $\mathscr{H} \cap \mathscr{H}'$ and D. As $|\mathscr{H} \cap \mathscr{H}'| \geqslant k - 6$ and $k - 6 > 7$, the cubics intersect in a double point and more than seven simple points; therefore they coincide. So $\mathscr{F}_3$ contains P_3 and the possible further intersection of P_1P_2 with $\mathscr{K}$. □

COROLLARY. *In $PG(2,q)$ with q odd and $q > 11$, a 3-curve $(\mathscr{K},\mu)$ with a double point D and $k > q - \sqrt{q}/4 + 19/4$ is contained in an absolutely irreducible cubic $\mathscr{F}_3$ with double point D. If $(\mathscr{K},\mu)$ is also complete, then it coincides with $\mathscr{F}_3$.* □

12.5.9. THEOREM. *In $PG(2,q)$ with q odd and $q > 11$, a 4-curve $(\mathscr{K},\mu)$ with three double points P_1,P_2,P_3 satisfying $k > q - \sqrt{q}/4 + 19/4$ is contained in an absolutely irreducible quartic $\mathscr{F}_4$ with double points P_1,P_2,P_3. If $(\mathscr{K},\mu)$ is also complete, it coincides with $\mathscr{F}_4$.*

Proof. By lemma 12.5.5, the points of $(\mathscr{K},\mu)$ other than P_1, P_2, and P_3 are all simple. By (12.32), the only points of $(\mathscr{K},\mu)$ on the sides of the triangle $P_1P_2P_3$ are the vertices. Let $\mathscr{H} = \mathscr{K}\backslash\{P_1,P_2,P_3\}$; then $|\mathscr{H}| = k - 3$.

Consider the standard quadratic transformation ϕ with triangle of reference $P_1P_2P_3$. By (12.32), every conic through P_1, P_2, and P_3 meets $\mathscr{H}$ in at most two points, as does any line through one P_i. Since ϕ transforms conics through P_1, P_2, and P_3 into lines containing no P_i and transforms the pencil with centre P_i into itself, $\mathscr{H}\phi$ meets every line of the plane in at most two points. As ϕ restricted to the set of points of the plane off the sides of the triangle $P_1P_2P_3$ is bijective, $\mathscr{H}\phi$ is a $(k - 3)$-arc. But $k - 3 > q - \sqrt{q}/4 + 7/4$; so, by theorem 10.4.4, $\mathscr{H}\phi$ is contained in a conic $\mathscr{F}_2$.

The conic $\mathcal{F}_2$ contains none of the P_i. For, if $\mathcal{F}_2$ contains P_1 only, then $\mathcal{F}_2\phi^{-1}$ would be an irreducible cubic with double point P_1 and would contain $\mathcal{H}$. So $\mathcal{F}_2\phi^{-1}$ would meet $(\mathcal{K},\mu)$ in at least $k - 3$ simple points as well as the double point P_1. Then, by (12.32), $k - 3 + 2 < 12$ giving $k < 13$, which contradicts the initial hypothesis. If $\mathcal{F}_2$ contains P_1 and P_2 but not P_3, $\mathcal{F}_2\phi^{-1}$ would be a conic containing P_1, P_2, and $\mathcal{H}$. So, by (12.32), $k - 3 + 4 < 8$ giving $k < 7$ and again a contradiction. Finally, if $\mathcal{F}_2$ contains P_1, P_2, and P_3, $\mathcal{F}_2\phi^{-1}$ would be a line containing $\mathcal{H}$. So, by (12.32), $k - 3 < 4$ giving $k < 7$ and another contradiction.

Since $\mathcal{F}_2$ contains none of P_1, P_2, and P_3, $\mathcal{F}_2\phi^{-1}$ is an irreducible quartic $\mathcal{F}_4$ containing $\mathcal{H}$ with P_1, P_2, and P_3 as double points.□

12.6. Notes and references

§12.1. Lemmas 12.1.2, 12.1.3 were given by Tallini Scafati [1966a]. Lemmas 12.1.4, 12.1.5 were given by Haĭmulin [1966a]. Lemmas 12.1.6, 12.1.7 were given by Lunelli and Sce [1964]. See also Barlotti [1965].

§12.2. Theorem 12.2.1 is due to Cossu [1961]. Theorem 12.2.2 is due to Denniston [1969]. Theorem 12.2.3 and its corollaries are due to Barlotti [1956b]. Lemma 12.2.4 and theorem 12.2.5 are due to Thas [1975a]. See also D'Orgeval [1960], [1967a], Thas [1974b].

§12.3. This is taken from Tallini Scafat [1966b], apart from theorem 12.3.9, which is due to Hubaut [1970]. See also Halder [1976]. Buekenhout [1976] gave an example of a Hermitian arc other than $\mathcal{U}_{2,q}$ for certain even q. The method has been extended by R. Metz to produce an example of a Hermitian arc other than $\mathcal{U}_{2,q}$ ("non-classical initial" is a frequent terminology) for every square q.

§12.4. This section is based on Lunelli and Sce [1964]. Theorem 12.4.6 has also been proved by Haĭmulin [1966a] for $q \equiv 1 \pmod 4$ and asserted for $q \equiv 3 \pmod 4$. The important corollary has been given a shorter proof by Wilson [1974]. The classification of complete arcs in $PG(2,4)$ is due to F. Conti [1968].

For examples of $(k;n)$-arcs, see Barlotti [1965], Bramwell and Wilson [1973b], Casse [1972], Ceccherini [1969], Lunelli and Sce [1964], Stein and Retkin [1965].

See also Basile and Brutti [1971], [1973], Haĭmulin [1966b], Halder [1974].

§12.5. Theorems 12.5.1 - 12.5.3 come from Tallini Scafati [1966a] and the rest of the section follows Tallini Scafati [1971].

13
BLOCKING SETS

13.1. Definitions and examples

A *blocking set* $\mathscr{B}$ in $\Pi = PG(2,q)$ is a subset of Π which meets every line but contains no line completely; that is, $1 \leqslant |\mathscr{B} \cap l| \leqslant q$ for every line l in Π. So $\mathscr{B}$ is a blocking set if and only if $\Pi \backslash \mathscr{B}$ is. If $|\mathscr{B}| = b$, we will speak of $\mathscr{B}$ as a *blocking* b*-set*. Such a set for which b has the minimum value is a *committee*. A blocking set $\mathscr{B}$ is *minimal* if $\mathscr{B}\backslash\{P\}$ is not a blocking set for every P in $\mathscr{B}$. We may note that, in terms of Chapter 12, a blocking set is merely a $(k;n)$-arc with $n \leqslant q$ and no 0-secants.

Firstly the existence of blocking sets is considered and then the size of committees.

13.1.1. LEMMA. *A blocking set* $\mathscr{B}$ *exists in* Π *if and only if* $q > 2$.

Proof. Let $P_1P_2P_3$ be a triangle and P a point other than a vertex on P_2P_3. Let $\mathscr{B}$ be the set of $2q$ points on the sides P_1P_2 and P_2P_3, excluding P_2 and P_3, plus the point P. Then any line not through P contains one, two, or q points of $\mathscr{B}$ and any line through P contains two or three points of $\mathscr{B}$. So, if $q > 2$, $\mathscr{B}$ is a blocking $(2q)$-set.

When $q = 2$, a blocking set is a k-arc since it contains more than one point and does not contain three collinear points. A 2-arc has two external lines, a 3-arc and a 4-arc just one. There is no 5-arc. So a blocking set does not exist.□

Before giving more examples of blocking sets, we first relate them to complete k-arcs.

13.1.2. THEOREM. *Let* $\mathscr{K}$ *be a complete* k*-arc in* $\Pi = PG(2,q)$ *with* $k < q + 2$. *The dual in* Π *of the set of bisecants of* $\mathscr{K}$ *is a blocking* $[k(k - 1)/2]$*-set* $\mathscr{B}$.

Proof. Since every point of Π lies on a bisecant of $\mathscr{K}$ every line of Π meets $\mathscr{B}$. Since no point of Π lies on more than $k - 1 < q + 1$ bisecants of $\mathscr{K}$, no line of Π has all its points in $\mathscr{B}$.□

A blocking set obtained from a k-arc as in the theorem will be called *k-arc derived*. By way of converse to the above theorem, there is the following.

13.1.3. THEOREM. *A blocking b-set $\mathscr{B}$ is k-arc derived if and only if*

(i) $b \leqslant k(k - 1)/2$;

(ii) *the number of $(k - 1)$-secants of $\mathscr{B}$ is* $\geqslant k$;

(iii) *no three $(k - 1)$-secants of $\mathscr{B}$ are concurrent.*

Proof. If a blocking b-set $\mathscr{B}$ is k-arc derived, then the three conditions follow immediately from the definition of a k-arc with, in fact, equality in (i) and (ii).

Conversely, assume that a blocking b-set $\mathscr{B}$ satisfies the three conditions. Let $\mathscr{L}_{k-1}$ denote the set of $(k - 1)$-secants of $\mathscr{B}$ and $\mathscr{I}$ the set of incidences of points of $\mathscr{B}$ with lines of $\mathscr{L}_{k-1}$. By (ii), $|\mathscr{I}| \geqslant k(k - 1)$. The points of intersections of pairs of lines of $\mathscr{L}_{k-1}$ are distinct, by (iii). Let n of these points be in $\mathscr{B}$. This leaves $b - n \leqslant k(k - 1)/2 - n$ points of $\mathscr{B}$ that lie on one or no lines of $\mathscr{L}_{k-1}$. So $|\mathscr{I}| \leqslant 2n + [k(k - 1)/2 - n]$. Therefore $n + k(k - 1)/2 \geqslant$ $\geqslant k(k - 1)$ giving $n \geqslant k(k - 1)/2$. But, by definition, $n \leqslant b \leqslant k(k - 1)/2$. Thus $n = k(k - 1)/2$, $|\mathscr{I}| = k(k - 1)$, $|\mathscr{L}_{k-1}| = k$, $b = k(k - 1)/2$, and the points of $\mathscr{B}$ are precisely the intersections of pairs of lines $\mathscr{L}_{k-1}$. If $\mathscr{K}$ is the dual of $\mathscr{L}_{k-1}$, then $\mathscr{K}$ is a k-arc and the dual of $\mathscr{B}$ is the set of bisecants of $\mathscr{K}$. Since $\mathscr{B}$ is a blocking set, $\mathscr{K}$ is complete. So $\mathscr{B}$ is k-arc derived.□

13.1.4. LEMMA. *A blocking set $\mathscr{B}$ is minimal if and only if for every point P of $\mathscr{B}$, there is some line l such that $\mathscr{B} \cap l = \{P\}$.*

Proof. If $\mathscr{B}$ satisfies the condition and P is a point such that $\mathscr{B} \cap l = \{P\}$ for some line l, then $\mathscr{B}\backslash\{P\}$ is skew to l and so is not a blocking set. Thus $\mathscr{B}$ is minimal.

Conversely, if the condition is not satisfied, there is some point P in $\mathscr{B}$ such that all lines through P contain another point of $\mathscr{B}$. So $\mathscr{B}\backslash\{P\}$ is a blocking set and $\mathscr{B}$ is not minimal.□

Thus a minimal blocking set $\mathscr{B}$ is a $(k;n)$-arc with $\rho_1 > 0$ for all points P on $\mathscr{B}$, §12.1.

13.1.5. LEMMA. *The following are minimal blocking b-sets in $PG(2,q)$:*

(i) *a line l minus a point P plus a set of q points, one on each of the q lines through P other than l but not all collinear: $b = 2q$ (the example in lemma* 13.1.1 *is a particular case of this);*

(ii) *$(\mathscr{C} \cup l \cup \{P\})\backslash\{P_1,P_2\}$, where $\mathscr{C}$ is a conic, l a bisecant, $\mathscr{C} \cap l = \{P_1,P_2\}$, and P is the point of intersection of the tangents to $\mathscr{C}$ at P_1 and P_2 : $b = 2q - 1$.*□

13.1.6. LEMMA. *For q square, the following are minimal blocking b-sets in $PG(2,q)$:*

(i) *a subgeometry $PG(2,\sqrt{q})$: $b = q + \sqrt{q} + 1$;*

(ii) *a Hermitian curve $\mathscr{U}_2$: $b = q\sqrt{q} + 1$.*

Proof. (i) A line of $PG(2,q)$ containing two points of $PG(2,\sqrt{q})$ contains $\sqrt{q} + 1$ points of it. So the number of lines containing one or $\sqrt{q} + 1$ points of the subgeometry is $(q + \sqrt{q} + 1)(q - \sqrt{q}) + (q + \sqrt{q} + 1) = q^2 + q + 1$.

(ii) Every line meets $\mathscr{U}_2$ in one or $\sqrt{q} + 1$ points.□

We now prove some fundamental lemmas that will be required in the subsequent sections.

13.1.7. LEMMA. *If $\mathscr{B}$ is a blocking b-set,*

(i) *no line of Π contains more than $b - q$ points of $\mathscr{B}$;*

(ii) *if a line contains n points of $\mathscr{B}$, $b \geq n + q$;*

(iii) *there is some line containing at least three points of $\mathscr{B}$;*

(iv) $b \geq q + 3$.

Proof. Suppose $|l \cap \mathcal{B}| = n$ for a line l. Let $P \in l\backslash\mathcal{B}$. Then there are q lines through P other than l each containing a point of $\mathcal{B}$. So $b \geqslant n + q$, which proves (i) and (ii). Since $\mathcal{B}$ contains more than one point, there is a line meeting it in two points. If every line meets it in at most two points, it is a b-arc and so, by lemma 8.1.1, it has $q(q - 1)/2 + t(t - 1)/2$ external lines, where $t = q + 2 - b$. So $\mathcal{B}$ is not a b-arc. This proves (iii), and then (iv) follows by (ii).□

13.1.8. LEMMA. *Let* Π' *and* $\mathcal{L}'$ *be subsets of points and lines respectively of* Π *such that* $|\Pi'| = n$, $|\mathcal{L}'| = m$, $m \leqslant n \leqslant 2m$, *and the lines of* $\mathcal{L}'$ *contain all the points of* Π' *but no two meet in* Π'. *If* $\mathcal{I}$ *is the set of incidences of points of* Π' *with those lines* l *of* $\mathcal{L}'$ *satisfying* $|l \cap \Pi'| \geqslant 2$, *then* $|\mathcal{I}| \leqslant 2(n - m)$.

Proof. Let r_i be the number of lines of $\mathcal{L}'$ meeting Π' in exactly i points for $i = 1,2,\ldots, s$. Then

$$r_1 + r_2 + \ldots + r_s = m ,$$

$$r_1 + 2r_2 + \ldots + sr_s = n .$$

Hence

$$r_2 + 2r_3 + \ldots + (s - 1)r_s = n - m.$$

So

$$\begin{aligned} |\mathcal{I}| &= 2r_2 + 3r_3 + \ldots + sr_s \\ &= 2(n - m) - r_3 - 2r_4 \ldots - (s - 2)r_s \\ &\leqslant 2(n - m). \end{aligned}$$

We note that the maximum is attained when $r_1 = 2m - n$, $r_2 = n - m$, $r_3 = r_4 \ldots = r_s = 0$.□

We now prove a lemma which, in the subsequent sections, is

the main tool for counting arguments about blocking sets. Firstly a number of definitions are required.

(1) $\mathscr{B}$ is a blocking b-set and l a line containing the maximum number k of points of $\mathscr{B}$.

(2) $\mathscr{B}$ is such that $q \leqslant b - k \leqslant 2q$.

(3) $\mathscr{L}'$ is the set of lines containing a point of $l \backslash \mathscr{B}$ and at least two points of $\mathscr{B} \backslash l$.

(4) s is the maximum number of points of $\mathscr{B}$ on any line of $\mathscr{L}'$ and, for $i = 2,3,\ldots, s$, $r_i = r_i(Q)$ is the number of lines of $\mathscr{L}'$ through a point Q of $l \backslash \mathscr{B}$ containing exactly i points of $\mathscr{B}$.

(5) $m = m(P)$ is the number of lines of $\mathscr{L}'$ through a point P of $\mathscr{B} \backslash l$ and $k_1, k_2, \ldots, k_m$ are the respective numbers of points of $\mathscr{B}$ on these lines.

(6) $\mathscr{I}$ is the set of incidences of points of $\mathscr{B} \backslash l$ with lines of $\mathscr{L}'$.

13.1.9. LEMMA.

(i) $k + s \leqslant b - q + 1$;

(ii) $s \leqslant (b - q + 1)/2$;

(iii) $(s - 1)m(P) \geqslant \sum_1^m (k_i - 1) \geqslant b - 1 + k - k^2$;

(iv) $|\mathscr{I}| = m(P) \geqslant (b - k)(b - 1 + k - k^2)/(s - 1)$;

(v) $r(Q) = \Sigma i r_i \leqslant 2(b - k - q)$;

(vi) $|\mathscr{I}| = \Sigma r(Q) \leqslant 2(q + 1 - k)(b - k - q)$;

(vii) $2(s - 1)(q + 1 - k)(b - k - q)$
$\geqslant (b - k)(b - 1 + k - k^2)$.

Proof. (i) Let l' be a line of $\mathscr{L}'$ containing exactly s points of $\mathscr{B} \backslash l$ and let $l \cap l' = Q$; then, by definition of $\mathscr{L}'$, $Q \notin \mathscr{B}$. Since $\mathscr{B}$ is a blocking set, each line through Q contains at least one point of $\mathscr{B}$, while, in particular, l and l' contain k and s points of $\mathscr{B}$ respectively. So $k + s + q - 1 \leqslant b$, whence $k + s \leqslant b - q + 1$.

(ii) Since $s \leqslant k$, so $2s \leqslant b - q + 1$.

(iii) For the point P in $\mathscr{B} \backslash l$, we count the number $n(P)$ of points of $\mathscr{B} \backslash l$ other than P on lines of $\mathscr{L}'$ through P. $n(P) = \sum_1^m (k_i - 1)$. Since $k_i \leqslant s$ for each i, $n(P) \leqslant (s - 1)m$. Also, there are $b - 1 - k$ points of $\mathscr{B} \backslash l$ other than P, of which the lines joining P to the k points of $\mathscr{B} \cap l$ account for at most $k(k - 2)$. So there are at least $b - 1 - k - k(k - 2) = b - 1 + k - k^2$ points of $\mathscr{B}\backslash l$ other than P which are incident with lines of $\mathscr{L}'$ through P. So $n(P) \geqslant b - 1 + \quad - k^2$.

(iv) $|\mathscr{I}| = \Sigma m(P)$. As there are $b - k$ points in $\mathscr{B}\backslash l$, (iii) implies that $|\mathscr{I}| \geqslant (b - k)(b - 1 + k - k^2)/(s - 1)$.

(v) For each point Q in $l\backslash\mathscr{B}$, the points of $\mathscr{B}\backslash l$ are partitioned among the q lines other than l through Q. So, from lemma 13.1.8, the number $r(Q)$ of incidences in $\mathscr{I}$ of these lines with points of $\mathscr{B}\backslash l$ satisfies

$$r(Q) = \sum_{2}^{s} i r_i \leqslant 2(b - k - q).$$

(vi) Since there are $q + 1 - k$ points Q,

$$|\mathscr{I}| = \Sigma r(Q) \leqslant 2(q + 1 - k)(b - k - q).$$

(vii) Combining (iv) and (vi) gives the final result.□

COROLLARY. *Let $s = 3$. Then the upper and lower bounds for $|\mathscr{I}|$ can be improved by use of the following:*

(i) *if b is even, then, for all points P in $\mathscr{B}\backslash l$*

$$m(P) \geqslant (b + k - k^2)/2;$$

(ii) *if, for some P in $\mathscr{B}\backslash l$, $k_1 = 2$, then*

$$m(P) \geqslant (b + k - k^2)/2;$$

(iii) *for all points Q in $l\backslash\mathscr{B}$,*

$$r(Q) = 2(b - k - q) - r_3.$$

Proof. (i) From (iii) of the lemma,

$$m(P) \geqslant (b - 1 + k - k^2)/2.$$

As b is even, $b - 1 + k - k^2$ is odd. So, as $m(P)$ is an integer,

$$m(P) \geq (b + k - k^2)/2.$$

(ii) Since $k_1 = 2$, $\sum_1^m (k_i - 1) = 1 + \sum_2^m (k_i - 1) \leq 1 + (s - 1)(m - 1) = 2m - 1$. So, from (iii) of the lemma,

$$2m - 1 \geq b - 1 + k - k^2,$$

whence

$$m \geq (b + k - k^2)/2.$$

(iii) From the proof of lemma 13.1.8 with $n = b - k$ and $m = q$,

$$|\mathscr{I}| = 2(n - m) - r_3 \ldots -(s - 2)r_s$$
$$= 2(n - m) - r_3,$$

since $s = 3$.□

13.2. Bounds for blocking sets with q square

13.2.1. LEMMA. *If q is a square and $\mathscr{B}$ a blocking $(q + \sqrt{q} + 1)$-set in $\Pi = PG(2,q)$, there exists a line containing exactly $\sqrt{q} + 1$ points of $\mathscr{B}$.*

Proof. Let l be a line containing the maximum number k of points of $\mathscr{B}$. Then, by lemma 13.1.7, $3 \leq k \leq \sqrt{q} + 1$. Applying lemma 13.1.9(vii) with $b = q + \sqrt{q} + 1$ and $s = k$ gives

$$2(k - 1)(q + 1 - k)(\sqrt{q} + 1 - k) \geq (q + \sqrt{q} + 1 - k)(q + \sqrt{q} + k - k^2).$$

As $q + \sqrt{q} + k - k^2 = (\sqrt{q} - k + 1)(\sqrt{q} + k)$, this implies

$$2(k - 1)(q + 1 - k) \geq (q + \sqrt{q} + 1 - k)(\sqrt{q} + k).$$

If $k < \sqrt{q} + 1$, then $2(k - 1) < \sqrt{q} + k$. So

$$q + 1 - k > q + \sqrt{q} + 1 - k : \text{a contradiction.}$$

Hence $k = \sqrt{q} + 1$.□

13.2.2. THEOREM. *In $PG(2,q)$, q square,*

(i) a blocking $(q + \sqrt{q} + 1)$-set $\mathscr{B}$ is a subgeometry $PG(2,\sqrt{q})$;

(ii) a blocking $(q^2 - \sqrt{q})$-set $\mathscr{B}'$ is the complement of a subgeometry $PG(2,\sqrt{q})$.

Proof. (i) By the foregoing lemma, some line l contains precisely $\sqrt{q} + 1$ points of $\mathscr{B}$. If P and Q are any two points of $\mathscr{B}\backslash l$, the line PQ meets l in a point of $\mathscr{B} \cap l$. For, if $PQ \cap l \in l\backslash\mathscr{B}$, then the q lines through this point other than l have only $|\mathscr{B}\backslash l| = q$ points to share; so, none of these lines can contain two points of $\mathscr{B}$. Thus for any point P in $\mathscr{B}\backslash l$, the $\sqrt{q} + 1$ lines joining it to the points of $\mathscr{B} \cap l$ contain all q points of $\mathscr{B}\backslash l$; as each such line contains at most $\sqrt{q} + 1$ points of $\mathscr{B}$, each contains exactly $\sqrt{q} + 1$ points of $\mathscr{B}$. So a line through two points of $\mathscr{B}$ contains $\sqrt{q} + 1$ points of $\mathscr{B}$ and two such lines meet in $\mathscr{B}$. So $\mathscr{B}$ is a subgeometry $PG(2,\sqrt{q})$.

(ii) $\Pi\backslash\mathscr{B}'$ is a blocking $(q + \sqrt{q} + 1)$-set, and so (i) gives the result.□

COROLLARY 1. *If q is a square and $\mathscr{B}$ a blocking b-set, then $q + \sqrt{q} + 1 \leqslant b \leqslant q^2 - \sqrt{q}$.*

Proof. Suppose $b = q + \sqrt{q} + 1 - n$, $n > 0$. By lemma 13.1.7, no line contains more than $\sqrt{q} + 1 - n$ points of $\mathscr{B}$. Let $\mathscr{S}$ be any set of n points such that $\mathscr{S} \cap \mathscr{B} = \emptyset$ and $\mathscr{B}' = \mathscr{B} \cup \mathscr{S}$ is not a subgeometry $PG(2,\sqrt{q})$. Then, $\mathscr{B}'$ is a blocking set since no line contains more than $(\sqrt{q} + 1 - n) + n = \sqrt{q} + 1$ of its points. So, by the theorem, $\mathscr{B}'$ is a subgeometry, contradicting the choice of $\mathscr{S}$.

If $b > q^2 - \sqrt{q}$, then $|\Pi\backslash\mathscr{B}| < q + \sqrt{q} + 1$, contradicting the lower bound.□

COROLLARY 2. *If q is a square, a committee is a subgeometry $PG(2,\sqrt{q})$.*□

13.3. Bounds for blocking sets with q non-square

In this section, the same bounds for blocking sets as for the case with q square are established. Throughout this section

$$q = q_1^2 + e \quad \text{where} \quad 1 \leqslant e \leqslant 2q_1.$$

13.3.1. LEMMA. *If $\mathscr{B}$ is a blocking $(q + q_1 + 2)$-set, then some line l contains exactly $q_1 + 2$ points of $\mathscr{B}$.*

Proof. Let l be a line containing the maximum number k of points of $\mathscr{B}$. By lemma 13.1.7, $k \leqslant q_1 + 2$. Applying lemma 13.1.9(vii) with $b = q + q_1 + 2$ gives

$$2(s - 1)(q + 1 - k)(q_1 + 2 - k) \geqslant (q + q_1 + 2 - k)(q + q_1 + 1 + k - k^2).$$

Since $q + q_1 + 2 - k > q + 1 - k$,

$$2(s - 1)(q_1 + 2 - k) > q + q_1 + 1 + k - k^2. \tag{13.1}$$

By lemma 13.1.9(ii), $s \leqslant (q_1 + 3)/2$. So

$$(q_1 + 1)(q_1 + 2 - k) > q + q_1 + 1 + k - k^2.$$

Substituting $q = q_1^2 + e$ gives

$$F(k) > 0,$$

where

$$F(k) = k^2 - (2 + q_1)k + 2q_1 + 1 - e.$$

In particular,

$$\begin{aligned} F(0) &= 2q_1 + 1 - e > 0, \\ F(2) &= 1 - e \leqslant 0, \\ F(q_1) &= 1 - e \leqslant 0. \end{aligned}$$

There are three cases to consider:

(i) $q_1 = 1$. Since $1 \leqslant e \leqslant 2q_1$, $q = 2$ or 3. As there are no blocking sets for $q = 2$, q must be 3. By lemma 13.1.7, $3 \leqslant k \leqslant 3$. So $k = 3 = q_1 + 2$.

(ii) $q_1 = 2$. Then $k \geqslant 3 = q_1 + 1$.

(iii) $q_1 > 2$. Since $F(2) = F(q_1) \leqslant 0$ and $F(0) > 0$, it follows that $F(k) \leqslant 0$ for $2 \leqslant k \leqslant q_1$. Therefore, in order that $F(k) > 0$, $k \geqslant q_1 + 1$.

So there remains the undesirable possibility that $k = q_1 + 1$. Then lemma 13.1.9(i) implies that $s \leqslant 2$ and, as by definition $s \geqslant 2$, so $s = 2$. Hence, from (13.1),

$$2 > q + q_1 + 1 + (q_1 + 1+ - (q_1 + 1)^2 = e + 1.$$

So $e < 1$, contradicting that $e \geqslant 1$.□

13.3.2. THEOREM. *If $\mathscr{B}$ is a blocking b-set in $\Pi = PG(2,q)$, where $q = q_1^2 + e$ with $1 \leqslant e \leqslant 2q_1$, then $b \geqslant q + q_1 + 2$.*

Proof. Suppose $b = q + q_1 + 2 - n$ with $n > 0$. Then, by lemma 13.1.7, we may consider two cases:

(i) no line contains more than $q_1 + 1 - n$ points of $\mathscr{B}$;

(ii) some line l contains exactly $q_1 + 2 - n$ points of $\mathscr{B}$.

In case (i), let $\mathscr{B}' = \mathscr{B} \cup \mathscr{S}$, where $|\mathscr{S}| = n$ and $\mathscr{B} \cap \mathscr{S} = \emptyset$. Since $\mathscr{B}$ meets every line of Π, so does $\mathscr{B}'$. As no line contains more than $(q_1 + 1 - n) + n$ points of $\mathscr{B}'$, the latter is a blocking $(q + q_1 + 2)$-set, which contradicts the previous lemma.

Suppose therefore that case (ii) is true. If Q is any point of $l\backslash\mathscr{B}$, then each line other than l through Q contains exactly one point of $\mathscr{B}\backslash l$; that is, no two points of $\mathscr{B}\backslash l$ are collinear with a point of $l\backslash\mathscr{B}$. So, if P is any point of $\mathscr{B}\backslash l$, the $q_1 + 2 - n$ lines joining P to the points of $\mathscr{B} \cap l$ contain all the points of $\mathscr{B}$. Again, by lemma 13.1.7, no such line contains more than $q_1 + 2 - n$ points of $\mathscr{B}$. Hence, counting the points of $(\mathscr{B}\backslash l)\backslash\{P\}$, we have

$$(q_1 + 2 - n)(q_1 - n) \geqslant q - 1.$$

So

$$(q \quad + 1 - n)^2 \geqslant q. \tag{13.2}$$

By lemma 13.1.7, $q + q_1 + 2 - n \geqslant q + 3$; whence $n \leqslant q_1 - 1$. Also, by hypothesis, $n \geqslant 1$. For n satisfying $1 \leqslant n \leqslant q_1 - 1$, the maximum value of $(q_1 + 1 - n)^2$ is $q_1{}^2$. Thus, from (13.2), $q_1{}^2 \geqslant q$, contradicting that $q = q_1{}^2 + e$.□

13.3.3. THEOREM. *If $\mathscr{B}$ is a blocking b-set in PG(2,q), then*

$$q + \sqrt{q} + 1 \leqslant b \leqslant q^2 - \sqrt{q}.$$

Proof. Since this result was proved in theorem 13.2.2, corollary 1, for q square, it remains to be established for q non-square. If the lower bound is established for $\mathscr{B}$, then the upper bound follows when the lower bound is applied to $\Pi\backslash\mathscr{B}$.

However, the previous theorem asserts that $b \geqslant q + q_1 + 2$ where $q = q_1{}^2 + e$ with $1 \leqslant e \leqslant 2q_1$; so $q_1{}^2 < q < (q_1 + 1)^2$. Hence $q_1 > \sqrt{q} - 1$, which gives the result.□

13.4. Blocking sets in planes of small order

A projective triangle $\mathscr{B}$ of side n in $PG(2,q)$ is a set of $3(n - 1)$ points, n of which lie on each side of a triangle $P_0P_1P_2$ such that the vertices are in $\mathscr{B}$, and, if Q_0 on P_1P_2 and Q_1 on P_2P_0 are in $\mathscr{B}$, then $Q_2 = Q_0Q_1 \cap P_0P_1$ is in $\mathscr{B}$.

13.4.1. THEOREM. *In PG(2,q), q odd, there exists a projective triangle of side $(q + 3)/2$ which is a blocking $[3(q + 1)/2]$-set.*

Proof. The points $Q_0(a_0) = \mathbf{P}(0,1,a_0)$, $Q_1(a_1) = \mathbf{P}(1,0,a_1)$ and $Q_2(a_2) = \mathbf{P}(-a_2,1,0)$, non-vertices on the sides of the triangle of reference, are collinear if and only if $a_0 = a_1a_2$. Let $\mathscr{B}$ consist of all the points $Q_i(a_i)$ such that a_i is a non-zero square plus the vertices of the triangle. Then any line l not through a vertex meets the sides $\mathbf{u}_0,\mathbf{u}_1,\mathbf{u}_2$ in $Q_0(a_0)$, $Q_1(a_1)$, $Q_2(a_2)$ where either one or three of a_0,a_1,a_2 are squares: so $|l \cap \mathscr{B}| = 1$ or 3.□

For q even we can give a similar construction. As on previous occasions the occurrence of squares and non-squares for q odd invites the use of elements of category zero and one for q even. A *projective triad* $\mathscr{B}$ *of side* n is a set of $3n - 2$ points n of which lie on each of three concurrent lines l_0, l_1, l such that the vertex $P \in \mathscr{B}$, and, if Q_0 on l_0 and Q_1 on l_1 are in $\mathscr{B}$. then $Q = Q_0Q_1 \cap l$ is in $\mathscr{B}$.

13.4.2. THEOREM. *In $PG(2,q)$, q even, there exists a projective triad of side $(q + 2)/2$ which is a blocking $[(3q + 2)/2]$-set.*

Proof. Consider the points $Q_0(a_0) = \mathbf{P}(0,1,a_0)$ on $\mathbf{u}_0$, $Q_1(a_1) = \mathbf{P}(1,0,a_1)$ on $\mathbf{u}_1$ and $Q(b) = \mathbf{P}(1,1,b)$ on $\mathbf{V}(x_0 + x_1)$. They are collinear if and only if $b = a_0 + a_1$. So let $\mathscr{B}$ consist of $\mathbf{U}_2$ and the points $Q_0(a_0)$, $Q_1(a_1)$, $Q(b)$ such that a_0, a_1, and b are of category zero, §1.4. Then any line of the plane not through $\mathbf{U}_2$ meets the three lines of the triad in one or three points of $\mathscr{B}$.□

(I) $q = 3$

13.4.3. LEMMA. *In $PG(3,3)$, a blocking* 6-*set* $\mathscr{B}$ *is a projective triangle of side three.*

Proof. By lemma 13.1.7, some line contains three points of $\mathscr{B}$ and no line contains more than three points of $\mathscr{B}$. Let $P_0 \in \mathscr{B}$. Then at least one line through P_0 contains two other points Q_2 and P_1 of $\mathscr{B}$, as otherwise $\mathscr{B}$ could only have five points. If the other three lines through P_0 each contain another point of $\mathscr{B}$, then, by lemma 13.1.4, $\mathscr{B}$ is not minimal and so a blocking 5-set exists, which contradicts theorem 13.3.3. So the four lines through P_0 meet $\mathscr{B}$ in the respective sets $\{P_0,Q_2,P_1\}$, $\{P_0,Q_1,P_2\}$, $\{P_0,Q_0\}$, $\{P_0\}$. Considering the similar result for P_1, either P_1Q_1 or P_1P_2 meets $\mathscr{B}$ in Q_0. Suppose it is P_1P_2. Then, considering similarly Q_0, we have that Q_0,Q_1,Q_2 are collinear. So $\mathscr{B}$ is a projective triangle with vertices P_0,P_1,P_2 and with Q_0,Q_1,Q_2 the remaining points on the sides.□

13.4.4. THEOREM. *In $PG(3,3)$, a blocking b-set $\mathscr{B}$ is a projective triangle of side three or its complement.*

Proof. By theorem 13.3.3, $b = 6$ or 7. If $b = 6$, the previous lemma implies that $\mathscr{B}$ is a projective triangle of side three. If $b = 7$, $\Pi\backslash\mathscr{B}$ is a blocking 6-set and so also a projective triangle.□

(II) $q = 4$

13.4.5. LEMMA. *If $\mathscr{B}$ is a blocking b-set in $\Pi = PG(2,q)$ with $q \geq 4$ such that $|l \cap \mathscr{B}| \leq 3$ for all lines l, then*

(i) $b = 2q - 1$, *or* $2q + 1$, *or* $2q + 3$;

(ii) $|l' \cap \mathscr{B}| \neq 2$ *for all lines l' meeting a fixed trisecant l in $l\backslash\mathscr{B}$.*

Proof. First, bounds for b are obtained. By lemma 13.1.7, there is some line l meeting $\mathscr{B}$ in exactly three points. So, applying lemma 13.1.9(vii) with $k = s = 3$ gives

$$4(q - 2)(b - 3 - q) \geq (b - 3)(b - 7). \qquad (13.3)$$

Rearranging gives

$$4q^2 - 4(b - 1)q + b^2 - 2b - 3 \leq 0,$$

whence

$$(2q - b + 1)^2 \leq 4; \qquad (13.4)$$

so

$$2q - 1 \leq b \leq 2q + 3. \qquad (13.5)$$

The remainder of the proof consists of improving the upper and lower bounds for $|\mathscr{I}|$ as in the corollary to lemma 13.1.9.

When b is even, lemma 13.1.9, corollary (ii), gives that $m(P) \geq (b - 6)/2$ instead of $m(P) \geq (b - 7)/2$ used to obtain (13.3), which now becomes

$$4(q - 2)(b - 3 - q) \geqslant (b - 3)(b - 6). \qquad (13.6)$$

Then $b = 2q$ and $b = 2q + 2$ lead to a contradiction in (13.6). So $b = 2q - 1$, $2q + 1$, or $2q + 3$.

Suppose now that there exists some line l' such that $l' \cap \mathscr{B} = \{P_1, P_2\}$ and $l' \cap l = \{Q\} \notin \mathscr{B}$. Then. by lemma 13.1.9, corollary (ii), $m(P_i) \geqslant (b - 6)/2$; since b is odd, this implies that $m(P_i) \geqslant (b - 5)/2$. Thus as (13.3) was obtained by using $m(P_i) \geqslant (b - 7)/2$ in the lower bound for $|\mathscr{I}|$, we have the improvement

$$4(q - 2)(b - 3 - q) \geqslant (b - 3)(b - 7) + 4, \qquad (13.7)$$

whence

$$(2q - b + 1)^2 \leqslant 0. \qquad (13.8)$$

Consider the lines other than l through q and take, as in lemma 13.1.9, r_i to be the number of lines, other than l, through q meeting $\mathscr{B}$ in exactly i points $(i = 1,2,3)$. Then $r_1 + r_2 + r_3 = q$, $r_1 + 2r_2 + 3r_3 = 2q - 2$. If $r_2 = 1$ and $r_3 = 0$, then $r_1 = q - 1 = 2q - 4$, whence $q = 3$, a contradiction. So there is either a bisecant of $\mathscr{B}$ besides l', in which case there is a further increase in the lower bound for $|\mathscr{I}|$, or there is another trisecant of $\mathscr{B}$ besides l, in which case lemma 13.1.9, corollary (iii), decreases $r(Q)$ and the upper bound for $|\mathscr{I}|$. In either case, we have an immediate contradiction.□

13.4.6. THEOREM. *If, in $PG(2,q)$ with $q \geqslant 4$, a blocking b-set $\mathscr{B}$ is such that $|\mathscr{B} \cap l| \leqslant 3$ for all lines l, then*

(i) $q = 4$;

(ii) $|\mathscr{B} \cap l| = 1$ *or* 3 *for all lines* l;

(iii) $\mathscr{B}$ *is a subgeometry* $PG(2,2)$ *or a Hermitian curve* $\mathscr{U}_2$.

Proof. In terms of lemma 13.1.9, let l be a trisecant of $\mathscr{B}$ and $\mathscr{L}'$ the lines joining points of $l \backslash \mathscr{B}$ to at least two points of $\mathscr{B} \backslash l$. By the previous lemma, every line of $\mathscr{L}'$ is a trisecant of $\mathscr{B}$. If Q is any point of $l \backslash \mathscr{B}$, then every line through Q is a unisecant or a trisecant of $\mathscr{B}$. Then lemma 13.1.9, corollary (iii), becomes $r(Q) = 3r_3 = 2(b - 3 - q) - r_3$, whence $r_3 = (b - 3 - q)/2$, and $r(Q) = 3(b - 3 - q)/2$. So from part (vi) of lemma 13.1.9, $|\mathscr{I}| = 3(q - 2)(b - 3 - q)/2$. From part (iv) of the same lemma, $|\mathscr{I}| \geqslant (b - 3)(b - 7)/2$, whence

$$3(q - 2)(b - 3 - q) \geqslant (b - 3)(b - 7) \tag{13.9}$$

By the previous lemma, only the cases $b = 2q - 1$, $2q + 1$, and $2q + 3$ need to be considered.

(a) $b = 2q + 3$. (13.9) becomes

$$3q(q - 2) \geqslant 4q(q - 2),$$

a contradiction since $q \geqslant 4$.

(b) $b = 2q + 1$. (13.9) becomes

$$3(q - 2)^2 \geqslant 4(q - 1)(q - 3),$$

which holds only for $q = 4$.

(c) $b = 2q - 1$. (13.9) becomes

$$3(q - 2)(q - 4) \geqslant 4(q - 2)(q - 4),$$

which holds only for $q = 4$.

In case (c), $\mathscr{B}$ is a blocking 7-set in $PG(2,4)$ and therefore, by theorem 13.2.2, a subgeometry $PG(2,2)$. It is also the example given in lemma 13.1.5 (ii) and in theorem 13.4.2. In case (b), $\mathscr{B}$ is a blocking 9-set in $PG(2,4)$ and the inequality is an equality. So, from above, $|\mathscr{I}| = 6$ and $m(P) = 1$ for each P in $\mathscr{B}\backslash l$. This means that, for a point P not on a particular trisecant l of $\mathscr{B}$, the two lines joining P to the points of $l\backslash\mathscr{B}$ are a trisecant and a unisecant. So the lines joining P to the three points of $l \cap \mathscr{B}$ are all trisecants. Thus through P there are four trisecants and one unisecant of $\mathscr{B}$. As there is a trisecant through P meeting l outside $\mathscr{B}$. the three points of $\mathscr{B} \cap l$ have the same property as P. Hence $\mathscr{B}$ is a (9;3)-arc such that through each of its points there are four trisecants and one unisecant. So $\mathscr{B}$ is a Hermitian arc, which, by theorem 11.1.1, corollary, is a Hermitian curve $\mathscr{U}_2$.□

(III) $q = 5$

13.4.7. THEOREM. *In PG(2,5), a committee has nine points.*

Proof. By theorem 13.3.3, a blocking b-set has $b \geqslant 9$. Both lemma 13.1.5(ii) and theorem 13.4.1 give examples with $b = 9$. In this case these two examples are the same.□

(IV) $q = 7$

By theorem 13.3.3 a blocking b-set has $b \geqslant 11$.

13.4.8. THEOREM. *In PG(2,7), a committee has* 12 *points.*

Proof. Suppose there exists a blocking 11-set $\mathscr{B}$ and k is the largest number of points of $\mathscr{B}$ on any line. By lemma 13.1.7, $k < 5$ and, by theorem 13.4.5, $k > 3$; so $k = 4$. Therefore, there exists a line l_0 with $|l_0 \cap \mathscr{B}| = 4$, and, for any such line, every other line through a point of $l_0 \backslash \mathscr{B}$ meets $\mathscr{B}$ in exactly one point. Alternatively, the joins of the seven points of $\mathscr{B} \backslash l_0$ meet l_0 in the four points of $l_0 \cap \mathscr{B}$. So at least one set of three points of $\mathscr{B} \backslash l_0$ is collinear on the line l_1, say. If the remaining four points P_1, P_2, P_3, P_4 of $\mathscr{B} \backslash l_0$ are vertices of a tetrastigm $\mathscr{Q}$, the six sides of $\mathscr{Q}$ would pass through the four points of $l_0 \cap \mathscr{B}$ and similarly through the four points of $l_1 \cap \mathscr{B}$. Then both l_0 and l_1 contain two diagonal points of $\mathscr{Q}$. So $P_0 = l_0 \cap l_1$ is a diagonal point, say $P_0 = P_1P_2 \cap P_3P_4$ as in Fig. 18.

Each tetrad of points on a line is harmonic. For example,

$$(P_0, S_1, R_1, T_1) \overset{P_3}{\barwedge} (P_4, P_1, S_0, T_1) \overset{P_2}{\barwedge} (S_1, P_0, R_1, T_1) \barwedge (P_0, S_1, T_1, R_1).$$

Now consider the point $R_1 = l_1 \cap P_2P_3$. The line R_1P_1 must pass through a point of $\mathscr{B} \cap l_0$; the only possible one is $T_0 = l_0 \cap P_2P_4$.

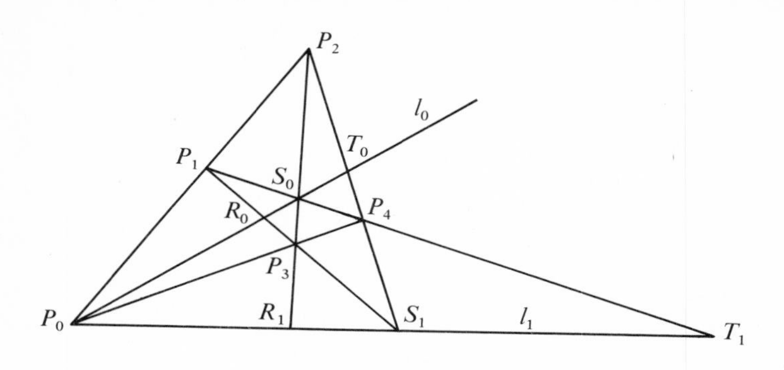

FIG. 18

But then $(P_0,S_1,R_1,T_1)\overset{P_1}{\bar{\barwedge}}(P_2,T_0,S_1,P_4)\overset{S}{\bar{\barwedge}}(R_1,P_0,S_1,T_1)\barwedge(P_0,R_1,T_1,S_1)$. So the tetrad of points in $\mathscr{B}\cap l_1$ is both equianharmonic and harmonic, which can only occur for characteristic three, §6.1. So P_1,P_2,P_3,P_4 are not the vertices of a tetrastigm.

Now suppose that P_2,P_3,P_4 are collinear on l_2. The point $l_2\cap l_1\in\mathscr{B}$. If it is not P_0, then $l_2\cap l_0 = P\notin\mathscr{B}$, since $|l_2\cap\mathscr{B}|<5$. Since l_2 and l_0 contain seven points of $\mathscr{B}$, only six lines through P can contain a point of $\mathscr{B}$. So l_0,l_1,l_2 are concurrent at P_0. The nine points of $\mathscr{B}\backslash\{P_0,P_1\}$ are also collinear in threes on lines through P_1, as no line through P_1 and a point of l_0, say, can meet l_1 or l_2 outside $\mathscr{B}$. For the same reason, any two of these nine points are collinear with a third. These nine points therefore form a tactical configuration $(9_4,12_3)$, which only exists when $q\equiv 0$ or $1 \pmod 3$, §11.1. However, only when $q\equiv 0 \pmod 3$ is a set of three lines containing the nine points concurrent; for $q\equiv 1 \pmod 3$, the three lines form a triangle. So, again the configuration cannot exist (Fig. 19).

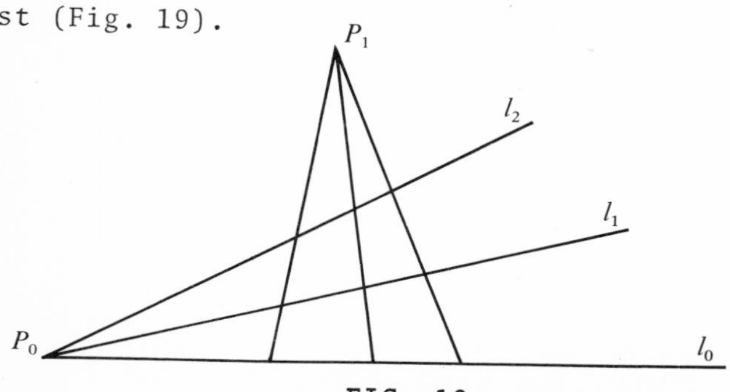

FIG. 19

Two distinct examples of 12-point committees are a projective triangle of side five and the dual of the $(9_4, 12_3)$ configuration. In the latter case, $\mathscr{B}$ is a set of 12 points such that nine lines are 4-secant to $\mathscr{B}$, three through each point, and that 12 lines are bisecant to $\mathscr{B}$, two through each point. So the number of lines meeting $\mathscr{B}$ is $9 + 12 \times 2/2 + 12 \times 3 = 57$. So $\mathscr{B}$ is a blocking set with at most four points on any line of the plane.□

(V) $q = 8$

By theorem 13.3.3, a blocking b-set has $b \geqslant 12$.

13.4.9. THEOREM. *In $PG(2,8)$, a committee has* 13 *points.*

Proof. By theorem 13.4.2, a projective triad of side five is a blocking 13-set. So we consider a blocking 12-set $\mathscr{B}$ and show that it cannot exist. By lemma 13.1.7, the maximum number k of points of $\mathscr{B}$ on a line satisfies $k \leqslant 4$. By theorem 13.4.6, $k > 3$. So $k = 4$.

In contrast to the geometrical arguments used in the previous theorem, we use the more combinatorial methods of Chapter 12. In that terminology, $\mathscr{B}$ is a (12;4)-arc with $\tau_0 = 0$, that is, it has no 0-secants. To recall the notation, τ_i is the number of i-secants of $\mathscr{B}$, ρ_i the number of i-secants through a point of $\mathscr{B}$, and σ_i the number of i-secants through a point of $\Pi \backslash \mathscr{B}$. With $\tau_0 = 0$, $k = 12$, and $n = 4$, eqns (12.1_i), $i = 1,2,3$, are

$$\begin{aligned} \tau_1 + \tau_2 + \tau_3 + \tau_4 &= 73, \\ \tau_1 + 2\tau_2 + 3\tau_3 + 4\tau_4 &= 108, \\ \tau_2 + 3\tau_3 + 6\tau_4 &= 66. \end{aligned}$$

In terms of τ_4,

$$\tau_1 = 69 - \tau_4, \; \tau_2 = 3\tau_4 - 27, \; \tau_3 = 31 - 3\tau_4. \qquad (13.10)$$

Since $\tau_0 = \sigma_0 = 0$, more information should be obtainable from

from the σ_i than the ρ_i. Equations (12.3_i), $i = 1,2$, are

$$\sigma_1 + \sigma_2 + \sigma_3 + \sigma_4 = 9,$$
$$\sigma_1 + 2\sigma_2 + 3\sigma_3 + 4\sigma_4 = 12.$$

The only possible solutions are as follows:

	σ_1	σ_2	σ_3	σ_4
(1)	8	0	0	1
(2)	7	1	1	0
(3)	6	3	0	0

Suppose there are s_1 points with solution (1), s_2 with solution (2), and s_3 with solution (3). Then eqns (12.9) and (12.8_i), $i \in N_4$, are $s_1 + s_2 + s_3 = 61$, $8s_1 + 7s_2 + 6s_3 = 8\tau_1$, $s_2 + 3s_3 = 7\tau_2$, $s_2 = 6\tau_3$, $s_1 = 5\tau_4$. Substitution for τ_1,τ_2,τ_3 from (13.10) in the last three equations gives

$$s_1 = 5\tau_4,\ s_2 = 186 - 18\tau_4,\ s_3 = 13\tau_4 - 125;$$

these values for s_1,s_2,s_3 satisfy the other two equations. Therefore, from s_2 and s_3, $\tau_4 = 10$. However, eqn (12.2_2) is

$$\rho_2 + 2\rho_3 + 3\rho_4 = 11,$$

whence $\rho_4 \leqslant 3$. Finally, eqn (12.4_4) is

$$4\tau_4 = \Sigma\rho_4.$$

As $\rho_4 \leqslant 3$ and $\mathscr{B}$ has 12 points, $\Sigma\rho_4 \leqslant 36$. However, $4\tau_4 = 40$ and we have the desired contradiction.□

13.5. Further examples

Let $q = p^h$ and $q_1 = p^d$, where $h = md$ and $d \neq h$. $\gamma' = GF(q_1)$ and $\Pi' = PG(2,q_1)$ is a subgeometry of $\Pi = PG(2,q)$. In each of the following cases, $\mathscr{B}$ is a blocking b-set in Π.

(i) $\mathscr{B} = \{P(1,t,t^{q_1}) \mid t \in \gamma\} \cup \{P(0,1,t^{q_1-1}) \mid t \in \gamma\}$:

$$b = (q_1^{m+1} - 1)/(q_1 - 1) =$$

$$= (p^{h+d} - 1)/(p^d - 1) = (q^{(h+d)/h} - 1)/(q^{d/h} - 1).$$

(ii) Every t in γ can be written

$$t = c_0 + c_1\alpha + \ldots + c_{m-1}\alpha^{m-1}, \quad \alpha \in \gamma, \ c_i \in \gamma'.$$

$$\mathscr{B} = \{\mathbf{P}(1,c,t)\} \cup \{\mathbf{P}(0,1,t)\} \cup \mathbf{U}_2,$$

where c varies in γ' and $t = c_1\alpha + c_2\alpha^2 + \ldots + c_{m-1}\alpha^{m-1}$ with α fixed in γ and the c_i varying in γ':

$$b = q_1{}^m + q_1{}^{m-1} + 1 = p^h + p^{h-d} + 1 = q + q^{(h-d)/h} + 1.$$

(iii) Let l_0, l_1, l_2 be three lines concurrent at a point of Π' and all having $q_1 + 1$ points in Π'.

$$\mathscr{B} = (l_0 \cap \Pi') \cup (l_1\backslash\Pi') \cup (l_2\backslash\Pi'):$$

$$b = 2q - q_1 + 1 = 2p^h - p^d + 1 = 2q - q^{d/h} + 1.$$

(iv) $q = 3^h$, $h \geq 2$, θ is a non-square in γ.

$$\mathscr{B} = \mathbf{V}((x_0{}^2x_2 - x_1{}^3)(x_0{}^2x_2 - \theta x_1{}^3)):$$

$$b = 2q.$$

$$|l \cap \mathscr{B}| \leq 4, \text{ all lines } l \text{ in } \Pi.$$

13.6. Notes and references

§13.1. This section is based on Bruen [1971a], Bruen and Fisher [1974]. See also Isbell [1958], Richardson [1956].

§13.2. This section comes from Bruen [1970].

§13.3. This section comes from Bruen [1971a].

§13.4. See Di Paola [1969], Bruen and Fisher [1973], [1974].

§13.5. See Bruen [1971a], Bruen and Fisher [1974], Ostrom [1968].

For the related concept of a k-block, see Tutte [1966], as well as Datta [1976a], [1976b], Tutte [1967], [1969].

14
PG (2, *q*) FOR SMALL *q*

14.1. $PG(2,2)$

In $PG(2,2)$, $\theta(1) = 3$ and $\theta(2) = 7$. The points are denoted P_i, $i \in \overline{\mathbf{N}}_6$, with the lines given in Fig. 20 and i written for P_i.

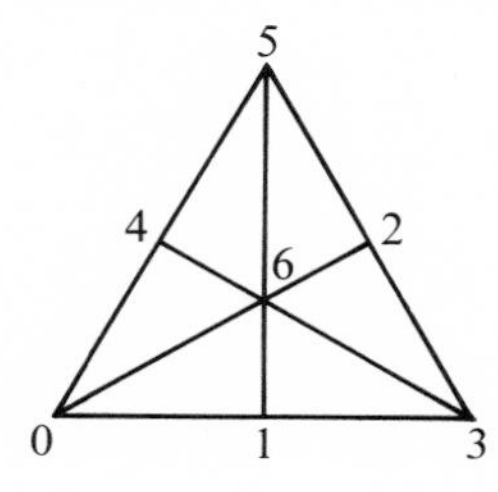

0	1	2	3	4	5	6
1	2	3	4	5	6	0
3	4	5	6	0	1	2

FIG. 20

Given two points in the plane, there are four not collinear with them. Given a 3-arc, its bisecants contain six points, leaving just one in the plane. Thus each 3-arc is in a unique 4-arc, and the maximum number $m(2,2)$ of points in $PG(2,2)$ no three of which are collinear is four. The number of 3-arcs or triangles is $7 \cdot 6 \cdot 4/(3 \cdot 2) = 28$ and the number of 4-arcs (ovals) or tetrastigms (and hence tetragrams) is $28/4 = 7$. Also, each point lies in four ovals.

Since a tetrastigm $\mathscr{Q}$ has four vertices and six sides, there remain in the plane the three diagonal points of $\mathscr{Q}$ and one line l. Through each vertex of $\mathscr{Q}$ there are three sides and through each diagonal point two sides. Hence the three diagonal points are collinear on l.

A conic is a 3-arc and vice versa. Hence every oval is a conic plus its nucleus in four ways. For example, if $\mathscr{O} = \{U_0, U_1, U_2, U\}$, then the four conics are

$$V(x_0^2 + x_1x_2),$$
$$V(x_1^2 + x_0x_2),$$
$$V(x_2^2 + x_0x_1),$$
$$V(x_0x_1 + x_0x_2 + x_1x_2).$$

Since a 3-arc determines a 4-arc, each projectivity is determined by the images of points of a 3-arc. From theorem 7.4.1, the number of distinct types of projectivity is $e(3,2) = 6$: there is one type of order one, two, three, and four and two types of order seven. Also $p(3,2) = 168$, and we now find the size of each conjugacy class in $PGL(3,2)$.

Let Q_1, Q_2, Q_3, Q_4 be vertices of a tetrastigm $\mathcal{Q}$ and let $D_1 = Q_1Q_2 \cap Q_3Q_4$, $D_2 = Q_1Q_3 \cap Q_2Q_4$, $D_3 = Q_1Q_4 \cap Q_2Q_3$ be the diagonal points.

(i) If a projectivity $\mathfrak{T}$ acts as a 3-cycle $(Q_1Q_2Q_3)$, then Q_4 is fixed, as is the line $D_1D_2D_3$: the diagonal points are permuted as $(D_1D_3D_2)$. As there is a projectivity determined by the cycle $(Q_1Q_3Q_2)$, there are $2\cdot 28 = 56$ projectivities of order three.

(ii) If $Q_1\mathfrak{T} = Q_2$, $Q_2\mathfrak{T} = Q_3$, $Q_3\mathfrak{T} = Q_4$, then $\mathcal{Q}$ is fixed and $\mathfrak{T}$ acts as $(Q_1Q_2Q_3Q_4)$, whence D_2 is fixed, but D_1 and D_3 are interchanged. As there are six distinct cycles $(Q_1Q_iQ_jQ_k)$ with $\{i,j,k\} = \{2,3,4\}$, there are $6\cdot 7 = 42$ projectivities of order four.

(iii) If there is an involution $\mathfrak{T}$ which interchanges the points P_0 and P_1 in Fig. 20, then P_3 is fixed. As $\mathfrak{T}$ must fix an odd number N of points, $N = 1$ or 3. Suppose $N = 3$: they cannot be non-collinear, as $\mathfrak{T}$ would be the identity. Suppose therefore that $\mathfrak{T}$ fixes P_2 and P_5, apart from P_3. Then P_4 and P_6 are interchanged, as are the lines $P_0P_4P_5$ and $P_1P_6P_5$, $P_0P_2P_6$ and $P_1P_2P_4$. Thus $\mathfrak{T}$ acts as $(P_0P_1)(P_4P_6)$, fixing a tetrastigm and leaving each of the diagonal points fixed. As there are three pairs of opposite vertices of a tetrastigm, there are $3\cdot 7 = 21$ involutions of this type, and each is the square of a projectivity of order four as described in (ii).

If $\mathfrak{T}$ only fixes P_3, then the lines $P_2P_3P_5$ and $P_3P_4P_6$ are interchanged, as otherwise the 4-arc $\{P_0, P_1, P_4, P_6\}$ would be fixed as above. But then, if P_2 and P_4 are interchanged, P_1 is fixed, and, if P_2 and P_6 are interchanged, P_0 is fixed: in either case, we have a contradiction.

(iv) Finally, we consider a projectivity $\mathfrak{T}$ not of order one, two, three, or four. If $Q_1\mathfrak{T} = Q_2$, $Q_2\mathfrak{T} = Q_3$, $Q_3\mathfrak{T} = P$, then P cannot be Q_1, as $\mathfrak{T}$ does not have order three; nor can $\{Q_1,Q_2,Q_3,P\}$ be a 4-arc. As $\{Q_1,Q_2,Q_3\}$ and $\{Q_2,Q_3,P\}$ are 3-arcs, either Q_1Q_2P or Q_1Q_3P is a line. Hence, if $P_0\mathfrak{T} = P_1$ and $P_1\mathfrak{T} = P_2$, there are the possible cycles $(P_0P_1P_2P_3P_4P_5P_6)$ and $(P_0P_1P_2P_6P_5P_3P_4)$. If the cycles are $(a_0a_1a_2a_3a_4a_5a_6)$, then, in the first cycle, a_i,a_{i+1},a_{i+3} are collinear, and, in the second, a_i,a_{i+2},a_{i+3} are collinear, where the indices are taken modulo seven in both cases. This gives two classes of projectivities of order seven, each with $6\cdot 4 = 24$ elements.

The conjugacy classes of $PGL(3,2)$ can therefore be summarized in Table 14.1.

TABLE 14.1

Order of Elements	Number of Elements in class	Typical Element	Canonical elementary divisors
1	1	$(P_0)(P_1)(P_2)(P_3)(P_4)(P_5)(P_6)$	$x+1$, $x+1$, $x+1$
2	21	$(P_0P_1)(P_4P_6)(P_2)(P_3)(P_5)$	$(x+1)^2$, $x+1$
3	56	$(P_0P_1P_2)(P_3P_4P_6)(P_5)$	x^2+x+1, $x+1$
4	42	$(P_0P_1P_2P_5)(P_3P_4)(P_6)$	$(x+1)^3$
7	24	$(P_0P_1P_2P_3P_4P_5P_6)$	x^3+x+1
7	24	$(P_0P_6P_5P_4P_3P_2P_1)$	x^3+x^2+1

Finally, the projective classification of k-arcs for all possible k is given in Table 14.2, with G the projective group.

TABLE 14.2

k	Total number of k-arcs	G	$\|G\|$
1	7	$\mathbf{S}_4$	24
2	21	$\mathbf{D}_4$	8
3	28	$\mathbf{S}_3$	6
4	7	$\mathbf{S}_4$	24

14.2. $PG(2,3)$

TABLE 14.3

0	1	2	3	4	5	6	7	8	9	10	11	12
1	2	3	4	5	6	7	8	9	10	11	12	0
3	4	5	6	7	8	9	10	11	12	0	1	2
9	10	11	12	0	1	2	3	4	5	6	7	8

In $PG(2,3)$, $\theta(1) = 4$ and $\theta(2) = 13$. The points P_i, $i \in \bar{\mathbf{N}}_{12}$, lie on lines as in the columns of Table 14.3 with i for P_i. Alternatively, the points and lines can be represented as in Fig. 21 below.

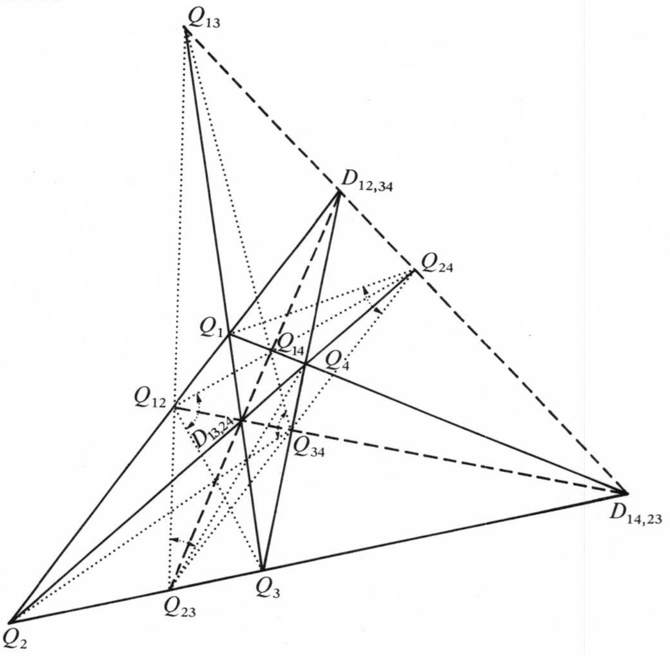

FIG. 21

Fig. 21 begins with a tetrastigm $\mathcal{Q}$ with vertices Q_1,Q_2,Q_3,Q_4, diagonal points $D_{12,34},D_{13,24},D_{14,23}$, and fourth points on the sides $Q_{12},Q_{13},Q_{14},Q_{23},Q_{24},Q_{34}$. There are

six black lines (sides of $\mathcal{Q}$)	$Q_i\ Q_j\ Q_{ij}\ D_{ij,kl}$
three dashed lines (sides of the diagonal triangle)	$Q_{ij}\ Q_{kl}\ D_{ik,jl}\ D_{il,jk}$
four dotted lines	$Q_i\ Q_{jk}\ Q_{jl}\ Q_{kl}$

Each dotted line is drawn in two segments joined by dotted connecting arrows.

The main property of the plane is that the four points on a line form a superharmonic set; that is, they are projectively equivalent in any order, or, equivalently, the set of four points is both harmonic and equianharmonic. For example,

$$(Q_1, Q_2, Q_{12}, D_{12,34}) \overset{D_{14,23}}{\bar{\bar{\wedge}}} (Q_{14}, Q_{23}, D_{13,24}, D_{12,34}), \tag{14.1}$$

$$(Q_{14}, Q_{23}, D_{13,24}, D_{12,34}) \overset{Q_{24}}{\bar{\wedge}} (Q_{12}, Q_1, Q_2, D_{12,34}), \tag{14.2}$$

$$(Q_{14}, Q_{23}, D_{13,24}, D_{12,34}) \overset{Q_3}{\bar{\wedge}} (Q_{12}, Q_2, Q_1, D_{12,34}). \tag{14.3}$$

From (14.1) and (14.2), the four points on Q_1Q_2 are equianharmonic: from (14.1) and (14.3), they are harmonic: see §6.1.

If a line and the four points on it are deleted from $PG(2,3)$ to give $AG(2,3)$, there remains a $(9_4, 12_3)$ figure of nine points on 12 lines with three points on a line and four lines through a point. This is the arrangement of the nine inflexions of a non-singular cubic curve in $PG(2,\mathbf{C})$: see §11.1. A difference from the classical figure is that each of the four sets of three lines through the nine points of $AG(2,3)$ is concurrent on the deleted line, instead of forming a triangle. Indeed, the nine points of $AG(2,3)$ are not the inflexions of a cubic, since there can be at most a pencil of cubics through the nine points and we have four cubics each consisting of three lines. So there is no absolutely irreducible cubic through the nine points. As $p(3,3) = 2^4 \cdot 3^3 \cdot 13$, so $p(3,3)/13 = 2^4 \cdot 3^3 = 432$, which is the order of the projective group fixing the nine points.

The number of conjugacy classes in $PGL(3,3)$ is $e(3,3) = 12$. The classes are given in Table 14.4.

The number of points on the sides of a triangle is nine. Hence each 3-arc lies in four 4-arcs. If the vertices of the **triangle** $\mathscr{T}$ are $\mathbf{U}_0, \mathbf{U}_1, \mathbf{U}_2$, then the 4-arcs are completed by $\mathbf{U}$, $\mathbf{P}(1,1,-1)$, $\mathbf{P}(1,-1,1)$, $\mathbf{P}(-1,1,1)$. These four points themselves form a 4-arc, which is a conic $\mathscr{C} = \mathbf{V}(x_0^2 + x_1^2 + x_2^2)$ for which $\mathscr{T}$ is self-polar. Thus, given a triangle $\mathscr{T}$,

TABLE 14.4

Order of element in class	Canonical elementary divisors
1	$x - 1, x - 1, x - 1$
2	$x - 1, x - 1, x + 1$
2	$x^2 + 1, x - 1$
3	$(x - 1)^3$
3	$(x - 1)^2, x - 1$
4	$x^2 + x - 1, x - 1$
6	$(x - 1)^2, x + 1$
8	$x^2 - x - 1, x - 1$
13	$x^3 - x - 1$
13	$x^3 + x - 1$
13	$x^3 - x^2 - 1$
13	$x^3 + x^2 - 1$

there is a unique conic with no point on a side of $\mathscr{T}$, and, equivalently, given a conic $\mathscr{C}$, there is a unique self-polar triangle with all sides external lines of $\mathscr{C}$; compare theorem 8.3.4(ii)(a).

For $k \geq 3$, the projective classification of k-arcs is given in Table 14.5, with G the group.

TABLE 14.5

k	Total number of k-arcs	G	$\lvert G\rvert$
3	234	$\mathbf{S}_4$	24
4	234	$\mathbf{S}_4$	24

14.3. $PG(2,4)$

In $PG(2,4)$, $\theta(1) = 5$ and $\theta(2) = 21$: see Table 14.6. A 4-arc has 19 points on its bisecants and a 5-arc 20 points. So there are exactly two points on no bisecant of a 4-arc and one point on no bisecant of a 5-arc. Therefore each 4-arc lies in a unique 6-arc, and $m(2,4) = 6$. Thus a 6-arc or oval $\mathscr{O}$ consists of a 5-arc or conic plus its nucleus in six ways.

TABLE 14.6

0	1	2	3	4	5	6	7	8	9	10	11	12	13	14	15	16	17	18	19	20
1	2	3	4	5	6	7	8	9	10	11	12	13	14	15	16	17	18	19	20	0
4	5	6	7	8	9	10	11	12	13	14	15	16	17	18	19	20	0	1	2	3
14	15	16	17	18	19	20	0	1	2	3	4	5	6	7	8	9	10	11	12	13
16	17	18	19	20	0	1	2	3	4	5	6	7	8	9	10	11	12	13	14	15

Let $\mathscr{K} = \{Q_1, Q_2, Q_3, Q_4\}$ be a 4-arc and let $\mathscr{Q}$ be the tetrastigm with the same points as vertices. Let $D_1 = Q_1Q_2 \cap Q_3Q_4$, $D_2 = Q_1Q_3 \cap Q_2Q_4$, $D_3 = Q_1Q_4 \cap Q_2Q_3$ be the diagonal points of $\mathscr{Q}$. If D_1, D_2, D_3 are not collinear, then D_1D_2 contains D_1, D_2, a point on Q_1Q_4, a point on Q_2Q_3, and one other point R_3, which lies on no bisecant of $\mathscr{K}$. Similarly D_1D_3 contains a point R_2 and D_2D_3 a point R_1, neither of which lies on a bisecant of $\mathscr{K}$. So $\mathscr{K} \cup \{R_i\}$ is a 5-arc for $i = 1,2,3$, contradicting that $\mathscr{K}$ only lies in two 5-arcs. Thus D_1, D_2, D_3 are collinear and the remaining two points on this diagonal line of $\mathscr{Q}$ form the unique oval $\mathscr{O}$ containing $\mathscr{K}$.

This applies to all fifteen 4-arcs contained in $\mathscr{O}$. Further, there are 15 ways in which the six points of $\mathscr{O}$ can be partitioned into three pairs. Hence, in each such partition, the three bisecants of $\mathscr{O}$ joining the pairs are concurrent.

A hexagon is a *Brianchon hexagon* if the joins of opposite vertices are concurrent. Thus $\mathscr{O}$ forms a Brianchon hexagon in 15 ways, and the 15 points of concurrency are called *Brianchon points* or *B-points* of $\mathscr{O}$.

The six points of $\mathscr{O}$ and its 15 B-points include all 21 points of the plane. Each B-point is a diagonal point of three of the tetrastigms with vertices in $\mathscr{O}$. Now, $\mathscr{O}$ has 15 bisecants all containing three B-points and no unisecants; also, through each B-point there are three bisecants. Hence there are six lines skew to $\mathscr{O}$, called *S-lines*, and through each B-point there are two S-lines. So the six S-lines have no three concurrent and thus form a figure $\mathscr{O}'$ dual to $\mathscr{O}$.

If $\mathcal{O} = \{Q_i \mid i \in \mathbf{N}_6\}$, then the six lines of $\mathcal{O}'$ are as follows, where we write $(ij, kl, mn) = Q_iQ_j \cap Q_kQ_l \cap Q_mQ_n$:

I.	(12,34,56),	(13,25,46),	(14,26,35),	(15,24,36),	(16,23,45)
II.	(12,34,56),	(13,26,45),	(14,25,36),	(15,23,46),	(16,24,35)
III.	(12,36,45),	(13,25,46),	(14,23,56),	(15,26,34),	(16,24,35)
IV.	(12,36,45),	(13,24,56),	(14,26,35),	(15,23,46),	(16,25,34)
V.	(12,35,46),	(13,26,45),	(14,23,56),	(15,24,36),	(16,25,34)
VI.	(12,35,46),	(13,24,56),	(14,25,36),	(15,26,34),	(16,23,45)

Since $\mathcal{O}$ is determined by the 4-arc $\mathcal{K}$, if a projectivity $\mathfrak{T}$ fixes each point of $\mathcal{K}$, it fixes the remaining two points of $\mathcal{O}$. As a 4-arc can be transformed to any other by a projectivity, any even permutation of the six points of $\mathcal{O}$ can be performed by a projectivity. Thus $\mathcal{O}$ has a projective group $\mathbf{A}_6$ of order 360 and a collineation group $\mathbf{S}_6$ of order 720.

The existence of $\mathcal{O}$ and $\mathcal{O}'$ also illustrates that $\mathbf{S}_6$ has 12 subgroups isomorphic to $\mathbf{S}_5$. If Q_i is fixed, then the collineations permuting the other five points of $\mathcal{O}$ form a group isomorphic to $\mathbf{S}_5$. This group does not fix any of the lines of $\mathcal{O}'$. So the collineations fixing one line of $\mathcal{O}'$ and permuting the others form a group isomorphic to $\mathbf{S}_5$ and fixing $\mathcal{O}$ but not any of its points. Thus the collineation group of $\mathcal{O}$ contains two sets of six conjugate subgroups all isomorphic to $\mathbf{S}_5$. The outer automorphisms of $\mathbf{A}_6$ and $\mathbf{S}_6$ are given by the correlations and reciprocities respectively that interchange $\mathcal{O}$ and $\mathcal{O}'$.

A hexagon is a *Pascal hexagon* if the intersections of pairs of opposite sides are collinear. Dual to the points of $\mathcal{O}$ as the vertices of a hexastigm, the lines of $\mathcal{O}'$ can be considered as sides of a hexagram. Hence $\mathcal{O}'$ forms a Pascal hexagon in 15 ways and the 15 lines formed are called *Pascal lines* of $\mathcal{O}'$. These lines are the 15 sides of the hexastigm with vertices the points of $\mathcal{O}$. The 15 Brianchon points of $\mathcal{O}$ are the vertices of the hexagram whose sides are the lines of $\mathcal{O}'$.

Each bisecant l of $\mathcal{O}$ contains three B-points of $\mathcal{O}$, which are the diagonal points of the tetrastigm with vertices in $\mathcal{O} \backslash l$. The fact that $\mathcal{O}$ has 15 B-points suffices to construct the plane. This means that if some $PG(2,q)$ contains a 6-arc with 15 B-points, then these 21 points form a $PG(2,4)$ and $q = 4^m$. In $PG(2,q)$ with q odd, a bisecant of a 6-arc contains at most two B-points, whence the 6-arc has at most $15 \cdot 2/3 = 10$ B-points. Compare the chapter on cubic surfaces in Volume II.

In the language of §9.1, the B-points of $\mathscr{O}$ are points of index three. So, from (9.1),

$$c_0 = 0,\ c_1 = 0,\ c_2 = 0,\ c_3 = 15.$$

Consider now the Hermitian curve $\mathscr{U}_{2,4} = \mathrm{V}(x_0{}^3 + x_1{}^3 + x_2{}^3)$. It comprises nine points, any two of which are collinear with a third. So, as in $PG(2,3)$, a $(9_4, 12_3)$ figure is formed. Here the figure differs in that three lines containing the nine points of $\mathscr{U}_{2,4}$ are not collinear. So we obtain four triangles of this type, and their 12 vertices complete the 21 points of the plane. The nine tangents to $\mathscr{U}_{2,4}$ with the 12 sides of the four triangles form the 21 lines of the plane.

Let $\mathscr{T}_1, \mathscr{T}_2, \mathscr{T}_3, \mathscr{T}_4$ be the four triangles. Then $\mathscr{U}_{2,4}$ and the four $\mathscr{T}_i$ form the pencil of cubic curves through the nine points. Further, the vertices of any two of the triangles form an oval; that is, the line joining any two of the 12 vertices of the triangles is a trisecant of $\mathscr{U}_{2,4}$. Thus $\mathscr{T}_1$ and $\mathscr{T}_2$, say, are in six-fold perspective from six points, which can only be the vertices of $\mathscr{T}_3$ and $\mathscr{T}_4$. This can be rephrased as follows. Let $\mathscr{O}$ be an oval and $\mathscr{K}_1, \mathscr{K}_2$ two 3-arcs with $\mathscr{O} = \mathscr{K}_1 \cup \mathscr{K}_2$. Then there exist two other 3-arcs $\mathscr{K}_3, \mathscr{K}_4$ such that $\mathscr{K}_i \cup \mathscr{K}_j$ is an oval for each pair $\{i,j\}$. The nine points in the plane in no $\mathscr{K}_i$ form a Hermitian curve.

With $\mathscr{U}_{2,4}$ as above, the vertices of the triangles are as follows:

$\mathscr{T}_1$: $\mathrm{U}_0, \mathrm{U}_1, \mathrm{U}_2$;

$\mathscr{T}_2$: U, $\mathbf{P}(1,\omega,\omega^2)$, $\mathbf{P}(1,\omega^2,\omega)$;

$\mathscr{T}_3$: $\mathbf{P}(1,1,\omega^2)$, $\mathbf{P}(1,\omega^2,1)$, $\mathbf{P}(\omega^2,1,1)$;

$\mathscr{T}_4$: $\mathbf{P}(1,1,\omega)$, $\mathbf{P}(1,\omega,1)$, $\mathbf{P}(\omega,1,1)$.

The sides of the triangles form the cubic curves $\mathscr{F}(\infty)$, $\mathscr{F}(1)$, $\mathscr{F}(\omega)$, $\mathscr{F}(\omega^2)$ as in §11.5.

From theorem 7.3.1, the number of Hermitian curves in $PG(2,4)$ is 280.

The projective classification of k-arcs is given in Table 14.7: $p(3,4) = 4^3(4^3 - 1)(4^2 - 1) = 2^6 \cdot 3^3 \cdot 5 \cdot 7$.

TABLE 14.7

k	Total number of k-arcs	G	$\|G\|$
3	1120	$\mathbf{S}_3\mathbf{Z}_3^2$	54
4	2520	$\mathbf{S}_4$	24
5	1008	$\mathbf{A}_5$	60
6	168	$\mathbf{A}_6$	360

We now investigate the set of 168 ovals. From above, there is only one oval containing a given 4-arc. Through a given point there are $20 \cdot 16 \cdot 9/(5 \cdot 4 \cdot 3) = 48$ ovals; through two points there are $16 \cdot 9/(4 \cdot 3) = 12$ ovals; through a 3-arc there are $9/3 = 3$ ovals.

Given the oval $\mathscr{O}$, let N_i for $i = 0,1,2,3$ be the number of ovals which have exactly i points in common with $\mathscr{O}$. Then, as there are two other ovals through each three points of $\mathscr{O}$, so $N_3 = 2\mathbf{c}(6,3) = 40$. Of the 11 other ovals through any pair of points of $\mathscr{O}$, there are $4 \cdot 2 = 8$ meeting $\mathscr{O}$ in a third point. So $N_2 = (11 - 8)\mathbf{c}(6,2) = 45$. Of the 47 other ovals through a point of $\mathscr{O}$, there are $5 \cdot 3 = 15$ meeting it in exactly one other point and $10 \cdot 2 = 20$ meeting it in exactly two other points. Hence $N_1 = (47 - 20 - 15)6 = 72$. Thus $N_0 = 168 - 72 - 45 - 40 - 1 = 10$. We also know how these ten ovals arise. For each of the ten partitions of $\mathscr{O}$ into 3-arcs $\mathscr{K}_1$ and $\mathscr{K}_2$, there are two 3-arcs $\mathscr{K}_3$ and $\mathscr{K}_4$ such that $\mathscr{K}_i \cup \mathscr{K}_j$ is an oval for all six pairs $\{i,j\}$: in particular, $\mathscr{K}_3 \cup \mathscr{K}_4$ is an oval with no points in $\mathscr{O}$.

The $1 + 10 + 45 = 56$ ovals having an even number of points in common with $\mathscr{O}$ form an interesting configuration with the following properties:

(i) $\mathcal{D}$ is the orbit of $\mathcal{O}$ under $PSL(3,4)$;

(ii) any two distinct ovals in $\mathcal{D}$ have zero or two points in common;

(iii) if two ovals in $\mathcal{D}$ have two points in common, there are exactly two ovals disjoint from them both;

(iv) $\mathcal{D}$ forms a symmetric 2-(56,11,2) design whose points are the 56 ovals and whose blocks are the sets of 11 ovals equal to or disjoint from a given oval: compare the chapter on Hermitian surfaces in Volume II.

14.4. *PG*(2,5)

In $PG(2,5)$, $\theta(1) = 6$ and $\theta(2) = 31$. The numbers of points of the plane on no bisecant of a 3-arc, a 4-arc, and a 5-arc are respectively 16, 6, and 1. Hence the conic containing a 5-arc $\mathcal{K}$ is the only 6-arc or oval containing $\mathcal{K}$; that is, every oval is a conic.

Let $\mathcal{C}$ be a conic and l one of its bisecants. Let $\mathcal{Q}$ be the tetrastigm with vertices the points of $\mathcal{C}\backslash l$. Each of the six sides of $\mathcal{Q}$ meets l in one of the four points of $l\backslash\mathcal{C}$. The six points of intersection of sides of $\mathcal{Q}$ with l cannot coincide in three pairs, for then the three diagonal points of would be collinear. So l must contain exactly two diagonal points of $\mathcal{Q}$. So $\mathcal{C}$ has $15\cdot 2/3 = 10$ B-points. There are also 15 points in the plane through which exactly two bisecants of $\mathcal{C}$ pass, and through each of these points there are two tangents to $\mathcal{C}$. These points are thus the external points of $\mathcal{C}$, and the B-points are the internal points. Apart from the six tangents and the 15 bisecants of $\mathcal{C}$, there are ten external lines, three of which pass through each B-point. Thus the ten B-points and the ten external lines form a Desargues figure, §7.5. The external lines of $\mathcal{C}$ are also its Pascal lines. In terms of (9.1), the parameters c_i for $\mathcal{C}$ are $c_0 = 0$, $c_1 = 0$, $c_2 = 15$, $c_3 = 10$. A conic is the only complete arc in $PG(2,5)$.

$p(3,5) = 5^3(5^3 - 1)(5^2 - 1) = 2^5\cdot 3\cdot 5^3\cdot 31$. For a given k, each k-arc is projectively unique. The projective classification is given in Table 14.8:

TABLE 14.8

k	Total number of k-arcs	G	$\|G\|$
3	$5^3\cdot 31$	$\mathbf{S}_3\mathbf{Z}_4^2$	96
4	$3\cdot 5^3\cdot 31$	$\mathbf{S}_4$	24
5	$2^3\cdot 3\cdot 5^2\cdot 31$	$\mathbf{Z}_4\mathbf{Z}_5$	20
6	$2^2\cdot 5^2\cdot 31$	$\mathbf{S}_5$	120

14.5. $PG(2,7)$

In $PG(2,7)$, $\theta(1) = 8$ and $\theta(2) = 57$. The numbers of points on no bisecant of a 3-arc, a 4-arc, and a 5-arc are respectively 36, 20, and 7. Also the number of points on no bisecant of a 6-arc $\mathscr{K}$ is $6 - c_3$, where c_3 is the number of B-points of $\mathscr{K}$, §9.2. As in theorem 9.2.2, every 7-arc lies on a conic, whence every 6-arc lies on a conic or is complete. If $\mathscr{K}$ lies on a conic $\mathscr{C}$, then the only points on none of its bisecants are the two points of $\mathscr{C}\backslash\mathscr{K}$; so $c_3 = 4$. If $\mathscr{K}$ is complete, then $c_3 = 6$. To classify all k-arcs, it remains to classify the complete 6-arcs.

Let $\mathscr{K}$ be a complete 6-arc. If each of its 15 bisecants contained no more than one B-point, then $\mathscr{K}$ has no more than $15/3 = 5$ B-points. So at least one bisecant l contains two B-points and hence $\mathscr{K}$ comprises the vertices of a tetrastigm $\mathscr{Q}$ and two points on the join l of two of its diagonal points; these diagonal points are the B-points of $\mathscr{K}$ on l.

Let the vertices of $\mathscr{Q}$ be

$$P_1 = \mathbf{P}(-1,1,1),\ P_2 = \mathbf{P}(1,1,1),\ P_3 = \mathbf{P}(1,-1,1),\ P_4 = \mathbf{P}(-1,-1,1).$$

So its diagonal triangle is that of reference. Also choose $l = \mathbf{u}_0$. The pencil $\mathscr{F}$ of quadrics through the vertices of $\mathscr{Q}$ is

$$\mathbf{V}(a_0x_0^2 + a_1x_1^2 + a_2x_2^2) \text{ with } a_0 + a_1 + a_2 = 0.$$

Of the eight quadrics in the pencil, three are line pairs and five are conics. $\mathscr{F}$ cuts out an involution on l with double points $\mathbf{U}_1$ and $\mathbf{U}_2$, and pairs $\mathbf{P}(0,t,1)$, $\mathbf{P}(0,-t,1)$ for $t \neq 0$.

The double points are vertices of the respective line pairs $V(x_0^2 - x_2^2)$ and $V(x_0^2 - x_1^2)$; the third line pairs meets l in $P(0,1,1)$ and $P(0,-1,1)$. So the only possible pairs of points on l for $\mathscr{K}$ are:

(i) $\left.\begin{matrix} P(0,2,1) \\ P(0,3,1) \end{matrix}\right\}$, (ii) $\left.\begin{matrix} P(0,2,1) \\ P(0,-3,1) \end{matrix}\right\}$, (iii) $\left.\begin{matrix} P(0,-2,1) \\ P(0,3,1) \end{matrix}\right\}$, (iv) $\left.\begin{matrix} P(0,-2,1) \\ P(0,-3,1) \end{matrix}\right\}$.

By the harmonic homology $P(x_0,x_1,x_2)\mathfrak{T} = P(x_0,-x_1,x_2)$, which fixes $\mathscr{F}$ and the involution on l, the 6-arcs obtained from (i) and (iv) are equivalent, as are those from (ii) and (iii). Thus there are only two projectively distinct complete 6-arcs, $\mathscr{K}_1$ and $\mathscr{K}_2$, which will be called type *I* and type *II* respectively:

$$\mathscr{K}_1 = \{P_1, P_2, P_3, P_4, P(0,2,1),\ P(0,3,1)\},$$

$$\mathscr{K}_2 = \{P_1, P_2, P_3, P_4, P(0,2,1),\ P(0,-3,1)\}.$$

Firstly consider $\mathscr{K}_1$ Write $A_1 = P_1$, $A_2 = P_2$, $A_3 = P(0,2,1)$, $A_4 = P_3$, $A_5 = P_4$, $A_6 = P(0,3,1)$. The six B-points of $\mathscr{K}_1$ are

$$A_1' = A_1A_4 \cap A_2A_5 \cap A_3A_6, \quad A_4' = A_1A_4 \cap A_2A_6 \cap A_3A_5,$$

$$A_2' = A_1A_5 \cap A_2A_6 \cap A_3A_4, \quad A_5' = A_1A_5 \cap A_2A_4 \cap A_3A_6,$$

$$A_3' = A_1A_6 \cap A_2A_4 \cap A_3A_5, \quad A_6' = A_1A_6 \cap A_2A_5 \cap A_3A_4.$$

Thus $\mathscr{K}_1$ consists of two 3-arcs $\{A_1,A_2,A_3\}$ and $\{A_4,A_5,A_6\}$, which are in sextuple perspective from the six B-points. Each bisecant A_iA_j of $\mathscr{K}_1$ contains zero or two B-points as A_i and A_j belong to the same or different 3-arcs. Thus nine bisecants contain two B-points and six contain one. Further, $\mathscr{K}_1' = \{A_i' \mid i \in N_6\}$ is also a 6-arc of type *I*, whose B-points are just the points of $\mathscr{K}_1$. The 3-arcs of $\mathscr{K}_1'$ in sextuple perspective are $\{A_1',A_2',A_3'\}$ and $\{A_4',A_5',A_6'\}$. Thus the 6-arcs of type *I* occur in pairs.

To consider $\mathscr{K}_2$, write $B_1 = P_1$, $B_2 = P_2$, $B_3 = P_3$, $B_4 = P_4$, $B_5 = P(0,2,1)$, $B_6 = P(0,-3,1)$. The six B-points of $\mathscr{K}_2$ are

$$B_1' = B_1B_2 \cap B_3B_5 \cap B_4B_6, \quad B_2' = B_1B_2 \cap B_3B_6 \cap B_4B_5,$$
$$B_3' = B_3B_4 \cap B_1B_5 \cap B_2B_6, \quad B_4' = B_3B_4 \cap B_1B_6 \cap B_2B_5,$$
$$B_5' = B_5B_6 \cap B_1B_3 \cap B_2B_4, \quad B_6' = B_5B_6 \cap B_1B_4 \cap B_2B_3.$$

Therefore $\mathscr{K}_2$ comprises the three pairs $\{B_1,B_2\}$, $\{B_3,B_4\}$, $\{B_5,B_6\}$ such that a bisecant B_iB_j contains two B-points or one as B_i and B_j belong to the same or different pairs. Thus three bisecants contain two B-points and nine contain one. Let the triangle $\mathscr{T}$ with sides B_1B_2, B_3B_4, B_5B_6 have vertices D,E,F. Then the tetragram with sides B_3B_5, B_3B_6, B_4B_5, B_4B_6 has $\mathscr{T}$ as its diagonal triangle (Fig. 22).

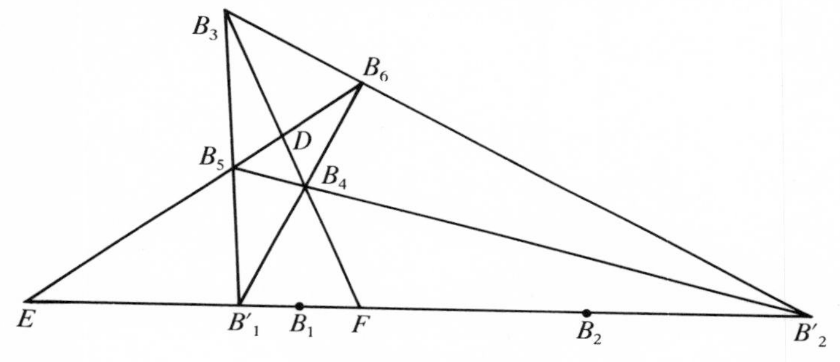

FIG. 22

From Fig. 22, we have that $H(B_3,B_4;\ D,F)$, $H(B_5,B_6;\ D,E)$, $H(B_1',B_2';\ E,F)$. Similarly, $H(B_1,B_2;\ E,F)$, $H(B_3',B_4';\ D,F)$, $H(B_5',B_6';\ D,E)$.

The six points B_i' form a 6-arc $\mathscr{K}_2'$, also of type II, separated into the three pairs $\{B_1',B_2'\}$, $\{B_3',B_4'\}$, $\{B_5',B_6'\}$. The B-points of $\mathscr{K}_2'$ must again lie on the sides of $\mathscr{T}$. However, they cannot be points of $\mathscr{K}_2$. For, if $P = B_3'B_4' \cap$ $\cap\ B_1'B_5' \cap B_2'B_6'$, say, were a point of $\mathscr{K}_2$, it would have to be B_3 or B_4. But since B_1',B_3,B_5 are collinear, B_1',B_3,B_5' cannot be: so $P \neq B_3$. Since B_1',B_4,B_6 are collinear, B_1',B_4,B_5' are not: so $P \neq B_4$. Thus, if the B-points of $\mathscr{K}_2'$ are B_i'', $i \in \mathrm{N}_6$, they form a 6-arc $\mathscr{K}_2''$ of type II with $B_1''B_2'' = EF$, $B_3''B_4'' = FD$, $B_5''B_6'' = DE$. The three pairs of harmonic conjugates of E and F are $\{B_1,B_2\}$, $\{B_1',B_2'\}$, $\{B_1'',B_2''\}$, and similarly for F and D, and for D and E. The B-points of $\mathscr{K}_2''$ must now be the points of $\mathscr{K}_2$. Thus the 6-arcs of type II occur in trios.

The numbers of 6-arcs of type I and II can now be calculated. Once the vertices of $\mathscr{Q}$ are chosen, there are $3 \cdot 2 = 6$ ways of selecting the remaining points of $\mathscr{K}$ on a side of the diagonal triangle of $\mathscr{Q}$ in both cases. However, for each 6-arc, the number of ways of selecting $\mathscr{Q}$ is the number of bisecants containing two B-points, which is nine for type I and three for type II. The number of tetrastigms is $57 \cdot 56 \cdot 49 \cdot 36/4! = 2^2 \cdot 2^2 \cdot 7^3 \cdot 19$. Hence the number of 6-arcs of type I is

$$2^2 \cdot 3^2 \cdot 7^3 \cdot 19 \cdot 6/9 = 2^3 \cdot 3 \cdot 7^3 \cdot 19$$

and of type II is

$$2^2 \cdot 3^2 \cdot 7^3 \cdot 19 \cdot 6/3 = 2^3 \cdot 3^2 \cdot 7^3 \cdot 19.$$

The total number of complete 6-arcs is therefore $2^5 \cdot 3 \cdot 7^3 \cdot 19$, agreeing with the number of 6-arcs not on a conic, calculated in theorem 7.2.3, corollary 2.

As $p(3,7) = 7^3(7^3 - 1)(7^2 - 1) = 2^5 \cdot 3^3 \cdot 7^3 \cdot 19$, the projective group of a complete 6-arc has order 36 or 12 as it is of type I of type II. The group G_1 of $\mathscr{K}_1$ comprises all even permutations of $A_1, \ldots, A_6$ fixing the pair of triads $\{A_1, A_2, A_3\}$, $\{A_4, A_5, A_6\}$. So $G_1 \cong \mathbf{Z}_4(\mathbf{Z}_3 \times \mathbf{Z}_3)$. The group G_2 of order 12 is isomorphic to $\mathbf{A}_4$.

The projective classification of k-arcs in $PG(2,7)$ is given in Table 14.9, where each row gives a projectively unique k-arc.

14.6. $PG(2,8)$

In $PG(2,8)$, $\theta(1) = 9$ and $\theta(2) = 73$. From theorem 9.2.5, a complete arc is either a regular oval or a 6-arc. Thus a 6-arc $\mathscr{K}$ is one of three types:

(i) $\mathscr{K}$ lies on a conic;

(ii) $\mathscr{K}$ comprises a 5-arc plus the nucleus N of the conic containing the 5-arc;

(iii) $\mathscr{K}$ is complete.

All 5-arcs are projectively equivalent; for, a 5-arc lies on a conic $\mathscr{C}$ and the residual set of four points can be transformed to any other set of four points on $\mathscr{C}$, as each set of four points on $\mathscr{C}$ admits as cross-ratios all six elements of γ_{01}.

TABLE 14.9

k	Total number of k-arcs	G	$\|G\|$	Description
3	$2^2 \cdot 7^3 \cdot 19$	$\mathbf{S}_3\mathbf{Z}_6^{\,2}$	108	*incomplete*
4	$2^2 \cdot 3^2 \cdot 7^3 \cdot 19$	$\mathbf{S}_4$	24	*incomplete*
5	$2^4 \cdot 3^2 \cdot 7^3 \cdot 19$	$\mathbf{S}_3$	6	*incomplete*
6	$2^3 \cdot 3^2 \cdot 7^3 \cdot 19$	$\mathbf{Z}_6\mathbf{Z}_2$	12	*incomplete, on a conic*
6	$2^3 \cdot 3 \cdot 7^3 \cdot 19$	$\mathbf{Z}_4(\mathbf{Z}_3 \times \mathbf{Z}_3)$	36	*complete , type I*
6	$2^3 \cdot 3^2 \cdot 7^3 \cdot 19$	$\mathbf{A}_4$	12	*complete, type II*
7	$2^4 \cdot 3^2 \cdot 7^2 \cdot 19$	$\mathbf{Z}_7\mathbf{Z}_6$	42	*incomplete, on a conic*
8	$2 \cdot 3^2 \cdot 7^2 \cdot 19$	$PGL(2,7)$	336	*complete, conic*

A conic through $\mathbf{U}_0, \mathbf{U}_1, \mathbf{U}_2, \mathbf{U}$ has the form $\mathscr{C} = \mathbf{V}(a_0x_1x_2 + a_1x_0x_2 + a_2x_0x_1)$ with $a_0 + a_1 + a_2 = 0$. The nucleus $N = \mathbf{P}(a_0, a_1, a_2)$, which lies on the line $\mathbf{u}$ of diagonal points of the tetrastigm $\mathscr{Q}$ with the four points as vertices. Thus N lies on the line of diagonal points of every tetrastigm inscribed in $\mathscr{C}$.

The number of points on no bisecant of a 6-arc $\mathscr{K}$ is $7 - c_3$, where c_3 is the number of B-points of $\mathscr{K}$, §9.2. If $\mathscr{K}$ is of type (i) or (ii) and lies on the oval $\mathscr{O}$, then the only points on no bisecant of $\mathscr{K}$ are the points of $\mathscr{O} \backslash \mathscr{K}$. So $7 - c_3 = 4$, whence $c_3 = 3$.

For $\mathscr{K}$ of type (ii), we can find where these three B-points are. Let $\mathscr{O} = \mathscr{C} \cup \{N\}$ and let $\mathscr{K} = \{P_1, P_2, P_3, P_4, P_5, N\}$; also, let d_i be the line of diagonal points of the tetrastigm with vertices $\mathscr{K} \backslash \{P_i, N\}$. The five lines d_i all pass through N and are therefore tangents to the conic $\mathscr{C}$. As there are only four points on $\mathscr{C}$ besides the five P_i, at least one line, say d_5, passes through the corresponding point P_5. So $d_5 = NP_5$ and contains the three B-points of $\mathscr{K}$, namely

$P_1P_2 \cap P_3P_4$, $P_1P_3 \cap P_2P_4$, $P_1P_4 \cap P_2P_3$.

Another way of looking at it is to recall from §6.4 that, on $\mathscr{C}$, one of the P_i, it will be P_5, is the associated point of the other four. Thus P_5 is the double point of the involution $\mathscr{I}_1$ on $\mathscr{C}$ with pairs $\{P_1,P_2\}$ and $\{P_3,P_4\}$. This means that $\mathscr{I}_1$ is defined by the pencil of lines through $D_1 = P_1P_2 \cap P_3P_4$. So D_1P_5 is tangent to $\mathscr{C}$ and hence passes through N. Similarly P_5 is the double point of the involutions $\mathscr{I}_2$ with pairs $\{P_1,P_3\}$ and $\{P_2,P_4\}$, and $\mathscr{I}_3$ with pairs $\{P_1,P_4\}$ and $\{P_2,P_3\}$. So $D_2 = P_1P_3 \cap P_2P_4$ and $D_3 = P_1P_4 \cap P_2P_3$ also lie on NP_5. This incidentally shows that the three diagonal points of a tetrastigm are collinear. It also follows that d_5 is the line of diagonal points of the tetrastigm with vertices, the four points of $\mathscr{C}$ not in $\mathscr{K}$.

The projective group $G(\mathscr{K})$ fixes both N and P_5 and so the set $\{P_1,P_2,P_3,P_4\}$. The number of tetrads of points on a conic is $9\cdot8\cdot7\cdot6/(4\cdot3\cdot2) = 9\cdot8\cdot7/4$, whence $|G(\mathscr{K})| = 4$. Thus

$$G(\mathscr{K}) = \{1,\ (P_1P_2)(P_3P_4),\ (P_1P_3)(P_2P_4),\ (P_1P_4)(P_2P_3)\}$$

$$\cong \mathbf{Z}_2 \times \mathbf{Z}_2.$$

Now let $\mathscr{K}$ be complete: so $c_3 = 7$. If no bisecant of $\mathscr{K}$ contained more than one B-point, then $c_3 \leqslant 5$. So some bisecant l of $\mathscr{K}$ contains two B-points and therefore three. Let $\mathscr{Q}$ be the tetrastigm with vertices $\mathscr{K}\backslash l = \{\mathbf{U}_0,\mathbf{U}_1,\mathbf{U}_2,\mathbf{U}\}$. Then the points of $l \cap \mathscr{K}$ lie on $\mathbf{u}$ and are not the diagonal points $\mathbf{P}(0,1,1)$, $\mathbf{P}(1,0,1)$, $\mathbf{P}(1,1,0)$ of $\mathscr{K}$. In the notation of §1.7, $GF(8)\backslash\{0\} = \{\varepsilon^i \mid i \in \overline{\mathbf{N}}_6\}$, where

$$\varepsilon^3 + \varepsilon^2 + 1 = \varepsilon^6 + \varepsilon^4 + 1 = \varepsilon^5 + \varepsilon + 1 = 0.$$

So the points of $l \cap \mathscr{K}$ comprise two from

$$P(\varepsilon),\ P(\varepsilon^2),\ P(\varepsilon^3),\ P(\varepsilon^4),\ P(\varepsilon^5),\ P(\varepsilon^6),$$

where $P(t) = \mathbf{P}(t,1+t,1)$.

The conic through the vertices of $\mathscr{Q}$ and $\mathbf{P}(b_0,b_1,b_2)$ where

$b_0 + b_1 + b_2 = 0$ is $\mathbf{V}(b_0{}^2 x_1 x_2 + b_1{}^2 x_0 x_2 + b_2{}^2 x_0 x_1)$, which has nucleus $\mathbf{P}(b_0{}^2, b_1{}^2, b_2{}^2)$. Thus the addition to the vertices of $\mathscr{Q}$ of any of the following pairs gives a 6-arc of type (ii):

$$\{P(\varepsilon),\ P(\varepsilon^2)\},\ \{P(\varepsilon^2),\ P(\varepsilon^4)\},\ \{P(\varepsilon^4),\ P(\varepsilon)\},$$

$$\{P(\varepsilon^3),\ P(\varepsilon^6)\},\ \{P(\varepsilon^6),\ P(\varepsilon^5)\},\ \{P(\varepsilon^5),\ P(\varepsilon^3)\}.$$

Let ϕ be the automorphism of $PG(2,8)$ given by $\mathbf{P}(x_0, x_1, x_2)\phi = \mathbf{P}(x_0{}^2, x_1{}^2, x_2{}^2)$. Let $\mathfrak{T}$ be the projectivity given by $\mathbf{P}(x_0, x_1, x_2)\mathfrak{T} = \mathbf{P}(x_2, x_0, x_1)$ so that $P(t)\mathfrak{T} = \mathbf{P}(1, t, 1 + t) = P(1/(1 + t))$. Then, for $\mathscr{K}$ complete, the points of $l \cap \mathscr{K}$ are one of the pairs

$$P(\varepsilon^3),\ P(\varepsilon^2);\ P(\varepsilon^5),\ P(\varepsilon^4);\ P(\varepsilon^6),\ P(\varepsilon);$$

$$P(\varepsilon^6),\ P(\varepsilon^4);\ P(\varepsilon^3),\ P(\varepsilon);\ \ P(\varepsilon^5),\ P(\varepsilon^2);$$

$$P(\varepsilon^5),\ P(\varepsilon);\ \ P(\varepsilon^6),\ P(\varepsilon^2);\ P(\varepsilon^3),\ P(\varepsilon^4).$$

These nine pairs are arranged so that ϕ operates down the columns and $\mathfrak{T}$ along the rows of the 3×3 array. This shows that all complete 6-arcs are equivalent under collineations, but that there may be three projectively distinct complete arcs.

To fix one complete arc, let $\mathscr{K} = \{P_i \mid i \in \mathbf{N}_6\}$, where $P_1 = \mathbf{U}_0$, $P_2 = \mathbf{U}_1$, $P_3 = \mathbf{U}_2$, $P_4 = \mathbf{U}$, $P_5 = P(\varepsilon^3)$, $P_6 = P(\varepsilon^2)$. Then $P_1P_2 = \mathbf{u}_2$, $P_3P_4 = \mathbf{V}(x_0 + x_1)$, and $P_5P_6 = \mathbf{u}$ are the lines of diagonal points of the residual tetrastigms, and each of the three lines contains three B-points of $\mathscr{K}$. The other 12 bisecants each contain one B-point of $\mathscr{K}$. Also $\mathscr{K}$ consists in three ways of the vertices of a tetrastigm and two points on the latter's line of diagonal points. The seven B-points of $\mathscr{K}$ form a subplane $PG(2,2)$ as in Fig. 23, where $(ij, kl, mn) = P_iP_j \cap P_kP_l \cap P_mP_n$.

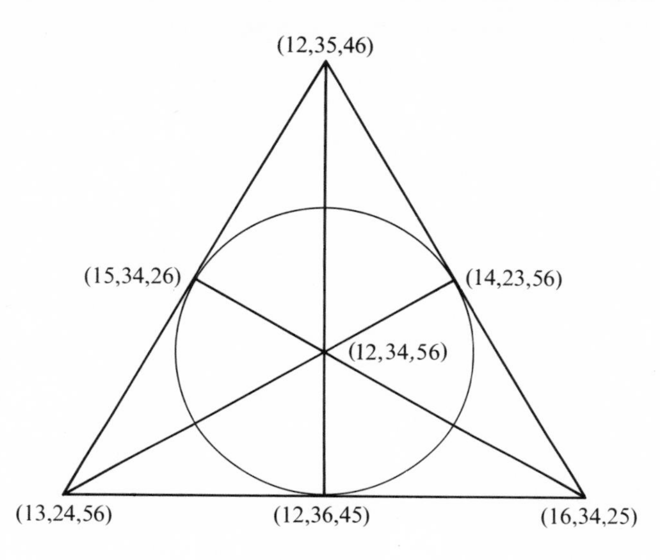

FIG. 23

The number of tetrastigms in $PG(2,8)$ is $73\cdot 72\cdot 64\cdot 49/4! = 2^6\cdot 3\cdot 7^2\cdot 73$, and the number of complete 6-arcs is therefore $(2^6\cdot 3\cdot 7^2\cdot 73)9/3 = 2^6\cdot 3^2\cdot 7^2\cdot 73$. However, $p(3,8) = 73\cdot 72\cdot 64\cdot 49 = 2^9\cdot 3^2\cdot 7^2\cdot 73$. So, either there are three projectively distinct complete 6-arcs each with a group of order 24, or only one with a group of order eight. The number of permutations of $P_1,\ldots,P_6$ fixing the set of three lines P_1P_2, P_3P_4, P_5P_6 is $3!\cdot 2^3 = 48$. The projective group $G(\mathscr{K})$ comprises all the even ones. As $G(\mathscr{K})$ operates on the four lines of $PG(2,2)$ not through $P_1P_2 \cap P_3P_4 \cap P_5P_6$, so $G(\mathscr{K}) \cong \mathbf{S}_4$. Hence there are three orbits of complete 6-arcs under $PGL(3,8)$ each of size $2^6\cdot 3\cdot 7^2\cdot 73$. From §14.1, the stabilizer in $PGL(3,2)$ of a point in $PG(2,2)$ is isomorphic to $\mathbf{S}_4$. From lemma 4.3.1, the number of subplanes in $PG(2,8)$ is $p(3,8)/p(3,2) = 2^6\cdot 3\cdot 7\cdot 73$. So there are 21 complete 6-arcs associated to each $PG(2,2)$.

The number of 6-arcs of type (ii) is the number of 5-arcs, which is $73\cdot 72\cdot 64\cdot 49\cdot 30/5! = 2^7\cdot 3^2\cdot 7^2\cdot 73$. Adding this to the number of complete 6-arcs gives $2^6\cdot 3^3\cdot 7^2\cdot 73$, which is the number of 6-arcs not on a conic, as in lemma 7.2.3, corollary 2.

Now we list all k-arcs in Table 14.10. Each row gives a

projectively unique k-arc except in the case of the complete 6-arcs, where although they are all equivalent under $P\Gamma L(3,8)$, there are three classes under $PGL(3,8)$. Also $\mathscr{C}_m$ denotes an m-arc on a conic with nucleus N.

TABLE 14.10

k	Total number of k-arcs	G	$\|G\|$	Description	
3	$2^8 \cdot 3 \cdot 73$	$\mathbf{S}_3\mathbf{Z}_7{}^2$	294	*incomplete*	
4	$2^6 \cdot 3 \cdot 7^2 \cdot 73$	$\mathbf{S}_4$	24	*incomplete*	
5	$2^7 \cdot 3^2 \cdot 7^2 \cdot 73$	$\mathbf{Z}_2 \times \mathbf{Z}_2$	4	*incomplete*	
6	$2^8 \cdot 3 \cdot 7^2 \cdot 73$	$\mathbf{S}_3$	6	*incomplete*	$\mathscr{C}_6$
6	$2^7 \cdot 3^2 \cdot 7^2 \cdot 73$	$\mathbf{Z}_2 \times \mathbf{Z}_2$	4	*incomplete*	$\mathscr{C}_5 \cup \{N\}$
6	$2^6 \cdot 3^2 \cdot 7^2 \cdot 73$	$\mathbf{S}_4$	24	*complete*	
7	$2^8 \cdot 3^2 \cdot 7 \cdot 73$	$\mathbf{Z}_7\mathbf{Z}_2$	14	*incomplete*	$\mathscr{C}_7$
7	$2^8 \cdot 3 \cdot 7^2 \cdot 73$	$\mathbf{S}_3$	6	*incomplete*	$\mathscr{C}_6 \cup \{N\}$
8	$2^6 \cdot 3^2 \cdot 7 \cdot 73$	$\mathbf{Z}_8\mathbf{Z}_7$	56	*incomplete*	$\mathscr{C}_8$
8	$2^8 \cdot 3^2 \cdot 7 \cdot 73$	$\mathbf{Z}_7\mathbf{Z}_2$	14	*incomplete*	$\mathscr{C}_7 \cup \{N\}$
9	$2^6 \cdot 7 \cdot 73$	$PGL(2,8)$	504	*incomplete*	$\mathscr{C}_9$
9	$2^6 \cdot 3^2 \cdot 7 \cdot 73$	$\mathbf{Z}_8\mathbf{Z}_7$	56	*incomplete*	$\mathscr{C}_8 \cup \{N\}$
10	$2^6 \cdot 7 \cdot 73$	$PGL(2,8)$	504	*complete*	$\mathscr{C}_9 \cup \{N\}$

14.7. $PG(2,9)$

In $PG(2,9)$, $\theta(1) = 10$ and $\theta(2) = 91$. From theorem 8.2.4, every 10-arc is a conic: from theorem 8.6.10, every 9-arc lies on a conic. From §1.7, $GF(9) = \{0,\pm 1,\pm\sigma,\pm\sigma^2,\pm\sigma^3\}$, where

$$\sigma^2 - \sigma - 1 = \sigma^3 - \sigma^2 - \sigma = \sigma^3 + \sigma^2 + 1 = \sigma^3 + \sigma - 1 = 0 .$$

First, we need to develop from §6.4 some more properties of $PG(1,9)$. There are $\mathbf{c}(10,3)/4 = 30$ superharmonic tetrads and $\mathbf{c}(10,4) - 30 = 180$ non-harmonic tetrads on the line. A superharmonic tetrad has projective group $\mathbf{S}_4$. Any other tetrad takes all six values of $GF(9)\backslash\{0,1,-1\}$ for its different cross-ratios. So the non-harmonic tetrads are all projectively equivalent and each has a projective group $\mathbf{Z}_2 \times \mathbf{Z}_2$. Alternatively, the six projectivities fixing the set of three points $\{\infty,0,1\}$ are given by $x \to x$, $1/x$, $1 - x$, $1/(1 - x)$, $x/(x - 1)$, $(x - 1)/x$. So, as the cross-ratio $\{\infty,0;1,t\} = t$, these six transformations take t, for $t \in \gamma^+\backslash\{\infty,0,1,-1\}$, to all six elements of this set.

If a pentad $\mathscr{P}$ contains a harmonic tetrad $\mathscr{H} = \{\infty,0,1,-1\}$, then the six projectivities above fix $\mathscr{H}$ and are transitive on the other six points of the line. So all pentads containing a harmonic tetrad are projectively equivalent and will be called *harmonic pentads*. As a triad is in only one harmonic tetrad, $\mathscr{P}$ contains only one harmonic tetrad. The number of harmonic pentads is therefore $6\cdot 30 = 180$, and each has a projective group $\mathbf{Z}_4$.

The number of non-harmonic pentads is $\mathbf{c}(10,5) - 180 = 72$. Given a non-harmonic tetrad, $\mathscr{T}$, there are four points on the line each of which added to $\mathscr{T}$ makes a harmonic pentad. So $\mathscr{T}$ lies in two non-harmonic pentads $\mathscr{T} \cup \{P\}$ and $\mathscr{T} \cup \{Q\}$. If, still in non-homogeneous coordinates, $\mathscr{T} = \{\infty,0,1,\sigma\}$, then $\{P,Q\} = \{\sigma^2,-\sigma^3\}$; so the projectivity $x \to \sigma + \sigma^3 x$ transforms $\{\infty,0,1,\sigma,\sigma^2\}$ to $\{\infty,\sigma,1,-\sigma^3,0\}$. Thus the non-harmonic pentads are projectively equivalent. As $p(2,9) = 10\cdot 9\cdot 8$, each has a projective G of order $10\cdot 9\cdot 8/72 = 10$; in fact, $G \cong \mathbf{D}_5$. Therefore, we may take as canonical harmonic and non-harmonic pentads respectively

$$\mathscr{P}_1 = \{\infty,0,1,-1,\sigma\}, \quad \mathscr{P}_2 = \{\infty,0,1,\sigma,\sigma^2\}.$$

All conics are projectively equivalent and, by the above, there are two projectively distinct pentads on a conic. As a 5-arc uniquely determines a conic, there are two projectively distinct 5-arcs, which may be called *harmonic* or *non-harmonic* according to the type of pentad they form on the unique conic containing them. These 5-arcs then have projective groups isomorphic to $\mathbf{Z}_4$ and $\mathbf{D}_5$ respectively.

Let $\mathscr{C} = \{P(t) = \mathbf{P}(1,\sigma^3/(t - \sigma), 1/t) \mid t \in \gamma^+\}$. Let

$\mathscr{F}_1 = \{P(t) \mid t \in \mathscr{P}_1\}$, $\mathscr{F}_2 = \{P(t) \mid t \in \mathscr{P}_2\}$. So the harmonic 5-arc

$$\mathscr{F}_1 = \{U_0, U_1, U_2, U, P(1, -\sigma, -1)\}$$

and the non-harmonic 5-arc

$$\mathscr{F}_2 = \{U_0, U_1, U_2, U, P(1, \sigma^3, -\sigma^2)\}.$$

From lemma 7.1.2, five of the diagonal points of the pentastigm with vertices $\mathscr{F}_2$ are collinear: they are the points given there by *II*.

From §9.2, the number of points on no bisecant of a 5-arc $\mathscr{F}$ is $c_0 = 21$; of these five lie on the conic through $\mathscr{F}$. Thus through each of $\mathscr{F}_1$ and $\mathscr{F}_2$ the number of 6-arcs not on $\mathscr{C}$ is 16. It takes only a little calculation to show that the 6-arcs can have ten, four, three, or two *B*-points and the number of each type through $\mathscr{F}_1$ and $\mathscr{F}_2$ is as follows:

B-points	10	4	3	2
Number of 6-arcs containing $\mathscr{F}_1$	0	4	8	4
Number of 6-arcs containing $\mathscr{F}_2$	1	10	0	5

(14.4)

The existence of these 6-arcs also follows from §20.3, IV and §20.4 in Volume II.

The 6-arc with ten *B*-points is complete and projectively unique. The numbers of harmonic and non-harmonic 5-arcs are respectively

$$N_1 = p(3,9)/4, \quad N_2 = p(3,9)/10\,. \tag{14.5}$$

As there is one complete 6-arc through $\mathscr{F}_2$ and six non-harmonic 5-arcs in a complete 6-arc, there are $p(3,9)/60$ complete 6-arcs, each of which therefore has a projective group of order 60. Of the 15 sets of three bisecants containing all six points of the 6-arc, ten are concurrent at the

B-points and the remaining five form triangles. The group G_{10} of the 6-arc operates faithfully on these five triangles. Hence $G_{10} \cong \mathbf{A}_5$. Each bisecant of the complete 6-arc contains two B-points, the maximum possible, whence the 6-arc is symmetric.

In fact, the 6-arcs with two and three B-points are also projectively unique, whereas there are two types with four B-points. Let $\mathscr{K}_r$ be a 6-arc with r B-points. The following are projectively distinct and, with $\mathscr{K} = (A_1, A_2, A_3, A_4, A_5, A_6)$, we write $A_iA_j \cap A_kA_l \cap A_mA_n = (ij, kl, mn)$ for a B-point:

$$\mathscr{K}_2 = (\mathbf{U}_0, \mathbf{U}_1, \mathbf{U}_2, \mathbf{U}, \mathbf{P}(1,-\sigma,-1), \mathbf{P}(1,\sigma,-\sigma^3))$$
$$(15,24,36),\ (15,34,26);$$

$$\mathscr{K}_3 = (\mathbf{U}_0, \mathbf{U}_1, \mathbf{U}_2, \mathbf{U}, \mathbf{P}(1,-\sigma,-1), \mathbf{P}(1,-1,\sigma))$$
$$(13,24,56),\ (15,23,46),\ (14,25,36);$$

$$\mathscr{K}_4 = (\mathbf{U}_0, \mathbf{U}_1, \mathbf{U}_2, \mathbf{U}, \mathbf{P}(1,-\sigma,-1), \mathbf{P}(1,-1,-\sigma))$$
$$(14,35,26),\ (12,35,46),\ (14,25,36),\ (13,45,26);$$

$$\mathscr{K}'_4 = (\mathbf{U}_0, \mathbf{U}_1, \mathbf{U}_2, \mathbf{U}, \mathbf{P}(1,-\sigma,-1), \mathbf{P}(1,\sigma^2,-\sigma))$$
$$(14,35,26),\ (16,35,24),\ (14,23,56),\ (13,45,26);$$

$$\mathscr{K}_{10} = (\mathbf{U}_0, \mathbf{U}_1, \mathbf{U}_2, \mathbf{U}, \mathbf{P}(1,\sigma^3,-\sigma^2), \mathbf{P}(1,-\sigma,\sigma^3))$$
$$(12,35,46),\ (12,45,36),\ (13,24,56),\ (13,25,46),\ (14,23,56),$$
$$(14,26,35),\ (15,24,36),\ (15,34,26),\ (16,23,45),\ (16,25,34).$$

The projective groups are generated as follows:

$$\begin{aligned} G(\mathscr{K}_2) &= \langle (15)(2436) \rangle \cong \mathbf{Z}_4\,, \\ G(\mathscr{K}_3) &= \langle (126)(354) \rangle \cong \mathbf{Z}_3\,, \\ G(\mathscr{K}_4) &= \langle (156)(243),\ (15)(34) \rangle \cong \mathbf{S}_3\,, \\ G(\mathscr{K}'_4) &= \langle (125)(346),\ (12)(46) \rangle \cong \mathbf{S}_3\,. \end{aligned} \tag{14.6}$$

There is only one class of 6-arcs with four B-points under collineations. For example, $\mathscr{F}_1$ is fixed by a collineation

$\mathfrak{T}$ effecting (14)(35), namely $P(x_0,x_1,x_2)\mathfrak{T} = \mathbf{P}(x_0^3 + x_2^3, x_0^3 - \sigma^3 x_1^3 - \sigma x_2^3, x_0^3 - x_2^3)$, which sends $\mathscr{K}_4$ to $\mathscr{K}_4'$.

Let m_2, m_3, m_4 be the total number of 6-arcs not on a conic with two, three, four B-points respectively, and, for $i = 2,3,4$, let r_i and s_i be the respective numbers of harmonic and non-harmonic 5-arcs in a 6-arc with i B-points. Then

$$r_2 + s_2 = r_3 + s_3 = r_4 + s_4 = 6.$$

Also, from (14.4),

$$m_2 = 4N_1/r_2 = 5N_2/s_2,$$

whence, from (14.5), $r_2 = 2s_2$. Hence

$$r_2 = 4,\ s_2 = 2,\ m_2 = N = p(3,9)/4.$$

Similarly, $r_3 = 6$, $s_3 = 0$, whence $m_3 = 8N_1/6 = p(3,9)/3$. Finally,

$$m_4 = 4N_1/r_4 = 10N_2/s_4,$$

whence

$$r_4 = s_4 = 3,\ m_4 = 4N_1/3 = p(3,9)/3.$$

The values of r_i and s_i also follow from the group structure of $\mathscr{K}_i$.

On a conic, there are two projectively distinct 6-arcs, residual to a harmonic and non-harmonic tetrad respectively. On $\mathscr{C}$,

$$\mathscr{K}_6 = (\mathbf{U}_0,\mathbf{U}_1,\mathbf{U}_2,\mathbf{U},\mathbf{P}(1,-\sigma,-1),\mathbf{P}(1,\sigma,-\sigma))$$

is of the former type. Its six B-points are

(12,34,56), (13,26,45), (14,26,35)
(15,23,46), (15,24,36), (16,25,34).

So the four points of index zero for $\mathscr{K}_6$ are the points of $\mathscr{C}\backslash\mathscr{K}_6$. Residual to a non-harmonic tetrad on $\mathscr{C}$ is

$$\mathscr{K}_2' = (U_0, U_1, U_2, U, P(1,-\sigma,-1), P(1,\sigma^2,\sigma^3)).$$

It has the two B-points (13,24,56), (13,25,46), but is distinguished from $\mathscr{K}_2$ by having $G(\mathscr{K}_2')$ isomorphic to $Z_2 \times Z_2$.

As a 9-arc lies on a conic, the 6-arcs $\mathscr{K}_2, \mathscr{K}_3, \mathscr{K}_4$ either lie in complete 7-arcs or in 7-arcs contained in complete 8-arcs. If a 6-arc has c_3 B-points, then from §9.2 it has $10 - c_3$ points of index 0. So the number of 7-arcs in which $\mathscr{K}_2, \mathscr{K}_3$, and $\mathscr{K}_4$ respectively lie are eight, seven, and six. They separate as follows:

	$\mathscr{K}_2$	$\mathscr{K}_3$	$\mathscr{K}_4$
incomplete 7-arcs	8	6	6
complete 7-arcs	0	1	0

The incomplete 7-arcs all have 20 points of index three. As, for a 7-arc, $c_0 = 21 - c_3$, so each of these incomplete 7-arcs lies in a unique 8-arc, which is complete. The complete 7-arc obtained is

$$\begin{aligned}\mathscr{S}_1 &= \mathscr{K}_3 \cap \{P(1,-\sigma^3,\sigma^3)\}\\ &= (U_0, U_1, U_2, U, P(1,-\sigma,-1), P(1,-1,\sigma), P(1,-\sigma^3,\sigma^3)).\end{aligned}$$

It is symmetric since each bisecant contains three points of index three; these 21 points are

(12,34,67), (15,24,67), (13,45,67), (14,23,57), (12,46,57), (26,34,57), (13,26,47), (16,35,47), (23,56,47), (16,25,37), (14,56,37), (12,45,37), (15,36,27), (16,45,27), (35,46,27), (25,34,17), (26,35,17), (24,36,17), (13,24,56), (15,23,46), (14,25,36).

The number of complete 7-arcs is $m_3/7 = p(3,9)/21$. Hence $|G(\mathscr{S}_1)| = 21$ and $G(\mathscr{S}_1) \cong Z_3 Z_7$.

The other six 7-arcs through $\mathscr{K}_3$ fall into two classes of three under the group of $\mathscr{K}_3$; they consist of $\mathscr{K}_3$ and the following points:

(i) $\mathbf{P}(1,\sigma,\sigma^3)$, $\mathbf{P}(1,\sigma^3,-\sigma^2)$, $\mathbf{P}(1,\sigma^3,-\sigma^3)$;
(ii) $\mathbf{P}(1,\sigma,-\sigma^3)$, $\mathbf{P}(1,-\sigma^3,-\sigma)$, $\mathbf{P}(1,\sigma^2,-\sigma)$.

Let $\mathscr{S}_2 = \mathscr{K}_3 \cup \{\mathbf{P}(1,\sigma,\sigma^3)\}$
$= (\mathbf{U}_0,\mathbf{U}_1,\mathbf{U}_2,\mathbf{U},\mathbf{P}(1,-\sigma,-1),\mathbf{P}(1,-1,\sigma),\mathbf{P}(1,\sigma,\sigma^3))$.
The 20 points of index three for $\mathscr{S}_2$ are

(14,23,67), (15,34,67), (24,35,67), (12,36,57), (13,26,57),
(16,34,57), (12,35,47), (15,26,47), (16,25,47), (15,24,37),
(14,26,37), (12,56,37), (15,36,27), (16,45,27), (35,46,27),
(23,45,17), (25,46,17), (13,24,56), (15,23,46), (14,25,36).

The bisecant A_1A_5 contains five points of index three, the bisecants A_1A_3, A_1A_7, A_5A_4, A_5A_6, A_3A_4 all contain two points of index three, and the other 15 A_iA_j contain three points of index three. Thus $\{A_2,A_3,A_4,A_6,A_7\}$ is a non-harmonic five-arc, and A_1A_5 contains five diagonal points of the associated pentastigm. Also the only possible element of $G(\mathscr{S}_2)$ other than the identity acts as $(A_1A_5)(A_3A_4)(A_7A_6)$, and indeed there is such a projectivity. Hence $G(\mathscr{S}_2) = \mathbf{Z}_2$.

Let $\mathscr{S}'_2 = \mathscr{K}_3 \cup \{\mathbf{P}(1,\sigma,-\sigma^3)\}$
$= (\mathbf{U}_0,\mathbf{U}_1,\mathbf{U}_2,\mathbf{U},\mathbf{P}(1,-\sigma,-1),\mathbf{P}(1,-1,\sigma),\mathbf{P}(1,\sigma,-\sigma^3))$.

Then there is no projectivity but there is a collineation transforming $\mathscr{S}_2$ to $\mathscr{S}'_2$. All the 7-arcs containing $\mathscr{K}_2$ or $\mathscr{K}_4$ are projectively equivalent to $\mathscr{S}_2$ or $\mathscr{S}'_2$. Of the seven 6-arcs in $\mathscr{S}_2$, two have four B-points, two have three B-points, and three have two B-points: exactly one of the last 6-arcs lies on a conic.

There are three complete 8-arcs through $\mathscr{K}_3$:

$\mathscr{E}_1 = \mathscr{S}_2 \cup \{\mathbf{P}(1,\sigma^2,-\sigma)\} = \mathscr{K}_3 \cup \{\mathbf{P}(1,\sigma,\sigma^3),\mathbf{P}(1,\sigma^2,-\sigma)\}$,
$\mathscr{E}_2 = \mathscr{S}'_2 \cup \{\mathbf{P}(1,\sigma^3,-\sigma^2)\} = \mathscr{K}_3 \cup \{\mathbf{P}(1,\sigma,-\sigma^3),\mathbf{P}(1,\sigma^3,-\sigma^2)\}$,
$\mathscr{E}_3 = \mathscr{K}_3 \cup \{\mathbf{P}(1,\sigma^3,-\sigma^3),\mathbf{P}(1,-\sigma^3,-\sigma)\}$.

The generator of $G(\mathscr{K}_3)$ in (14.6) induces $(\mathscr{E}_1\mathscr{E}_2\mathscr{E}_3)$. So all complete 8-arcs are projectively equivalent.

It was shown above that $m_3 = p(3,9)/3$. As there are six incomplete 7-arcs through $\mathscr{K}_3$ and as there are two 6-arcs with three B-points in such a 7-arc, there are $6(p(3,9)/3)/2 = p(3,9)$ incomplete 7-arcs not on a conic and hence $p(3,9)/8$ complete 8-arcs. Therefore $|G(\mathscr{E}_1)| = 8$. In fact, with

$$\mathscr{E}_1 = (\mathbf{U}_0, \mathbf{U}_1, \mathbf{U}_2, \mathbf{U}, \mathbf{P}(1,-\sigma,-1), \mathbf{P}(1,-1,\sigma), \mathbf{P}(1,\sigma,\sigma^3), \mathbf{P}(1,\sigma^2,-\sigma)),$$

$G(\mathscr{E}_1) = \langle(15)(34)(67),\ (16)(23)(48)\rangle \cong \mathbf{D}_4$: four of the 7-arcs in $\mathscr{E}_1$ belong to each projective class. If $\mathscr{E}_1$ were symmetric, then as there are 28 bisecants, the constants c_1, c_2, c_3, c_4 of (9.3) would all be divisible by seven, as would their sum. So, from (9.1_1), the total number of points off $\mathscr{E}_1$, namely $91 - 8 = 83$, would also be divisible by seven. Hence $\mathscr{E}_1$ is not symmetric.

Table 14.11 gives the complete list of k-arcs in $PG(2,9)$. The complete ones are indicated, $p(3,9) = 9^3(9^3 - 1)(9^2 - 1) = 2^7\cdot 3^6\cdot 5\cdot 7\cdot 13$.

14.8. $PG(2,2^h)$ for $h = 4,5,6,7,8$

The object of this section is to find all ovals of the form

$$\mathscr{D}(k) = \{\mathbf{P}(t^k, t, 1) \mid t \in \gamma^+\} \cup \{\mathbf{U}_1\}$$

for the given fields. It was shown in theorem 8.4.2, corollary 2, that there are three conditions for $\mathscr{D}(k)$ to be an oval:

$$\begin{aligned} &\text{(i) } (k, q-1) = 1;\\ &\text{(ii) } (k-1, q-1) = 1;\\ &\text{(iii) } E_k(x) = [(x+1)^k + 1]/x \in \mathscr{P}(q;x). \end{aligned} \tag{14.7}$$

By corollary 1 of the same theorem, k must be even. In theorem 8.4.3, it is shown that, if $\mathscr{D}(k)$ is an oval, then, from (i) and (ii) of (14.7),

$$\mathscr{D}(k) \sim \mathscr{D}(k_1) \sim \mathscr{D}(k_2) \sim \mathscr{D}(k_3),$$

TABLE 14.11

k	Total number of k-arcs	G	$\|G\|$	Number of projectively distinct types	Description
3	$3^5 \cdot 5 \cdot 7 \cdot 13$	$S_3 Z_8{}^2$	384	1	
4	$2^4 \cdot 3^5 \cdot 5 \cdot 7 \cdot 13$	S_4	24	1	
5	$2^5 \cdot 3^6 \cdot 5 \cdot 7 \cdot 13$	Z_4	4	1	*harmonic*
5	$2^6 \cdot 3^6 \cdot 7 \cdot 13$	D_5	10	1	*non-harmonic*
6	$2^4 \cdot 3^5 \cdot 5 \cdot 7 \cdot 13$	S_4	24	1	*on a conic, residual of harmonic tetrad, six B-points*
6	$2^5 \cdot 3^6 \cdot 5 \cdot 7 \cdot 13$	$Z_2 \times Z_2$	4	1	*on a conic, residual of non-harmonic tetrad, two B-points*
6	$2^5 \cdot 3^5 \cdot 7 \cdot 13$	A_5	60	1	*complete, symmetric, ten B-points*
6	$2^7 \cdot 3^5 \cdot 5 \cdot 7 \cdot 13$	S_3	6	2	*not on a conic, four B-points*
6	$2^7 \cdot 3^5 \cdot 5 \cdot 7 \cdot 13$	Z_3	3	1	*not on a conic, three B-points*
6	$2^5 \cdot 3^6 \cdot 5 \cdot 7 \cdot 13$	Z_4	4	1	*not on a conic, two B-points*
7	$2^6 \cdot 3^5 \cdot 5 \cdot 7 \cdot 13$	S_3	6	1	*on a conic*
7	$2^7 \cdot 3^5 \cdot 5 \cdot 13$	$Z_3 Z_7$	21	1	*complete, symmetric*
7	$2^7 \cdot 3^6 \cdot 5 \cdot 7 \cdot 13$	Z_2	2	2	*in a complete* 8-*arc*
8	$2^3 \cdot 3^6 \cdot 5 \cdot 7 \cdot 13$	$Z_2 Z_8$	16	1	*on a conic*
8	$2^4 \cdot 3^6 \cdot 5 \cdot 7 \cdot 13$	D_4	8	1	*complete*
9	$2^4 \cdot 3^4 \cdot 5 \cdot 7 \cdot 13$	$Z_8 Z_9$	72	1	*on a conic*
10	$2^3 \cdot 3^4 \cdot 7 \cdot 13$	$PGL(2,9)$	720	1	*conic, complete*

where

$$\begin{aligned} kk_1 &\equiv 1 \quad (\text{mod } q-1), \\ (k-1)(k_2-1) &\equiv 1 \quad (\text{mod } q-1), \\ k + k_3 &= q, \end{aligned} \tag{14.8}$$

and, for definiteness, $1 < k_1 < q - 1$ and $1 < k_2 < q - 1$.

For the given values of q, we compile lists of k, k_1, k_2, k_3, in Table 14.12, where $2 \leqslant k \leqslant q/2$ and k satisfies (i) and (ii) of (14.7). Since $k_3 = q - k$, we need not consider $k > q/2$.

TABLE 14.12

$q = 16 = 2^4$

k	2	8
k_1	8	2
k_2	2	14
k_3	14	8

$q = 32 = 2^5$

k	2	4	6	8	10	12	14	16
k_1	16	8	26	4	28	13	20	2
k_2	2	22	26	10	8	18	13	30
k_3	30	28	26	24	22	20	18	16

$q = 64 = 2^6$

k	2	20	26	32
k_1	32	41	17	2
k_2	2	11	59	62
k_3	62	44	38	32

$q = 128 = 2^7$

k	2	4	6	8	10	12	14	16	18	20	22
k_1	64	32	106	16	89	53	118	8	120	108	52
k_2	2	86	52	110	114	105	89	18	16	108	122
k_3	126	124	122	120	118	116	114	112	110	108	106
k	24	26	28	30	32	34	36	38	40	42	44
k_1	90	44	59	72	4	71	60	117	54	124	26
k_2	117	62	81	93	42	78	99	104	115	32	66
k_3	104	102	100	98	96	94	92	90	88	86	84
k	46	48	50	52	54	56	58	60	62	64	
k_1	124	26	58	45	94	22	40	93	46	36	
k_2	49	101	71	6	13	98	79	29	26	126	
k_3	82	80	78	76	74	72	70	68	66	64	

TABLE 14.12 (continued)

$q = 256 = 2^8$

k	2	8	14	32	38	44	62	74	92	98	104	122	128
k_1	128	32	164	8	47	29	218	224	158	242	172	23	2
k_2	2	74	158	182	194	173	47	8	242	164	53	197	254
k_3	254	248	242	224	218	212	194	182	164	158	152	134	128

14.9.1. THEOREM. *In $PG(2,2^h)$ with $4 \leqslant h \leqslant 8$, the projectively distinct ovals of the form $\mathscr{D}(k)$ are as follows:*

(i) $h = 4$: $\mathscr{D}(2)$;
(ii) $h = 5$: $\mathscr{D}(2)$, $\mathscr{D}(4)$, $\mathscr{D}(6)$;
(iii) $h = 6$: $\mathscr{D}(2)$;
(iv) $h = 7$: $\mathscr{D}(2)$, $\mathscr{D}(4)$, $\mathscr{D}(6)$, $\mathscr{D}(8)$, $\mathscr{D}(20)$;
(v) $h = 8$: $\mathscr{D}(2)$, $\mathscr{D}(8)$.

The group of $\mathscr{D}(2)$ is given by all transformations $t \to (at + b)/(ct + d)$ with $ad - bc \neq 0$ and is isomorphic to $PGL(2,q)$. The group of $\mathscr{D}(4)$ and $\mathscr{D}(8)$ is given by all transformations $t \to ct + d$ with $c \neq 0$ and is isomorphic to $\mathbf{Z}_q\mathbf{Z}_{q-1}$. The group of $\mathscr{D}(6)$ and $\mathscr{D}(20)$ is given by all transformations $t \to ct$ with $c \neq 0$ and is isomorphic to $\mathbf{Z}_{q-1}$.

Proof. From Table 14.12, the only possibilities for projectively distinct ovals $\mathscr{D}(k)$ are the following, where k has been chosen as low as possible:

(i) $h = 4$: $\mathscr{D}(2)$;
(ii) $h = 5$: $\mathscr{D}(2)$, $\mathscr{D}(4)$, $\mathscr{D}(6)$;
(iii) $h = 6$: $\mathscr{D}(2)$;
(iv) $h = 7$: $\mathscr{D}(2)$, $\mathscr{D}(4)$, $\mathscr{D}(6)$, $\mathscr{D}(8)$, $\mathscr{D}(20)$, $\mathscr{D}(26)$;
(v) $h = 8$: $\mathscr{D}(2)$, $\mathscr{D}(8)$, $\mathscr{D}(14)$.

In each case, $\mathscr{D}(2^n)$ with $(n,h) = 1$ is an oval by theorem 8.4.2, corollary 3. With h odd, $\mathscr{D}(6)$ is an oval by theorem 8.4.4(iii). This leaves the cases $\mathscr{D}(20)$ and $\mathscr{D}(26)$ for $h = 7$, and $\mathscr{D}(14)$ for $h = 8$.

In each of these three cases it remains to consider whether or not E_k is in $\mathscr{P}(q;x)$. Tables for calculating in $GF(128)$ and $GF(256)$ are given in Appendix II.

TABLE 14.13

i'	i	i'	i	i'	i	i'	i	i'	i	i'	i
1	15	22	35	43	57	64	71	85	92	106	46
2	30	23	117	44	70	65	103	86	114	107	123
3	79	24	124	45	101	66	19	87	119	108	2
4	60	25	36	46	107	67	89	88	13	109	86
5	38	26	39	47	111	68	69	89	4	110	53
6	31	27	64	48	121	69	104	90	75	111	65
7	51	28	77	49	26	70	105	91	45	112	54
8	120	29	84	50	72	71	55	92	87	113	109
9	11	30	93	51	8	72	88	93	106	114	94
10	76	31	66	52	78	73	34	94	95	115	40
11	81	32	99	53	23	74	28	95	3	116	82
12	62	33	73	54	1	75	122	96	115	117	125
13	83	34	98	55	90	76	18	97	108	118	43
14	102	35	116	56	27	77	32	98	52	119	96
15	110	36	44	57	47	78	42	99	91	120	118
16	113	37	14	58	41	79	33	100	17	121	20
17	49	38	9	59	85	80	100	101	61	122	126
18	22	39	21	60	59	81	58	102	16	123	48
19	68	40	50	61	63	82	7	103	80	124	10
20	25	41	67	62	5	83	74	104	29	125	24
21	56	42	112	63	6	84	97	105	37	126	12

$$GF(128)_0 = \{\beta^i \mid i \in \bar{\mathrm{N}}_{126},\ \beta^7 = \beta + 1\},$$

Then $E_{26}(\beta^5) = E_{26}(\beta^9) = \beta^{123}$. So E_{26} is not in $\mathscr{P}(128;x)$ and $\mathscr{D}(26)$ is not an oval.

Write $\beta^{i'} = E_{20}(\beta^i)$. Then the pairs (i',i) are listed in Table 14.13 showing that i' takes all values in N_{126} as i does. So E_{20} is in $\mathscr{P}(128;x)$.

$$GF(256)_0 = \{\rho^i \mid i \in \bar{\mathrm{N}}_{254},\ \rho^8 = \rho^4 + \rho^3 + \rho^2 + 1\}.$$

Then $E_{14}(\rho^4) = E_{14}(\rho^{15}) = \rho^{190}$. So E_{14} is not in $\mathscr{P}(256;x)$ and

$\mathscr{D}(14)$ is not an oval in $PG(2,256)$.

To show that the ovals found are projectively distinct, it almost suffices to determine their projective groups. These are as in the statement of the theorem. The group of $\mathscr{D}(2)$ fixes $\mathbf{U}_1$; the groups of $\mathscr{D}(4)$ and $\mathscr{D}(8)$ fix $\mathbf{U}_0$ and $\mathbf{U}_1$; the groups of $\mathscr{D}(6)$ and $\mathscr{D}(20)$ fix $\mathbf{U}_0$, $\mathbf{U}_1$, and $\mathbf{U}_2$. So the ovals given are distinct except perhaps in the case $h = 7$.

If $\mathscr{D}(4)$ is mapped to $\mathscr{D}(8)$ by $\mathfrak{T}$, then $\mathfrak{T}$ fixes the set $\{\mathbf{U}_0,\mathbf{U}_1\}$: a calculation shows that no such projectivity $\mathfrak{T}$ exists. Similarly if $\mathscr{D}(6)\mathfrak{T} = \mathscr{D}(20)$, then $\mathfrak{T}$ fixes $\{\mathbf{U}_0,\mathbf{U}_1,\mathbf{U}_2\}$ and again no such $\mathfrak{T}$ exists.□

14.9. Notes and references

§14.1. This is based on Edge [1970a].

§14.2. See also Abdul-Elah [1974], Coxeter [1956]. For the embedding of diagrams of $PG(2,q)$ in the real Euclidean plane, see Lehti [1957].

§14.3. This is based on Edge [1965]. See also Abdul-Elah [1974], Freudenthal [1972], Hall, Lane, and Wales [1970], Havlíček and Tietze [1970], [1971], Hirschfeld [1964], [1967a], Jónsson [1972], [1973], Lüneberg [1969], Schulz [1975].

§14.4. See also Coxeter [1974] Chapter 10, Cundy [1952], Edge [1955b], Rickart [1940].

§14.5. See Hirschfeld [1967a], as well as Hall [1953], [1954], Martinov and Lozanov [1976].

§14.6. See also D'Orgeval [1975a], Hall, Swift, and Walker [1956], Segre [1957a], Tallini Scafati [1958].

§14.7. See also Hy [1971], Room and Kirkpatrick [1971], Segre [1959a], Yff [1977].

§14.8. See Hirschfeld [1971], [1975].

For the structure of small arcs, see Sce [1960].

See also Berman [1952], MacInnes [1907], MacNeish [1942].

APPENDIX I
ORDERS OF THE SEMI-LINEAR GROUPS
$DX(n, q)$, $q=p^h$, $n \geq 2$

AI.1. Definitions

(a) We define the groups $DX(n,q)$ for $n \geq 2$, where

$$D = I,\ S,\ G,\ \Gamma,\ \Gamma S,\ P,\ PS,\ PG,\ P\Gamma,\ P\Gamma S,$$

and
$$X = L,\ O,\ O_+,\ O_-,\ U,\ Sp,\ Ps,\ Ps^*.$$

For $D = I, S, G, \Gamma, \Gamma S$, the group $DX(n,q)$ is a group of semi-linear transformations in $V(n,q)$. For $D = P, PS, PG, P\Gamma, P\Gamma S$, the group $DX(n,q)$ is a group of collineations in $PG(n-1,q)$. The symbol X refers to the form (quadratic, Hermitian, or bilinear) involved. In particular, the full linear and semi-linear groups of varieties and polarities in $V(n,q)$ are given by the respective rows $GX(n,q)$ and $\Gamma X(n,q)$ in Table AI.1. The full projective and collineation groups of varieties or polarities in $PG(n-1,q)$ are given by the respective rows $PGX(n,q)$ and $P\Gamma X(n,q)$: these have already been met in the text.

(b) $IX(n,q) = X(n,q)$.

$X(n,q)$ = group of linear transformations leaving invariant the canonical form specified in the table.

$SX(n,q)$ = subgroup of $X(n,q)$ of linear transformations of determinant one.

$GX(n,q)$ = group of linear transformations leaving the canonical form invariant up to a scalar multiple.

$\Gamma X(n,q)$ = group of semi-linear transformations leaving the canonical form invariant up to an automorphism of $GF(q)$ and a scalar multiple.

$\Gamma SX(n,q)$ = group of semi-linear transformations that are products of an element of $SX(n,q)$ and an automorphism of $GF(q)$.

(c) The groups $DX(n,q)$ beginning with the symbol "P" are the corresponding projective groups.

The centre of $\Gamma L(n,q)$ is $Z = \{\gamma I_n \mid \gamma \in \gamma_0,\ I_n$ identity of $\Gamma L(n,q)\}$. Then PX, PSX, PGX, $P\Gamma X$, $P\Gamma SX$ are the respective canonical images of X, SX, GX, ΓX, ΓSX under $\Gamma L(n,q) \to \Gamma L(n,q)/Z$. Thus,

TABLE

$|DX(n,q)|$, q =

	L	O	O_+	O_-
I	$\lambda_+(n,q)$	$(q-1,2)q^{(n-1)/2}\lambda_+((n-1)/2,q^2)$	$2q^{n-2}(q^{n/2}-1)\lambda_+((n-2)/2,q^2)$	$2q^{n-2}(q^{n/2}+1)\lambda_+((n-2)/$
S	$(q-1)^{-1}\lambda_+(n,q)$	$q^{(n-1)/2}\lambda_+((n-1)/2,q^2)$	$(q,2)q^{n-2}(q^{n/2}-1)\lambda_+((n-2)/2,q^2)$	$(q,2)q^{n-2}(q^{n/2}+1)\lambda_+((n-$
G	$\lambda_+(n,q)$	$q^{(n-1)/2}(q-1)\lambda_+((n-1)/2,q^2)$	$2q^{n-2}(q^{n/2}-1)(q-1)\lambda_+((n-2)/2,q^2)$	$2q^{n-2}(q^{n/2}+1)(q-1)\lambda_+(($
Γ	$h\lambda_+(n,q)$	$hq^{(n-1)/2}(q-1)\lambda_+((n-1)/2,q^2)$	$2hq^{n-2}(q^{n/2}-1)(q-1)\lambda_+((n-2)/2,q^2$	$2hq^{n-2}(q^{n/2}+1)(q-1)\lambda_+($
ΓS	$h(q-1)^{-1}\lambda_+(n,q)$	$hq^{(n-1)/2}\lambda_+((n-1)/2,q^2)$	$h(q,2)q^{n-2}(q^{n/2}-1)\lambda_+((n-2)/2,q^2)$	$h(q,2)q^{n-2}(q^{n/2}+1)\lambda_+(($
P	$(q-1)^{-1}\lambda_+(n,q)$	$q^{(n-1)/2}\lambda_+((n-1)/2,q^2)$	$(q,2)q^{n-2}(q^{n/2}-1)\lambda_+((n-2)/2,q^2)$	$(q,2)q^{n-2}(q^{n/2}+1)\lambda_+((n-$
PS	$(n,q-1)^{-1}(q-1)^{-1}\lambda_+(n,q)$	$q^{(n-1)/2}\lambda_+((n-1)/2,q^2)$	$2^{(-1)^q}q^{n-2}(q^{n/2}-1)\lambda_+((n-2)/2,q^2)$	$2^{(-1)^q}q^{n-2}(q^{n/2}+1)\lambda_+(($
PG	$(q-1)^{-1}\lambda_+(n,q)$	$q^{(n-1)/2}\lambda_+((n-1)/2,q^2)$	$2q^{n-2}(q^{n/2}-1)\lambda_+((n-2)/2,q^2)$	$2q^{n-2}(q^{n/2}+1)\lambda_+((n-2)/$
$P\Gamma$	$h(q-1)^{-1}\lambda_+(n,q)$	$hq^{(n-1)/2}\lambda_+((n-1)/2,q^2)$	$2hq^{n-2}(q^{n/2}-1)\lambda_+((n-2)/2,q^2)$	$2hq^{n-2}(q^{n/2}+1)\lambda_+((n-2)/$
$P\Gamma S$	$h(n,q-1)^{-1}(q-1)^{-1}\lambda_+(n,q)$	$hq^{(n-1)/2}\lambda_+((n-1)/2,q^2)$	$h2^{(-1)^q}q^{n-2}(q^{n/2}-1)\lambda_+((n-2)/2,q^2)$	$h2^{(-1)^q}q^{n-2}(q^{n/2}+1)\lambda_+(($
n	*all*	*odd*	*even*	*even*
q	*all*	*all*	*all*	*all*
invariant in $PG(n-1,q)$	$PG(n-1,q)$	$\mathscr{P}_{n-1}$	$\mathscr{H}_{n-1}$	$\mathscr{E}_{n-1}$
canonical form of invariant		$x_1^2+x_2x_3+\ldots+x_{n-1}x_n$	$x_1x_2+x_3x_4+\ldots+x_{n-1}x_n$	$f(x_1,x_2)+x_3x_4+\ldots+x_{n-1}x$ f *irreducible*
name of group	*linear*	*orthogonal*	*orthogonal*	*orthogonal*

$\lambda_\varepsilon(m,r) = r^{m(m-1)/2}\prod_{i=1}^{m}(r^i-\varepsilon^i)$; write $\lambda_+(m,r)$ when $\varepsilon = 1$, and $\lambda_-(m,r)$ when $\varepsilon = -1$. $p_{ij} = x_iy_j - x_jy_i$.

U	Sp	Ps	Ps^*
$\lambda_-(n,\sqrt{q})$	$q^{n/2}\lambda_+(n/2,q^2)$	$q^{(n-1)/2}\lambda_+((n-1)/2,q^2)$	$q^{(3n-4)/2}\lambda_+((n-2)/2,q^2)$
$(\sqrt{q}+1)^{-1}\lambda_-(n,\sqrt{q})$	$q^{n/2}\lambda_+(n/2,q^2)$	$q^{(n-1)/2}\lambda_+((n-1)/2,q^2)$	$q^{(3n-4)/2}\lambda_+((n-2)/2,q^2)$
$(\sqrt{q}-1)\lambda_-(n,\sqrt{q})$	$q^{n/2}(q-1)\lambda_+(n/2,q^2)$	$q^{(n-1)/2}(q-1)\lambda_+((n-1)/2,q^2)$	$q^{(3n-4)/2}(q-1)\lambda_+((n-2)/2,q^2)$
$h(\sqrt{q}-1)\lambda_-(n,\sqrt{q})$	$hq^{n/2}(q-1)\lambda_+(n/2,q^2)$	$hq^{(n-1)/2}(q-1)\lambda_+((n-1)/2,q^2)$	$hq^{(3n-4)/2}(q-1)\lambda_+((n-2)/2,q^2)$
$h(\sqrt{q}+1)^{-1}\lambda_-(n,\sqrt{q})$	$hq^{n/2}\lambda_+(n/2,q^2)$	$hq^{(n-1)/2}\lambda_+((n-1)/2,q^2)$	$hq^{(3n-4)/2}\lambda_+((n-2)/2,q^2)$
$(\sqrt{q}+1)^{-1}\lambda_-(n,\sqrt{q})$	$(q-1,2)^{-1}q^{n/2}\lambda_+(n/2,q^2)$	$q^{(n-1)/2}\lambda_+((n-1)/2,q^2)$	$q^{(3n-4)/2}\lambda_+((n-2)/2,q^2)$
$(n,\sqrt{q}+1)^{-1}(\sqrt{q}+1)^{-1}\lambda_-(n,\sqrt{q})$	$(q-1,2)^{-1}q^{n/2}\lambda_+(n/2,q^2)$	$q^{(n-1)/2}\lambda_+((n-1)/2,q^2)$	$q^{(3n-4)/2}\lambda_+((n-2)/2,q^2)$
$(\sqrt{q}+1)^{-1}\lambda_-(n,\sqrt{q})$	$q^{n/2}\lambda_+(n/2,q^2)$	$q^{(n-1)/2}\lambda_+((n-1)/2,q^2)$	$q^{(3n-4)/2}\lambda_+((n-2)/2,q^2)$
$h(\sqrt{q}+1)^{-1}\lambda_-(n,\sqrt{q})$	$hq^{n/2}\lambda_+(n/2,q^2)$	$hq^{(n-1)/2}\lambda_+((n-1)/2,q^2)$	$hq^{(3n-4)/2}\lambda_+((n-2)/2,q^2)$
$h(n,\sqrt{q}+1)^{-1}(\sqrt{q}+1)^{-1}\lambda_-(n,\sqrt{q})$	$h(q-1,2)^{-1}q^{n/2}\lambda_+(n/2,q^2)$	$hq^{(n-2)/2}\lambda_+((n-1)/2,q^2)$	$hq^{(3n-4)/2}\lambda_+((n-2)/2,q^2)$
all	*even*	*odd*	*even*
square	*all*	*even*	*even*
$\mathcal{U}_{n-1}$	*null polarity*	*pseudo polarity*	*pseudo polarity*
$x_1^{\sqrt{q}+1}+\ldots+x_n^{\sqrt{q}+1}$	$p_{12}+p_{34}+\ldots+p_{n-1,n}$	$x_1y_1+p_{23}+p_{45}+\ldots+p_{n-1,n}$	$x_1y_1+p_{12}+p_{34}+\ldots+p_{n-1,n}$
unitary	*symplectic*	*pseudo symplectic*	*pseudo symplectic*

(n_1,n_2) = greatest common divisor of n_1 and n_2.

$$PX(n,q) = X(n,q)/(Z \cap X(n,q)),$$
$$PSX(n,q) = SX(n,q)/(Z \cap SX(n,q)),$$
$$PGX(n,q) = GX(n,q)/(Z \cap GX(n,q)),$$
$$P\Gamma X(n,q) = \Gamma X(n,q)/(Z \cap \Gamma X(n,q)),$$
$$P\Gamma SX(n,q) = \Gamma SX)n,q)/(Z \cap \Gamma SX(n,q)).$$

(d) For q square, it is sometimes opportune to consider only the involutory and identity automorphisms of γ instead of all automorphisms. In this case, we write γX instead of ΓX and $P\gamma X$ instead of $P\Gamma X$. In particular, $|P\gamma U(n,q)| = 2|PGU(n,q)|$.

(e) Commonly, one sees ΣX for ΓSX and $P\Sigma X$ for $P\Gamma SX$. Also O^+ and O^- are written elsewhere for O_+ and O_- respectively. In other works, when q is even, the orthogonal groups $SO^+(n,q)$ and $SO^-(n,q)$ are the kernels, not of the determinant map, but taken with respect to the Dickson invariant. In that case $|SO^{\pm}(n,q)| = |O^{\pm}(n,q)|/2$. See Dieudonné [1955], [1971].

AI.2. Comparative orders and isomorphisms in the same column of Table AI.1.

For all X,

$$|GX|/(q-1) = |\Gamma X|/[h(q-1)] = |P\Gamma X|/h = |PGX|.$$

Now we consider the rows I, S, P, PS, PG.

(a) $X = L$

Orders: $|L| = |GL|$, $|PGL| = |SL| = |PL| = |L|/(q-1)$,
$|PSL| = |SL|/(n,q-1)$.

Isomorphisms:

$L = GL$, $PL = PGL$,

$PGL \cong SL \cong PSL \Leftrightarrow (n,q-1) = 1$.

(b) $X = O$

Orders: $|PGO| = |PSO| = |PO| = |SO| = |O|/(q-1,2)$.

Isomorphisms:

q even, $PGO \cong PSO \cong PO \cong SO \cong O$;

q odd, $PGO \cong PSO \cong PO \cong SO$.

(c) $X = O_+, O_-$

Orders: q even, $|PGX| = |PSX| = |PX| = |SX| = |X|$;

q odd, $|PGX| = 4|PSX| = 2|PX| = 2|SX| = |X|$.

Isomorphisms:

q even, $PGX \cong PSX \cong PX \cong SX \cong X$.

(d) $X = U$

Orders: $|PGU| = |PU| = |SU| = |U|/(\sqrt{q} + 1)$,

$|PSU| = |SU|/(n, \sqrt{q} + 1)$.

Isomorphisms:

$PGU \cong PSU \cong SU \cong PU \Leftrightarrow (n, \sqrt{q} + 1) = 1$.

(e) $X = Sp$

Orders: $|PGSp| = |SSp| = |Sp|$,

$|PSSp| = |PSp| = |Sp|/(q - 1, 2)$.

Isomorphisms:

q even, $PGSp \cong PSSp \cong PSp \cong SSp \cong Sp$;

q odd, $SSp \cong Sp$, $PSSp \cong PSp$.

(f) $X = Ps, Ps^*$

Orders: $|PGX| = |PSX| = |PX| = |SX| = |X|$.

Isomorphisms:

$PGX \cong PSX \cong PX \cong SX \cong X$.

Remarks on proof of (b) - (f)

(b) $|PGL(n,q)|/|PGO(n,q)| = |GL(n,q)|/|GO(n,q)|$

$=$ number of parabolic quadrics $\mathscr{P}_{n-1}$

$=$ (number of n-ary quadratic forms)$/(q - 1)$

$= (|L(n,q)|/|O(n,q)|)(q - 1, 2)/(q - 1))$.

Note that, under linear transformations, $x_0^2 + x_1x_2$ is not equivalent to $\nu(x_0^2 + x_1x_2)$ in the case that q is odd and ν a non-square. So for q and n odd, L acting on the set of n-ary quadratic forms has two orbits.

(c) $|PGL(n,q)|/|PGO_+(n,q)| = |GL(n,q)|/|GO_+(n,q)|$

$=$ number of hyperbolic quadrics $\mathscr{H}_{n-1}$

$=$ (number of hyperbolic quadratic forms)$/(q - 1)$

$= (|L(n,q)|/|O_+(n,q)|)/(q - 1)$.

$|PGL(n,q)|/|PGO_-(n,q)| = |GL(n,q)|/|GO_+(n,q)|$

$=$ number of elliptic quadrics $\mathscr{E}_{n-1}$

$=$ (number of elliptic quadratic forms)$/(q - 1)$

$= (|L(n,q)|/|O_-(n,q)|)/(q - 1)$.

(d) $|PGL(n,q)|/|PGU(n,q)| = |GL(n,q)|/|GU(n,q)|$

$=$ number of Hermitian varieties $\mathscr{U}_{n-1}$

$=$ (number of Hermitian forms)$/(\sqrt{q} - 1)$

$= (|L(n,q)|/|U(n,q)|)/(\sqrt{q} - 1)$.

(e) $|PGL(n,q)|/|PGSp(n,q)| = |GL(n,q)|/|GSp(n,q)|$
= number of null polarities in $PG(n-1,q)$
= (number of alternating skew-symmetric bilinear forms)/$(q - 1)$
$= (|L(n,q)|/|Sp(n,q)|)/(q - 1)$.

(f) For $X = Ps$, Ps^*,
$|PGL(n,q)|/|PGX(n,q)| = |GL(n,q)|/|GX(n,q)|$
= number of pseudo polarities in $PG(n-1,q)$
= (number of non-alternating (skew)-symmetric bilinear forms)/$(q - 1)$
$= (|L(n,q)|/|X(n,q)|)/(q - 1)$.

AI.3. Isomorphisms proved in the text

$\mathbf{S}_3 \cong PGL(2,2) \cong PSL(2,2)$	§6.4 (i)
$\mathbf{A}_4 \cong PSL(2,3)$	§6.4 (ii)
$\mathbf{S}_4 \cong PGL(2,3)$	§6.4 (ii)
$\mathbf{A}_5 \cong PGL(2,4) \cong PSL(2,4)$	§6.4 (iii)
$\mathbf{A}_5 \cong PSL(2,5)$	§6.4 (iv)
$\mathbf{S}_5 \cong P\Gamma L(2,4)$	§6.4 (iii)
$\mathbf{S}_5 \cong PGL(2,5)$	§6.4 (iv)
$\mathbf{A}_6 \cong PSL(2,9)$	§6.4 (vii)
$\mathbf{S}_6 \cong P\Gamma SL(2,9)$	§6.4 (vii)
$PGL(3,2) \cong PSL(3,2) \cong PSL(2,7)$	§6.4 (v)
$PGL(2,q) \cong PGSp(2,q)$	lemma 6.2.2
$PGL(2,q) \cong PGU(2,q^2)$	lemma 6.2.1, corollary
$PSL(2,q) \cong PSU(2,q^2)$	lemma 6.2.1, corollary
$PGPs^*(2,q) \cong \mathbf{Z}_q$	lemma 6.2.2
$PGO(3,q) \cong PGL(2,q)$	lemma 7.2.3, corollary 7
$PGPs(3,q) \cong PGSp(2,q)$	lemma 8.3.6

A full list of all isomorphisms among the semi-linear groups as well as geometrical proofs will be given in Volume II.

On semi-linear groups, see Artin [1955a], [1955b], [1957], Dickson [1901], Dieudonné [1954], [1971], Huppert [1967], van der Waerden [1935].

APPENDIX II
ADDITION TABLES FOR $GF(2^h)$, $h \leq 8$

Addition tables for $GF(2^h)$, $h \leqslant 8$.

In all of Table AII.1, $GF(2^h)$ is considered as an extension of degree h of $GF(2)$.

$$GF(4)_0 = \{\omega^i \mid i \in \mathbf{N}_2, \quad \omega^2 = \omega + 1\}$$
$$GF(8)_0 = \{\varepsilon^i \mid i \in \mathbf{N}_6, \quad \varepsilon^3 = \varepsilon^2 + 1\}$$
$$GF(16)_0 = \{\eta^i \mid i \in \mathbf{N}_{14}, \quad \eta^4 = \eta + 1\}$$
$$GF(32)_0 = \{\chi^i \mid i \in \mathbf{N}_{30}, \quad \chi^5 = \chi^2 + 1\}$$
$$GF(64)_0 = \{\zeta^i \mid i \in \mathbf{N}_{62}, \quad \zeta^6 = \zeta + 1\}$$
$$GF(128)_0 = \{\beta^i \mid i \in \mathbf{N}_{126}, \quad \beta^7 = \beta + 1\}$$
$$GF(256)_0 = \{\rho^i \mid i \in \mathbf{N}_{254}, \quad \rho^8 = \rho^4 + \rho^3 + \rho^2 + 1\}$$

For each $GF(q)$, we give the function $\mathfrak{A} : \mathbf{N}_{q-2} \to \mathbf{N}_{q-2}$, where if α is the primitive element of the field, as above, $1 + \alpha^i = \alpha^{i\mathfrak{A}}$.

TABLE AII.1

	i	$i\mathfrak{A}$		i	$i\mathfrak{A}$	i	$i\mathfrak{A}$
$GF(8)$	1	5	$GF(16)$	1	4	8	2
	2	3		2	8	9	7
	3	2		3	14	10	5
	4	6		4	1	11	12
	5	1		5	10	12	11
	6	4		6	13	13	6
				7	9	14	3

TABLE AII.1 (continued)

	i	$i\mathfrak{A}$		i	$i\mathfrak{A}$	i	$i\mathfrak{A}$
$GF(32)$	1	18	$GF(64)$	1	6	32	3
	2	5		2	12	33	16
	3	29		3	32	34	31
	4	10		4	24	35	13
	5	2		5	62	36	54
	6	27		6	1	37	44
	7	22		7	26	38	49
	8	20		8	48	39	43
	9	16		9	45	40	55
	10	4		10	61	41	28
	11	19		11	25	42	21
	12	23		12	2	43	39
	13	14		13	35	44	37
	14	13		14	52	45	9
	15	24		15	23	46	30
	16	9		16	33	47	17
	17	30		17	47	48	8
	18	1		18	27	49	38
	19	11		19	56	50	22
	20	8		20	59	51	53
	21	25		21	42	52	14
	22	7		22	50	53	51
	23	12		23	15	54	36
	24	15		24	4	55	40
	25	21		25	11	56	19
	26	28		26	7	57	58
	27	6		27	18	58	57
	28	26		28	41	59	20
	29	3		29	60	60	29
	30	17		30	46	61	10
				31	34	62	5

TABLE AII.1 (continued)

	i	$i\mathfrak{A}$	i	$i\mathfrak{A}$	i	$i\mathfrak{A}$	i	$i\mathfrak{A}$
GF(128)	1	7	32	97	63	3	95	65
	2	14	33	77	64	67	96	111
	3	63	34	86	65	95	97	32
	4	28	35	109	66	27	98	117
	5	54	36	106	67	64	99	103
	6	126	37	46	68	45	100	39
	7	1	38	58	69	107	101	84
	8	56	39	100	70	91	102	80
	9	90	40	51	71	79	103	99
	10	108	41	75	72	85	104	59
	11	87	42	114	73	78	105	25
	12	125	43	17	74	92	106	36
	13	55	44	94	75	41	107	69
	14	2	45	68	76	116	108	10
	15	31	46	37	77	33	109	35
	16	112	47	22	78	73	110	26
	17	43	48	119	79	71	111	96
	18	53	49	122	80	102	112	16
	19	29	50	83	81	118	113	115
	20	89	51	40	82	23	114	42
	21	57	52	93	83	50	115	113
	22	47	53	18	84	101	116	76
	23	82	54	5	85	72	117	98
	24	123	55	13	86	34	118	81
	25	105	56	8	87	11	119	48
	26	110	57	21	88	61	120	121
	27	66	58	38	89	20	121	120
	28	4	59	104	90	9	122	49
	29	19	60	124	91	70	123	24
	30	62	61	88	92	74	124	60
	31	15	62	30	93	52	125	12
					94	44	126	6

TABLE AII.1 (continued)

GF(256)	i	$i\mathfrak{A}$	i	$i\mathfrak{A}$	i	$i\mathfrak{A}$	i	$i\mathfrak{A}$	i	$i\mathfrak{A}$	i	$i\mathfrak{A}$
	1	25	37	179	73	236	109	71	145	16	181	29
	2	50	38	184	74	103	110	126	146	217	182	163
	3	223	39	106	75	199	111	246	147	53	183	123
	4	100	40	84	76	113	112	7	148	206	184	38
	5	138	41	157	77	228	113	76	149	188	185	249
	6	191	42	20	78	212	114	166	150	143	186	61
	7	112	43	121	79	174	115	243	151	178	187	204
	8	200	44	215	80	168	116	214	152	226	188	149
	9	120	45	31	81	160	117	122	153	119	189	219
	10	21	46	137	82	59	118	164	154	201	190	97
	11	245	47	101	83	57	119	153	155	159	191	6
	12	127	48	253	84	40	120	9	156	169	192	247
	13	99	49	197	85	170	121	43	157	41	193	28
	14	224	50	2	86	242	122	117	158	93	194	125
	15	33	51	238	87	167	123	183	159	155	195	72
	16	145	52	141	88	175	124	180	160	81	196	23
	17	68	53	147	89	203	125	194	161	108	197	49
	18	240	54	208	90	62	126	110	162	65	198	26
	19	92	55	63	91	209	127	12	163	182	199	75
	20	42	56	131	92	19	128	140	164	118	200	8
	21	10	57	83	93	158	129	239	165	227	201	154
	22	235	58	107	94	202	130	69	166	114	202	94
	23	196	59	82	95	176	131	56	167	87	203	89
	24	254	60	132	96	251	132	60	168	80	204	187
	25	1	61	186	97	190	133	250	169	156	205	207
	26	198	62	90	98	139	134	177	170	85	206	148
	27	104	63	55	99	13	135	144	171	211	207	205
	28	193	64	70	100	4	136	34	172	229	208	54
	29	181	65	162	101	47	137	46	173	232	209	91
	30	66	66	30	102	221	138	5	174	79	210	241
	31	45	67	216	103	74	139	98	175	88	211	171
	32	35	68	17	104	27	140	128	176	95	212	78
	33	15	69	130	105	248	141	52	177	134	213	233
	34	136	70	64	106	39	142	218	178	151	214	116
	35	32	71	109	107	58	143	150	179	37	215	44
	36	225	72	195	108	161	144	135	180	124	216	67

TABLE AII.1 (continued)

	i	iz
$GF(256)$	217	146
	218	142
	219	189
	220	252
	221	102
	222	237
	223	3
	224	14
	225	36
	226	152
	227	165
	228	77
	229	172
	230	231
	231	230
	232	173
	233	213
	234	244
	235	22
	236	73
	237	222
	238	51
	239	129
	240	18
	241	210
	242	86
	243	115
	244	23
	245	11
	246	111
	247	192
	248	105
	249	185
	250	133
	251	96
	252	220
	253	48
	254	24

BIBLIOGRAPHY

This is an attempt to give a comprehensive bibliography of projective spaces *PG(n,q)*. Some adjacent topics, particularly algebraic geometry over an arbitrary field, coding theory, finite linear groups, generalized quadrangles, miscellaneous properties of finite fields, permutation polynomials, projective planes, and t-designs are only represented. It is also intended to suffice for Volume II.

The abbreviations for periodicals follow Mathematical Reviews almost entirely. Occasionally the Mathematical Reviews abbreviation has been curtailed. When a work has several volumes, the year given refers to the first volume. The Bibliography contains 819 items.

ABDUL-ELAH, M.S. (1974). Some configurational aspects of π_{13} and π_{21}. *Math. Student* **42**, 60-64.

AGOU, S. (1971). Polynômes sur un corps fini. *Bull. Sci. Math.* **95**, 327-330.

— (1976). Critères d'irreductibilité des polynômes composés à coefficients dans un corps fini. *Acta Arith.* **30**, 213-223.

ALBERT, A.A. (1955). Leonard Eugene Dickson. *Bull. Amer. Math. Soc.* **61**, 331-345.

— (1956). *Fundamental concepts of higher algebra*. University of Chicago.

— and SANDLER, R. (1968). *An introduction to finite projective planes*. Holt, Rinehart, and Winston.

AL-DHAHIR, M.W. (1972). A theorem in projective 3-space equivalent to commutativity. *Progr. Math. (Allahabad)* **6**, 51-55.

ALLTOP, W.O. (1971). 5-designs in affine spaces. *Pacific J. Math.* **39**, 547-552.

ANDRÉ, J. (1954). Uber nicht-Desarguessche Ebenen mit transitiver Translationsgruppe. *Math. Z.* **60**, 156-186.

ANDREWS, G.E. (1977). Partitions, q-series and the Lusztig-Macdonald-Wall conjectures. *Inventiones Math.* **41**, 91-102.

ARCHBOLD, J.W. (1960). A metric for plane affine geometry over GF(2^λ). *Mathematika* **7**, 145-148.

— (1966). Permutation mappings in finite projective planes. *Mathematika* **13**, 45-48.

ARF, C. (1940). Untersuchungen über quadratische Formen in Körpern der Charakteristik 2 (Teil I). *J. Reine Angew. Math.* **183**, 148-167.

ARTIN, E. (1955*a*). The orders of the linear groups. *Comm. Pure Appl. Math.* **8**, 355-366.

— (1955*b*). The orders of the classical simple groups. *Comm. Pure Appl. Math.* **8**, 455-472.

— (1957). *Geometric algebra*. Interscience, New York.

— (1959). *Galois Theory*. University of Notre Dame.

ARTZY, R. (1965). *Linear geometry*. Addison Wesley.

— (1968). Pascal's theorem on an oval. *Amer. Math. Monthly* **75**, 143-146.

ASSMUS, E.F. and MATTSON, H.F. (1966). Perfect codes and the Mathieu groups. *Arch. Math.* **17**, 121-135.

— — (1967). On tactical configurations and error-correcting codes. *J. Combinatorial Theory* **2**, 243-257.

ASSMUS, E.F. and MATTSON, H.F. (1974). Coding and combinatorics. *SIAM Review* **16**, 349-388.

— and SACHER, H.E. (1977). Ovals from the point of view of Coding Theory. *Higher Combinatorics (Berlin, 1976)*. Reidel, 213-216.

AX, J. (1964). Zeros of polynomials over finite fields. *Amer. J. Math.* **86**, 255-261.

— (1968). The elementary theory of finite fields. *Ann. of Math.* **88**, 239-271.

BAER, R. (1952). Linear algebra and projective geometry. Academic Press.

BAKER, H.F. (1921). Principles of geometry (six volumes). Cambridge University Press (Ungar, 1960).

BARLOTTI, A. (1955). Un' estensione del teorema di Segre-Kustaanheimo. *Boll. Un. Mat. Ital.* **10**, 96-98.

— (1956*a*). Un' osservazione sulle k-calotte degli spazi lineari finiti di dimensione tre. *Boll. Un. Mat. Ital.* **11**, 248-252.

— (1956*b*). Su {k;n}-archi di un piano lineare finito. *Boll. Un. Mat. Ital.* **11**, 553-556.

— (1957). Un limitazione superiore per il numero dei punti appartenenti a una k-calotta C(k,0) di uno spazio lineare finito. *Boll. Un. Mat. Ital.* **12**, 67-70.

— (1965). Some topics in finite geometrical structures. *Institute of Statistics mimeo series no. 439*, Univ. of North Carolina.

— (1966*a*). Un' osservazione intorno ad un teorema di B. Segre sui q-archi. *Matematiche (Catania)* **21**, 23-29.

— (1966*b*). Una caratterizzazione grafica delle ipersuperficie hermitiane non singolari in uno spazio lineare finito di ordine quattro. *Matematiche (Catania)* **21**, 287-395.

— (1967). Sulle 2-curve nei piani grafici. *Rend. Sem. Mat. Univ. Padova* **37**, 91-97.

— and BOSE, R.C. (1971). Linear representation of a class of projective planes in a four dimensional projective space. *Ann. Mat. Pura Appl.* **88**, 9-31.

BASILE, A. and BRUTTI, P. (1971). Alcuni risultati sui {q(n-1)+1;n} - archi di un piano proiettivo finito. *Rend. Sem. Mat. Univ. Padova* **46**, 107-125.

— — (1973). On the completeness of regular {q(n-1)+m;n}-arcs in a finite projective plane. *Geometriae Dedicata* **1**, 340-343.

BASU, N.K. (1973). Harmonic inversions in PG(4,3) and some related configurations. *J. Austral. Math. Soc.* **25**, 54-59.

BAUMERT, L.D. (1971). Cyclic difference sets. *Lecture Notes in Mathematics* **182**, Springer.

BEARD, J.T.B. (1972*a*). Matrix fields over prime fields. *Duke Math. J.* **39**, 313-322.

— (1972*b*). Matrix fields over finite extensions of prime fields. *Duke Math. J.* **39**, 475-484.

— (1974). Computing in GF(q). *Math. Comp.* **28**, 1159-1166.

— and WEST, K.I. (1974*a*). Some primitive polynomials of the third kind. *Math. Comp.* **28**, 1166-1167.

— — (1974*b*). Factorization tables for $x^n - 1$ over GF(q). *Math. Comp.* **28**, 1167-1168.

BELTRAMETTI, E.G. and BLASI, A. (1969). On rotations and Lorentz transformations in a Galois space-time. *Atti Accad. Naz. Lincei Rend.* **46**, 390-394.

BENSON, C.J. (1970). On the structure of generalized quadrangles. *J. Algebra* **15**, 443-454.

BERLEKAMP, E.R. (1968). Algebraïc Coding Theory. McGraw-Hill.

— (1970). Factoring polynomials over large finite fields. *Math. Comp.* **24**, 713-735.

BERLEKAMP, E.R. (1976). An analog to the discriminant over fields of characteristic two. *J. Algebra* **38**, 315-317.

—, RUMSEY, H., and SOLOMON, G. (1967). On the solution of algebraic equations over finite fields. *Information and Control* **10**, 553-564.

BERMAN, G. (1952). Finite projective geometries. *Canad. J. Math.* **4**, 302-313.

BERTHELOT, P. (1971). Une formule de rationalité pour la fonction des schémas propres et lisses sur un corps fini. *C.R. Acad. Sci. Paris* **272**, 1574-1577.

BEUTELSPACHER, A. (1974). On parallelisms of finite projective spaces. *Geometriae Dedicata* **3**, 35-40.

— (1975). Partial spreads in finite projective spaces and partial designs. *Math. Z.* **145**, 211-230.

— (1976). Correction to 'Partial spreads in finite projective spaces and partial designs'. *Math. Z.* **147**, 303.

BIGGS, N. (1971). Finite groups of automorphisms. *London Mathematical Society Lecture Note Series* **6**. Cambridge University Press.

BIRKHOFF, G. and BARTEE, T.C. (1970). Modern Applied Algebra. McGraw-Hill.

BLAKE, I.F. and MULLIN, R.C. (1975). The mathematical theory of coding. Academic Press.

BLICHFELDT, H.J. (1917). Finite collineation groups. University of Chicago Press.

BONARDI, M.T. (1964). Intorno a certe superficie cubiche dello spazio di Galois. *Atti Accad. Naz. Lincei Rend.* **37**, 396-400.

— (1967). Sopra i monoidi cubici di un $S_{3,q}$ con speciale riguardo al caso q = 5. *Atti Accad. Naz. Lincei Rend.* **42**, 792-796.

BORHO, W. (1973). Kettenbrüche im Galoisfeld. *Abh. Math. Sem. Univ. Hamburg* **39**, 76-82.

BOSE, R.C. (1938). On the application of the properties of Galois fields to the problem of the construction of hyper Graeco-Latin squares. *Sankhyā* **3**, 323-338.

— (1947). Mathematical theory of the symmetrical factorial design. *Sankhyā* **8**, 107-166.

— (1959). On the application of finite projective geometry for deriving a certain series of balanced Kirkman arrangements. Golden Jubilee Commemoration Volume (1958-9). Calcutta Math. Soc. 341-354.

— (1971). Selfconjugate tetrahedra with respect to the hermitian variety $x_0^3 + x_1^3 + x_2^3 + x_3^3 = 0$ in PG$(3,2^2)$ and a representation of PG(3,3). *Proc. Sympos. Pure Math.* **19**, 27-37.

— (1976). On a representation of the Baer subplanes of the Desarguesian plane PG$(2,q^2)$ in a projective five dimensional space. *Teorie Combinatorie* (Rome, 1973), volume I, Accad. Naz. dei Lincei, 381-391.

— and BURTON, R.C. (1966). A characterization of flat spaces in a finite geometry and the uniqueness of the Hamming and the MacDonald codes. *J. Combinatorial Theory* **1**, 96-104.

— and CHAKRAVARTI, I.M. (1966). Hermitian varieties in a finite projective space PG(N,q^2). *Canad. J. Math.* **18**, 1161-1182.

— and SRIVASTAVA, J.N. (1964). On a bound useful in the theory of factorial designs and error correcting codes. *Ann. Math. Statist.* **35**, 408-414.

BOTTEMA, O. (1940). Self-projective point-sets. *Nederl. Akad. Wetensch. Proc. Ser. A* **43**, 591-598.

BOUGHON, P. (1955). Formule de Taylor pour un polynome à plusiers indeterminées sur un anneau de caractéristique p > 0. *C.R. Acad. Sci. Paris* **240**, 720-722.

—, NATHAN, J., and SAMUEL, P. (1955*a*). Courbes planes en caractéristique 2. *Bull. Soc. Math. France*. **83**, 275-278.

BOUGHON, P., NATHAN, J., and SAMUEL, P. (1955*b*). Une classe de séries formelles transcendantes. *Acad. Roy. Belg. Bull. Cl. Sc.* **41**, 93-96.
BRAMWELL, D.L. and WILSON, B.J. (1973*a*). Cubic caps in three dimensional Galois space. *Proc. Royal Irish Acad.* **73**A, no. 20, 279-283.
— — (1973*b*). The (11,3)-arcs of the Galois plane of order 5. *Proc. Cambridge Philos. Soc.* **74**, 247-250.
BRAWLEY, J.V., CARLITZ, L., and VAUGHAN, T. (1973). Linear permutation polynomials with coefficients in a subfield. *Acta Arith.* **24**, 193-199.
— and LEVINE, J. (1972). Equivalence classes of linear mappings with applications to algebraic cryptography I, II. *Duke Math. J.* **39**, 121-132; 133-142.
BRETAGNOLLE-NATHAN, J. (1958). Cubiques définies sur un corps de caractéristique quelconque. *Ann. Fac. Sci. Toulouse* **22**, 175-234.
BRUCK, R.H. (1969). Construction problems of finite projective planes. Conference on Combinatorial Mathematics and its Applications (University of North Carolina, 1967), University of North Carolina Press, 426-514.
— (1970). Some relatively unknown ruled surfaces in projective space. *Archives, Nouvelle Serie, Section des Sciences, Institut Grand-Ducal de Luxembourg* **34**, 361-376.
— (1973*a*). Circle geometry in higher dimensions. *A Survey of Cominatorial Theory* (Colorado State University, 1971). North Holland, 69-77.
— (1973*b*). Circle geometry in higher dimensions II. *Geometriae Dedicata* **2**, 133-188.
— (1973*c*). Construction problems in finite projective spaces. *Finite Geometric Structures and their Applications* (Bressanone, 1972). Cremonese, 105-188.
— and BOSE, R.C. (1964). The construction of translation planes from projective spaces. *J. Algebra* **1**, 85-102.
— — (1966). Linear representations of projective planes in projective spaces. *J. Algebra* **4**, 117-172.
BRUEN, A.A. (1970). Baer subplanes and blocking sets. *Bull. Amer. Math. Soc.* **76**, 342-344.
— (1971*a*). Blocking sets in finite projective planes. *SIAM J. Appl. Math.* **21**, 380-392.
— (1971*b*). Partial spreads and replaceable nets. *Canad. J. Math.* **23**, 381-391.
— (1972*a*). Unimbeddable nets of small deficiency. *Pacific J. Math.* **43**, 51-54.
— (1972*b*). Spreads and a conjecture of Bruck and Bose. *J. Algebra* **23**, 1-19.
— (1975*a*). Collineations and extensions of translation nets. *Math. Z.* **145**, 243-250.
— (1975*b*). Subregular spreads and indicator sets. *Canad. J. Math.* **27**, 1141-1148.
— (1977). Some new replaceable translation nets. *Canad. J. Math.* **29**, 225-237.
— and FISHER, J.C. (1969). Spreads which are not dual spreads. *Canad. Math. Bull.* **12**, 801-803.
— — (1972). Arcs and ovals in derivable planes. *Math. Z.* **125**, 122-128.
— — (1973). Blocking sets, k-arcs and nets of order 10. *Advances in Math.* **10**, 317-320.
— — (1974). Blocking sets and complete k-arcs. *Pacific J. Math.* **53**, 73-84.
— and LEVINGER, B. (1973). A theorem on permutations of a finite field. *Canad. J. Math.* **25**, 1060-1065.
— and SILVERMAN, R. (1974). Switching sets in PG(3,q). *Proc. Amer. Math. Soc.* **43**, 176-180.
— and THAS, J.A. (1975). Flocks, chains and configurations in finite geometries. *Atti Accad. Naz. Lincei Rend.* **59**, 744-748.

BRUEN, A.A. and THAS, J.A. (1976). Partial spreads, packings and Hermitian manifolds in PG(3,q). *Math. Z.* **151**, 207-214.

BRUNO, S.R. (1969). Points interior and exterior to a quadric on a subspace in P_{2m} [GF(q)]. (Spanish) *Math. Notae* **21**, 137-158.

BUCKHIESTER, P.G. (1972). The number of n × n matrices of rank r and trace α over a finite field. *Duke Math. J.* **39**, 695-700.

BUEKENHOUT, F. (1966*a*). Ovales et ovales projectifs. *Atti Accad. Naz. Lincei Rend.* **40**, 46-49.

— (1966*b*). Plans projectifs à ovoides Pascaliens. *Arch. Math.* **17**, 89-93.

— (1966*c*). Étude intrinsèque des ovales. *Rend. Mat. e Appl.* **25**, 1-61.

— (1969*a*). Ensembles quadratiques des espaces projectifs. *Math. Z.* **110**, 306-318.

— (1969*b*). Une charactérization des espaces affins basée sur la notion de droite. *Math. Z.* **111**, 367-371.

— (1976). Characterizations of semi quadrics. A survey. *Teorie Combinatorie* (Rome, 1973), volume I, Accad. Naz. dei Lincei, 393-421.

— (1976). Existence of unitals in finite translation planes of order q^2 with a kernel of order q. *Geometriae Dedicata* **5**, 189-194.

— and HUBAUT, X. (1977). Locally polar spaces and related rank 3 groups. *J. Algebra* **45**, 391-434.

— and LEFÈVRE, C. (1974). Generalized quadrangles in projective spaces. *Arch. Math.* **25**, 540-552.

— and SHULT, E. (1974). On the foundations of polar geometry. *Geometriae Dedicata* **3**, 155-170.

BÜKE, M. (1974). Sui q-archi di un piano di Galois d'ordine q. *Atti Accad. Naz. Lincei Rend.* **57**, 355-359.

BUMCROT, R.J. (1969). Modern projective geometry. Holt, Rinehart, and Winston.

— (1973). An index to papers in 'Finite Geometries'. *Proc. International Conference on Projective Planes* (Washington State University, 1973). Washington State University Press, 7-25.

BURNSIDE, W. (1911). Theory of groups of finite order. 2nd edition, Cambridge (Dover, 1955).

BUSH, K.A. (1952). Orthogonal arrays of index unity. *Ann. Math. Statist.* **23**, 426-434.

BUSSEMAKER, J.C. and SEIDEL, J.J. (1970). Symmetric Hadamard matrices of order 36. *Ann. New York Acad. Sci.* **175**, 66-79.

BUSSEY, W. (1910). Tables of Galois fields of order less than 1000. *Bull. Amer. Math. Soc.* **16**, 188-206.

CALABI, E. and WILF, H.S. (1977). On the sequential and random selection of subspaces over a finite field. *J. Combinatorial Theory A* **22**, 107-109.

CAMERON, P.J. (1975). Partial quadrangles. *Quart. J. Math. Oxford Ser.* **26**, 61-74.

— (1976). Parallelisms of complete designs. *London Mathematical Society Lecture Note Series* **23**, Cambridge University Press.

— and SEIDEL, J.J. (1973). Quadratic forms over GF(2). *Nederl. Akad. Wetensch. Proc. Ser. A* **76**, 1-8.

— and Van LINT, J.H. (1975). Graph theory, coding theory and block designs. *London Mathematical Society Lecture Note Series* **19**, Cambridge University Press.

CAMPBELL, A.D. (1926). Plane cubic curves in the Galois fields of order 2^n. *Ann. of Math.* **27**, 395-406.

— (1927*a*). Three-parameter and four-parameter linear families of conics in the Galois fields of order 2^n. *Bull. Amer. Math. Soc.* **33**, 608-612.

— (1927*b*). Pencils of conics in the Galois fields of order 2^n. *Amer. J. Math.* **49**, 401-406.

— (1928*a*). The polar curves of plane algebraic curves in Galois fields. *Bull. Amer. Math. Soc.* **34**, 361-363.

CAMPBELL, A.D. (1928*b*). Nets of conics in the Galois fields of order 2^n. *Bull. Amer. Math. Soc.* **34**, 481-489.
— (1928*c*). Note on the Plücker equations for plane algebraic curves in the Galois fields. *Bull. Amer. Math. Soc.* **34**, 718-720.
— (1928*d*). The discriminant of the m-ary quadratic in the Galois fields of order 2^n. *Ann. of Math.* **29**, 395-398.
— (1928*e*). Nets of conics in the real domain. *Amer. J. Math.* **50**, 459-466.
— (1929). Note on linear transformations of n-ics in m variables. *Bull. Amer. Math. Soc.* **35**, 718-720.
— (1931). Pencils of quadrics in the Galois fields of order 2^n. *Tôhoku Math. J.* **34**, 236-248.
— (1932). Apolarity in the Galois fields of order 2^n. *Bull. Amer. Math. Soc.* **38**, 52-56.
— (1933*a*). Pseudo-covariants of an n-ic in m variables in a Galois field that consists of terms of this n-ic. *Bull. Amer. Math. Soc.* **39**, 252-256.
— (1933*b*). Note on cubic surfaces in the Galois fields of order 2^n. *Bull. Amer. Math. Soc.* **39**, 406-410.
— (1933*c*). Plane quartic curves in the Galois fields of order 2^n. *Tôhoku Math. J.* **37**, 88-93.
— (1937). Pseudo-covariants of n-ics in a Galois field. *Tôhoku Math. J.* **43**, 17-29.
CAR, M. (1971*a*). Le problème de Waring pour l'anneau des polynômes sur un corps fini. *C.R. Acad. Sc. Paris* **273**, 141-144.
— (1971*b*). Le problème de Goldbach pour l'anneau des polynômes sur un corps fini. *C.R. Acad. Sc. Paris* **273**, 201-204.
CARLITZ, L. (1953*a*). Permutations in a finite field. *Proc. Amer. Math. Soc.* **4**, 538.
— (1953*b*). Invariantive theory of equations in a finite field. *Trans. Amer. Math. Soc.* **75**, 405-427.
— (1956*a*). A note on nonsingular forms in a finite field. *Proc. Amer. Math. Soc.* **7**, 27-29.
— (1956*b*). Resolvents of certain linear groups in a finite field. *Canad. J. Math.* **8**, 568-579.
— (1957). The number of certain cubic surfaces over a finite field. *Boll. Un. Mat. Ital.* **12**, 19-21.
— (1960). A theorem on permutations in a finite field. *Proc. Amer. Math. Soc.* **11**, 456-459.
— (1962*a*). Some theorems on permutation polynomials. *Bull. Amer. Math. Soc.* **68**, 120-122.
— (1962*b*). A note on permutation functions over a finite field. *Duke Math. J.* **29**, 325-332.
— (1963). Permutations in finite fields. *Acta Sci. Math. Szeged.* **24**, 196-203.
— (1966). A note on quadrics over finite fields. *Duke Math. J.* **33**, 453-458.
— (1975*a*). Correspondences in a finite field, I. *Acta Arith.* **27**, 101-123.
— (1975*b*). Correspondences in a finite field, II. *Indiana Univ. Math. J.* **24**, 785-811.
— (1977). Some theorems on polynomials over a finite field. *Amer. Math. Monthly* **84**, 29-32.
— and HAYES, D.R. (1972). Permutations with coefficients in a subfield. *Acta Arith.* **21**, 131-135.
CARMICHAËL, R.D. (1937). Introduction to the theory of groups of finite order. Boston (Dover, 1956).
CASSE, L.R.A. (1969). A solution to Beniamino Segre's 'Problem $I_{r,q}$' for q even. *Atti Accad. Naz. Lincei Rend.* **46**, 13-20.

CASSE, L.R.A. (1972). (15;3)-arcs of $S_{2,7}$ *Proc. First Australian Conference on Combinatorial Math.*, Newcastle, 193-196.

— (1976). Concerning bitangents of irreducible plane quartic curves over GF(2^h). *Teorie Combinatorie* (Rome, 1973), volume II, Accad. Naz. dei Lincei, 381-387.

CATES, M., ERDÖS, P., HINDMAN, N., and ROTHSCHILD, B. (1976). Partition theorems for subsets of vector spaces. *J. Combinatorial Theory A* **20**, 279-291.

— and HINDMAN, N. (1975). Partition theorems for subspaces of vector spaces. *J. Combinatorial Theory A* **19**, 13-25.

CECCHERINI, P.V. (1969). Su certi {k,n}-archi dedotti da curve piane e sulle {m,n}-calotte di tipo (0,n) di un $S_{r,q}(r\geq 2)$. *Rend. Mat.* 2, 185-196.

CHAKRAVARTI, I.M. (1971). Some properties and applications of Hermitian varieties in a finite projective space PG(N,q^2) in the construction of strongly regular graphs and block designs. *J. Combinatorial Theory B* **11**, 268-283.

CHÂTELET, F. (1946). Essais de géométrie Galoisienne. *Bull. Soc. Math. France* **74**, 69-86.

— (1947). Les courbes de genre 1 dans un champ de Galois. *C.R. Acad. Sci. Paris* **224**, 1616-1618.

— (1948). Formes quadratiques dans un corp arbitraire. *C.R. Acad. Sci. Paris* **226**, 1233-1235.

CHEUNG, A. and CRAPO, H. (1976). A combinatorial perspective on algebraic geometry. *Advances in Math.* **20**, 388-414.

CICCHESE, M. (1962). Sulle cubiche di un piano di Galois. *Atti Accad. Naz. Lincei Rend.* **32**, 38-42.

— (1965). Sulle cubiche di un piano di Galois. *Rend. Mat. e Appl.* **24**, 291-330.

— (1970). Sulle cubiche di un piano lineare $S_{2,q}$ con $q \equiv 1$ (mod 3). *Atti Accad. Naz. Lincei Rend.* **48**, 584-588.

— (1971). Sulle cubiche di un piano lineare $S_{2,q}$ con $q \equiv 1$ (mod 3). *Rend. Mat.* **4**, 249-283.

COBLE, A.B. (1908). A configuration in finite geometry isomorphic with that of the twenty-seven lines of a cubic surface. *Johns Hopkins Univ. Circ. no. 7*, 80-88.

— (1912). The linear complex in the finite geometry (mod 2) of an S_5 as related to the double tangents of a quartic. *Johns Hopkins Univ. Circ. no. 2*, 43-46.

— (1913). An application of finite geometry to the characteristic theory of the odd and even theta functions. *Trans. Amer. Math. Soc.* **14**, 241-276.

— (1929). Algebraic geometry and theta functions. *Amer. Math. Soc. Colloquium Publications* **10**, Amer. Math. Soc. (revised edition 1961).

— (1936). Collineation groups in a finite space with a linear and a quadratic invariant. *Amer. J. Math.* **58**, 15-34.

COHEN, E. (1965). Linear and quadratic equations in a Galois field with applications to geometry. *Duke Math. J.* **32**, 633-641.

COLE, F.N. (1922). Kirkman parades. *Bull. Amer. Math. Soc.* **28**, 435-437.

CONWELL, G.M. (1910). The 3-space PG(3,2) and its group. *Ann. of Math.* **11**, 60-76.

CONTI, F. (1968). Determinazione dei (k;n)-archi completi del piano grafico finito di ordine quattro. *Matematiche (Catania)* **23**, 425-433.

CONTI, G. (1975). Piani proiettivi dotati di un ovale pascaliano. *Boll. Un. Mat. Ital.* **11**, 330-338.

CORAY, D.F. (1976). Algebraic points on cubic hypersurfaces. *Acta Arith.* **30**, 267-296.

CORDES, C.M. (1975*a*). A note on Pall partitions over finite fields. *Linear Algebra and Appl.* **12**, 81-85.

CORDES, C.M. (1975*b*). Some results on totally isotropic subspaces and five-dimensional quadratic forms over GF(q). *Canad. J. Math.* **27**, 271-275.
COSSU, A. (1960). Sulle ovali in un piano proiettivo sopra un corpo finito. *Atti Accad. Naz. Lincei Rend.* **28**, 342-344.
— (1961). Su alcune proprietà dei {k,n}-archi di un piano proiettivo sopra un corpo finito. *Rend. Mat. e Appl.* **20**, 271-277.
COXETER, H.S.M. (1940). The polytope 2_{21}, whose twenty-seven vertices correspond to the lines on a general cubic surface. *Amer. J. Math.* **62**, 457-486.
— (1950). Self-dual configurations and regular graphs. *Bull. Amer. Math. Soc.* **56**, 413-455.
— (1956). The collineation groups of the finite affine and projective planes with four lines through each point. *Abh. Math. Sem. Univ. Hamburg* **20**, 165-177.
— (1958*a*). The chords of the non-ruled quadric in PG(3,3). *Canad. J. Math.* **10**, 484-488.
— (1958*b*). Twelve points in PG(5,3) with 95040 self-transformations. *Proc. Roy. Soc. London Ser. A* **427**, 279-293.
— (1959). Polytopes over GF(2) and their relevance for the cubic surface group. *Canad. J. Math.* **11**, 646-650.
— (1962). Projective line geometry. *Math. Notae* **1**, 197-216.
— (1969). Introduction to geometry, second edition, Wiley.
— (1974). Projective geometry, second edition, University of Toronto Press.
— (1975*a*). Regular complex polytopes. Cambridge University Press.
— (1975*b*). Desargues configurations and their collineation groups. *Math. Proc. Cambridge Philos. Soc.* **78**, 227-246.
— (1976). The Pappus configuration and its groups. *Nederl. Akad. Wetensch. Verslag Afd. Natuurk.* **85**, 44-46.
CRAPO, H. (1976). A combinatorial perspective on algebraic geometry. *Teorie Combinatorie* (Rome, 1973), volume II, Accad. Naz. dei Lincei, 343-357.
CULLEN, C.G. (1966). Matrices and linear transformations. Addison-Wesley.
CUNDY, H.M. (1952). 25-point geometry. *Math. Gaz.* **36**, 158-166.
CURTES, C.W., KANTOR, W.M., and SEITZ, G.M. (1976). The 2-transitive permutation representations of the finite Chevalley groups. *Trans. Amer. Math. Soc.* **218**, 1-59.
DAI, Zong-Duo (1966). On transitivity of subspaces in orthogonal geometry over fields of characteristic 2. *Chinese Math.* **8**, 569-584.
— and FENG, Xu-Ning (1964). Notes on finite geometries and the construction of PBIB designs. IV. Some 'Anzahl' theorems in orthogonal geometry over finite fields of characteristic not 2. *Sci. Sinica* **13**, 2001-2004.
— — (1965). Studies in finite geometries and the construction of incomplate block designs. IV. Some 'Anzahl' theorems in orthogonal geometry over finite fields of characteristic $\neq$ 2. *Chinese Math.* **7**, 265-279.
DATTA, B.T. (1976*a*). On tangential 2-blocks. *Discrete Math.* **15**, 1-22.
— (1976*b*). Nonexistence of six-dimensional tangential 2-blocks. *J. Combinatorial Theory* B**21**, 171-193.
DAVIS, C. (1969). A bibliographic survey of simple groups of finite order, 1900-1965. New York University.
De GROOTE, R. (1973). Les cubiques dans un plan projectif sur un corps de caractéristique trois. *Acad. Roy. Belg. Bull. A. Sci.* **59**, 1140-1155.
— (1974). Les cubiques dans un plan projectif sur un corps fini de caractéristique 3. *Acad. Roy. Belg. Bull. A. Sci.* **60**, 43-57.
DELIGNE, P. (1974). La conjecture de Weil I. *Inst. Hautes Études Sci. Publ. Math.* **43**, 273-307.
DELSARTE, P. (1969). A geometric approach to a class of cyclic codes. *J. Combinatorial Theory* **6**, 340-358.
— and GOETHALS, J.-M. (1970). Codes correcteurs d'erreurs et décodage par décision majoritaire. *Revue MBLE* **13**, 23-25.

DELSARTE, P. and GOETHALS, J.-M. (1975). Alternating bilinear forms over GF(q). *J. Combinatorial Theory A* **19**, 26-50.
— and McELIECE, R.J. (1976). Zeros of functions in finite abelian group algebras. *Amer. J. Math.* **98**, 197-224.
DEMBOWSKI, P. (1968). Finite geometries. Springer.
DENNISTON, R.H.F. (1969). Some maximal arcs in finite projective planes. *J. Combinatorial Theory* **6**, 317-319.
— (1972). Some packings of projective spaces. *Atti Accad. Naz. Lincei Rend.* **52**, 36-40.
— (1973*a*). Spreads which are not subregular. *Glasnik Mat.* **8**, 3-5.
— (1973*b*). Cyclic packings of the projective space of order 8. *Atti Accad. Naz. Lincei Rend.* **54**, 373-377.
— (1973*c*). Packings of PG(3,q). Finite Geometric Structures and their Applications (Bressanone, 1972), Cremonese, 195-199.
— (1976). Some spreads which contain reguli without being subregular. *Teorie Combinatorie* (Rome, 1973), volume II, Accad. Naz. dei Lincei, 367-371.
De RESMINI, M.J. (1970). Sulle quartiche piane sopra un campo di caratteristica due. *Ricerche Mat.* **19**, 133-160.
— (1971). On quartics in a plane over a field of characteristic 2. *Atti del Convegno di Geometria Combinatoria e sue Applicazioni* (Università di Perugia, 1970), Università di Perugia, 187-197.
DICKEY, L.J. (1971). Embedding the complement of a unital in a projective plane. *Atti del Convegno di Geometria Combinatoria e sue Applicazioni* (Università di Perugia, 1970), Università di Perugia, 199-202.
DICKSON, L.E. (1897). The analytic representation of substitutions on a power of a prime number of letters with a discussion of the linear group I;II. *Ann. of Math.* **11**, 65-120; 161-183.
— (1901). Linear groups with an exposition of the Galois field theory. Teubner, Leipzig (Dover, 1958).
— (1905). Determination of all subgroups of the known simple group of order 25920. *Proc. Amer. Math. Soc.* **5**, 126-166.
— (1907*a*). Invariants of binary forms under modular transformations. *Trans. Amer. Math. Soc.* **8**, 205-232.
— (1907*b*). Invariants of the general quadratic form modulo 2. *Proc. London Math. Soc.* **5**, 301-324.
— (1907*c*). On quadratic forms in a general field. *Bull. Amer. Math. Soc.* **14**, 108-115.
— (1908). On families of quadratic forms in a general field. *Quarterly Journal of Pure and Applied Mathematics* **39**, 316-333.
— (1909). Modular invariants of a general system of linear forms. *Proc. London Math. Soc.* **7**, 430-444.
— (1914*a*). On invariants and the theory of numbers. Amer. Math. Soc. (Dover, 1966).
— (1914*b*). The invariants, seminvariants and linear covariants of the binary quadratic form modulo 2. *Ann. of Math.* **15**, 114-117.
— (1914*c*). Modular invariants of the system of a binary cubic, quadratic and linear form. *Quarterly Journal of Pure and Applied Mathematics* **45**, 373-384.
— (1915*a*). Recent progress in the theories of modular and formal invariants and in modular geometry. *Proc. Nat. Acad. Sci. U.S.A.* **1**, 1-4.
— (1915*b*). Invariantive theory of plane cubic curves modulo 2. *Amer. J. Math.* **37**, 107-116.
— (1915*c*). Quartic curves modulo 2. *Trans. Amer. Math. Soc.* **16**, 111-120.
— (1915*d*). Classification of quartic curves, modulo 2. *Messenger of Mathematics* **44**, 189-192.
— (1915*e*). Geometrical and invariantive theory of quartic curves modulo 2. *Amer. J. Math.* **37**, 337-354.

DICKSON, L.E. (1915*f*). Projective classification of cubic surfaces modulo 2. *Ann. of Math.* **16**, 139-157.
— (1915*g*). The straight lines on modular cubic surfaces. *Proc. Nat. Acad. Sci. U.S.A.* **1**, 248-253.
— (1915*h*). Invariantive classification of pairs of conics modulo 2. *Amer. J. Math.* **37**, 337-354.
— (1915*i*). Invariants, seminvariants and covariants of the ternary and quaternary form modulo 2. *Bull. Amer. Math. Soc.* **21**, 174-179.
— (1926). Modern algebraic theories. Sanborn (*Algebraic theories*, Dover, 1959).
Di COMITE, C. (1962). Su k-archi deducibili da cubiche piane. *Atti Accad. Naz. Lincei Rend.* **33**, 429-435.
— (1963). Su k-archi contenuti in cubiche piane. *Atti Accad. Naz. Lincei Rend.* **35**, 274-278.
— (1964). Intorno a certi (q+9)/2-archi di $S_{2,q}$. *Atti Accad. Naz. Lincei Rend.* **36**, 819-824.
— (1965). Intorno a certa k-calotte di $S_{3,q}$ (q dispari). *Atti Accad. Naz. Lincei Rend.* **39**, 249-254.
— (1969*a*). Calotte complete di un $S_{3,q}$, con q pari. *Atti Accad. Naz. Lincei Rend.* **46**, 385-389.
— (1969*b*). Alcuni *k*-archi completi di un piano di Galois di caratteristica due. *Atti Accad. Naz. Lincei Rend.* **47**, 240-244.
— and PELUSO, R. (1972). Sulla determinazione di alcuni k-archi completi di una calcolatrice elettronica. *Atti Accad. Naz. Lincei Rend.* **53**, 94-99.
DIENST, K.J. (1977). Zu Sekanten symmetrische Halbovoide. *Arch. Math.* **28**, 325-329.
— and MAÜRER, H. (1974). Zwei charakteristische Eigenschaften hermitescher Quadriken. *Geometriae Dedicata* **3**, 131-138.
DIEUDONNÉ, J.A. (1954). Les isomorphismes exceptionnels entre les groupes classiques finis. *Canad. J. Math.* **6**, 305-315.
— (1955). Pseudodiscriminant and Dickson invariant. *Pacific J. Math.* **5**, 907-910.
— (1971). La géometrie des groupes classiques, third edition, Springer.
— and CARRELL, J.B. (1971). Invariant theory, old and new. Academic Press.
Di PAOLA, J.W. (1969). On minimum blocking coalitions in small projective plane games. *SIAM J. Appl. Math.* **17**, 378-392.
D'ORGEVAL, B. (1960). Sur certains (k,3)-arcs en géométrie di Galois. *Acad. Roy. Belg. Bull. Cl. Sci.* **46**, 597-603.
— (1963). Sur quelques courbes en géométrie di Galois. *J. Math. Pures Appl.* **42**, 97-102.
— (1964). A propos des courbes d'ordre q + 2 qui contiennent tous les points d'un plan de Galois $S_{2,q}$. *Bull. Soc. Roy. Sci. Liège* **33**, 264-269.
— (1967*a*). Sur certains (k,n)-arcs. *Acad. Roy. Belg. Bull. Cl. Sci.* **53**, 178-182.
— (1967*b*). Sur les cubiques gauches d'un espace S_3 construit sur un champ de Galois de caractéristique 2. *Bul. Inst. Politehn. Iaşi* (N.S.) **13**(17), 105-106.
— (1969). Sur une configuration de points, droites et quadriques. *Bul. Inst. Politehn. Iaşi* (N.S.) **15**(19), 35-38.
— (1975*a*). Sur certaines proprietés des k-arcs. *Ann. Mat. Pura Appl.* **102**, 91-102.
— (1975*b*). Les noyaux des courbes planes dans la geometrie sur un champ de caracteristique P. *Acad. Roy. Belg. Bull. Cl. Sci.* **61**, 91-95.
DORWART, H.L. (1966). The geometry of incidence. Prentice-Hall.
DOWLING, T.A. (1971). Codes, packings and the critical problem. *Atti del Convegno di Geometria Combinatoria e sue Applicazioni* (Università di Perugia, 1970), Università di Perugia, 209-224.

DUGUÉ, D. (1966). Géométrie algébrique finie et plans d'expêriences. *Rend. Mat. e Appl.* **25**, 139-146.

DWORK, B.M. (1960). On the rationality of the zeta function of an algebraic variety. *Amer. J. Math.* **82**, 631-648.

— (1965). Analytic theory of the zeta function of algebraic varieties. *Arithmetical Algebraic Geometry* (Purdue, 1963), Harper and Row, 18-32.

DYE, R.H. (1968). The simple group FH(8,2) of order 2^{12}. 3^5. 5^2. 7 and the associated geometry of triality. *J. London Math. Soc.* **18**, 521-562.

— (1970*a*). On the transitivity of the orthogonal and symplectic groups in projective space. *Proc. Cambridge Philos. Soc.* **68**, 33-43.

— (1970*b*). Maximal subgroups of index 960 in the group of bitangents. *J. London Math. Soc.* **2**, 746-748.

— (1971*a*). On the conjugacy classes of involutions of the simple orthogonal groups over perfect fields of characteristic two. *J. Algebra* **18**, 414-425.

— (1971*b*). On the conjugacy classes of involutions of the orthogonal groups over perfect fields of characteristic 2. *Bull. London Math. Soc.* **3**, 61-66.

— (1971*c*). Some geometry of triality with applications to involutions of certain orthogonal groups. *Proc. London Math. Soc.* **22**, 217-234.

— (1972). On the involution classes of the linear groups $GL_n(K)$, $SL_n(K)$, $PGL_n(K)$, $PSL_n(K)$ over fields of characteristic two. *Proc. Cambridge Philos. Soc.* **72**, 1-6.

— (1973*a*). On the conjugacy classes of involutions of the unitary groups $U_m(K)$, $SU_m(K)$, $PU_m(K)$, $PSU_m(K)$, over perfect fields of characteristic 2. *J. Algebra* **24**, 453-459.

— (1973*b*). The anomalous involution classes of $Sp_{2n}(2^p)$. *J. London Math. Soc.* **6**, 459-463.

— (1976). Elementary 2-subgroups of the classical groups over perfect fields of characteristic 2 containing all involution types. *J. Algebra* **38**, 398-406.

EDGE, W.L. (1931). Ruled surfaces. Cambridge University Press.

— (1953). Geometry in three dimensions over GF(3). *Proc. Roy. Soc. London* A**222**, 262-286.

— (1954). The geometry of the linear fractional group LF(4,2). *Proc. London Math. Soc.* **4**, 317-342.

— (1955*a*). Line geometry in three dimensions over GF(3), and the allied geometry of quadrics in four and five dimensions. *Proc. Roy. Soc. London* A**228**, 129-146.

— (1955*b*). 31-point geometry. *Math. Gaz.* **39**, 113-121.

— (1955*c*). The isomorphism between $LF(2,3^2)$ and A_6. *J. London Math. Soc.* **30**, 172-185.

— (1955*d*). The conjugate classes of the cubic surface group in an orthogonal representation. *Proc. Roy. Soc. London* A**233**, 126-146.

— (1956). Conics and orthogonal projectivities in a finite plane. *Canad. J. Math.* **8**, 362-382.

— (1958*a*). The geometry of an orthogonal group in six variables. *Proc. London Math. Soc.* **8**, 416-446.

— (1958*b*). The partitioning of an orthogonal group in six variables. *Proc. Roy. Soc. London* A**247**, 539-549.

— (1959). Quadrics over GF(2) and their relevance for the cubic surface group. *Canad. J. Math.* **11**, 625-645.

— (1960*a*). The simple group of order 6048. *Proc. Cambridge Philos. Soc.* **56**, 189-204.

— (1960*b*). A setting for the group of bitangents. *Proc. London Math. Soc.* **10**, 583-603.

— (1961). A permutation representation of the group of bitangents. *J. London Math. Soc.* **36**, 340-344.

EDGE, W. L. (1963*a*). A second note on the simple group of order 6048. *Proc. Cambridge Philos. Soc.* **59**, 1-9.
— (1963*b*). An orthogonal group of order 2^{13}. 3^5. 5^2. 7. *Ann. Mat. Pura Appl.* **61**, 1-96.
— (1964). Fundamental figures, in four and six dimensions over GF(2). *Proc. Cambridge Philos. Soc.* **60**, 183-195.
— (1965). Some implications of the geometry of the 21-point plane. *Math. Z.* **87**, 348-362.
— (1970*a*). The seven-point plane. *Inaugural lecture* no. 48, University of Edinburgh.
— (1970*b*). Double six. *Eureka* **33**, 10-12.
— (1973). An operand for a group of order 1512. *J. London Math. Soc.* **7**, 101-110.
— (1975). A footnote on the mystic hexagram. *Math. Proc. Cambridge Philos. Soc.* **77**, 29-42.
EICH, M.M. and PAYNE, S.E. (1972). Nonisomorphic symmetric block designs derived from generalized quadrangles. *Atti Accad. Naz. Lincei Rend.* **52**, 893-902.
EICHLER, M. (1966). Introduction to the theory of algebraic numbers and functions. Academic Press.
ENRIQUES, F. (1915). Lezioni sulla teoria geometrica delle equazioni e delle funzioni algebriche (edited by O. Chisini), four volumes, Zanichelli.
FANO, G. (1892). Sui postulati fondamentali della geometria proiettiva. *Giornale di Matematiche* **30**, 106-132.
— (1937). Osservazioni su alcune 'geometrie finite' I; II. *Atti Accad. Naz Lincei Rend.* **26**, 55-60; 129-134.
FEIT, W. and FINE, N.J. (1960). Pairs of commuting matrices over a finite field. *Duke Math. J.* **27**, 91-94.
FELLEGARA, G. (1962). Gli ovaloidi di uno spazio tridimensionale di Galois di ordine 8. *Atti Accad. Naz. Lincei Rend.* **32**, 170-176.
FENDEL, D. (1967). The number of classes of linearly equivalent functions. *J. Combinatorial Theory* **3**, 48-53.
FENG, Xu-Ning and DAI, Zong-Duo (1964). Notes on finite geometries and the construction of PBIB designs. V. Some 'Anzahl' theorems in orthogonal geometry over finite fields of characteristic 2. *Sci. Sinica* **13**, 2005-2008.
— — (1965). Studies in finite geometries and the construction of incomplete block designs. V. Some 'Anzahl' theorems in orthogonal geometry over finite fields of characteristic 2. *Chinese Math.* **7**, 392-410.
FERRI, O. (1968). Sulle quintiche irreducibili con un punto triplo che invadono un piano di Galois S_{2,γ_2}. *Rend. Acc. Sc. Fis. Mat. Napoli* **35**, 413-420.
— (1970). Intorno alle curve C^{q+3} irreducibili con un punto triplo che invadono un piano di Galois $S_{2,q}$. *Rend. Acc. Sc. Fis. Mat. Napoli* **36**, 167-173.
FLESNER, D.E. (1975*a*). Finite symplectic geometry in dimension four and characteristic two. *Illinois J. Math.* **19**, 41-47.
— (1975*b*). The geometry of subgroups of $PSp_4(2^n)$. *Illinois J. Math.* **19**, 48-70.
— (1975*c*). Maximal subgroups of $PSp_4(2^n)$ containing central elations or noncentred skew elations. *Illinois J. Math.* **19**, 247-268.
FOULSER, D.E. (1964). The flag-transitive collineation groups of the finite Desarguesian affine planes. *Canad. J. Math.* **16**, 443-472.
— and SANDLER, R. (1967). Certain properties of orbits under collineation groups. *J. Combinatorial Theory* **2**, 546-570.
FRAME, J.S. (1938). A symmetric representation of the twenty-seven lines on a cubic surface by lines in a finite geometry. *Bull. Amer. Math. Soc.* **44**, 658-661.

FRAME, J.S. (1951). The classes and representations of the groups of 27 lines and 28 bitangents. *Ann. Mat. Pura Appl.* **32**, 83-119.
FREDMAN, M.L. (1972). The distribution of absolutely irreducible polynomials in several indeterminates. *Proc. Amer. Math. Soc.* **31**, 387-390.
FREEMAN, J.W. (1973). A representation of the Baer subplanes of PG(2,q^2) in PG(5,q) and properties of a regular spread of PG(5,q). *Proc. International Conference on Projective Planes* (Washington State University, 1973), Washington State University Press, 91-97.
— (1974). On the Grassmannian of lines in PG(4,q) and R(1,2) reguli. *J. Geometry* **5**, 159-183.
FREUDENTHAL, H. (1972). Une interprétation géométrique des automorphismes exterieurs du groupe symétrique S_6. *Rend. Sem. Mat. Fis. Milano* **42**, 47-56.
— (1975). Une étude de quelques quadrangles generalisés. *Ann. Mat. Pura Appl.* **102**, 109-133.
FRIED, M. and SACERDOTE, G. (1976). Solving diophantine problems over all residue class fields and all finite fields. *Ann. of Math.* **104**, 203-233.
FULTON, J.D. (1972). Stochastic invoulutions over a finite field. *Duke Math. J.* **39**, 391-400.
— (1976). Representations by quadratic forms of arbitrary rank in a finite field of characteristic two. *Linear and Multilinear Algebra* **4**, 89-101.
— (1977). Representations by Hermitian forms in a finite field of characteristic two. *Canad. J. Math.* **29**, 169-179.
FULTON, W. (1969). Algebraic curves. Benjamin.
FURSTENBERG, H. (1967). Algebraic functions over finite fields. *J. Algebra* **7**, 271-277.
GAUTHIER, L. (1959). L'invariant modulaire dans la géometrie sur un corps de caractéristique 3. *J. Math. Pures Appl.* **38**, 117-163.
— (1966*a*). Géométrie hermitienne généralisée. *Acad. Roy. Belg. Bull. Cl. Sci.* **52**, 421-449.
— (1966*b*). Semirationalité des variétés sesquipolaries. *Rend. Mat e Appl.* **25**, 233-238.
— (1969). La géométrie sur les corps de caractéristique non nulle. *Enseignement Math.* **15**, 123-145.
GEWIRTZ, A. (1969). The uniqueness of G(2,2,10,56). *Trans. New York Acad. Sci.* **31**, 656-675.
— (1969). Graphs with maximal even girth. *Canad. J. Math.* **21**, 915-934.
GIOVAGNOLI, A. (1970). On some finite structures. *Atti Accad. Naz. Lincei Rend.* **49**, 231-235.
GOETHALS, J.-M. and SEIDEL, J.J. (1970). Strongly regular graphs and combinatorial designs. *Canad. J. Math.* **22**, 597-614.
GORENSTEIN, D. (1974). Reviews on finite groups. American Mathematical Society.
GRAVES, W.H. (1971). A categorical approach to combinatorial geometry. *J. Combinatorial Theory A* **11**, 222-232.
GREEN, J.A. (1955). The characters of the finite general linear groups. *Trans. Amer. Math. Soc.* **80**, 402-447.
GRUENBERG, K.W. and WEIR, A.J. (1967). Linear Geometry. Van Nostrand.
GULATI, B.R. (1971). On maximal (k,t)-sets. *Ann. Inst. Statist. Math.* **23**, 279-292; correction **23**, 527-529.
— and KOUNIAS, E.G. (1970). On bounds useful in the theory of symmetrical factorial designs. *J. Roy. Statist. Soc. Ser. B* **32**, 123-133.
— — (1973). Maximal sets of points in finite projective spaces, no t linearly dependent. *J. Combinatorial Theory A* **15**, 54-65.
—, JOHNSON, B.McK., and KOEHN, U. (1973). On maximal t-linearly independent sets. *J. Combinatorial Theory A* **15**, 45-53.
HAIMULIN, J.N. (1966*a*). Some properties of $\{k;n\}_q$-arcs in Galois planes. *Soviet Math. Dokl.* **7**, 1100-1103.

HAĬMULIN, J.N. (1966*b*). Barlotti's formula (Russian). *Cuvas. Gos. Ped. Inst. Učen. Zap. Vyp.* **27**, 58-63.
HALDER, H.-R. (1974). Über symmetrische (k,n)-Kurven in endlichen Ebenen. *J. Geometry* **5**, 147-157.
— (1976). Über Kurven vom Typ (m;n) und Beispiele total m-regulärer (k,n)-Kurven. *J. Geometry* **8**, 163-170.
— and HEISE, W. (1974). On the existence of finite chain-m-structures and k-arcs in finite projective space. *Geometriae Dedicata* **3**, 483-486.
HALL, M. (1953). Uniqueness of the projective plane with 57 points. *Proc. Amer. Math. Soc.* **4**, 912-916.
— (1954). Correction to 'Uniqueness of the projective plane with 57 points'. *Proc. Amer. Soc.* **5**, 994-997.
— (1959). The theory of groups. Macmillan, New York.
— (1967). Combinatorial theory. Blaisdell.
— (1975). Ovals in the Desarguesian plane of order 16. *Ann. Mat. Pura Appl.* **102**, 159-176.
—, LANE, R., and WALES, D. (1970). Designs derived from permutation groups. *J. Combinatorial Theory* **8**, 12-22.
—, SWIFT, J.D., and WALKER, R.J. (1956). Uniqueness of the projective plane or order eight. *Math. Tables Aids Comput.* **10**, 186-194.
HAMADA, N. (1968). The rank of the incidence matrix of points and d-flats in finite geometries. *J. Sci. Hiroshima Univ.* **32**, 381-396.
— (1973). On the p-rank of the incidence matrix of a balanced or partially balanced incomplete block design and its application to error correcting codes. *Hiroshima Math. J.* **3**, 153-226.
— and FUKUDA, T. (1967). A note on the cyclical generation of disjoint spreads. *J. Sci. Hiroshima Univ.* **31**, 191-194.
HARTLEY, R.W. (1926). Determination of the ternary collineation groups whose coefficients lie in the GF(2^n). *Ann. of Math.* **27**, 140-158.
HARTSHORNE, R. (1967). Foundations of projective geometry. Benjamin.
HASSE, H. (1934). Abstrakte Begründung der komplexen Multiplikation und Riemannsche Vermutung in Funktionenkörpern. *Abh. Math. Sem. Univ. Hamburg* **10**, 325-348.
— (1943). Punti razionali sopra curve algebriche a congruenze. *Atti dei Convegni* (1939), Reale Accademia d'Italia, Rome.
HAVLÍCEK, K. and TIETZE, J. (1970). Zur Geometrie der endlichen Ebene der Ordnung n=4. *Comment. Math. Univ. Carolinae* **11**, 593-594.
— — (1971). Zur Geometrie der endlichen Ebene der Ordnung n = 4. *Czechoslovak Math. J.* **21**, 157-164.
HAYES, D.R. (1967). A geometric approach to permutation polynomials over finite fields. *Duke Math. J.* **34**, 293-305.
HEISLER, J. (1974). Diagonal forms over finite fields. *J. Number Theory* **6**, 50-51.
HERING, C. (1970). A new class of quasifields. *Math. Z.* **118**, 56-57.
— (1971). Uber die Translationsebenen auf denen die Gruppe SL(2,q) operiert. *Atti del Convegno di Geometria Combinatoria e sue Applicazioni* (Università di Perugia, 1970), Università di Perugia, 259-261.
HESSENBERG, G. (1902/3). Über die projective Geometrie. *Sitzungsberichte der Berliner Mathematischen Gesellschaft*, 36-40.
HIGMAN, D.G. (1962). Flag transitive collineation groups of finite projective spaces. *Illinois J. Math.* **6**, 434-446.
— (1964). Finite permutation groups of rank 3. *Math. Z.* **86**, 145-156.
— and McLAUGHLIN, J.E. (1965). Rank 3 subgroups of finite symplectic and unitary groups. *J. Reine Angew. Math.* **218**, 174-189.
HIBERT, D. (1899). Grundlagen der Geometrie. Teubner, Leipzig (10th edition translated, Open Court, 1971).
HILL, R. (1973). On the largest size of cap in $S_{5,3}$. *Atti Accad. Naz. Lincei Rend.* **54**, 378-384.

HILL, R. (1976). Caps and groups. *Teorie Combinatorie* (Rome, 1973), volume II, Accad. Naz. dei Lincei, 389-394.
HILTON, H. (1920). Plane algebraic curves. Oxford University Press.
HIRSCHFELD, J.W.P. (1964). The double-six of lines over PG(3,4). *J. Austral. Math. Soc.* **4**, 83-89.
— (1967*a*). Classical configurations over finite fields: I. The double-six and the cubic surface with 27 lines. *Rend. Mat. e Appl.* **26**, 115-152.
— (1967*b*). Classical configurations over finite fields: II. Grace's extension of the double-six. *Rend. Mat. e Appl.* **26**, 349-374.
— (1967*c*). A curve over a finite field, the number of whose points is not increased by a quadratic extension of the field, and sub-Hermitian forms. *Atti Accad. Naz. Lincei Rend.* **42**, 365-367.
— (1969). A projective configuration. *Computational Problems in Abstract Algebra*, Pergamon, 321-323.
— (1971). Rational curves on quadrics over finite fields of characteristic two. *Rend. Mat.* **3**, 772-795.
— (1972). A configuration of lines in three dimensions. *Proc. Edinburgh Math. Soc.* **18**, 105-123.
— (1975). Ovals in Desarguesian planes of even order. *Ann. Mat. Pura Appl.* **102**, 79-89.
— (1976). Cyclic projectivities in PG(n,q). *Teorie Combinatorie* (Rome, 1973), volume I, Accad. Naz. dei Lincei, 201-211.
HODGE, W.V.D. and PEDOE, D. (1947). Methods of algebraic geometry (three volumes). Cambridge University Press.
HOFFER, A.R. (1972). On unitary collineation groups. *J. Algebra* **22**, 211-218.
HORADAM, A.F. (1970). A guide to undergraduate projective geometry. Pergamon.
HSIEH, W.N. (1975*a*). Families of intersecting finite vector spaces. *J. Combinatorial Theory A* **18**, 252-261.
— (1975*b*). Intersection theorems for systems of finite vector spaces. *Discrete Math.* **12**, 1-16.
HUA, L.K. and VANDIVER, H.S. (1949). On the nature of solutions of certain equations in a finite field. *Proc. Nat. Acad. Sci. U.S.A.* **35**, 481-487.
HUBAUT, X. (1968). Sur le groupe des 27 droites d'une surface cubique. *Acad. Roy. Belg. Bull. Cl. Sci.* **54**, 265-271.
— (1970). Limitation du nombre de points d'un (k,n)-arc regulier d'un plan projectif fini. *Atti Accad. Naz. Lincei Rend.* **48**, 490-493.
— (1972). Designs et graphes de Schläfli. *Bull. Acad. Roy. Belg.* **58**, 622-624.
— (1975). Strongly regular graphs. *Discrete Math.* **13**, 357-381.
HUGHES, D.R. (1959). Review of some results in collineation groups. *Proc. Sympos. Pure Math.* **1**, 42-55.
— and PIPER, F. (1973). Projective planes. Springer.
HUPPERT, B. (1967). Endliche Gruppen. I. Springer.
HY, N.M. (1971). On the regular polygons of the order nine affine Galois plane. *Mat. Lapok* **22**, 323-329.
ILKKA, S. (1974). On betweenness-relations and conics in affine planes over fields. *Math. Balkanica* **4**, 261-265.
IRELAND, K. and ROSEN, M.J. (1972). Elements of number theory. Bogden and Quigley.
ISBELL, J.R. (1958). A class of simple games. *Duke Math. J.* **25**, 425-436.
JAMIESON, R.E. (1977). Covering finite fields with cosets of subspaces. *J. Combinatorial Theory A* **22**, 253-266.
JÄRNEFELT, G. (1949). A plane geometry with a finite number of elements. *Veröf. Finn. Geod. Inst.* No. 36.
— (1951). Reflections on a finite approximation to Euclidean geometry. Physical and astronomical prospects. *Ann. Acad. Sci. Fenn. Ser. A,* No. 96.

JÄRNEFELT, G. and KUSTAANHEIMO, P. (1949). An observation on finite geometries. *Comptes Rendus XI Congr. Math. Scand.* Trondheim, 166-182.

—— (1955). Die Isomorphie eines elementargeometrischen und eines Galois-Gitterpunkte modells. *Ann. Acad. Sci. Fenn. Ser. A*, No. 201.

JOLY, J.-R. (1971). Nombre de solutions de certaines équations diagonales sur un corps fini. *C.R. Acad. Sci. Paris* 272, 1549-1552.

— (1973). Équations et variétés algébriques sur un corps fini. *Enseignement Math.* **19**, 1-117.

JÓNSSON, W. (1972). On the Mathieu groups M_{22}, M_{23}, M_{24} and the uniqueness of the associated Steiner systems. *Math. Z.* **125**, 193-214.

— (1973). The (56,11,2) design of Hall, Lane, and Wales. *J. Combinatorial Theory A* **14**, 113-118.

JORDAN, C. (1870). Traité des substitutions et des équations algébriques. Paris.

— (1906). Réduction d'un réseau de formes quadratiques ou bilineaires. *J. de Math.* **2**, 403-438.

JUNG, F.R. (1972*a*). On the exterior and interior points of quadrics over a fininite field. *Duke Math. J.* **39**, 183-188.

— (1972*b*). Solutions of some systems of equations over a finite field with applications to geometry. *Duke Math. J.* 39, 189-202.

— (1974). On conics over a finite field. *Canad. J. Math.* **26**, 1281-1288.

KANTOR, W.M. (1972). On incidence matrices of finite projective and affine spaces. *Math. Z.* **124**, 315-318.

— (1976). Generalized quadrangles having a prime parameter. *Israel J. Math.* **23**, 8-18.

KAPLANSKY, I. (1974). Linear algebra and geometry, second edition, Chelsea.

KÁRTESZI, F. (1976). Introduction to finite geometries. North Holland/American Elsevier.

KARZEL, H. and SÖRENSEN, K. (1971). Projektive Ebenen mit einem pascalschen Oval. *Abh. Math. Sem. Univ. Hamburg* **36**, 123-125.

KATZ, N.M. (1971). On a theorem of Ax. *Amer. J. Math.* **93**, 485-499.

KEEDWELL, A.D. (1963). On the order of projective planes with characteristic. *Rend. Mat. e Appl.* **22**, 498-530.

— (1964). A class of configurations associated with projective planes with characteristic. *Arch. Math.* **15**, 470-480.

— (1975). Self-collineations of Desarguesian projective planes. *Amer. Math. Monthly* **82**, 59-63.

KIMENYEMBO, J. (1971). Application des invariants de P. Buzano à la démonstration de l'existence du foyer pour les coniques d'un plan de caractéristique 2. *Bull. Soc. Roy. Sci. Liège* **40**, 5-6.

KOCH, G.G. (1969). A class of covers for finite projective geometries which are related to the design of combinatorial filing systems. *J. Combinatorial Theory* **7**, 215-220.

KOMMERELL, K. (1941). Die Pascalsche Konfiguration 9_3. *Deutsche Math.* **6**, 16-32.

KORCHMÁROS, G. (1974*a*). Osservazioni sui resultati di B. Segre relativi ai k-archi contenenti k - 1 punti di un ovale. *Atti Accad. Naz. Lincei Rend.* **56**, 690-697.

— (1974*b*). Poligoni affin-regolari dei piani di Galois d'ordine dispari. *Atti Accad. Naz. Lincei Rend.* **56**, 690-697.

— (1975/6). Su una classificazione delle ovali dotate di automorfismi. *Rend. Accad. Naz. dei XL* **1/2**, 1-10.

KUSTAANHEIMO, P. (1950). A note on a finite approximation of the Euclidean plane geometry. *Soc. Sci. Fenn. Comm. Phys. Math.* 15, No. 19.

— (1952). On the fundamental prime of a finite world. *Ann. Acad. Sci. Fenn. Ser. A*, No. 129.

— (1954). Über die Versuche, ein logisch finites Weltbild aufzubauen. *Proc. Second International Congress of the International Union for the Philosophy of Science*, Zurich, 60-65.

KUSTAANHEIMO, P. (1957*a*). On the relation of congruence in finite geometries. *Rend. Mat. e Appl.* **16**, 286-291.
— (1957*b*). On the relation of order in finite geometries. *Rend. Mat. e Appl.* **16**, 292-296.
— (1957*c*). On the relation of order in geometries over a Galois field. *Soc. Sci. Fenn. Comm. Phys. Math.* **20**, No. 8.
— (1957*d*). On the relation of congruence in finite geometries. *Math. Scand.* **5**, 197-201.
— and QVIST, B. (1954). Finite geometries and their application. *Nordisk Mat. Tidskr.* 2, 137-155.
LANG, S. and WEIL, A. (1954). Number of points of varieties in finite fields. *Amer. J. Math.* **76**, 819-827.
LAUSCH, H. and NÖBAUER, W. (1973). Algebra of Polynomials. North Holland/ American Elsevier.
LEFÈVRE, C. (1975). Tallini sets in projective spaces. *Atti Accad. Naz. Lincei Rend.* **59**, 392-400.
LEHTI, R. (1957). On the construction of elementary models for finite plane geometries. *13th Cong. Math. Scand.*
LENZ, H. (1954). Zur Begründung der analytischen Geometrie. *Bayer Akad. Wiss. Math.-Natur. Kl. S.-B.*, 17-72.
LEONARD, P.A. (1969). On factoring quartics (mod p). *J. Number Theory* **1**, 113-155.
— and WILLIAMS, K.S. (1972). Quartics over GF(2^n). *Proc. Amer. Math. Soc.* **36**, 347-350.
LEVI, F.W. (1942). Finite geometrical systems. University of Calcutta.
LEWIS, D.J. and SCHUUR, S.E. (1973). Varieties of small degree over finite fields. *J. Reine Angew. Math.* **262/263**, 293-306.
LIDL, R. (1972*a*). Über die Darstellung von Permutationen durch Polynome. *Abh. Math. Sem. Univ. Hamburg* **37**, 108-111.
— (1972*b*). Über Permutationsfunktionen in mehreren Unbestimmten. *Acta Arith.* **20**, 291-296.
LOMBARDO-RADICE, L. (1956). Sul problema dei k-archi completi di $S_{2,q}$. *Boll. Un. Mat. Ital.* **11**, 178-181.
— (1962). La decomposizione tattica di un piano grafico finito associata a un k-arco. *Ann. Mat. Pura Appl.* **60**, 37-48.
LONG, A.F. (1973). Factorization of irreducible polynomials over a finite field with the substitution $x^{q^r} - x$ for x. *Acta Arith.* **25**, 65-80.
LONGUET-HIGGINS, M.S. (1972). Clifford's chain and its analogues in relation to the higher polytopes. *Proc. Roy. Soc. London* A**330**, 443-466.
LÜNEBERG, H. (1965). Die Suzukigruppen und ihre Geometrien. *Lecture Notes in Mathematics* **10**, Springer.
— (1968). Über die Gruppen von Mathieu. *J. Algebra* **10**, 194-210.
— (1969). Transitive Erweiterungen endlicher Permutationsgruppen. *Lecture Notes in Mathematics* **84**, Springer.
LUNELLI, L., LUNELLI, M., and SCE, M. (1959). Calcolo di omografie cicliche di piani desarguesiani finiti. *Rend. Mat. e Appl.* **18**, 361-374.
— and SCE, M. (1958*a*). k-archi completi nei piani proiettivi desarguesiani di rango 8 e 16. *Centro di Calcoli Numerici*, Politecnico di Milano.
— — (1958*b*). Sulla ricerca dei k-archi completi mediante una calcolatrice elettronica. *Convegno internazionale: Reticoli e Geometrie Proiettive* (Palermo, Messina 1957), Cremonese, 81-86.
— — (1964). Considerazione aritmetiche e risultati sperimentali sui $\{K;n\}_q$ - archi. *Ist. Lombardo Accad. Sci. Rend.* A**98**, 3-52.
MacINNES, C.R. (1907). Finite planes with less than eight points on a line. *Amer. Math. Monthly* **14**, 171-174.
MacNEISH, H.F. (1942). Four finite geometries. *Amer. Math. Monthly* **49**, 15-23.
MacWILLIAMS, F.J. (1961). Error-correcting codes for multiple-level transmission. *Bell System Tech. J.* **40**, 281-308.

MacWILLIAMS, F.J. (1971). Orthogonal circulant matrices over finite fields, and how to find them. *J. Combinatorial Theory* **10**, 1-17.

— and SLOANE, N.J.A. (1977). The theory of error-correcting codes. North-Holland.

MAKOWSKI, A. (1963). Remarks on a paper of Tallini. *Acta Arith.* **8**, 469-470.

MAISANO, F. (1959). Sulle k-calotte contenute in una quadrica rigata di $S_{3,q}$. *Atti Accad. Naz. Lincei Rend.* **26**, 35-38.

— (1960). Sul massimo numero dei punti di una quartica gobba di 1^a specie in $S_{3,q}$ (q dispari). *Atti Accad. Sci. Lett. Palermo Parte I* **19**, 155-163.

MANERI, C. and SILVERMAN, R. (1966). A vector-space packing problem. *J. Algebra* **4**, 321-330.

— — (1971). A combinatorial problem with applications to geometry. *J. Combinatorial Theory A* **11**, 118-121.

MANIN, J.I. (1974). Cubic forms. North Holland/American Elsevier.

MANN, H.B. (1974). The solution of equations by radicals. *J. Algebra* **29**, 551-554.

MARTIN, G.E. (1967). On arcs in a finite projective plane. *Canad. J. Math.* **19**, 376-393.

— (1971*a*). Gobos in a finite projective plane. *J. Combinatorial Theory A* **10**, 92-96.

— (1971*b*). Symmetric gobos in a finite projective plane. *J. Combinatorial Theory A* **11**, 201-207.

MARTINOV, R. and LOZANOV, C. (1976). Equivalent and isomorphic complete 6-arcs in Desarguesian projective planes. *C.R. Acad. Bulgare Sci.* **29**, 1583-1584.

MAZUMDAR, S. (1967). On the construction of cyclic collineations for obtaining a balanced set of L-restrictional prime-powered lattice designs. *Ann. Math. Statist.* **38**, 1293-1295.

MAZUR, B. (1975). Eigenvalues of Frobenius acting on algebraic varieties over finite fields. *Proc. Sympos. Pure Math.* **29**, Algebraic Geometry (1974), 231-261.

MAZZOCCA, F. (1974*a*). Immergibilità in $S_{4,q}$ di certi sistemi rigati di seconda specie. *Atti Accad. Naz. Lincei Rend.* **56**, 189-196.

— (1974*b*). Caratterizzazione dei sistemi rigati isomorfi ad una quadrica ellittica dello $S_{5,q}$, con q dispari. *Atti Accad. Naz. Lincei Rend.* **57**, 360-368.

McCONNEL, R. (1972). Functions over finite fields satisfying coordinate ψ-conditions. *Duke Math. J.* **39**, 297-312.

McDONALD, B.R. (1972*a*). Involutory matrices over finite local rings. *Canad. J. Math.* **24**, 369-378.

— (1972*b*). Diagonal equivalence of matrices over a finite local ring. *J. Combinatorial Theory A* **13**, 100-104.

MESNER, D.M. (1967). Sets of disjoint lines in PG(3,q). *Canad. J. Math.* **19**, 273-280.

MEYER, K. (1972). Äquivalenz von quadratischen Formen über endlichen Körpern. *Abh. Math. Sem. Univ. Hamburg* **37**, 79-85.

MIELANTS, W. (1971). Pappus-Pascal configurations in Pascalian planes. *Atti del Convegno di Geometria Combinatoria e sue Applicazioni* (Università di Perugia, 1970), Università di Perugia, 339-350.

— (1972). On caps and ovoids in affine spaces. *Bull. Soc. Math. Belg.* **24**, 351-369.

MIGLIORI, G. (1975). Sulle k-calotte complete di $S_{3,q}$. *Atti Accad. Naz. Lincei Rend.* **59**, 737-743.

— (1976). Calotte di specie s in uno spazio r-dimensionale di Galois. *Atti Accad. Naz. Lincei Rend.* **60**, 789-792.

MITCHELL, H.H. (1911). Determination of the ordinary and modular ternary linear groups. *Trans. Amer. Math. Soc.* **12**, 207-242.

MITCHELL, H.H. (1913). Determination of the finite quaternary linear groups. *Trans. Amer. Math. Soc.* **14**, 123-142.
MONTAGUE, S. (1970). On rank 3 groups with a multiply transitive constituent. *J. Algebra* **13**, 506-522.
MORLAYE, B. (1971). Equations diagonales non homogènes sur un corps fini. *C.R. Acad. Sci. Paris* **272**, 1545-1548.
MULLEN, G.L. (1976*a*). Permutation polynomials in several variables over finite fields. *Acta Arith.* **31**, 107-111.
— (1976*b*). Equivalence classes of polynomials over finite fields. *Acta Arith.* **31**, 113-123.
MUMFORD, D. (1976). Algebraic geometry I. Complex projective varieties. Springer.
NAGAHARA, J. and TOMINAGA, H. (1974). Elementary proofs of a theorem of Wedderburn and a theorem of Jacobson. *Abh. Math. Sem. Univ. Hamburg* **41**, 72-74.
NIEDERREITER, H. (1970). Permutation polynomials in several variables over finite fields. *Proc. Japan Acad.* **46**, 1001-1005.
— (1972). Permutation polynomials in several variables. *Acta Sci. Math. Szeged* **33**, 53-58.
ORR, W.J. (1976). A characterization of subregular spreads in finite 3-space. *Geometriae Dedicata* **5**, 43-50.
OSTROM, T.G. (1955). Ovals, dualities, and Desargues's theorem. *Canad. J. Math.* 7, 417-431.
— (1960). Translation planes and configurations in Desarguesian planes. *Arch. Math.* **11**, 457-464.
— (1968). Vector spaces and construction of finite projective planes. *Arch. Math.* **19**, 1-25.
— (1969). Translation planes of order p^2. *Conference on Combinatorial Mathematics and its Applications* (University of North Carolina, 1967), University of North Carolina Press, 416-425.
— (1970). Linear transformations and collineations of translation planes. *J. Algebra* **14**, 405-416.
PALL, G. (1972). Partitionings by means of maximal isotropic subspaces. *Linear Algebra and Appl.* **5**, 173-180.
PANELLA, G. (1955). Caratterizzazione delle quadriche di uno spazio (tridimensionale) lineare sopra un corpo finito. *Boll. Un. Mat. Ital.* **10**, 507-513.
— (1967). Una classe di fibrazioni di uno spazio proiettivo. *Atti Accad. Naz. Lincei Rend.* 42, 611-615.
— (1970). Alcune considerazioni riguardanti le fibrazioni proiettive. *Rend. Circ. Mat. Palermo* **19**, 11-26.
PAYNE, S.E. (1970*a*). Collineations of affinely represented generalized quadrangles. *J. Algebra* **16**, 496-508.
— (1970*b*). Affine representations of generalized quadrangles. *J. Algebra* **16**, 473-485.
PATTERSON, N.J. (1976). A four-dimensional Kerdock set over GF(3). *J. Combinatorial Theory A* **20**, 365-366.
PAYNE, S.E. (1971*a*). A complete determination of translation ovoids in finite Desarguian planes. *Atti Accad. Naz. Lincei Rend.* **51**, 328-331.
— (1971*b*). Nonisomorphic generalized quadrangles. *J. Algebra* **18**, 201-212.
— (1971*c*). The equivalence of certain generalized quadrangles. *J. Combinatorial Theory A* **10**, 284-289.
— (1971*d*). A geometric representation of certain generalized hexagons in PG(3,s). *J. Combinatorial Theory B* **11**, 181-192.
— (1971*e*). Linear transformations of a finite field. *Amer. Math. Monthly* **78**, 659-660.
— (1972*a*). Nonisomorphic symmetric block designs derived from generalized quadrangles. *Atti Accad. Naz. Lincei* **52**, 893-902.

PAYNE, S.E. (1972*b*). Quadrangles of order (s-1, s+1). *J. Algebra* **22**, 97-119.
— (1972*c*). Generalized quadrangles as amalgamations of projective planes. *J. Algebra* **22**, 120-136.
PEDOE, D. (1963). An introduction to projective geometry. Pergamon.
PELLEGRINO, G. (1970). Sul massimo ordine delle calotte in $S_{4,3}$. *Matematiche (Catania)* **25**, 1-9.
— (1972). Procedimenti geometrici per la costruzione di alcune classi di calotte complete $S_{r,3}$. *Boll. Un. Mat. Ital.* **5**, 109-115.
— (1973). Sulle proprietà della 11-calotta completa di $S_{4,3}$ e su un BIB-disegno ad essa collegato. *Boll. Un. Mat. Ital.* **7**, 463-470.
— (1974). Su una interpretazione geometrica dei gruppi M_{11} ed M_{12} di Mathieu e su alcuni t-(v,k,λ)- disegni deducibili da una $(12)^4_{5,3}$ calotta completa. *Atti Sem. Mat. Fis. Univ. Modena* **23**, 103-117.
PERMUTTI, R. (1963). Sul teorema di Hesse per forme sopra un campo a caratteristica qualungue. *Matematiche (Catania)* **18**, 116-128.
PETERSON, W.W. (1961). Error-correcting codes. M.I.T. Press.
— (1969). Some new results on finite fields and their application to the theory of BCH codes. *Conference on Combinatorial Mathematics and its Applications* (University of North Carolina, 1967), University of North Carolina Press, 329-334.
PICASSO, E. (1971). Proprietà grafico-combinatorie di (q+1)-archi di un $S_{2,q}$ di ordine dispari. *Rend. Sem. Fac. Sc. Univ. Cagliari* **41**, 115-130.
PICKERT, G. (1952). Der Satz vom vollständigen Viereck bei kollinearen Diagonalpunkten. *Math. Z.* **55**, 355-363.
— (1955). Projektive Ebenen. Springer.
PIPER, F.C. (1963). Elations of finite projective planes. *Math. Z.* **82**, 247-258.
— (1965). Collineation groups containing elations. I. *Math. Z.* **89**, 181-191.
— (1966*a*). Collineation groups containing elations. II. *Math. Z.* **92**, 281-287.
— (1966*b*). On elations of finite projective spaces of odd order. *J. London Math. Soc.* **41**, 641-648.
— (1967). Collineation groups containing homologies. *J. Algebra* **6**, 256-269.
— (1968*a*). The orbit structure of collineation groups of finite projective planes. *Math. Z.* **103**, 318-332.
— (1968*b*). On elations of finite projective spaces of even order. *J. London Math. Soc.* **43**, 459-464.
— (1973). On elations of finite projective spaces. *Geometriae Dedicata* **2**, 13-27.
— and BEESLEY, F. (1969). Line orbits of collineation groups. *Matematiche (Catania)* **24**, 166-174.
— and NORMAN, C. (1971). On fixed points of collineation groups. *Bull. London Math. Soc.* **3**, 297-302.
— and WAGNER, A. (1968). Faithful orbits of collineation groups. *Math. Z.* **107**, 212-220.
PLESS, V. (1964). On Witt's theorem for nonalternating symmetric bilinear forms over a field of characteristic 2. *Proc. Amer. Math. Soc.* **15**, 979-983.
— (1965*a*). The number of isotropic subspaces in a finite geometry. *Atti Accad. Naz. Lincei Rend.* **39**, 418-421.
— (1965*b*). On the invariants of a vector subspace of a vector space over a field of characteristic two. *Proc. Amer. Math. Soc.* **16**, 1062-1067.
— (1968). On the uniqueness of the Golay codes. *J. Combinatorial Theory* **5**, 215-228.

PLESS, V. (1969). On a new family of symmetry codes and related new five-designs. *Bull. Amer. Math. Soc.* **75**, 1339-1342.

PREŠIĆ, M.D. (1970). A method for solving equations in finite fields. *Mat. Vesnik* **7**(22), 507-509.

PRIMROSE, E.J.F. (1951). Quadrics in finite geometries. *Proc. Cambridge Philos. Soc.* **47**, 299-304.

QVIST, B. (1952). Some remarks concerning curves of the second degree in a finite plane. *Ann. Acad. Sci. Fenn. Ser. A*, no. 134.

RABER, N.C. (1975). A geometric approach to counting the distribution of squares in a finite field. *Geometriae Dedicata* **4**, 297-303.

RAKTOE, B.L. (1967). Application of cyclic collineations to the construction of balanced L-restrictional prime-powered lattice designs. *Ann. Math. Statist.* **38**, 1127-1141.

— (1969). Combining elements from distinct finite fields in mixed factorials. *Ann. Math. Statist.* **40**, 498-504.

— (1970). Generalized combining of elements from finite fields. *Ann. Math. Statist.* **41**, 1763-1767.

RAMAMURTI, B. (1933). Desargues configurations admitting a collineation group. *J. London Math. Soc.* **8**, 34-39.

RAMANUJACHARYULU, C. (1965). Classification of canonical forms of quadrics in finite projective geometry. *J. Indian Statist. Assoc.* **3**, 91-96.

RAO, C.R. (1945). Finite geometries and certain derived results in theory of numbers. *Proc. Nat. Inst. Sci. India* **11**, 136-149.

— (1946). Difference sets and combinatorial arrangements derivable from finite geometries. *Proc. Nat. Inst. Sci. India* **12**, 123-135.

— (1969). Cyclical generation of linear subspaces in finite geometries. *Conference on Combinatorial Mathematics and its Applications* (University of North Carolina, 1967), University of North Carolina Press, 515-535.

RAY-CHAUDHURI, D.K. (1962*a*). Some results on quadrics in finite geometries based on Galois fields. *Canad. J. Math.* **14**, 129-138.

— (1962*b*). Application of the geometry of quadrics for constructing PBIB designs. *Ann. Math. Statist.* **33**, 1175-1186.

— (1965). Some configurations in finite projective spaces and PBIB designs. *Canad. J. Math.* **17**, 114-123.

RÉDEI, L. (1946*a*). Zur theorie der Gleichungen in endlichen Körpern. *Acta Sci. Math. Szeged* **11**, 63-70.

— (1946*b*). Über die Gleichungen dritten und vierten Grades in endlichen Körpern. *Acta Sci. Math. Szeged* **11**, 96-105.

— (1947). Bemerkung zu meiner Arbeit 'Über die Gleichungen dritten und vierten Grades in endlichen Körpern'. *Acta Sci. Math. Szeged* **11**, 184-190.

— (1967). Algebra. Vol. 1. Pergamon.

— (1973). Lacunary polynomials over finite fields. North Holland.

— and TURAN, P. (1959). Zur Theorie der algebraischen Gleichungen über endlichen Körpern. *Acta Arith.* **5**, 223-225.

RICHARDSON, M. (1956). On finite projective games. *Proc. Amer. Math. Soc.* **7**, 453-465.

RICKART, C.E. (1940). The Pascal configuration in a finite projective plane. *Amer. Math. Monthly* **47**, 89-96.

RIGBY, J.F. (1965). Affine subplanes of finite projective planes. *Canad. J. Math.* **17**, 977-1009.

— (1967). Collineations, correlations, polarities, and conics. *Canad. J. Math.* **19**, 1027-1041.

— (1969). Pascal ovals in projective planes. *Canad. J. Math.* **21**, 1462-1476.

ROGERS, K. (1971). An elementary proof of a theorem of Jacobson. *Abh. Math. Sem. Univ. Hamburg* **35**, 223-229.

— (1972). Correction. *Abh. Math. Sem. Univ. Hamburg* **37**, 268.

ROOM, T.G. (1932). The Schur quadrics of a cubic surface (I; II). *J. London Math. Soc.* **7**, 147-154; 154-160.
— (1938). Geometry of determinantal manifolds. Cambridge University Press.
— and KIRKPATRICK, P.B. (1971). Miniquaternion geometry. Cambridge University Press.
ROSATI, L.A. (1956). Sul numero dei punti di una superficie cubica in uno spazio lineare finito. *Boll. Un. Mat. Ital.* **11**, 412-418.
— (1957). L'equazione delle 27 rette della superficie cubica generale in un corpo finito: Nota I. *Boll. Un. Mat. Ital.* **12**, 612-626.
— (1958). Nota II. *Boll. Un. Mat. Ital.* **13**, 84-99.
— (1962*a*). Su certe varietà dello spazio proiettivo r-dimensionale sopra un corpo non commutativo. *Atti Accad. Naz. Lincei Rend.* **32**, 907-912.
— (1962*b*). Su alcune varietà dello spazio proiettivo sopra un corpo non commutativo. *Ann. Mat. Pura Appl.* **59**, 213-227.
— (1968). Gruppi di collineazioni strettamente transitivi sulle bandiere. *Atti Sem. Mat. Fis. Univ. Modena* **17**, 23-28.
ROSENBAUM, R.A. (1963). Introduction to projective geometry and modern algebra. Addison-Wesley.
ROTA, G.-C. and STEIN, J. (1976). Applications of Cayley Algebras. *Teorie Combinatorie* (Rome, 1973), volume II, Accad. Naz. dei Lincei, 71-97.
ROTHAUS, O.S. (1976). On 'bent' functions. *J. Combinatorial Theory A* **20**, 300-305.
ROTHSCHILD, B L. and van LINT, J.H. (1974). Characterizing finite subspaces. *J. Combinatorial Theory A* **16**, 97-110.
— and SINGHI, N.M. (1976). Characterizing k-flats in geometric designs. *J. Combinatorial Theory A* **20**, 398-403.
RUSSO, A. (1971). Calotte hermitiane di un $S_{r,4}$. *Ricerche Mat.* **20**, 297-307.
RYSER, H.J. (1963). Combinatorial mathematics. Wiley.
SAAVEDRO RIVANO, N. (1976). Finite geometries in the theory of theta characteristics. *Enseignement Math.* **22**, 191-218.
SALMON, G. (1879). A treatise on the higher plane curves, third edition, Hodges, Foster, and Figgis, Dublin (Chelsea, 1960).
— (1914). A treatise on the analytic geometry of three dimensions, sixth edition (edited by R.A.F. Rogers), two volumes, Longmans, Green and Co., London (Chelsea, 1958).
SCARPIS, U. (1914). Intorno all' interpretazione della Teoria di Galois in un campo di razionalità finito. *Ann. Mat. Pura Appl.* **23**, 41-60.
SCE, M. (1958*a*). Sulla completezza degli archi nei piani proiettivi. *Atti Accad. Naz. Lincei Rend.* **25**, 43-51.
— (1958*b*). Sui k_q-archi di indice h. *Convegno internazionale: Reticoli e Geometrie Proiettive* (Palermo, Messina, 1957), Cremonese, 133-135.
— (1960). Preliminari ad una teoria aritmetico-gruppale dei k-archi. *Rend. Mat. e Appl.* **19**, 241-291.
SCHMIDT, W.M. (1973). Zur Methode von Stepanov. *Acta Arith.* **24**, 347-367.
— (1974). A lower bound for the number of solutions of equations over finite fields. *J. Number Theory* **6**, 448-480.
— (1976). Equations over finite fields, an elementary approach. *Lecture Notes in Mathematics* **536**, Springer.
SCHRÖDER, E.M. (1976). Zur Characterisierung von Quadriken. *J. Geometry* **8**, 75-77.
SCHULZ, R.-H. (1975). Geometrie über GF(4). *Math.-Phys. Semesterber.* **22**, 158-184.
SEGRE, B. (1942). The non-singular cubic surfaces. Oxford.
— (1948). Lezioni di geometria moderna, vol. I, Zanichelli.
— (1949). Le rette delle superficie cubiche nei corpi commutativi. *Boll. Un. Mat. Ital.* **4**, 223-228.
— (1951). Arithmetical questions on algebraic varieties. The Athlone Press, London.

SEGRE, B. (1954). Sulle ovali nei piani lineari finiti. *Atti Accad. Naz. Lincei Rend.* **17**, 1-2.
— (1955*a*). Ovals in a finite projective plane. *Canad. J. Math.* **7**, 414-416.
— (1955*b*). Curve razionali normali e k-archi negli spazi finiti. *Ann. Mat. Pura. Appl.* **39**, 357-379.
— (1956). Intorno alla geometria sopra un campo di caratteristica due. *Istanbul Üniv. Fen Fak. Mec. Ser. A* **21**, 97-123.
— (1957*a*). Sui k-archi nei piani finiti di caratteristica due. *Rev. Math. Pures Appl.* **2**, 289-300.
— (1957*b*). Sulle geometrie proiettive finite. *Convegno internazionale: Reticoli e Geometrie Proiettive* (Palermo, Messina, 1957), Cremonese, 46-61.
— (1958*a*). Elementi di geometria non lineare sopra un corpo sghembo. *Rend. Circ. Mat. Palermo* **7**, 81-122.
— (1958*b*). Sulla geometria sopra un campo a caratteristica. *Archimede* **10**, 53-60.
— (1959*a*). Le geometrie di Galois. *Ann. Mat. Pura Appl.* **48**, 1-97.
— (1959*b*). On complete caps and ovaloids in three-dimensional Galois spaces of characteristic two. *Acta Arith.* **5**, 315-332.
— (1959*c*). Le geometrie di Galois: archi ed ovali, calotte ed ovaloidi. *Conf. Sem. Mat. Univ. Bari*, nos. 43, 44.
— (1959*d*). Intorno alla geometria di certi spazi aventi un numero finito dei punti. *Archimede* **11**, 1-15.
— (1960*a*). Lectures on modern geometry (with an appendix by L. Lombardo-Radice). Cremonese.
— (1960*b*). On Galois geometries. *Proc. Intern. Congress Math. 1958*, Cambridge, 488-499.
— (1960*c*). Sul numero delle soluzioni di un qualsiasi sistema di equazioni algebriche sopra un campo finito. *Atti Accad. Naz. Lincei Rend.* **28**, 271-277.
— (1960*d*). Sistemi di equazioni nei campi di Galois. *Convegno Teoria Gruppi Finiti* (Florence), 66-80.
— (1960*e*). Gli spazi grafici. *Rend. Sem. Mat. Fis. Milano* **30**, 223-241.
— (1961). Alcune questioni su insiemi finiti di punti in geometria algebrica. *Univ. e Politec. Torino Rend. Sem. Mat.* **20**, 67-85.
— (1962*a*). Geometry and algebra in Galois spaces. *Abh. Math. Sem. Univ. Hamburg* **25**, 129-139.
— (1962*b*). Ovali e curve σ nei piani di Galois di caratteristica due. *Atti Accad. Naz. Lincei Rend.* **32**, 785-790.
— (1963). Intorno ad una congettura di Lang e Weil. *Atti Accad. Naz. Lincei Rend.* **34**, 337-339.
— (1964*a*). Arithmetische Eigenschaften von Galois-Raumen. I. *Math. Ann.* **154**, 195-256.
— (1964*b*). Teoria di Galois, fibrazioni proiettive e geometrie non desarguesiane. *Ann. Mat. Pura Appl.* **64**, 1-76.
— (1964*c*). Galois spaces and non-Desarguesian geometries. *Rend. Mat. e Appl.* **23**, 1-3.
— (1965*a*). Forme e geometrie hermitiane, con particolare riguardo al caso finito. *Ann. Mat. Pura Appl.* **70**, 1-201.
— (1965*b*). Hermitian geometries, with special regard to the finite case. *Actas Col. Intern. Geom. Alg.* (Madrid, 1965), 9-15.
— (1965*c*). Algebra and geometry in Galois spaces. *Sitzungsber. d. Berliner Math. Ges.*, 1961-1964, 26.
— (1965*d*). Intorno ad alcune forme algebriche su di un campo a caratteristica positiva. *Atti Accad. Naz. Lincei Rend.* **39**, 219-223.
— (1966). Finite structures and algebraic geometry. *Rend. Mat. e Appl.* **25**, 46-50.

SEGRE, B. (1967). Introduction to Galois geometries (edited by J.W.P. Hirschfeld). *Atti Accad. Naz. Lincei Mem.* **8**, 133-236.

— (1972*a*). Prodromi di geometria algebrica (with an appendix by U. Bartocci and M. Lorenzani). Cremonese.

— (1972*b*). Uber Galoissche Geometrien. *Geometrie, Wissenschaftliche Buchgesellschaft,* Darmstadt, 100-117.

— (1973). Proprietà elementari relative ai segmenti ed alle coniche sopra un campo qualsiasi ed una congettura di Seppo Ilkka per il caso di campi di Galois. *Ann. Mat. Pura Appl.* **96**, 290-337.

— (1974). Sur les correspondances involutoires régulières dans les ensembles finis. *C.R. Acad. Sci. Paris* **279**, 309-311.

— (1975). Sugli insiemi finiti di punti di uno spazio grafico. *Rend. Mat.* **8**, 37-64.

— (1975/6). Un approccio al problema dei quattro colori. *Rend. Accad. Naz. dei XL* **1/2**, 1-60.

— (1976). Incidence structures and Galois geometries, with special regard to regular involutive correspondences, arcs, caps, segments and conics. *Teorie Combinatorie* (Rome, 1973), Vol. I, Accad. Naz. dei Lincei, 331-369.

— and BARTOCCI, U. (1971). Ovali ed altre curve nei piani di Galois di caratteristica due. *Acta Arith.* **8**, 423-449.

— and KORCHMÁROS, G. (1977). Una proprietà degli insiemi di punti di un piano di Galois caratterizzante quelli formati dai punti delle singole rette esterne ad una conica. *Atti Accad. Naz. Lincei Rend.* **62**, 1-7.

SEIDEN, E. (1950). A theorem in finite projective geometry and an application to statistics. *Proc. Amer. Math. Soc.* **1**, 282-286.

— (1973). On the problem of construction and uniqueness of saturated 2_R^{K-p} designs. *A Survey of Combinatorial Theory* (Colorado State University, 1971), North Holland, 397-401.

SEIDENBERG, A. (1968). Elements of the theory of algebraic curves. Addison-Wesley.

SEITZ, G.M. (1973). Flag-transitive subgroups of Chevalley groups. *Ann. of Math.* **97**, 27-56.

SEMPLE, J.G. and KNEEBONE, G.T. (1952). Algebraic projective geometry. Oxford University Press.

— and ROTH, L. (1949). Introduction to algebraic geometry. Oxford University Press.

SERRE, J.-P. (1975). Valeurs propres des endomorphismes de Frobenius. *Seminaire Bourbaki no. 446* (1973/4), *Lecture notes in Mathematics* **431**, Springer, 190-204.

SHAFAREVICH, I.R. (1974). Basic algebraic geometry. Springer.

SHERK, F.A. and PABST, G. (1977). Indicator sets, reguli, and a new class of spreads. *Canad. J. Math.* **29**, 132-154.

SHRIKHANDE, S.S. (1963). A note on finite Euclidean plane over GF(2^n). *J. Indian Statist. Assoc.* **1**, 48-49.

SINGER, J. (1938). A theorem in finite projective geometry and some applications to number theory. *Trans. Amer. Math. Soc.* **43**, 377-385.

SLOANE, N.J.A. (1975). A short course on error-correcting codes. *CISM course* **188**, Springer.

— (1977). Error-correcting codes and invariant theory: new applications of a nineteenth-century technique. *Amer. Math. Monthly* **84**, 82-107.

SMIT GHINELLI, D. (1969). Varietà hermitiane e strutture finite. *Rend. Mat.* **2**, 23-62.

— (1976). On the application of Hermitian geometries for the construction of association schemes and designs. *Teorie Combinatorie* (Rome, 1973), vol. I, Accad. Naz. dei Lincei, 213-228.

SMITH, K.J.C. (1969). On the p-rank of the incidence matrix of points and hyperplanes in a finite projective geometry. *J. Combinatorial Theory* **7**, 122-129.

SNAPPER, E. (1950). Periodic linear transformations of affine and projective geometries. *Canad. J. Math.* **2**, 149-151.
SOKOLOVA, J.G. (1967). Projective Galois spaces (Russian). *Moskov. Oblast. Ped. Inst. Učen. Zap.* **173**, 57-66.
STAHNKE, W. (1973). Primitive binary polynomials. *Math. Comput.* **27**, 977-980.
STARK, H.M. (1973). On the Riemann hypothesis in hyperelliptic function fields. *Proc. Sympos. Pure Math.* **24**, 285-302.
STEIN, E. and RETKIN, H. (1965). On the symmetric (15; 3) arcs of the finite projective plane of order 7. *Rend. Mat. e Appl.* **24**, 392-399.
STEPANOV, S.A. (1970). Elementary method in the theory of congruences for a prime modulus. *Acta Arith.* **17**, 231-247.
— (1972). An elementary proof of the Hasse-Weil theorem for hyperelliptic curves. *J. Number Theory* **4**, 118-143.
STEVENSON, F.W. (1972). Projective planes. Freeman.
STORER, T. (1967). On the unique determination of the cyclotomic numbers for Galois fields and Galois domains. *J. Combinatorial Theory* **2**, 296-300.
STREET, A.P. and WALLIS, W.D. (1977). Combinatorial theory: an introduction. Charles Babbage Research Centre, Manitoba.
SWAN, R.G. (1962). Factorization of polynomials over finite fields. *Pacific J. Math.* **12**, 1099-1106.
SWINNERTON-DYER, H.P.F. (1967). The zeta function of a cubic surface over a finite field. *Proc. Cambridge Philos. Soc.* **63**, 55-71.
— (1971). Applications of algebraic geometry to number theory. *Proc. Sympos. Pure Math.* **20**, *Institute on Number Theory* (1969), 1-52.
TALLINI, G. (1956*a*). Sulle k-calotte di uno spazio lineare finito. *Ann. Mat. Pura Appl.* **42**, 119-164.
— (1956*b*). Sulle k-calotte degli spazi lineari finiti (Note I; II). *Atti Accad. Naz. Lincei Rend.* **20**, 311-317; 442-446.
— (1957*a*). Caratterizzazione grafica delle quadriche ellittiche negli spazi finiti. *Rend. Mat. e Appl.* **16**, 328-351.
— (1957*b*). Sui q-archi di un piano lineare finito di caratteristica p = 2. *Atti Accad. Naz. Lincei Rend.* **23**, 242-245.
— (1958*a*). Una proprietà grafica caratteristica delle superficie di Veronese negli spazi finiti (Note I; II). *Atti Accad. Naz. Lincei Rend.* **24**, 19-23; 135-138.
— (1958*b*). Una proprietà grafica caratteristica della superficie di Veronese negli spazi finiti. *Convegno internazionale: Reticoli e Geometrie Proiettive* (Palermo, Messina 1957). Cremonese, 136-139.
— (1959). Caratterizzazione grafica di certe superficie di $S_{3,q}$ (Note I; II). *Atti Accad. Naz. Lincei Rend.* **26**, 484-489; 644-648.
— (1960). Le geometrie di Galois e le loro applicazioni alla statistica e alla teoria delle informazioni. *Rend. Mat. e Appl.* **19**, 379-400.
— (1961*a*). On caps of kind s in a Galois r-dimensional space. *Acta Arith.* **7**, 19-28.
— (1961*b*). Sulle ipersuperficie irriducibili d'ordine minimo che contengono tutti i punti di uno spazio di Galois $S_{r,q}$. *Rend. Mat. e Appl.* **20**, 431-479.
— (1961*c*). Le ipersuperficie irriducibili d'ordine minimo che invadono uno spazio di Galois. *Atti Accad. Naz. Lincei Rend.* **30**, 706-712.
— (1962). Intorno alle forme di uno spazio di Galois ed agli spazi subordinati giacenti su esse. *Atti Accad. Naz. Lincei Rend.* **35**, 421-428.
— (1963). Un'applicazione delle geometrie di Galois a questioni di statistica. *Atti Accad. Naz. Lincei Rend.* **35**, 479-485.
— (1964*a*). Calotte complete di $S_{4,q}$ contenenti due quadriche ellittiche quali sezioni iperpiane. *Rend. Mat. e Appl.* **23**, 108-123.

TALLINI, G. (1964*b*). Recent results on Galois geometries (Hungarian). *Magyar Tud. Akad. Mat. Fiz. Oszt. Közl.* **14**, 183-192.

— (1971*a*). Ruled graphic systems. *Atti del Convegno di Geometria Combinatoria e sue Applicazioni* (Università di Perugia, 1970), Università di Perugia, 403-411.

— (1971*b*). Topologia associata ad uno spazio grafico. *Ricerche Mat.* **20**, 253-259.

— (1976). Graphic characterization of algebraic varieties in a Galois space. *Teorie Combinatorie* (Rome, 1973), volume II, Accad. Naz. dei Lincei, 153-165.

— and BRUNO, S. (1970). Geometria iperbolica in un piano di Galois $S_{2,q}$ con q dispari. *Ricerche Mat.* **19**, 48-78.

TALLINI SCAFATI, M. (1958). Sui 6-archi completi di un piano lineare $S_{2,8}$. *Convegno internazionale: Reticoli e Geometrie Proiettive* (Palermo, Messina, 1957), Cremonese, 128-132.

— (1964). Archi completi in un $S_{2,q}$, con q pari. *Atti Accad. Naz. Lincei Rend.* 37, 48-51.

— (1966*a*). Sui {k,n}-archi di un piano grafico finito. *Atti Accad. Naz. Lincei Rend.* 40, 373-378.

— (1966*b*). {k,n}-archi di un piano grafico finito, con particolare riguardo a quelli con due caratteri (Note I; II). *Atti Accad. Naz. Lincei Rend.* **40**, 812-818; 1020-1025.

— (1967*a*). Caratterizzazione grafica delle forme hermitiane di un $S_{r,q}$. *Rend. Mat. e Appl.* **26**, 273-303.

— (1967*b*). Una proprietà grafica caratteristica delle forme hermitiane in uno spazio di Galois. *Univ. e Politec. Torino Rend. Sem. Mat.* **26**, 33-41.

— (1971). Graphic curves on a Galois plane. *Atti del Convegno di Geometria Combinatoria e sue Applicazioni* (Università di Perugia, 1970), Università di Perugia, 413-419.

— (1973). Calotte di tipo (m,n) in uno spazio di Galois $S_{r,q}$. *Atti Accad. Naz. Lincei Rend.* **53**, 71-81.

— (1976*a*). The k-sets of type (m,n) in a Galois space $S_{r,q}(r \geq 2)$. *Teorie Combinatorie* (Rome, 1973), volume II, Accad. Naz. dei Lincei, 459-463.

— (1976*b*). Sui k-insiemi di uno spazio di Galois $S_{r,q}$ a due soli caratteri nella dimensione d. *Atti Accad. Naz. Lincei Rend.* **60**, 782-788.

TAYLOR, D.E. (1974). Some classical theorems on division rings. *Enseignement Math.* **20**, 293-298.

THAS, J.A. (1968*a*). Normal rational curves and k-arcs in Galois spaces. *Rend. Mat.* **1**, 331-334.

— (1968*b*). Contribution to the theory of k-caps in projective Galois spaces (Flemish). *Med. Konink. Vlaamse Acad. Wetensch.* **30**, No. 13.

— (1969*a*). Connection between the Grassmannian $G_{k-1;n}$ and the set of the k-arcs of the Galois space $S_{n,q}$. *Rend. Mat.* **2**, 121-134.

— (1969*b*). Normal rational curves and (q+2)-arcs in a Galois space $S_{q-2,q}$ $(q = 2^h)$. *Atti Accad. Naz. Lincei Rend.* **47**, 249-252.

— (1970). Connection between the n-dimensional affine space $A_{n,q}$ and the curve C, with equation $y = x^q$, of the affine plane A_{2,q^n}. *Rend. Ist. di Mat. Univ. Trieste* **2**, 146-151.

— (1971*a*). The m-dimensional projective space $S_m(M_n(GF(q)))$ over the total matrix algebra $M_n(GF(q))$ of the (n×n)-matrices with elements in the Galois field GF(q). *Rend. Mat.* **4**, 459-532.

— (1971*b*). An equivalent to Pappus' theorem in the affine plane $A_2(K)$ over a corpus K with characteristic different from two. *Simon Stevin* **44**, 186-187.

— (1972). Ovoidal translation planes. *Arch. Math.* **23**, 110-112.

— (1973*a*). Deduction of properties, valid in the projective space $S_{3n-1}(K)$, using the projective plane over the total matrix algebra $M_n(K)$. *Simon Stevin* **46**, 3-16.

THAS, J.A. (1973*b*). Construction of partial geometries. *Simon Stevin* **46**, 95-98.
— (1973*c*). A combinatorial problem. *Geometriae Dedicata* **1**, 236-240.
— (1973*d*). On 4-gonal configurations. *Geometriae Dedicata* **2**, 317-326.
— (1973*e*). 4-gonal subconfigurations of a given 4-gonal configuration. *Atti Accad. Naz. Lincei Rend.* **53**, 520-530.
— (1973*f*). Flocks of finite egglike inversive planes. *Finite Geometric Structures and their Applications* (Bressanone, 1972), Cremonese, 189-191.
— (1973*g*). 4-gonal configurations. *Finite Geometric Structures and their Applications* (Bressanone, 1972), Cremonese, 251-263.
— (1974*a*). Geometric characterization of the [n-1]-ovaloids of the projective space PG(4n-1,q). *Simon Stevin* **47**, 97-106.
— (1974*b*). Construction of maximal arcs and partial geometries. *Geometriae Dedicata* **3**, 61-64.
— (1974*c*). On semi ovals and semi ovoids. *Geometriae Dedicata* **3**, 229-231.
— (1975*a*). Some results concerning {(q+1)(n-1);n}-arcs and {(q+1)(n-1)+1;n}-arcs in finite projective planes of order q. *J. Combinatorial Theory A* **19**, 228-232.
— (1975*b*). Flocks of non-singular ruled quadrics in PG(3,q). *Atti Accad. Naz. Lincei Rend.* **59**, 83-85.
— (1977). Two infinite classes of perfect codes in metrically regular graphs. *J. Combinatorial Theory B* **23**, 236-238.
— (1977) Combinatorics of partial geometries and generalized quadrangles. *Higher Combinatorics* (Berlin, 1976), Reidel, 183-199.
— and PAYNE, S.E. (1976). Classical finite generalized quadrangles, a combinatorial study. *Ars Combinatoria* **2**, 57-110.
— and PUYSTJENS, R. (1970). Number of regular homographies of a Galois space $S_{n,q}$. *Rend. Mat.* **3**, 258-289.
— — (1971). The number of Desargues-configurations of the projective plane $S_{2,q}$ over the Galois field GF(q) with characteristic two or three. *Simon Stevin* **44**, 160-170.
TIETÄVÄINEN, A. (1973). On the nonexistence of perfect codes over finite fields. *SIAM J. Appl. Math.* **24**, 88-96.
TIETZE, U.P. (1968). Zur Theorie quadratischen Formen über Körpern der Characteristik 2. *J. Reine Angew. Math.* **268/269**, 388-390.
TITS, J. (1959). Sur la trialité et certains groupes qui s'en déduisent. *Inst. Hautes Etudes Sci. Publ. Math.* **2**, 13-60.
— (1960). Les groupes simples de Suzuki et de Ree. *Sém. Bourkaki No. 210*.
— (1962*a*). Ovoides et groupes du Suzuki. *Arch. Math.* **13**, 187-198.
— (1962*b*). Ovoides à translations. *Rend. Mat. e Appl.* **21**, 37-59.
— (1963). Groupes simples et géométries associées. *Proc. Intern. Math. Congress* (Stockholm, 1962), 197-221.
— (1966). Une propriété caractéristique des ovoides associés aux groupes de Suzuki. *Arch. Math.* **17**, 136-153.
TODD, J.A. (1946). Projective and analytical geometry. Pitman, London.
— (1950). The invariants of a finite collineation group in five dimensions. *Proc. Cambridge Philos. Soc.* **46**, 73-90.
— (1959). On representations of the Mathieu groups as collineation groups. *J. London Math. Soc.* **34**, 406-416.
— (1966). A representation of the Mathieu group M_{24} as a collineation group. *Ann. Mat. Pura Appl.* **71**, 199-238.
TUTTE, W.J. (1958). The chords of the non-ruled quadric in PG(3,3). *Canad. J. Math.* **10**, 481-483.
— (1966). On the algebraic theory of graph colourings. *J. Combinatorial Theory* **1**, 15-50.
— (1967). A correction to: 'On the algebraic theory of graph colorings'. *J. Combinatorial Theory* **3**, 102.

TUTTE, W.T. (1969). Projective geometry and the 4-color problem. *Recent Progress in Combinatorics* (Waterloo, 1968), Academic Press, 199-207.
VAJDA, S. (1967). Patterns and configurations in finite spaces. Griffin, London.
van der WAERDEN, B.L. (1935). Gruppen von linearen transformationen. Springer (Chelsea, 1948).
— (1973). Einführung in die algebraische geometrie, second edition, Springer.
van LINT, J.H. (1971). Coding theory. *Lecture Notes in Mathematics* **201**, Springer.
VANSTONE, S.A. and SCHELLENBERG, P.J. (1974). Vector spaces, geometries, and designs. *Utilitas Math.* **6**, 337-341.
VAUGHAN, T.P. (1974). Polynomials and linear transformations over finite fields. *J. Reine Angew. Math.* **267**, 179-206.
VEBLEN, O. and BUSSEY, N.J. (1906). Finite projective geometries. *Trans. Amer. Math. Soc.* **7**, 241-259.
— and YOUNG, J.W. (1916). Projective geometry, two volumes, Ginn and Co. (Blaisdell, 1965).
von STAUDT, K.G.C. (1856). Beitrage zur Geometrie der Lage, vol. I, Nürnberg.
WAGNER, A. (1961). On collineation groups of finite projective spaces I. *Math. Z.* **76**, 411-426.
— (1969*a*). Collineations of finite projective spaces as permutations on the sets of dual subspaces. *Math. Z.* **111**, 249-254.
— (1969*b*). Involutions of projective spaces. *Math. Z.* **112**, 37-51.
— (1971). Correction to 'Collineations of finite projective spaces as permutations on the sets of dual spaces'. *Math. Z.* **121**, 190.
— (1974). Groups generated by elations. *Abh. Math. Sem. Univ. Hamburg* **41**, 190-205.
— (1976). The faithful linear representation of least degree of S_n and A_n over a field of characteristic 2. *Math. Z.* **151**, 57-70.
— (1977). The faithful linear representations of least degree of S_n and A_n over a field of odd characteristic. *Math. Z.* **154**, 103-114.
WALES, D. (1969). Uniqueness of the graph of a rank 3 group. *Pacific J. Math.* **30**, 271-277.
WALKER, A.G. (1947). Finite projective geometry. *Edinburgh Math. Notes* **36**, 12-17.
WALKER, M. (1976*a*). The collineation groups of derived translation planes II. *Math. Z.* **148**, 1-6.
— (1976*b*). A class of translation planes. *Geometriae Dedicata* **5**, 135-146.
WALKER, R.J. (1950). Algebraic curves. Princeton University Press (Dover, 1962).
WALL, C.T.C. (1977). Nets of conics. *Math. Proc. Cambridge Philos. Soc.* **81**, 351-364.
WALL, G.E. (1963). On the conjugacy classes in the unitary, symplectic and orthogonal groups. *J. Austral. Math. Soc.* **3**, 1-62.
WALMSLEY, J.W. (1971). On a condition for commutativity of rings. *J. London Math. Soc.* **4**, 331-332.
WAN, Zhe-Xian (1964*a*). Notes on finite geometries and the construction of PBIB designs. I. Some 'Anzahl' theorems in symplectic geometry over finite fields. *Sci. Sinica* **13**, 515-516.
— (1964*b*). Notes on finite geometries and the construction of PBIB designs. II. Some PBIB designs with two associate classes based on the symplectic geometry over finite fields. *Sci. Sinica* **13**, 516-517.
— (1964*c*). Notes on finite geometries and the construction of PBIB designs. VI. Some association schemes and PBIB designs based on finite geometries. *Sci. Sinica* **13**, 1872-1876.

WAN, Zhe-Xian (1965*a*). Studies in finite geometries and the construction of incomplete block designs. I, Some 'Anzahl' theorems in symplectic geometry over finite fields. *Chinese Math.* **7**, 55-62.
— (1965*b*). Studies in finite geometries and the construction of incomplete block designs. II, Some PBIB designs based on symplectic geometry over finite fields. *Chinese Math.* **7**, 63-73.
— (1965*c*). Studies in finite geometries and the construction of incomplete block designs. VI. An association scheme with many associate classes constructed from subspaces in a vector space over finite fields (Chinese). *Progress in Math.* **8**, 293-302.
— and YANG, Ben-Fu (1964). Notes on finite geometries and the construction of PBIB designs. III. Some 'Anzahl' theorems in unitary geometry over finite fields and their applications. *Sci. Sinica* **13**, 1006-1007.
— — (1965). Studies in finite geometries and the construction of incomplete block designs. III. Some 'Anzahl' theorems in unitary geometry over finite fields and their applications. *Chinese Math.* **7**, 252-264.
WATERHOUSE, W.C. and MILNE, J.S. (1971). Abelian varieties over finite fields. *Proc. Sympos. Pure Math.* **20**, Institute on Number Theory (1969), 53-64.
WEBB, W.A. (1973). On the representation of polynomials over finite fields as sums of powers and irreducibles. *Rocky Mountain J. of Math.* **3**, 23-30.
WEIL, A. (1948*a*). Sur les courbes algébriques et les variétés qui s'en déduisent. Hermann.
— (1948*b*). Variétés abéliennes et courbes algébriques. Hermann.
— (1949). Numbers of solutions of equations in finite fields. *Bull. Amer. Math. Soc.* **55**, 497-508.
WELLS, C. (1968). Generators for groups of permutation polynomials over finite fields. *Acta Sci. Math. Szeged* **29**, 167-176.
— (1969). The degrees of permutation polynomials over finite fields. *J. Combinatorial Theory* **7**, 49-55.
— (1974). Polynomials over finite fields which commute with translations. *Proc. Amer. Math. Soc.* **46**, 347-350.
WHITELAW, T.A. (1967). Janko's group as a collineation group in PG(6,11). *Proc. Cambridge Philos. Soc.* **63**, 663-677.
WILLIAMS, K.S. (1972). Exponential sums over GF(2^n). *Pacific J. Math.* **40**, 511-519.
WILSON, A.H. (1914). The canonical types of nets of modular conics. *Amer. J. Math.* **36**, 187-210.
WILSON, B.J. (1974). A note on (k,n)-arcs. *Proc. Cambridge Philos. Soc.* **76**, 57-59.
WITT, E. (1937). Theorie der quadratischen Formen in beliebigen Körpern. *J. Reine Angew. Math.* **176**, 31-44.
WOLFMANN, J. (1977*a*). Codes projectifs à deux poids, 'caps' complets et ensembles de differences. *J. Combinatorial Theory A* 23, 208-222.
— (1977*b*). Combinatorial properties deduced from the study of two-weight codes. *C.R. Acad. Sci. Paris* **284**, 641-642.
YANG Ben-Fu (1965*a*). Studies in finite geometries and the construction of incomplete block designs. VII, An association scheme with many associate classes constructed from maximal completely isotropic subspaces in symplectic geometry over finite fields. *Chinese Math.* **7**, 547-560.
— (1965*b*). Studies in finite geometries and the construction of incomplete block designs. VIII, An association scheme with many associate classes constructed from maximal completely isotropic subspaces in unitary geometry over finite fields. *Chinese Math.* **7**, 561-576.
YANG, C.-T. (1947). Certain chains in a finite projective geometry. *Acad. Sinica Science Record* **2**, 44-46.
— (1948*a*). Projective collineations in a finite projective plane. *Acad. Sinica Science Record* **2**, 157-164.

YANG, C.-T. (1948*b*). Certain chains in a finite projective geometry. *Duke Math. J.* **15**, 37-47.

— (1949*a*). A theorem in finite projective geometry. *Bull. Amer. Math. Soc.* **55**, 930-933.

— (1949*b*). Fixed subplanes of a projective collineation in a finite projective plane. *Acad. Sinica Science Record* **2**, 262-269.

YFF, P. (1977). On subplane partitions of a finite projective plane. *J. Combinatorial Theory A* **22**, 118-122.

ZEITLER, H. (1971). Über Ovale in endlichen Ebenen. *Math.-Phys. Semesterber.* **18**, 69-86.

— (1975). Ovoide in endlichen projektiven Räumen der dimension 3. *Math.-Phys. Semesterber.* **22**, 109-124.

ZIERLER, N. (1973). On the MacWilliams identity. *J. Combinatorial Theory A* **15**, 333-337.

ZIMMER, H.G. (1971). An elementary proof of the Riemann hypothesis for an elliptic curve over a finite field. *Pacific J. Math.* **36**, 267-278.

— (1970). Die Néron-Tate'schen quadratischen formen auf der rationalen Punkt-gruppe einer elliptischen Kurve. *J. Number Theory* **2**, 459-499.

ZIRILLI, F. (1973). Su una classe di k-archi di un piano di Galois. *Atti Accad. Naz. Lincei Rend.* **54**, 393-397.

INDEX OF NOTATION

Operations on sets and groups

$\lvert X \rvert$	number of elements in the set X
$X \cap Y$	the intersection of X and Y
$X \cap Y$	the union of X and Y
$X \backslash Y$	the set of elements of X not in Y
$X \subset Y$	X is a subset of Y
$\emptyset$	the empty set
$x \in X$	x is an element of X
$x \notin X$	x is not an element of X
$G \cong H$	the groups G and H are isomorphic
$G < H$	G is a subgroup of H
$G \lhd H$	G is a normal subgroup of H
$G \times H$	the direct product of the groups G and H
GH	a semi-direct product of the groups G and H
$\langle g_1, g_2, \ldots, g_n \rangle$	the group generated by $g_1, g_2, \ldots, g_n$

Miscellaneous sets and numbers

$\mathbf{N}$	the natural numbers $\{1,2,3,\ldots\}$
$\mathbf{N}_r$	$\{1,2,\ldots,r\}$
$\overline{\mathbf{N}}_r$	$\{0,1,\ldots,r\}$
$\theta(n)$	$(q^{n+1} - 1)/(q - 1)$ (§1.6)
$\phi(n)$	$\lvert\{m \in \mathbf{N} \mid 1 \leq m \leq n,\ (m,n) = 1\}\rvert$ (§1.6)
$[r,s]_-$	$\prod_{i=r}^{i=s} (q^i - 1)$ for $s \geq r$ (§2.3)
$[r,s]_-$	1 for $s < r$ (§2.3)
$[r,s]_+$	$\prod_{i=r}^{i=s} (q^i + 1)$ for $s \geq r$
$[r,s]_+$	1 for $s < r$
$\mathbf{c}(n,r)$	$n(n - 1)\ldots(n - r + 1)/r!$
$\mathbf{h}(n,r)$	$n(n + 1)\ldots(n + r - 1)/r!$
$(m_1,\ldots,m_r)$	greatest common divisor of $m_1,\ldots,m_r$
$\lvert \alpha_1,\ldots,\alpha_r \rvert$	smallest positive integer such that $\alpha_1{}^k = \ldots = \alpha_r{}^k$

Fields: subsets, functions, polynomials

K	an arbitrary field (§1.1)

K_0	$K\backslash\{0\}$ (§1.1)
$\mathbf{C}$	the complex numbers
$\mathbf{R}$	the real numbers
$\gamma, GF(q)$	the Galois field of $q = p^h$ elements (§1.1)
γ_0	$\gamma\backslash\{0\}$ (§1.1)
γ_{01}	$\gamma\backslash\{0,1\}$
γ^+	$\gamma \cup \{\infty\}$ (§6.1)
q	order of γ (§1.1)
p	characteristic of γ (§1.1)
h	order of the automorphism group of γ (§§1.1, 1.2)
$\overline{\gamma}$	algebraic closure of γ
$o(t)$	order of the element t
$K[X], K[x_0, x_1, \ldots, x_n]$	polynomial ring in $x_0, \ldots, x_n$ over K
$\gamma[X], \gamma[x_0, x_1, \ldots, x_n]$	polynomial ring in $x_0, \ldots, x_n$ over γ
$\Gamma[x]$	$\gamma[x]/(x^q - x)$ (§1.3)
$G[x]$	$\{F \in \gamma[x] \mid \deg F < q\}$ (§1.3)
$\mathscr{P}(q;x)$	permutation polynomials in x over γ (§1.3)
$D(t),\ D_2(t)$	$t + t^2 + t^4 + \ldots + t^{2^{h-1}}$, $q = 2^h$ (§1.4)
$D_p(t)$	$t + t^p + \ldots + t^{p^{h-1}}$, $q = p^h$ (§1.9)
$D_{q'}(t)$	$t + t^{q'} + \ldots + t^{q'^{m-1}}$, $q = q'^m$ (§1.9)
$\mathscr{C}_0$	$\{t \in \gamma \mid D(t) = 0\}$ (§1.4)
$\mathscr{C}_1$	$\{t \in \gamma \mid D(t) = 1\}$ (§1.4)
$C(t)$	$D_2(t)$ when $p = 2$ (§1.8)
$C(t)$	$(1 - t^{(q-1)/2})/2$ when $p > 2$ (§1.8)
$\Delta, \Delta(F)$	discriminant of the polynomial F (§1.8)
$R, R(F,G)$	resultant of the polynomials F and G (§1.8)
$\partial, \partial(F)$	S-invariant of the polynomial F (§1.8)
$C(F)$	$C(\Delta)$ when $p > 2$ (§1.8)
$C(F)$	$C(\partial)$ when $p = 2$ (§1.8)
$\chi, \chi(\gamma)$	character of γ (§1.10)
$\mathscr{N}(m,q)$	set of monic irreducible polynomials of degree m over γ (§1.6)
$N(m,q)$	$\lvert\mathscr{N}(m,q)\rvert$ (§1.6)
$\mathscr{N}^e(m,q)$	elements of $\mathscr{N}(m,q)$ of exponent e (§1.6)

$P(m,q)$	number of primitive polynomials in $\mathcal{N}(m,q)$ (§1.6)
$R(m,q)$	number of subprimitive polynomials in $\mathcal{N}(m,q)$ (§1.6)
$\{\alpha_1,\alpha_2;\alpha_3,\alpha_4\}$	$(\alpha_1 - \alpha_3)(\alpha_2 - \alpha_4)(\alpha_1 - \alpha_4)^{-1}(\alpha_2 - \alpha_3)^{-1}$ (§1.11)
$\Sigma f(i_1,\ldots,i_r)$	sum of all terms $f(i_1,\ldots,i_r)$
$\Sigma' f(i_1,\ldots,i_r)$	sum of all terms $f(i_1,\ldots,i_r)$ with $i_1 \leqslant i_2 \leqslant\ldots\leqslant i_r$
$\Sigma'' f(i_1,\ldots,i_r)$	sum of all terms $f(i_1,\ldots,i_r)$ with $i_1 < i_2 <\ldots< i_r$
$F_{(i)}$	$\partial F/\partial x_i$ (§11.2)
$F_{(ij)}$	$\partial^2 F/\partial x_i \partial x_j$ (§11.2)
$H^{(i)}$	$F_{(j)}^2 F_{(kk)} + F_{(k)}^2 F_{(jj)} - 2F_{(j)}F_{(k)}F_{(jk)}$ (§11.2)

Spaces, subspaces, numbers

$V(n,K)$	n-dimensional vector space over the field K (§2.1)
$V(n,q)$	$V(n,K)$ when $K = \gamma$ (§2.1)
$PG(n,K)$	n-dimensional projective space over the field K (§2.1)
$PG(n,q)$	$PG(n,K)$ when $K = \gamma$ (§2.1)
$AG(n,K)$	n-dimensional affine space over the field K (§2.2)
$AG(n,q)$	$AG(n,K)$ when $K = \gamma$ (§2.2)
$PG^{(r)}(n,q)$	set of r-spaces of $PG(n,q)$ (§3.1)
$\mathbf{P}(X), \mathbf{P}(x_0,x_1,\ldots,x_n)$	point of $PG(n,K)$ with vector $X = (x_0,x_1,\ldots,x_n)$ (§2.1)
$\pi(U), \pi(u_0,u_1,\ldots,u_n)$	prime of $PG(n,K)$ with vector $U = (u_0,u_1,\ldots,u_n)$ (§2.1)
$\mathbf{U}_i$	$\mathbf{P}(0,\ldots,0,1,0,\ldots,0)$ with 1 in the $(i+1)$-th place (§2.1)
$\mathbf{U}$	$\mathbf{P}(1,1,\ldots,1)$
$\mathbf{u}_i$	$\pi(0,\ldots,0,1,0,\ldots,0)$ with 1 in the $(i+1)$-th place (§2.1)
$\mathbf{u}$	$\pi(1,1,\ldots,1)$ (§2.1)

Π_r	an r-dimensional subspace of $PG(n,K)$, specific or generic (§2.1)
$\Pi_r \Pi_s$	the join of Π_r and Π_s (§2.1)
$\Pi_r \cap \Pi_s$	the meet of Π_r and Π_s (§2.1)
Π_{-1}	the empty subspace
$\phi(r;n,q)$	$\lvert PG^{(r)}(n,q)\rvert$ (§3.1)
$\phi(0;n,q),\theta(n,q)$	$\lvert PG(n,q)\rvert$ (§§3.1, 4.1)
$\chi(s,r;n,q)$	$\lvert\{\Pi_r$ through a fixed Π_s in $PG(n,q)\}\rvert$ (§3.1)
$\psi_{12}(t,s,r;n,q)$	$\lvert\{\Pi_r$ meeting a fixed Π_s in a fixed Π_t in $PG(n,q)\}\rvert$ (§3.1)
$\psi_2(t,s,r;n,q)$	$\lvert\{\Pi_r$ meeting a fixed Π_s in some Π_t in $PG(n,q)\}\rvert$ (§3.1)
$\psi_1(t,s,r;n,q)$	$\lvert\{(\Pi_r,\Pi_s)\mid\Pi_r \cap \Pi_s =$ a fixed Π_t in $PG(n,q)\}\rvert$ (§3.1)
$\psi(t,s,r;n,q)$	$\lvert\{(\Pi_r,\Pi_s)\mid\Pi_r \cap \Pi_s =$ some Π_t in $PG(n,q)\}\rvert$ (§3.1)
$m(l;r,s;n,q)$	maximum value of k of a $(k,l;r,s;n,q)$-set (§3.3)
$m_2(n,q)$	$m(0;2,1;n,q)$ (§3.3)
$m(n,q)$	$m(0;n,n-1;n,q)$ (§3.3)
$s(n,q,q^k)$	number of $PG(n,q)$ in $PG(n,q^k)$ (§4.3)
$\theta(n,q,q^k)$	number of $PG(n,q)$ in a partition of $PG(n,q^k)$ (§4.3)
$p(n,q,q^k)$	number of partitions of $PG(n,q^k)$ by subgeometries $PG(n,q)$ (§4.3)
$\rho(n,q,q^k)$	number of partitions of $PG(n,q^k)$ containing a given $PG(n,q)$ (§4.3)
$H(P_1,P_2;P_3,P_4)$	P_1 and P_2 harmonically separate P_3 and P_4 (§6.1)
$(P_1,P_2,\ldots)\overline{\wedge}(Q_1,Q_2,\ldots)$	there is a projectivity $\mathfrak{T}$ taking the collinear points $P_1,P_2,\ldots$ to the collinear points $Q_1,Q_2,\ldots$ (§7.5)
$(P_1,P_2,\ldots)\overset{R}{\overline{\overline{\wedge}}}(Q_1,Q_2,\ldots)$	$P_1,P_2,\ldots$ and $Q_1,Q_2,\ldots$ are in perspective from R (§7.5)
$\{P_1,P_2;P_3,P_4\}$	the cross-ratio of (P_1,P_2,P_3,P_4) (§6.1)

Varieties

$\mathbf{V}(F_1,\ldots,F_r)$, $\mathbf{V}_n(F_1,\ldots,F_r)$, $\mathbf{V}_{n,K}(F_1,\ldots,F_r)$	$\{\mathbf{P}(X) \in PG(n,K) \mid F_1(X)=\ldots=F_r(X) = 0,$ F_i forms in $K[X]\}$ (§2.6)
$\mathbf{V}(F_1,\ldots,F_r)$, $\mathbf{V}_n(F_1,\ldots,F_r)$, $\mathbf{V}_{n,q}(F_1,\ldots,F_r)$	$V_{n,K}(F_1,\ldots,F_r)$ when $K = \gamma$ (§2.6)
$m_P(l,\mathscr{F})$	intersection multiplicity of the line l and variety $\mathscr{F}$ at P (§2.6)
$m_P(\mathscr{F})$	multiplicity of P on the variety $\mathscr{F}$ (§2.6)
$T_P(\mathscr{F})$	tangent space to the variety $\mathscr{F}$ at P (§2.6)
$m_P(\mathscr{F},\mathscr{G})$	intersection multiplicity of the plane curves $\mathscr{F}$ and $\mathscr{G}$ at P (§10.1)
$\mathscr{Q}_n$	a non-singular quadric in $PG(n,q)$ (§§3.3, 5.1)
$\mathscr{U}_n$, $\mathscr{U}_{n,q}$	a non-singular Hermitian variety in $PG(n,q)$ (§5.1)
$\mathscr{P}_{2s}$, $\mathscr{P}_{2s,q}$	a non-singular (parabolic) quadric in $PG(2s,q)$ (§5.2)
$\mathscr{H}_{2s-1}$, $\mathscr{H}_{2s-1,q}$	a non-singular hyperbolic quadric in $PG(2s-1,q)$ (§5.2)
$\mathscr{E}_{2s-1}$, $\mathscr{E}_{2s-1,q}$	a non-singular elliptic quadric in $PG(2s-1,q)$ (§5.2)
$\mu(n,q)$	$\lvert\mathscr{U}_n\rvert$ (§5.1)
$\psi(2s,q)$	$\lvert\mathscr{P}_{2s}\rvert$ (§5.2)
$\psi_+(2s-1,q)$	$\lvert\mathscr{H}_{2s-1}\rvert$ (§5.2)
$\psi_-(2s-1,q)$	$\lvert\mathscr{E}_{2s-1}\rvert$ (§5.2)
$g, g(\mathscr{Q})$	projective index of the quadric $\mathscr{Q}$ (§5.2)
$w, w(\mathscr{Q})$	character of the quadric $\mathscr{Q}$ (§5.2)
$\kappa, \kappa(\mathscr{F})$	class of the curve $\mathscr{F}$ (§11.2)
$\mathscr{N}_i^j$	singular cubic curve with i real inflexions and j real tangents at the double point (§11.3)
$\mathscr{N}^j$	singular cubic curve with j real tangents at the double point (§11.3)
$\mathscr{F}_n^r$	non-singular cubic curve with n real inflexions and r real inflexional triangles (§11.4)

$\mathscr{G}_n{}^r$	general cubic curve (§11.4)
$\mathscr{E}_n{}^r$	equianharmonic cubic curve (§11.4)
$\mathscr{H}_n{}^r$	harmonic cubic curve (§11.4)
$\mathscr{F}_1 \sim \mathscr{F}_2$	the varieties $\mathscr{F}_1, \mathscr{F}_2$ are projectively equivalent
$\mathscr{U}_2', \mathscr{U}_{2,q}'$	a sub-Hermitian curve (§7.3)

Transformations

Mappings are written on the right of the elements being mapped.

$A \sim B$	the matrices A and B are equivalent (§2.3)
$A \simeq B$	the matrices A and B are similar (§2.3)
T^*	transpose of the matrix T
$\mathrm{M}(T)$	projectivity defined by the matrix T (§2.1)
$C(F)$	companion matrix of the polynomial F (§2.3)
$H(F)$	hypercompanion matrix of the polynomial F (§2.3)
$[(r_1^{(1)},\ldots,r_{k_1}^{(1)})m_1,\ldots]$	symbol of a projectivity or linear transformation (§2.3)
$[\lambda_1(r_1^{(1)},\ldots,r_{k_1}^{(1)})m_1,\ldots]$	extended symbol of a projectivity or linear transformation (§2.3)
$\mathfrak{U}_n$	polarity induced by $\mathscr{U}_n$ (§5.3)
$\mathfrak{P}_{2s}$	polarity induced by $\mathscr{P}_{2s}$ (§5.3)
$\mathfrak{H}_{2s-1}$	polarity induced by $\mathscr{H}_{2s-1}$ (§5.3)
$\mathfrak{E}_{2s-1}$	polarity induced by $\mathscr{E}_{2s-1}$ (§5.3)
$o(T)$	order of the linear transformation T (§2.5)
$o(\mathfrak{T})$	order of the projectivity $\mathfrak{T}$ (§2.5)
$\mathscr{M}(n,R)$	ring of $n \times n$ matrices over the ring R (§2.3)
$\mathscr{S}_1 \sim \mathscr{S}_2$	the subsets $\mathscr{S}_1$ and $\mathscr{S}_2$ of $PG(n,q)$ are projectively equivalent
$p(n,q)$	$\lvert PGL(n,q)\rvert$ (§2.3)
$\sigma(n,q)$	$\lvert\{$cyclic projectivities in $PG(n,q)\}\rvert$ (§§2.4, 4.2)

$\sigma(n,q,q^k)$	$\|$\{projectivities which act cyclically on a $PG(n,q)$ in $PG(n,q^k)$\}$\|$ (§4.3)
$E(n,q)$	$\|$\{conjugacy classes of $GL(n,q)$\}$\|$ (§2.3)
$e(n,q)$	$\|$\{conjugacy classes of $PGL(n,q)$\}$\|$ (§2.3)
$e^*(n,q)$	$\|$\{conjugacy classes of $SL(n,q)$\}$\|$ (§2.9)
I, I_n	identity in $GL(n,q)$
$\mathfrak{I}$	identity in $PGL(n,q)$
$\mathrm{diag}(a_1,a_2,\ldots,a_n)$	the matrix with diagonal elements $a_1,a_2,\ldots,a_n$ and zeros elsewhere

Groups

$\mathbf{S}_n$	the symmetric group of degree n
$\mathbf{A}_n$	the alternating group of degree n
$\mathbf{D}_n$	the dihedral group of order $2n$
$\mathbf{Z}_n$	the cyclic group of order n
$\mathbf{Z}$	the integers
$GL(n,q)$	non-singular linear transformations of $V(n,q)$ (§2.3, Appendix I)
$\Gamma L(n,q)$	non-singular semi-linear transformations of $V(n,q)$ (§2.3, Appendix I)
$SL(n,q)$	elements of $GL(n,q)$ of determinant one (§2.3, Appendix I)
$PGL(n,q)$	projectivities of $PG(n-1,q)$ (§2.3, Appendix I)
$P\Gamma L(n,q)$	collineations of $PG(n-1,q)$ (§2.3, Appendix II)
$PSL(n,q)$	projectivities of $PG(n-1,q)$ of "determinant one" (§2.3, Appendix I)
$P\Gamma SL(n,q)$	collineations of $PG(n-1,q)$ of "determinant one" (Appendix I)
$PGU(n,q)$	subgroup of $PGL(n,q)$ fixing $\mathscr{U}_{n-1}$ (§5.2, Appendix I)
$PGO(2s+1,q)$	subgroup of $PGL(2s+1,q)$ fixing $\mathscr{P}_{2s}$ (§5.2, Appendix I)
$PGO_+(2s,q)$	subgroup of $PGL(2s,q)$ fixing $\mathscr{H}_{2s-1}$ (§5.2, Appendix I)
$PGO_-(2s,q)$	subgroup of $PGL(2s,q)$ fixing $\mathscr{E}_{2s-1}$ (§5.2, Appendix I)

$PGSp(2s,q)$	subgroup of $PGL(2s,q)$ fixing a null polarity in $PG(2s-1,q)$ (§5.3, Appendix I)
$PGPs(2s+1,q)$	subgroup of $PGL(2s+1,q)$ fixing a pseudo-polarity in $PG(2s,q)$ (§5.3, Appendix I)
$PGPs^*(2s,q)$	subgroup of $PGL(2s,q)$ fixing a pseudo-polarity in $PG(2s-1,q)$ (§5.3, Appendix I)
$PSU(n,q)$	elements of $PGU(n,q)$ of "determinant one" (Appendix I)
$G(\mathscr{F})$	subgroup of $PGL(n+1,q)$ fixing a subset $\mathscr{F}$ of $PG(n,q)$

Constants for a k-arc $\mathscr{K}$ in $PG(2,q)$

$l(n,q)$	Total number of points on the bisecants of an n-arc (§7.1)
$l^*(n,q)$	$\theta(2,q) - l(n,q)$ (§7.1)
$L(n,q)$	Total number of n-arcs in $PG(2,q)$ (§7.1)
$m(2,q)$	maximum value of k (§§3.3, 8.1)
$t, t(P)$	number of unisecants through P in $\mathscr{K}$ (§8.1)
τ_i	number of i-secants to $\mathscr{K}$ (§8.1)
$\sigma_i, \sigma_i(Q)$	number of i-secants through Q in $PG(2,q)\backslash\mathscr{K}$ (§8.1)
c_i	number of points of $PG(2,q)\backslash\mathscr{K}$ through which i bisecants pass (§9.1)
t_i	number of points of $PG(2,q)\backslash\mathscr{K}$ through which i unisecants pass (§9.1)
e_i	number of points of $PG(2,q)\backslash\mathscr{K}$ through which i external lines pass (§9.1)
α	smallest i for which $c_i \neq 0$ (§9.1)
β	largest i for which $c_i \neq 0$ (§9.1)
k'	$[k/2]$ (§9.1)
l_i	number of points on bisecant l of index i (§9.1)
m_i	number of points on unisecant m of grade i (§9.1)
n_i	number of points on external line n of rank i (§9.1)
$\mathscr{D}(F)$	$\{\mathbf{P}(F(t),t,1) \mid t \in \gamma^+\} \cup \{\mathbf{U}_1\}$ (§8.4)
$\mathscr{D}(k)$	$\mathscr{D}(F)$ with $F(t) = t^k$ (§8.4)

Constants for a $(k;n)$-arc $\mathscr{K}$ in $PG(2,q)$

τ_i	number of i-secants to $\mathscr{K}$ (§12.1)
$\rho_i, \rho_i(P)$	number of i-secants to $\mathscr{K}$ through P in $\mathscr{K}$ (§12.1)
$\sigma_i, \sigma_i(Q)$	number of i-secants to $\mathscr{K}$ through Q not in $\mathscr{K}$ (§12.1)
m	smallest i for which $\tau_i \neq 0$ (§12.1)
ν_{id}	number of points of $PG(2,q)$ through which i d-secants pass (§12.1)
ν'_{id}	number of points of $PG(2,q)\backslash\mathscr{K}$ through which i d-secants pass (§12.1)
μ_{ij}	number of points of a d-secant through which i j-secants pass (§12.1)
θ_j	maximum number of j-secants through a point of $PG(2,q)$ (§12.1)
θ	θ_0 (§12.1)
μ	multiplicity map (§12.5)

Other symbols

$\mathscr{A}\mathscr{B}\mathscr{C}\mathscr{D}\mathscr{E}\mathscr{F}\mathscr{G}\mathscr{H}\mathscr{I}\mathscr{J}\mathscr{K}\mathscr{L}\mathscr{M}\mathscr{N}\mathscr{O}\mathscr{P}\mathscr{Q}\mathscr{R}\mathscr{S}\mathscr{T}\mathscr{U}\mathscr{V}\mathscr{W}\mathscr{X}\mathscr{Y}\mathscr{Z}$	script alphabet
$\mathfrak{A}\mathfrak{B}\mathfrak{C}\mathfrak{D}\mathfrak{E}\mathfrak{F}\mathfrak{G}\mathfrak{H}\mathfrak{I}\mathfrak{J}\mathfrak{K}\mathfrak{L}\mathfrak{M}\mathfrak{N}\mathfrak{O}\mathfrak{P}\mathfrak{Q}\mathfrak{R}\mathfrak{S}\mathfrak{T}\mathfrak{U}\mathfrak{V}\mathfrak{W}\mathfrak{X}\mathfrak{Y}\mathfrak{Z}$	German alphabet
$\square$	end of the proof of theorem or lemma

GENERAL INDEX